P9-AQN-453

Random Matrices

REVISED AND ENLARGED

SECOND EDITION

Random Matrices

REVISED AND ENLARGED

SECOND EDITION

MADAN LAL MEHTA
Centre d'Études Nucleaires de Saclay
Gif-sur-Yvette Cedex
France

Centre National de Recherche Scientifique
France

ACADEMIC PRESS, INC.

Harcourt Brace Jovanovich, Publishers

Boston San Diego New York
London Sydney Tokyo Toronto

This book is printed on acid-free paper. ♾

ACADEMIC PRESS, INC.
1250 Sixth Avenue, San Diego, CA 92101

United Kingdon Edition published by
ACADEMIC PRESS LIMITED
24-28 Oval Road, London NW1 7DX

Library of Congress Cataloging-in-Publication Data

Mehta, M. L.
Random matrices / Madan Lal Mehta.—Rev. and enl 2nd ed.
p. cm.
Includes bibliographical references.
ISBN 0-12-488051-7 (alk. paper)
1. Energy levels (Quantum mechanics)—Statistical methods.
2. Random matrices. I. Title.
QC174.45.M444 1990 90-257
530.1′2—dc20 CIP

Printed in the United States of America
90 91 92 93 9 8 7 6 5 4 3 2 1

Contents

9 / Circular Ensembles

10 / Circular Ensembles (Continued)

11 / Circular Ensembles. Thermodynamics

12 / Asymptotic Behavior of $E_\beta(0, s)$ for Large s

13 / Gaussian Ensemble of Antisymmetric Hermitian Matrices

14 / Another Gaussian Ensemble of Hermitian Matrices

15 / Matrices with Gaussian Element Densities but with No Unitary or Hermitian Conditions Imposed

16 / Statistical Analysis of a Level Sequence

17 / Selberg's Integral and Its Consequences

Preface to the Second Edition

The contemporary textbooks on classical or quantum mechanics deal with systems governed by differential equations that are simple enough to be solved in closed terms (or eventually perturbatively). Hence, the entire past and future of such systems can be deduced from a knowledge of their present state (initial conditions). Moreover, these solutions are stable in the sense that small changes in the initial conditions result in small changes in their time evolution. Such systems are called integrable. Physicists and mathematicians now realize that most of the systems in nature are not integrable. The forces and interactions are so complicated that either we cannot write the corresponding differential equation, or when we can, the whole situation is unstable: a small change in the initial conditions produces a large difference in the final outcome. They are called chaotic. The relation of chaotic to integrable systems is something like that of transcendental to rational numbers.

For chaotic systems, it is meaningless to calculate the future evolution starting from an exactly given present state, because a small error or change at the beginning will make the whole computation useless. One should rather try to determine the statistical properties of such systems.

The theory of random matrices makes the hypothesis that the characteristic energies of chaotic systems behave locally as if they were the eigenvalues of a matrix with randomly distributed elements. Random matrices were first encountered in mathematical statistics by Hsu, Wishart, and others in the 1930s, but an intensive study of their properties in connection with nuclear physics only began with the work of Wigner in the 1950s. In 1965, C. E. Porter edited a reprint volume of all important papers on the subject, with a critical and detailed introduction, that even today is very instructive. The first edition of the present book appeared in 1967. During the last two decades there have been many new results, and a larger number of physicists and mathematicians became interested in the subject owing to various potential applications. Consequently, it was felt that this book had to be revised even though a nice review article by Brody *et al.* has appeared in the meantime (*Rev. Mod. Phys.*, 1981).

Among the important new results, one notes the theory of matrices with quaternion elements, which serves to compute some multiple integrals, the evaluation of n-point spacing probabilities, the derivation of the asymptotic behavior of nearest neighbor spacings, the computation of a few hundred millions of zeros of the Riemann zeta function and the analysis of their statistical properties, the rediscovery of Selberg's 1944 paper, giving rise to hundreds of recent publications, the use of the diffusion equation to evaluate an integral over the unitary group, thus allowing the analysis of non-invariant Gaussian ensembles, and the numerical investigation of various systems with deterministic chaos.

After a brief survey of the symmetry requirements, the Gaussian ensembles of random Hermitian matrices are introduced in Chapter 2. In Chapter 3 the joint probability density of the eigenvalues of such matrices is derived. In Chapter 5 we give a detailed treatment of the simplest of the matrix ensembles, the Gaussian unitary one, deriving the n-point correlation functions and the n-point spacing probabilities. Here we explain how the Fredholm theory of integral equations can be used to derive the limits of large determinants. In Chapter 6 we study the Gaussian orthogonal ensemble, which in most cases is appropriate for applications but is mathematically more complicated. Here we introduce matrices with quaternion elements and their determinants as well as the method of integration over alternate variables. The short Chapter 8 introduces a Brownian motion model of Gaussian Hermitian matrices. Chapters 9, 10, and 11 deal with ensembles of unitary random matrices, the mathematical methods being the same as in Chapters 5 and 6. In Chapter 12 we derive the asymptotic series for the nearest neighbor spacing probability. In Chapter 14 we study a non-invariant Gaussian Hermitian ensemble, deriving its n-point correlation and cluster functions; it is a good example of the use of mathematical tools developed in Chapters 5 and 6. Chapter 16 describes a number of statistical quantities useful for the analysis of experimental data. Chapter 17 gives a detailed account of Selberg's integral and its consequences. Other chapters deal with questions or ensembles less important either for applications or for the mathematical methods used. Numerous appendices treat secondary mathematical questions, list power series expansions, and present numerical tables of various functions useful in applications.

The methods explained in Chapters 5 and 6 are basic; they are necessary to understand most of the material presented here. However, Chapter 17 is independent. Chapter 12 is the most difficult one, since it uses results from the asymptotic analysis of differential equations, Toeplitz determinants, and the inverse scattering theory, for which, in spite of a few nice references, we are unaware of a royal road. The rest of the material is self-contained and hopefully quite accessible to any diligent reader with modest mathematical background. Contrary to the general tendency these days, this book contains no exercises.

Acknowledgments

By way of education and inspiration I owe much to my late father, P. C. Mehta, to my late Professor, C. Bloch, and to Professors E. P. Wigner and F. J. Dyson. I had free access to published and unpublished works of Wigner, Dyson, and my colleagues M. Gaudin, R. A. Askey, A. Pandey, and J. des Cloizeaux. I had the privilege to attend the lectures on zeta functions by H. Cohen, H. Stark, and D. Zagier at Les Houches in March 1989. E. A. van Doorn and A. Gervois read the manuscript separately almost entirely while R. A. Askey, H. L. Montgomery, and G. Mahoux read portions of it, thus helping me to avoid black spots and inaccuracies. It is my misfortune that trying to do a complete job, A. Pandey gave me his comments too late. O. Bohigas kindly supplied me with a number of graphs. A. Odlyzko sent me his preprint on the zeros of the Riemann zeta. It is my pleasant duty to thank all of them.

In contrast to the first edition, this edition contains a large umber of figures, especially in Chapters 1, 5, 6, and 16. Encouraged by my colleague, O. Bohigas, this became possible by the generous permissions of various authors and scientific publications. It is my pleasant duty to thank all of them.

The major part of the manuscript was typed on an IBM-PC by the author himself who used the very convenient mathematical word processor "MATHOR" and then translated it into TeX using the same processor. I hope that no serious errors persist.

Preface to the First Edition

Though random matrices were first encountered in mathematical statistics by Hsu, Wishart, and others, intensive study of their properties in connection with nuclear physics began with the work of Wigner in the 1950s. Much material has accumulated since then, and it was felt that it should be collected. A reprint volume to satisfy this need had been edited by C. E. Porter with a critical introduction (see References); nevertheless, the feeling was that a book containing a coherent treatment of the subject would be welcome.

We make the assumption that the local statistical behavior of the energy levels of a sufficiently complicated system is simulated by that of the eigenvalues of a random matrix. Chapter 1 is a rapid survey of our understanding of nuclear spectra from this point of view. The discussion is rather general, in sharp contrast to the precise problems treated in the rest of the book. In Chapter 2 an analysis of the usual symmetries that a quantum system might possess is carried out, and the joint probability density function for the various matrix elements of the Hamiltonian is derived as a consequence. The transition from matrix elements to eigenvalues is made in Chapter 3, and the standard arguments of classical statistical mechanics are applied in Chapter 4 to derive the eigenvalue density. An unproved conjecture is also stated. In Chapter 5 the method of integration over alternate variables is presented, and an application of the Fredholm theory of integral equations is made to the problem of eigenvalue spacings. The methods developed in Chapter 5 are basic to an understanding of most of the remaining chapters. Chapter 6 deals with the correlations and spacings for less useful cases. A Brownian motion model is described in Chapter 7. Chapters 8 to 11 treat circular ensembles; Chapters 8 to 10 repeat calculations analogous to those of Chapters 4 to 7. The integration method discussed in Chapter 11 originated with Wigner and is being published here for the first time. The theory of non-Hermitian random matrices, though not applicable to any physical problems, is a fascinating subject and must be studied for its own sake. In this direction an impressive effort by Ginibre is described in Chapter 12. For the Gaussian ensembles the level density in regions where it is very low is discussed in Chapter 13. The investigations of Chapter 16 and Appendices A.29 and A.30 were recently carried out in collaboration with Professor Wigner at Princeton University. Chapters 14, 15, and 17 treat a number of other topics. Most of the material in the appendices is either well known or was published elsewhere and is collected here for ready reference. It was surprisingly difficult to obtain the proof contained in A.21, while A.29, A.30, and A.31 are new.

October, 1967 M. L. MEHTA
Saclay, France

1 / Introduction

In the theory of random matrices one is concerned with the following question. Consider a large matrix whose elements are random variables with given probability laws. Then what can one say about the probabilities of a few of its eigenvalues or of a few of its eigenvectors? This question is of pertinence for the understanding of the statistical behavior of slow neutron resonances in nuclear physics, where it was first proposed and intensively studied. The question gained importance in other areas of physics and mathematics, such as the characterization of chaotic systems, elastodynamic properties of structural materials, conductivity in disordered metals, or the distribution of the zeros of the Riemann zeta function. The reasons of this pertinence are not yet clear. The impression is that some sort of a law of large numbers is in the background. In this chapter we will try to give reasons why one should study random matrices.

1.1. Random Matrices in Nuclear Physics

Figure 1.1 shows a typical graph of slow neutron resonances. There one sees various peaks with different widths and heights located at various places. Do they have any definite statistical pattern? The locations of the peaks are called nuclear energy levels, their widths the neutron widths and their heights are called the transition strengths.

The experimental nuclear physicists have collected vast amounts of data concerning the excitation spectra of various nuclei such as those shown on Figure 1.1 (Garg et al., 1964, where a detailed description of the experimental work on thorium and uranium energy levels is given; Rosen et al., 1960; Camarda et al., 1973; Liou et al., 1972a, b). The ground state and low lying excited states have been impressively

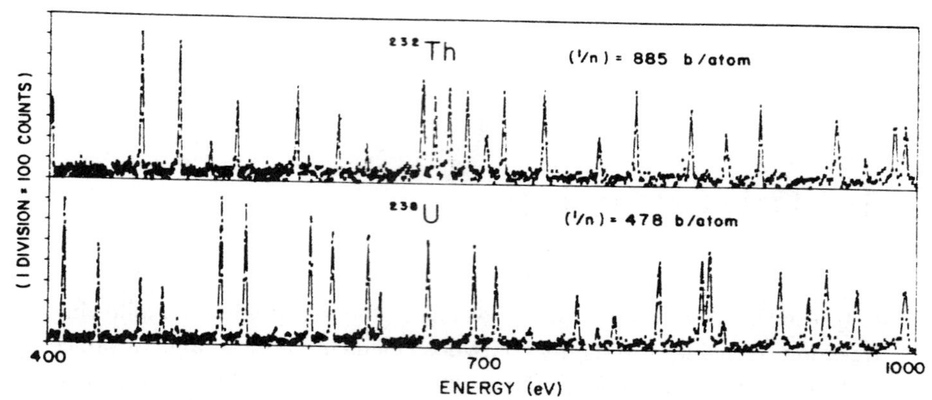

FIG. 1.1. Slow neutron resonance cross sections on thorium 232 and uranium 238 nuclei. Reprinted with permission from The American Physical Society, Rahn et al. Neutron resonance spectroscopy, X, *Phys. Rev. C*, 6, 1854 1869 (1972).

explained in terms of an independent particle model where the nucleons are supposed to move freely in an average potential well (Mayer and Jensen, 1955; Kisslinger and Sorenson, 1960). As the excitation energy increases, more and more nucleons are thrown out of the main body of the nucleus, and the approximation of replacing their complicated interactions with an average potential becomes more and more inaccurate. At still higher excitations the nuclear states are so dense and the intermixing is so strong that it is a hopeless task to try to explain the individual states; but when the complications increase beyond a certain point the situation becomes hopeful again, for we are no longer required to explain the characteristics of every individual state but only their average properties, which is much simpler.

The average behavior of the various energy levels is of prime importance in the study of nuclear reactions. In fact, nuclear reactions may be put into two major classes—fast and slow. In the first case a typical reaction time is of the order of the time taken by the incident nucleon to pass through the nucleus. The wavelength of the incident nucleon is much smaller than the nuclear dimensions, and the time it spends inside the nucleus is so short that it interacts with only a few nucleons inside the nucleus. A typical example is the head-on collision with one nucleon in which the incident nucleon hits and ejects a nucleon, thus giving it almost all its momentum and energy. Consequently, in such cases the coherence and the interference effects between incoming and outgoing nucleons is strong.

Another extreme is provided by the slow reactions in which the typical reaction times are two or three orders of magnitude larger. The incident nucleon is trapped and all its energy and momentum are quickly distributed among the various constituents of the target nucleus. It takes a long time before enough energy is again concentrated on a single nucleon to eject it. The compound nucleus lives long enough to forget the manner of its formation, and the subsequent decay is therefore independent of the way in which it was formed.

In the slow reactions, unless the energy of the incident neutron is very sharply defined, a large number of neighboring energy levels of the compound nucleus are involved, hence the importance of an investigation of their average properties, such as the distribution of neutron and radiation widths, level spacings, and fission widths. It is natural that such phenomena, which result from complicated many body interactions, will give rise to statistical theories. We shall concentrate mainly on the average properties of nuclear levels such as level spacings.

According to quantum mechanics, the energy levels of a system are supposed to be described by the eigenvalues of a Hermitian operator H, called the Hamiltonian. The energy level scheme of a system consists in general of a continuum and a certain, perhaps large, number of discrete levels. The Hamiltonian of the system should have the same eigenvalue structure and therefore must operate in an infinite dimensional Hilbert space. To avoid the difficulty of working with an infinite dimensional Hilbert space, we make approximations amounting to a truncation, keeping only the part of the Hilbert space that is relevant to the problem at hand and either forgetting about the rest or taking its effect in an approximate manner on the part considered. Because we are interested in the discrete part of the energy level schemes of various quantum systems, we approximate the true Hilbert space by one having a finite, though large, number of dimensions. Choosing a basis in this space, we represent our Hamiltonians by finite dimensional matrices. If we can solve the eigenvalue equation,

$$H\Psi_i = E_i\Psi_i,$$

we shall get all the eigenvalues and eigenfunctions of the system, and any physical information can then be deduced, in principle, from this knowledge. In the case of the nucleus, however, there are two difficulties. First, we do not know the Hamiltonian and, second, even if we did, it would be far too complicated to attempt to solve the corresponding equation.

Therefore, from the very begining we shall be making statistical hypotheses on H, compatible with the general symmetry properties. Choosing a complete set of functions as a basis, we represent the Hamiltonian operators H as matrices. The elements of these matrices are random variables whose distributions are restricted only by the general symmetry properties we might impose on the ensemble of operators. And the problem is to get information on the behavior of its eigenvalues. "The statistical theory will not predict the detailed level sequence of any one nucleus, but it will describe the general appearance and the degree of irregularity of the level structure that is expected to occur in any nucleus which is too complicated to be understood in detail." (Dyson, 1962a-I).

In classical statistical mechanics a system may be in any of the many possible states, but one does not ask in which particular state a system is. Here we shall renounce knowledge of the nature of the system itself. "We picture a complex nucleus as a black box in which a large number of particles are interacting according to unknown laws. As in orthodox statistical mechanics we shall consider an ensemble of Hamiltonians, each of which could describe a different nucleus. There is a strong logical expectation, though no rigorous mathematical proof, that an ensemble average will correctly describe the behavior of one particular system which is under observation. The expectation is strong, because the system might be one of a huge variety of systems, and very few of them will deviate much from a properly chosen ensemble average. On the other hand, our assumption that the ensemble average correctly describes a particular system, say the U^{239} nucleus, is not compelling. In fact, if this particular nucleus turns out to be far removed from the ensemble average, it will show that the U^{239} Hamiltonian posseses specific properties of which we are not aware. This, then will prompt one to try to discover the nature and the origin of these properties." (Dyson, 1962a-I).

Wigner was the first to propose in this connection the hypothesis alluded to, namely that the local statistical behavior of levels in a simple sequence is identical with the eigenvalues of a random matrix. A simple sequence is one whose levels all have the same spin, parity, and other strictly conserved quantities, if any, which result from the symmetry of the system. The corresponding symmetry requirements are to be imposed on the random matrix. There being no other restriction on the matrix, its elements are taken to be random with, say, a Gaussian distribution. Porter and Rosenzweig (1960a, b) were the early workers in the field who analyzed the nuclear experimental data made available by Harvey, Hughes (1958) and by Rosen and co-workers (1960) and the atomic data compiled by C. E. Moore (1949, 1958). They found that the occurence of two levels close to each other in a simple sequence is a rare event. They also used the computer to generate and diagonalize a large number of random matrices. This Monte-Carlo analysis indicated the correctness of Wigner's hypothesis. In fact it indicated more; the density and the spacing distribution of eigenvalues of real symmetric matrices are independent of many details of the distribution of individual matrix elements. From a group theoretical analysis Dyson found that an irreducible ensemble of matrices, invariant under a symmetry group G, necessarily belongs to one of three classes, named by him orthogonal, unitary, and symplectic. We shall not go into these elegant group theoretical arguments but shall devote enough space to the study of the circular ensembles introduced by Dyson. As we will see, Gaussian ensembles are equivalent to the circular ensembles for large orders. In other words, when the order of the matrices goes to infinity, the limiting correlations extending to any finite number of eigenvalues are identical in the two cases. The spacing distribution, which depends on an infinite number of correlation functions, is also identical in the two cases. It is remarkable that standard thermodynamics can be applied to obtain certain results which otherwise require long and difficult analysis to derive. A theory of Brownian motion of matrix elements has also been created by Dyson, thus rederiving a few known results.

Various numerical Monte-Carlo studies indicate, as Porter and Rosenzweig noted earlier, that a few level correlations of the eigenvalues depend only on the overall symmetry requirements that a matrix should satisfy, and they are independent of all other details of the distribution of individual matrix elements. The matrix has to be Hermitian to have real eigenvalues, the diagonal elements should have the same distribution, and the off-diagonal elements should be distributed symmetrically about the zero mean and the same mean square deviation for all independent parts entering in their definition. What is then decisive is whether the matrix is symmetric or self-dual or something else or none of these. In the limit of large orders other details are not seen. Similarly, in the circular ensembles, the matrices are taken to be unitary to have the eigenvalues on the circumference of the unit circle; what counts then is whether they are symmetric or self-dual or none of these. Other details are washed out in the limit of large matrices.

This independence is expected; but apart from the impressive numerical evidence, some heuristic arguments of Wigner and the equivalence of Gaussian and circular ensembles, no rigorous derivation of this fact has yet been found. Its generality seems something like that of the central limit theorem.

1.2. Random Matrices in Other Branches of Knowledge

The physical properties of metals depend characteristically on their excitation spectra. In bulk metal at high temperatures the electronic energy levels lie very near to one another and are broad enough to overlap and form a continuous spectrum. As the sample gets smaller, this spectrum becomes discrete, and as the temperature decreases the widths of the individual levels decrease. If the metallic particles are minute enough and at low enough temperatures, the spacings of the electronic energy levels may eventually become larger than the other energies, such as the level widths and the thermal energy, kT. Under such conditions the thermal and the electromagnetic properties of the fine metallic particles may deviate considerably from those of the bulk metal. This circumstance has already been noted by Fröhlich (1937) and proposed by him as a test of the quantum mechanics. Because it is difficult to control the shapes of such small particles while they are being experimentally produced, the electronic energy levels are seen to be random and the theory for the eigenvalues of the random matrices may be useful in their study.

In the mathematical literature the Riemann zeta function is quite important. It is suspected that all its nonreal zeros lie on a line parallel to the imaginary axis; it is also suspected (Montgomery, 1973, 1975) that the local fluctuation properties of these zeros on this line are identical to those of the eigenvalues of matrices from a unitary ensemble.

Random matrices are also encountered in other branches of physics. For example, glass may be considered as a collection of random nets, that is, a collection of particles with random masses exerting random mutual forces, and it is of interest to determine the distribution of frequencies of such nets (Dyson, 1953). The most studied model of glass is the so called random Ising model or spin glass. On each site of a 2- or 3-dimensional regular lattice a spin variable σ_i is supposed to take values $+1$ or -1, each with a probability equal to $1/2$. The interaction between neighboring spins σ_i and σ_j is $J_{ij}\sigma_i\sigma_j$, and that between any other pair of spins is zero. If $J_{ij} = J$ is fixed, we have the Ising model, whose partition function was first calculated by Onsager (*cf.* McCoy and Wu, 1973). If J_{ij} is a random variable, with a symmetric distribution around zero mean, we have the random Ising model, the calculation of the partition function of which is still an open problem.

A problem much studied during the last few years is that of characterizing a chaotic system. Classically, a system is chaotic if small differences in the initial conditions result in large differences in the final outcome. A polygonal billiard table with incommensurate sides and angles is such an example; two billiard balls starting on nearby paths have diverging trajectories. According to their increasing chaoticity, systems are termed classically as ergodic, mixing, a K-system, or a Bernoulli shift. Quantum mechanically, one may consider, for example, a free particle confined to a finite part of the space (billiard table). Its possible energies are discrete; they are the eigenvalues of the Laplace operator in the specified finite space. Given the sequence of these discrete energy values, what can one say about its chaoticity; whether it is ergodic or mixing or ...; i.e., is there any correspondence with the classical notions? A huge amount of numerical evidence tells us that these energies behave as if they were the eigenvalues of a random matrix taken from the Gaussian orthogonal ensemble (GOE). Finer recognition of the chaoticity has not yet been possible.

Recently, Weaver (1989) found that the ultrasonic resonance frequencies of structural materials, such as aluminium blocks, do behave as the eigenvalues of a matrix from the GOE. This is expected of the vibrations of complex structures at frequencies well above the frequencies of their lowest modes. Thus random matrix theory may be of relevance in the non-destructive evaluation of materials, architectural acoustics, and the decay of ultrasound in heterogeneous materials.

The series of nuclear energy levels, or any sequence of random numbers for that matter, can be thought to have two distinct kinds of statistical properties, which may be called global and local. A global property varies slowly, its changes being appreciable only on a large interval. The average number of levels per unit of energy or the mean level density, for example, is a global property; it changes appreciably only on intervals containing thousands of levels. Locally it may be treated as a constant. A local property on the other hand fluctuates from level to level. The distance between two successive levels, for example, is a local property. Moreover, global and local properties of a complex system seem to be quite disconnected. Two systems having completely different global properties may have identical fluctuation properties, and, inversely, two systems having the same global properties may have different fluctuation properties.

The random matrix models studied here will have quite different global properties, none of them corresponding exactly to the nuclear energy levels or to the sequence of zeros on the critical line of the Riemann zeta function. One may even choose the global properties at will (Balian, 1968)! However, the nice and astonishing thing about them is that their local fluctuation properties are always the same and determined only by the over-all symmetries of the system. From the extended numerical experience (*cf.* Appendix A.1) one might state the kind of central limit theorem, referred to earlier, as follows.

Conjecture 1.2.1. *Let H be an $N \times N$ real symmetric matrix, its off-diagonal elements H_{ij}, for $i < j$, being independent identically distributed (i.i.d.) random variables with mean zero and variance $\sigma > 0$, i.e., $\langle H_{ij} \rangle = 0$, and $\langle H_{ij}^2 \rangle = \sigma^2 \neq 0$. Then in the limit of large N the statistical properties of n eigenvalues of H become independent of the probability density of the H_{ij}, i.e., when $N \to \infty$, the joint probability density (j.p.d.) of arbitrarily chosen n eigenvalues of H tends, for every finite n, with probability one, to the n-point correlation function of the Gaussian orthogonal ensemble studied in Chapter* 6.

Note that the off-diagonal elements H_{ij} for $i > j$ are determined from symmetry, and the diagonal elements H_{ii} may have any distributions.

For Hermitian matrices we suspect the following.

Conjecture 1.2.2. *Let H be an $N \times N$ Hermitian matrix with complex numbers as elements. Let the real parts of H_{ij} for $i < j$ be i.i.d. random variables with mean zero and variance $\sigma_1 > 0$, while let the imaginary parts of H_{ij} for $i < j$ be i.i.d. random variables with mean zero and variance $\sigma_2 > 0$. Then as the order $N \to \infty$, the j.p.d. of n arbitrarily chosen eigenvalues of H, for every finite n, tends with probability one to the n-point correlation function of the Gaussian unitary ensemble studied in Chapter* 5.

Note that, as for real symmetric matrices, the elements H_{ij} for $i > j$ are determined from Hermiticity and the distributions of the digonal elements H_{ii} seem irrelevant.

A similar result is suspected for self-dual Hermitian quaternion matrices; all finite correlations being identical to those for the Gaussian symplectic ensemble studied in Chapter 7.

In other words, the local statistical properties of a few eigenvalues of a large random matrix seem to be independent of the distributions of individual matrix elements. What matters is whether the matrix is real symmetric, or self-dual (quaternion) Hermitian or only Hermitian. The rest does not seem to matter. And this seems to be true even under less restrictive conditions; for example, the probability law for different matrix elements (their real and imaginary parts) may be different.

For level density, i.e., the case $n = 1$, we have some arguments to say that it follows Wigner's "semi-circle" law (*cf.* Chapter 4). Except for this special case we have only numerical evidence in favor of the above conjectures.

Among the Hermitian matrices, the case of the Gaussian distributions of matrix elements is still the only one treated analytically by Hsu, Selberg, Wigner, Mehta, Gaudin, Dyson, Rosenzweig, Bronk, Ginibre, Pandey, des Cloizeaux, and others. Circular ensembles of unitary matrices have similarly been studied. We will describe these developments in great detail in the following pages.

1.3. A Summary of Statistical Facts about Nuclear Energy Levels

1.3.1. LEVEL DENSITY

As the excitation energy increases, the nuclear energy levels occur on the average at smaller and smaller intervals. In other words, level density increases with the excitation energy. The first question we might ask is how fast does this level density increase for a particular nucleus and what is the distribution of these levels with respect to spin and parity? This is an old problem treated by Bethe (1937). Even a simple model in which the nucleus is taken as a degenerate Fermi gas with equidistant single-particle levels gives an adequate result. It amounts to determining the number of partitions $\lambda(n)$ of a positive integer n into smaller positive integers $\nu_1, \nu_2, \ldots$:

$$n = \nu_1 + \nu_2 + \cdots + \nu_\ell, \qquad \nu_1 \geq \nu_2 \geq \cdots \geq \nu_\ell > 0.$$

For large n this number, according to the Hardy-Ramanujan formula, is given by

$$\lambda(n) \sim \exp\left[\left(\theta\pi^2 n/3\right)^{1/2}\right],$$

where θ is equal to 1 or 2 according to whether the ν_i are all different or whether some of them are allowed to be equal. With a slight modification due to later work (Lang and Lecouteur, 1954; Cameron, 1956). Bethe's result gives the level density as

$$\rho(E, j, \pi) \propto (2j+1)(E-\Delta)^{-5/4} \\ \times \exp\left[-j(j+1)/2\sigma^2\right] \exp\left[2a(E-\Delta)^{1/2}\right],$$

where E is the excitation energy, j is the spin and π is the parity. The dependence of the parameters σ, a, and Δ on the neutron and proton numbers is complicated and only imperfectly understood. However, for any particular nucleus a few measurements will suffice to determine them all; the formula will then remain valid for a wide range of energy that contains thousands and even millions of levels.

1.3.2. DISTRIBUTION OF NEUTRON WIDTHS

An excited level may decay in many ways: for example, by neutron ejection or by giving out a quantum of radiation. These processes are characterized by the corresponding decay widths of the levels. The neutron reduced widths $\Gamma_n^0 = \Gamma_n / E^{1/2}$, in which Γ_n is the neutron width and E is the excitation energy of the level, show large fluctuations from level to level. From an analysis of the available data Scott (1954) and later Porter and Thomas (1956) concluded that they had a χ^2 distribution with $\nu = 1$ degree of freedom:

$$\begin{aligned} P(x) &= (\nu/2)\left[\Gamma(\nu/2)\right]^{-1}(\nu x/2)^{(\nu/2)-1} e^{-\nu x/2} \\ &= (2\pi x)^{-1/2} e^{-x/2}, \\ x &= \Gamma_n^0 / \bar{\Gamma}_n^0 , \end{aligned}$$

where $\bar{\Gamma}_n^0$ is the average of Γ_n^0 and $P(x)\,dx$ is the probability that a certain reduced width will lie in an interval dx around the value x. This indicates a Gaussian distribution for the reduced width amplitude

$$\left(\frac{2}{\pi}\right)^{1/2} \exp\left[-\frac{1}{2}\left(\sqrt{x}\right)^2\right] d\left(\sqrt{x}\right)$$

expected from the theory. In fact, the reduced width amplitude is proportional to the integral of the product of the compound nucleus wave function and the wave function in the neutron-decay channel over the channel surface. If the contributions from the various parts of the channel surface are supposed to be random and mutually independent, their sum will have a Gaussian distribution with zero mean.

1.3.3. RADIATION AND FISSION WIDTHS

The total radiation width is almost a constant for particular spin states of a particular nucleus. The total radiation width is the sum of partial radiation widths,

$$\Gamma = \sum_{i=1}^{m} \Gamma_i .$$

If we assume that each of these $\Gamma_i/\bar{\Gamma}_i$ ($\bar{\Gamma}_i$ denoting the average of Γ_i) has a χ^2 distribution with one degree of freedom like the neutron widths and all the $\bar{\Gamma}_i$ are the same, then $\Gamma/\bar{\Gamma}$, ($\bar{\Gamma}$ being the average of Γ), will have a χ^2 distribution with m degrees of freedom. For (moderately) large m, this is a narrow distribution. This conclusion remains valid even when the $\bar{\Gamma}_i$ are not all equal.

It is difficult to measure the partial radiation widths. Little is known about the fission-width distributions. Some known fission widths of U^{235} have been analyzed (Bohr, 1956) and a χ^2 distribution with 2 to 3 degrees of freedom has been found to give a satisfactory fit.

From now on we shall no longer consider neutron, radiation, or fission widths.

1.3.4. LEVEL SPACINGS

Let us regard the level density as a function of the excitation energy as known and consider an interval of energy δE centered at E. This interval is much smaller compared with E, whereas it is large enough to contain many levels; that is,

$$E \gg \delta E \gg D,$$

where D is the mean distance between neighboring levels. How are the levels distributed in this interval? On Figure 1.2 a few examples of level series are shown. In all these cases the level density is taken to be the same, i.e., the scale in each case is chosen so that the average distance between the neighboring levels is unity. It is evident that these different level sequences do not look similar. There are many coincident pairs or sometimes even triples of levels as well as large gaps when the levels have no correlations, i.e., the Poisson distribution; whereas the zeros of the Riemann zeta function are more or less equally spaced. The cases of prime numbers, the slow neutron resonance levels of the erbium 166 nucleus, and the possible energy values of a free particle confined to a billiard table of a specified shape are far from either regularly spaced uniform series or the completely random Poisson series with no correlations.

Although the level density varies strongly from nucleus to nucleus, the fluctuations in the precise positions of the levels seem not to depend

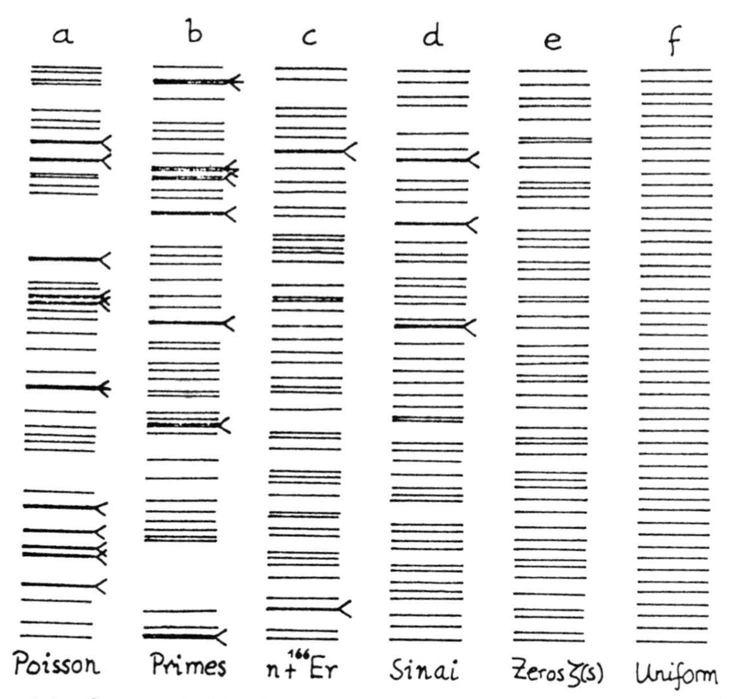

FIG. 1.2. Some typical level sequences. From Bohigas et al. (1983). (a) Random levels with no correlations, Poisson series. (b) Sequence of prime numbers. (c) Slow neutron resonance levels of the erbium 166 nucleus. (d) Possible energy levels of a particle free to move inside the area bounded by 1/8 of a square and a circular arc whose center is the midpoint of the square; i.e., the area specified by the inequalities, $y \geq 0$, $x \geq y$, $x \leq 1$, and $x^2 + y^2 \geq r$ (Sinai's billiard table). (e) The zeros of the Riemann zeta function on the line Re $z = 1/2$. (f) A sequence of equally spaced levels. Reprinted with permission from Kluwer Academic Publishers, Bohigas, O. Haq, R. U., and Pandey A., Fluctuation properties of nuclear energy levels and widths: comparison of theory with experiment, in *Nuclear Data for Science and Technology*, ed. K. H. Böckhoff, 809–814 (1983).

on the nucleus and not even on the excitation energy. As the density of the levels is nearly constant in this interval, we might think that they occur at random positions without regard to one another, the only condition being that their density be a given constant. However, as we see on Figure 1.2(c), such is not the case. It is true that nuclear levels with different spin and parity or atomic levels with different sets of good quantum numbers seem to have no influence on each other. However, levels with the same set of good quantum numbers show a large degree of regularity. For instance, they rarely occur close together.

A more detailed discussion of the experimental data regarding the above quantities as well as the strength functions may be found in the review article by Brody et al. (1981).

1.4. Definition of a Suitable Function for the Study of Level Correlations

To study the statistical properties of the sequence of eigenvalues one defines suitable functions such as level spacings or correlation and cluster functions, or some suitable quantity called a statistic, such as the Δ, Q, F or the Λ statistic of the literature. We will consider a few of them in due course. For the spacings let $E_1, E_2, ...$ be the positions of the successive levels in the interval $\delta E \, (E_1 \leq E_2 \leq \cdots)$ and let $S_1, S_2, ...$ be their distances apart, $S_i = E_{i+1} - E_i$. The average value of S_i is the mean spacing D. We define the relative spacing $s_i = S_i/D$. The probability density function $p(s)$ is defined by the condition that $p(s)ds$ is the probability that any s_i will have a value between s and $s + ds$.

For the simple case in which the positions of the energy levels are not correlated the probability that any E_i will fall between E and $E + dE$ is independent of E and is simply $\rho \, dE$, where $\rho = D^{-1}$ is the average number of levels in a unit interval of energy. Let us determine the probability of a spacing S; that is, given a level at E, what is the probability of having no level in the interval $(E, E + S)$ and one level in the interval $(E + S, E + S + dS)$. For this we divide the interval S into m equal parts.

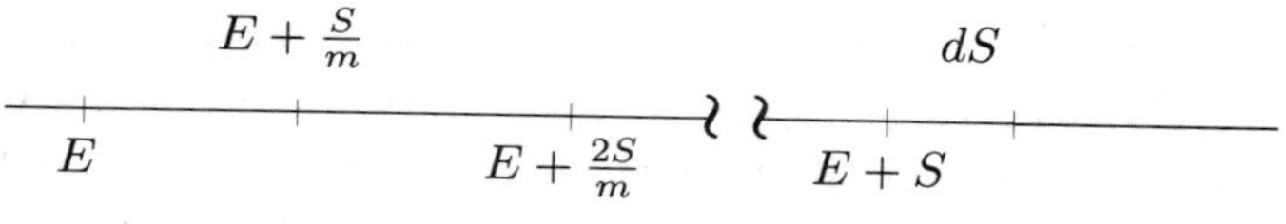

$E + \frac{S}{m}$ $\quad dS$

$E \quad E + \frac{2S}{m} \quad E + S$

Because the levels are independent, the probability of having no level in $(E, E + S)$ is the product of the probabilities of having no level in any of these m parts. If m is large, so that S/m is small, we can write this as $[1 - \rho(S/m)]^m$, and in the limit $m \to \infty$,

$$\lim_{m \to \infty} \left(1 - \rho \frac{S}{m}\right)^m = e^{-\rho S}.$$

Moreover, the probability of having a level in dS at $E+S$ is ρdS. Therefore, given a level at E, the probability that there is no level in $(E, E+S)$ and one level in dS at $E+S$ is

$$e^{-\rho S} \rho \, dS,$$

or in terms of the variable $s = S/D = \rho S$

$$p(s) ds = e^{-s} ds. \tag{1.4.1}$$

This is known as the Poisson distribution or the spacing rule for random levels.

That (1.4.1) is not correct for nuclear levels of the same spin and parity or for atomic levels of the same parity and orbital and spin angular momenta is clearly seen by a comparison with the empirical evidence (Figures 1.3 and 1.4). It is not true either for the eigenvalues of a matrix from any of the Gaussian ensembles, as we will see.

1.5. Wigner Surmise

When the experimental situation was not yet conclusive, Wigner proposed the following rules for spacing distributions:

1. In the sequence of levels with the same spin and parity, called a simple sequence, the probability density function for a spacing is given by

$$p_W(s) = \frac{\pi s}{2} \exp\left(-\frac{\pi}{4} s^2\right), \qquad s = \frac{S}{D}. \tag{1.5.1}$$

2. Levels with different spin and parity are not correlated. The function $p(s)$ for a mixed sequence may be obtained by randomly superimposing the constituent simple sequences (*cf.* Appendix A.2).

Two simple arguments give rise to Rule 1. As pointed out by Wigner (1957b) and by Landau and Smorodinsky (1955), it is reasonable to expect that, given a level at E, the probability that another level will

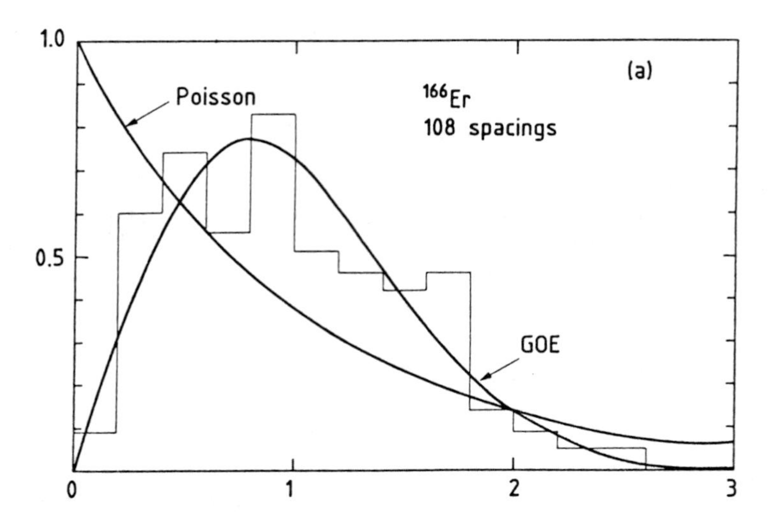

FIG. 1.3. The probability density for the nearest neighbor spacings in slow neutron resonance levels of erbium 166 nucleus. The histogram shows the first 108 levels observed. The solid curves correspond to the Poisson distribution, i.e., no correlations at all, and that for the eigenvalues of a real symmetric random matrix taken from the Gaussian orthogonal ensemble (GOE). Reprinted with permission from The American Physical Society, Liou, H. I. et al. Neutron resonance spectroscopy data, *Phys. Rev. C*, 5, 974–1001 (1972a).

lie around $E + S$ is proportional to S for small S. Now if we extrapolate this to all S and, in addition, assume that the probabilities in various intervals of length S/m obtained by dividing S into m equal parts are mutually independent, we arrive at

$$p(S/D)dS = \lim_{m \to \infty} \prod_{r=0}^{m-1} \left(1 - \frac{Sr}{m}\frac{S}{m}a\right) aS\, dS = aSe^{-aS^2/2}\, dS. \tag{1.5.2}$$

The constant a can be determined by the condition that the average value of $s = S/D$ is unity:

$$\int_0^\infty sp(s)ds = 1. \tag{1.5.3}$$

Let us, at this point, define the n-point correlation function $R_n(E_1, ..., E_n)$ so that $R_n dE_1 dE_2 \cdots dE_n$ is the probability of finding a level in each

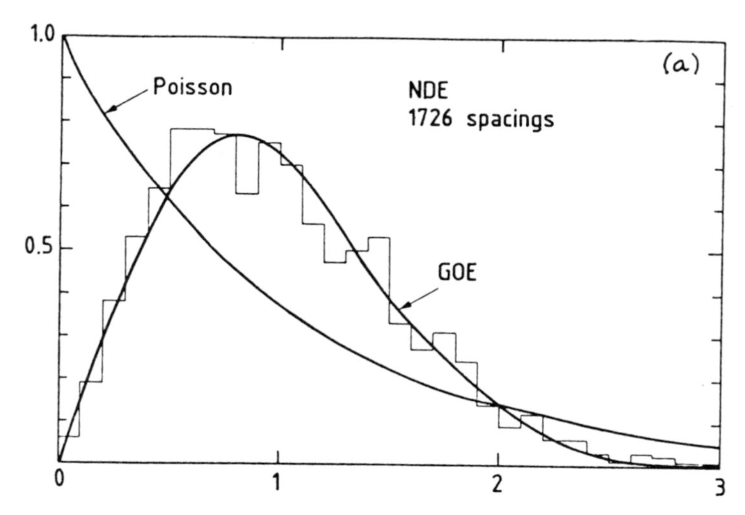

FIG. 1.4. Level spacing histogram for a large set of nuclear levels, often referred to as nuclear data ensemble. The data considered consist of 1407 resonance levels belonging to 30 sequences of 27 different nuclei: (i) slow neutron resonances of Cd(110, 112, 114), Sm(152, 154), Gd(154, 156, 158, 160), Dy(160, 162, 164), Er(166, 168, 170), Yb(172, 174, 176), W(182, 184, 186), Th(232), and U(238) (1146 levels); (ii) proton resonances of Ca(44) (J=1/2+), Ca(44) (J=1/2−), and Ti(48) (J=1/2+) (157 levels); and (iii) (n,γ)-reaction data on Hf(177) (J=3), Hf(177) (J=4), Hf(179) (J=4), and Hf(179) (J=5) (104 levels). The data chosen in each sequence is believed to be complete (no missing levels) and pure (the same angular momentum and parity). For each of the 30 sequences the average quantities (e.g., the mean spacing, spacing/mean spacing, number variance μ_2, etc.; see Chapter 16) are computed separately and their agregate is taken weighted according to the size of each sequence. The solid curves correspond to the Poisson distribution, i.e., no correlations at all, and that for the eigenvalues of a real symmetric random matrix taken from the Gaussian orthogonal ensemble (GOE). Reprinted with permission from Kluwer Academic Publishers, Bohigas, O. Haq, R. U., and Pandey A., Fluctuation properties of nuclear energy levels and widths: comparison of theory with experiment, in *Nuclear Data for Science and Technology*, ed. K. H. Böckhoff, 809–814 (1983).

unobserved. The two simple arguments of Wigner given in the derivation of Rule 1 are equivalent to the following. The two-point correlation function $R_2(E_1, E_2)$ is linear in the variable $|E_1 - E_2|$, and three and higher order correlation functions are negligibly small.

We shall see in Chapter 6 that both arguments are inaccurate, whereas Rule 1 is very near the correct result (*cf.* Figure 1.5). It is surprising that the two errors compensate so nearly each other.

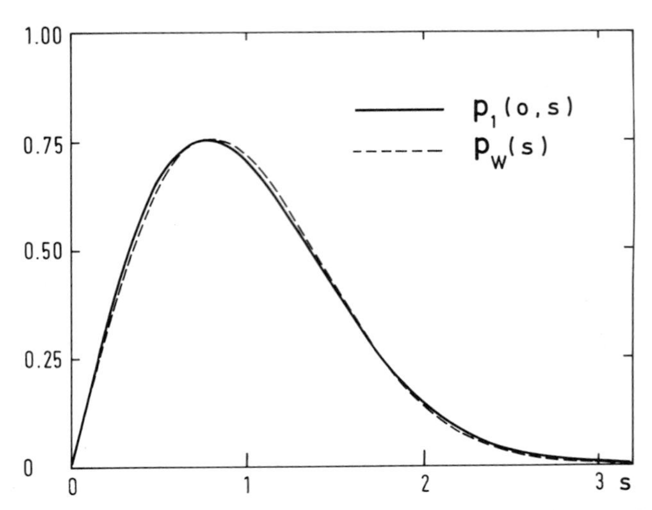

FIG. 1.5. The probability density $p_1(0; s)$ of the nearest neighbor spacings for the eigenvalues of a random real symmetric matrix from the Gaussian orthogonal ensemble, Eq. (6.6.18), and the Wigner surmise $p_W(s)$; Eq. (1.5.1). Reprinted with permission from Elsevier Science Publishers, Gaudin, M. Sur la loi limite de l'éspacement des valeurs propres d'une matrice aléatoire, *Nucl. Phys.*, 25, 447–458 (1961).

1.6. Electromagnetic Properties of Small Metallic Particles

Consider small metallic particles at low temperatures. The number of electrons in a volume V is $n \approx 4\pi p_0^3 V/(3h^3)$, where p_0 is the Fermi momentum and h is Planck's constant. The energy of an excitation near the Fermi surface is $E_0 \approx p_0^2/2m^*$, where m^* is the effective mass of the electron. The level density at zero excitation is therefore $\sigma = dn/dE_0 \approx 4\pi p_0 V m^*/h^3$, and the average level spacing is the inverse of this quantity $D \approx \sigma^{-1}$. For a given temperature we can easily estimate the size of the metallic particles for which $D \gg kT$, where k is Boltzmann's constant and T is the temperature in degrees Kelvin. For example, the number of electrons in a metallic particle of size 10^{-6}–10^{-7} cm may be as low as a few hundred and, at $T \approx 10K, D \approx 1eV$, whereas $kT \approx 10^{-3}$ eV. It is possible to produce particles of this size experimentally and then to sort them out according to their size (e.g., by centrifuging and sampling at a certain radial distance). Thus we have a large number of metallic particles, each of which has a different shape and therefore a different set of electronic energy levels but the same average level spacing, for the volumes are equal. It would be desirable if we could separate (e.g., by applying a nonuniform magnetic field) particles containing an odd number of conduction electrons from those containing an even number. The energy-level schemes for these two types of particles have very different properties (see Chapters 2 and 3).

Given the position of the electron energies, we can calculate the partition function in the presence of a magnetic field and then use thermodynamic relations to derive various properties such as electronic specific heat and spin paramagnetism. Fröhlich (1937) assumed that the energies were equally spaced, and naturally obtained the result that all physical quantities decrease exponentially at low temperatures as $\exp(-D/kT)$ for $1 \ll D/kT$. Kubo (1969) repeated the calculation with the assumption that the energies were random without correlations and that their spacings therefore follow a Poisson law. He arrived at a linear law for the specific heat $\sim kT/D$. The constants are different for particles containing an odd number of electrons from those containing an even number. For spin paramagnetism even the dependence on temperature is different for the two sets of particles. Instead of Fröhlich's equal spacing rule or Kubo's Poisson law, it would perhaps be better to suppose with Gorkov and Eliashberg (1965), that these energies behave as the eigenvalues of a random matrix. This point of view may be justified as follows. The energies are the eigenvalues of a fixed Hamiltonian with random boundary conditions. We may incorporate these boundary conditions into the Hamiltonian by the use of fictitious potentials. The energies are thus neither equally spaced, nor follow the Poisson law, but they behave as the eigenvalues of a random matrix taken from a suitable ensemble. In contrast to nuclear spectra, we have the possibility of realizing in practice all three ensembles considered in various sections of this book. They apply in particular when (*a*) the number of electrons (in each of the metallic particles) is even and there is no external magnetic field, (*b*) the number of electrons (in each of the metallic particles) is odd and there is no external magnetic field, (*c*) there is an external magnetic field much greater than D/μ, where μ is the magnetic moment of the electron.

As to which of the three assumptions is correct should be decided

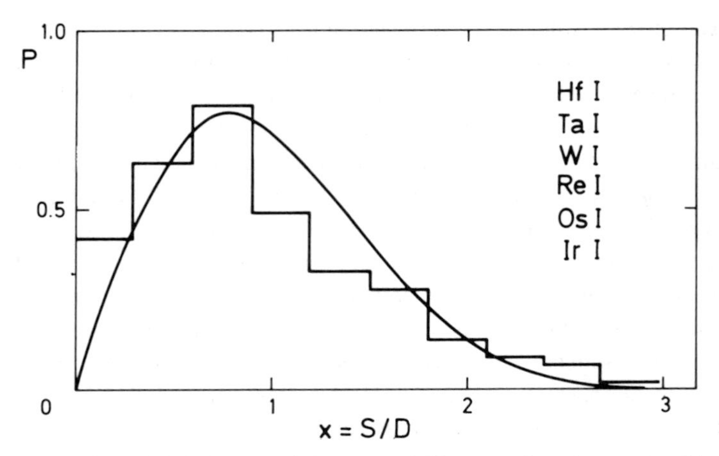

FIG. 1.6. Plot of the density of nearest neighbor spacings between odd parity atomic levels of a group of elements in the region of osmium. The levels in each element were separated according to angular momentum, and separate histograms were constructed for each level series, and then combined. The elements and the number of contributed spacings are Hf I, 74; Ta I, 180; W I, 262; Re I, 165; Os I, 145; Ir I, 131, which lead to a total of 957 spacings. The solid curve corresponds to the Wigner surmise, Eq. (1.5.1). Reprinted with permission from Annales Academiae Scientiarum Fennicae, Porter, C. E., and Rosenzweig, N. Statistical properties of atomic and nuclear spectra, *Annales Academiae Scientiarum Fennicae, Serie A, VI Physica*, 44, 1–66 (1960).

by the experimental evidence. Unfortunately, such experiments are difficult to perform neatly and no clear conclusion is yet available. See the discussion in Brody et al. (1981).

1.7. Analysis of Experimental Nuclear Levels

The enormous amount of available nuclear data was recently analyzed by French and coworkers, specially in view of deriving an upper bound for the time reversal violating part of the nuclear forces. Actually, if nuclear forces are time reversal invariant, then the nuclear energy levels should behave as the eigenvalues of a random real symmetric matrix; if they are not, then they should behave as those of a random Hermitian matrix. The level sequences in the two cases have different properties; for example, the level spacing curves are quite distinct. Figures 1.6 and 1.7 indicate that nuclear and atomic levels behave as the eigenvalues

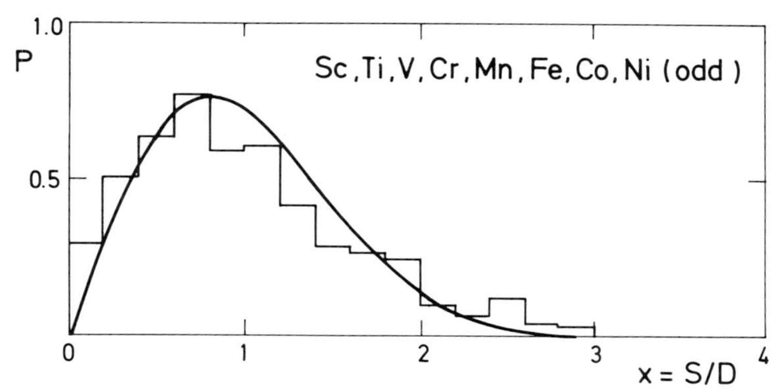

FIG. 1.7. Plot of the density of nearest neighbor spacings between odd parity levels of the first period corresponding to a separation of the levels into sequences each of which is labeled by definite values of S, L, and J. Comparision with the Wigner surmise (solid curve) shows a good fit if the (approximate) symmetries resulting from (almost) complete absence of spin-orbit forces are taken into consideration. Reprinted with permission from Annales Academiae Scientiarum Fennicae, Porter, C. E., and Rosenzweig, N. Statistical properties of atomic and nuclear spectra, *Annales Academiae Scientiarum Fennicae, Serie A, VI Physica*, 44, 1–66 (1960).

of a random real symmetric matrix and the admixture of a Hermitian anti-symmetric part, if any, is small. Figure 1.4 shows the level spacing histogram for a really big sample of available nuclear data from which French and coworkers (1985) could deduce an upper bound for this small admixture. Figure 1.8 shows how the probability density of the nearest neighbor spacings of atomic levels changes as one goes from one long period to another in the periodic table of elements. For further details on the analysis of nuclear levels and other sequences of levels, see Chapter 16.

1.8. The Zeros of the Riemann Zeta Function

The zeta function of Riemann is defined for $\operatorname{Re} z > 1$ by

$$\zeta(z) = \sum_{n=1}^{\infty} n^{-z} = \prod_{p} (1 - p^{-z})^{-1}, \tag{1.8.1}$$

and for other values of z by its analytical continuation. The product in Eq. (1.8.1) is taken over all primes p, and is known as the Euler product formula.

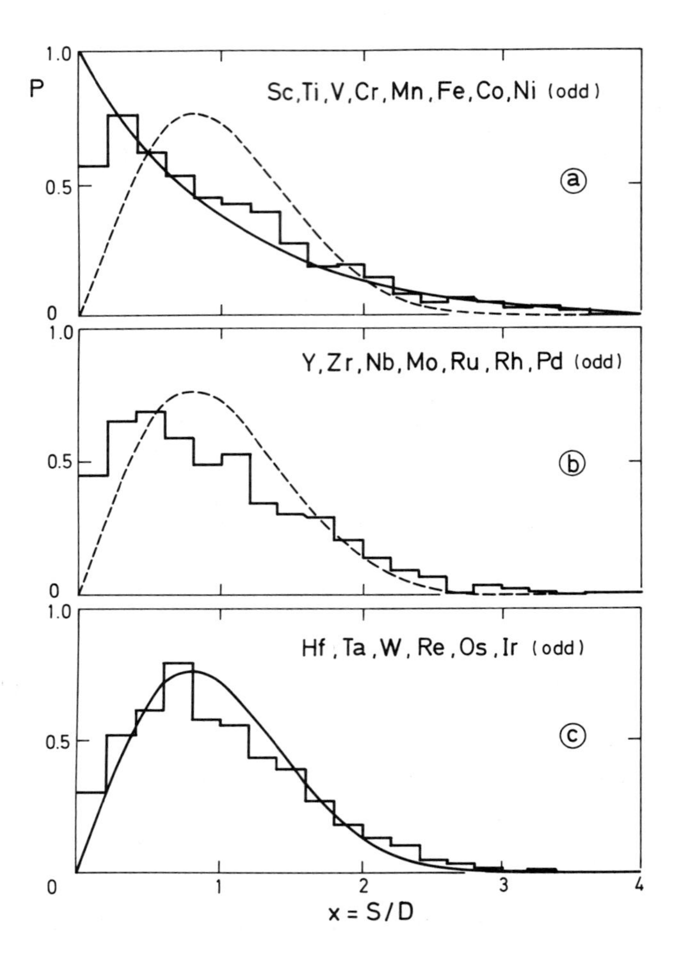

FIG. 1.8. Empirical density of nearest neighbor spacings between odd parity levels of elements in the first, second, and third long periods (histograms a, b, and c, respectively). To obtain these figures separate histograms were constructed for the J sequences of each element and then the results were combined. Comparision with the Poisson (or exponential) distribution and Wigner surmise (also shown) indicates that curves go from Poisson to Wigner curves as one goes from the first to the second and then finally to the third period. This variation can be understood in terms of the corresponding increase in strength of the spin dependent forces. Reprinted with permission from Annales Academiae Scientiarum Fennicae, Porter, C. E., and Rosenzweig, N. Statistical properties of atomic and nuclear spectra, *Annales Academiae Scientiarum Fennicae, Serie A, VI Physica*, 44, 1–66 (1960).

It is well known that $\zeta(z)$ is zero for $z = -2n$, $n = 1, 2, \ldots$; these are called the "trivial zeros." All the other zeros of $\zeta(z)$ lie in the strip $0 < \text{Re } z < 1$, and are symmetrically situated with respect to the critical line Re $z = 1/2$. It is conjectured that they actually lie on the critical line Re $z = 1/2$ (Riemann hypothesis, 1876). Since $\zeta(z^*) = \zeta^*(z)$, if z is a zero of $\zeta(z)$, then z^* is also a zero of it. Assuming the truth of the Riemann hypothesis (RH), let $z = 1/2 + i\gamma_n$, γ_n real, be the "nontrivial" zeros of $\zeta(z)$. How the γ_n are distributed on the real line?

These questions have their importance in number theory. Many, as yet unsuccessful, attempts have been made to prove or disprove the RH. With the advent of electronic computers of ever faster speeds and larger memories, a series of efforts have also been made to verify the truth of the RH and to evaluate some statistical properties of the γ_n.

An eventual proof of the RH is important for several reasons: (i) it has remained a challenge for such a long time; (ii) it will imply a significant improvement in the estimation of various arithmetic functions, such as $\pi(x)$, the number of primes $\leq x$; (iii) it will shed some light on the problem of determining effectively all imaginary quadratic fields $Q\sqrt{-d}$ having a given class number h. However, one needs some further knowledge of the distribution of the γ_n to answer other questions like how the largest gap between two consecutive primes $\leq x$ increases with x. Or can one approximate real numbers by rational numbers whose numerator and denominator are both primes; more specifically, is it true that for every irrational number θ there are infinitely many prime numbers p, q such that $|\theta - p/q| < q^{-2+\varepsilon}$. For such questions the RH alone is not sufficient; one has to know more about the γ_n.

Until the 1970s few persons were interested in the distribution of the γ_n. The main reason seems to be the feeling that if one cannot prove (or disprove!) the γ_n to be all real (RH), then there is no point in asking even harder questions. Montgomery (1973) took the lead in investigating questions about the distribution of the γ_n. Assuming the RH (i.e., that the γ_n are real), one can show (*cf.* Titchmarsh, 1972) that the number $N(T)$ of the γ_n with $0 < \gamma_n \leq T$ is

$$N(T) = \frac{T}{2\pi} \ln\left(\frac{T}{2\pi}\right) - \frac{T}{2\pi} + S(T) + \frac{7}{8} + O\left(T^{-1}\right), \qquad (1.8.2)$$

as $T \to \infty$, where

$$S(t) = \frac{1}{\pi} \arg \zeta \left(\frac{1}{2} + it \right). \tag{1.8.3}$$

Maximum order of $S(T)$ remains unknown. It is known that $S(T) = O(\ln T)$ and if one assumes the RH, then $S(T) = O(\ln T/\ln \ln T)$ (*cf.* Titchmarsh, 1972). Probably

$$S(T) = O\left(\left(\frac{\ln T}{\ln \ln T} \right)^{1/2} \right), \tag{1.8.4}$$

or more likely (Montgomery, private communication)

$$S(T) = O\left((\ln T \ln \ln T)^{1/2} \right). \tag{1.8.5}$$

Assuming the RH Montgomery studied $D(\alpha, \beta)$, the number of pairs γ, γ' such that $\zeta(1/2 + i\gamma) = \zeta(1/2 + i\gamma') = 0$, $0 < \gamma \leq T$, $0 < \gamma' \leq T$, and $2\pi\alpha/(\ln T) \leq \gamma - \gamma' \leq 2\pi\beta/(\ln T)$. Or taking the Fourier transforms, it amounts to evaluating the function

$$F(\alpha) = \frac{2\pi}{T \ln T} \sum_{0 < \gamma, \gamma' \leq T} T^{i\alpha(\gamma - \gamma')} \frac{4}{4 + (\gamma - \gamma')^2} \tag{1.8.6}$$

for real α. Since F is symmetric in γ, γ', it is real and even in α. Montgomery (1973) showed that if RH is true then $F(\alpha)$ is nearly non-negative, $F(\alpha) \geq -\varepsilon$ uniformly in α for $T > T_0(\varepsilon)$, and that

$$F(\alpha) = [1 + o(1)]\, T^{-2\alpha} \ln T + \alpha + o(1) \tag{1.8.7}$$

uniformly for $0 \leq \alpha \leq 1$. For $\alpha > 1$, the behavior changes. He also gave heuristic arguments to suggest that, for $\alpha \geq 1$,

$$F(\alpha) = 1 + o(1). \tag{1.8.8}$$

And from this conjecture he deduced that

$$\frac{2\pi}{T \ln T} D(\alpha, \beta) \approx \int_\alpha^\beta \left(1 - \left(\frac{\sin r}{r} \right)^2 + \delta(\alpha, \beta) \right) dr \tag{1.8.9}$$

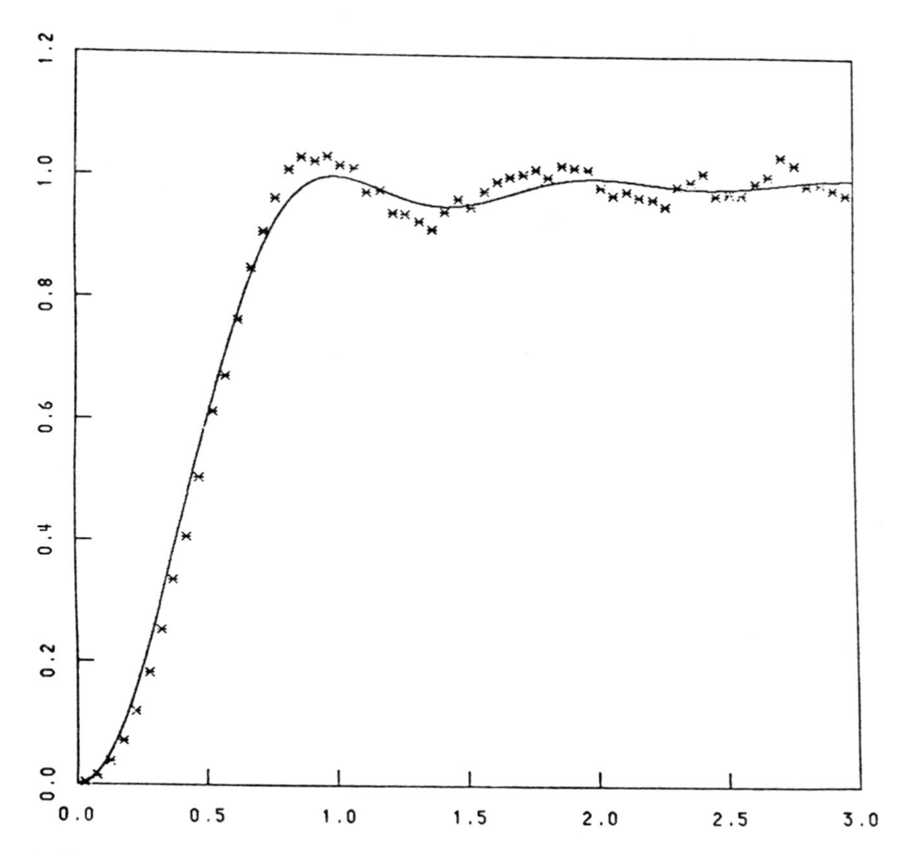

FIG. 1.9. Two point correlation function for the zeros $0.5 \pm i\gamma_n$, γ_n real, of the Riemann zeta function; $1 < n < 10^5$. The solid curve is Montgomery's conjecture, Eq. (1.8.10). Reprinted from "On the Distribution of Spacings Between Zeros of the Zeta Function," A. M. Odlyzko, *Mathematics of Computation*, (1987), pages 273–308, by permission of the American Mathematical Society.

for any real α, β with $\alpha < \beta$. Here $\delta(\alpha, \beta) = 1$ if $\alpha < 0 < \beta$, and $\delta(\alpha, \beta) = 0$ otherwise. This $\delta(\alpha, \beta)$ is there because for $\alpha < 0 < \beta$, $D(\alpha, \beta)$ includes terms with $\gamma = \gamma'$.

Equation (1.8.8) says that the two-point correlation function of the zeros of the zeta function $\zeta(z)$ on the critical line is

$$R_2(r) = 1 - [\sin(\pi r)/(\pi r)]^2 . \tag{1.8.10}$$

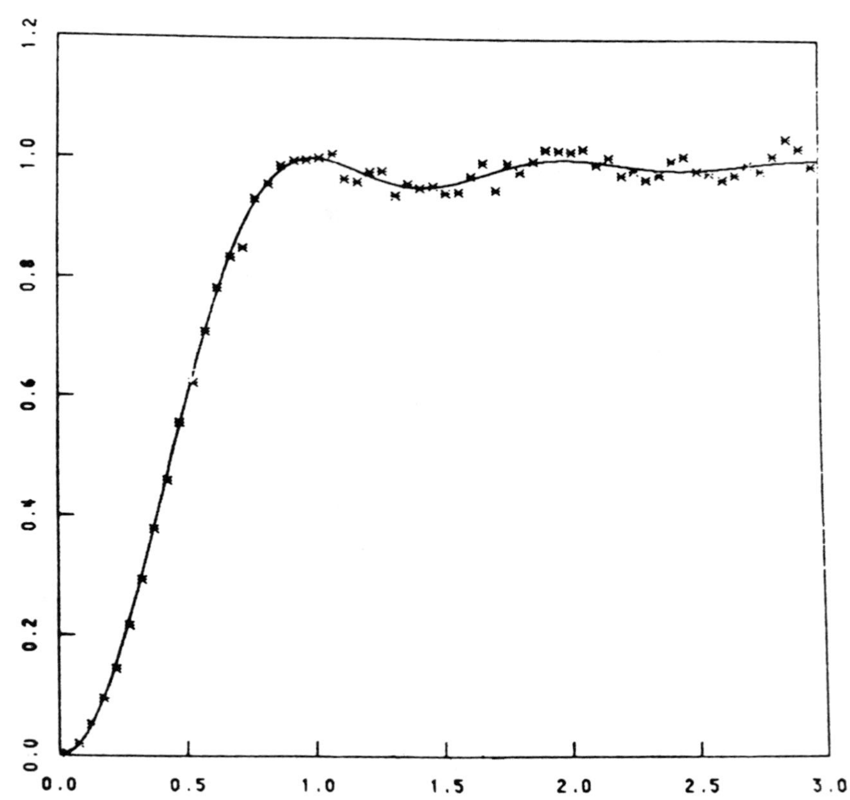

FIG. 1.10. The same as Figure 1.9 with $10^{12} < n < 10^{12} + 10^5$; Note that the curves fit much better. Reprinted from "On the Distribution of Spacings Between Zeros of the Zeta Function," A. M. Odlyzko, *Mathematics of Computation*, (1987), pages 273–308, by permission of the American Mathematical Society.

As we will see in Chapter 5.2, this is precisely the two-point correlation function of the eigenvalues of a random Hermitian matrix taken from the Gaussian unitary ensemble (GUE) (or that of a random unitary matrix taken from the circular unitary ensemble, *cf.* Chapter 10.1). This is consistent with the view (quoted to be conjectured by Polya and Hilbert) that the zeros of $\zeta(z)$ are related to the eigenvalues of a Hermitian operator. As the two-point correlation function seems to be the same for the zeros of $\zeta(z)$ and the eigenvalues of a matrix from the GUE, it is natural to think that other statistical properties also coincide. With a view

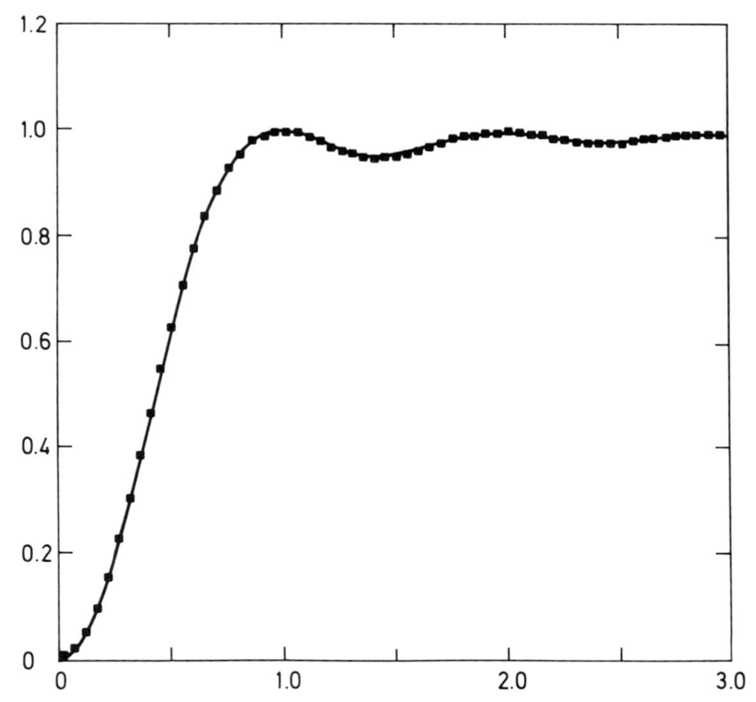

FIG. 1.11. The same as Figure 1.9, but for 79 million zeros around $n \approx 10^{20}$. From Odlyzko (1989). Copyright ©, 1989, American Telephone and Telegraph Company reprinted with permission.

to prove (or disprove!) this expectation, Odlyzko (1987) and Odlyzko and Schönhage, 1988) computed a large number of zeros $1/2 + i\gamma_n$ of $\zeta(z)$ with great accuracy. Taking γ_n to be positive and ordered, i.e., $0 < \gamma_1 \leq \gamma_2 \leq \cdots$ with $\gamma_1 = 14.134...$, $\gamma_2 = 21.022...$, etc., he computed sets of 10^5 zeros γ_n with a precision $\pm 10^{-8}$, and $N+1 \leq n \leq N+10^5$, for $N = 0$, $N = 10^6$, $N = 10^{12}$, $N = 10^{18}$, and recently for $N = 10^{20}$. (This computation gives, for example, $\gamma_n = 15202440115920747268.6290299...$ for $n = 10^{20}$; *cf.* Odlyzko, 1989 or Cipra, 1989) This huge numerical undertaking was possible only with the largest available supercomputer CRAY X-MP and fast efficient algorithms devised for the purpose. The numerical evidence so collected is shown on Figures 1.9 to 1.14.

Note that as one moves away from the real axis, the fit improves for

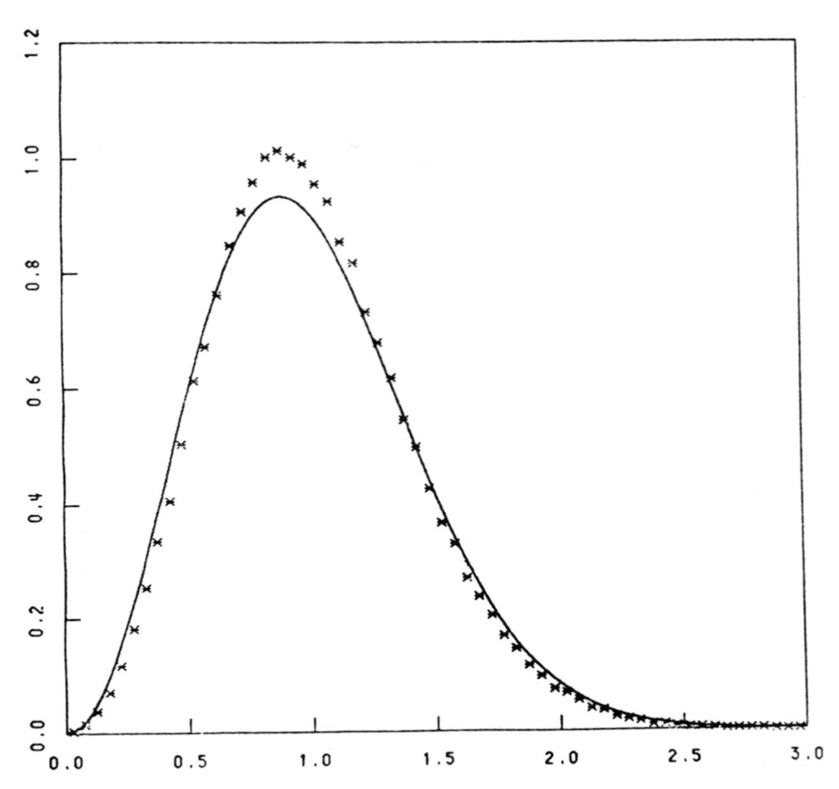

FIG. 1.12. Plot of the density of normalized spacings for the zeros $0.5 \pm i\gamma_n$, γ_n real, of the Riemann zeta function on the critical line. $1 < n < 10^5$. The solid curve is the spacing probability density for the Gaussian unitary ensemble, Eq. (5.4.32). From Odlyzko (1987). Reprinted from "On the Distribution of Spacings Between Zeros of the Zeta Function," A. M. Odlyzko, *Mathematics of Computation*, (1987), pages 273–308, by permission of the American Mathematical Society.

both the spacing as well as the two point correlation function. Surprisingly the convergence is very slow; the numerical curves become indistinguishable from the GUE curves only far away from the real axis, i.e., when $n \geq 10^{12}$. In contrast, for Hermitian matrices the limit is practically reached for matrix orders 20×20. It is hard to imagine the zeros of the zeta function as the eigenvalues of some unitary or Hermitian operator.

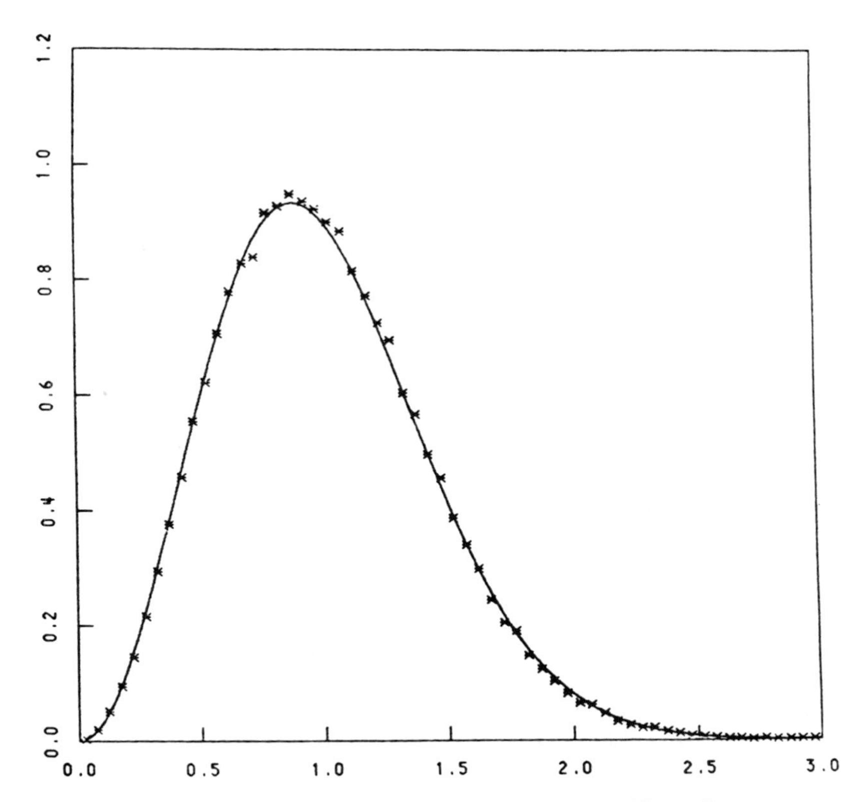

FIG. 1.13. The same as Figure 1.12 with $10^{12} < n < 10^{12} + 10^5$. Note the improvement in the fit. Reprinted from "On the Distribution of Spacings Between Zeros of the Zeta Function," A. M. Odlyzko, *Mathematics of Computation*, (1987), pages 273–308, by permission of the American Mathematical Society.

One may also compare other statistical quantities relating to the zeros of the Riemann zeta with the corresponding predictions of the GUE; see Chapter 16.

A generalization of the Riemann zeta function is the function $\zeta(z, a)$ defined for $\text{Re } z > 1$, by

$$\zeta(z, a) = \sum_{n=0}^{\infty} (n + a)^{-z}, \quad 0 < a \leq 1, \tag{1.8.11}$$

and by its analytical continuation for other values of z. For $a = 1/2$ and

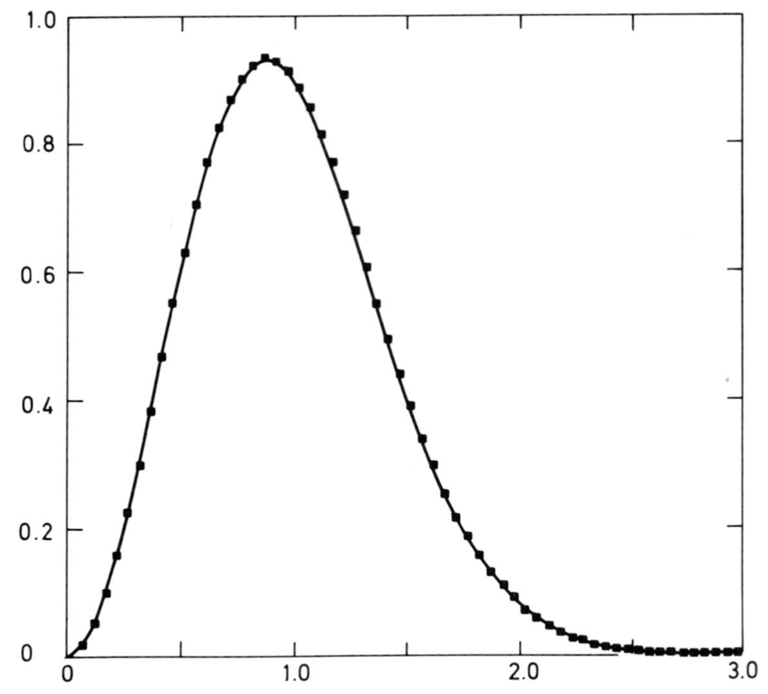

FIG. 1.14. The same as Figure 1.12 but for the 79 million zeros around the 10^{20}th zero. From Odlyzko (1989). Copyright ©, 1989, American Telephone and Telegraph Company reprinted with permission.

$a = 1$, one has

$$\zeta(z, 1/2) = (2^z - 1)\zeta(z), \quad \zeta(z, 1) = \zeta(z), \tag{1.8.12}$$

so there is nothing more about their zeros. For values of a other than 1/2 or 1 it is known that $\zeta(z, a)$ has an infinity of zeros with Re $z > 1$. (Davenport and Heilbronn, 1936; Cassels, 1961.)

For any positive definite quadratic form

$$Q(x, y) = a^2 + bxy + cy^2, \qquad a, b, c \quad \text{integers}, \tag{1.8.13}$$

the Epstein zeta function is defined for Re $z > 1$ by

$$\zeta(z, Q) = \sum_{(m,n) \neq (0,0)} Q(m, n)^{-z}, \tag{1.8.14}$$

and for other values of z by its analytical continuation. This Epstein zeta function satisfies the functional equation

$$\left(\frac{2\pi}{\sqrt{-d}}\right)^{-z}\Gamma(z)\zeta(z) = \left(\frac{2\pi}{\sqrt{-d}}\right)^{z-1}\Gamma(1-z)\zeta(1-z), \tag{1.8.15}$$

where d is the discriminant of Q, $d = b^2 - 4ac < 0$. But it has no Euler product formula. Let $h(d)$ be the class function, i.e., $h(d)$ is the number of such different quadratic forms Q having the the same discriminant d. If $h(d) = 1$, then the RH seems to be true, i.e., all the nontrivial zeros of $\zeta(z, Q)$ seem to lie on the critical line Re $z = 1/2$. If $h(d) > 1$, then it is known that $\zeta(z, Q)$ has an infinity of zeros with Re $z > 1$. (Davenport and Heilbronn, 1936.)

In case $h(d) > 1$, it is better to consider the sum

$$\zeta_d(z) = \sum_{r=1}^{h(d)} \zeta(z, Q_r), \tag{1.8.16}$$

where the quadratic forms Q_1, Q_2,...,$Q_{h(d)}$ all have the same discriminant d. This function is proportional to a Dedekind zeta function, satisfies a functional equation, and has an Euler product formula. It has no zeros with Re $z > 1$, and it is believed that all its nontrival zeros lie on the critical line Re $z = 1/2$ (Potter and Titchmarsh, 1935).

For example, $Q_1 = x^2 + y^2$ is the only quadratic form with $d = -4$, while $Q_2 = x^2 + 5y^2$ and $Q_3 = 3x^2 + 4xy + 3y^2$ both have $d = -20$. Thus, it is known that each of the functions

$$\zeta_2 = \sum_{(m,n)\neq(0,0)} (m^2 + 5n^2)^{-z}$$

and

$$\zeta_3 = \sum_{(m,n)\neq(0,0)} (3m^2 + 4mn + 3n^2)^{-z}$$

has an infinity of zeros in the half plane Re $z > 1$, while their sum $\zeta_2 + \zeta_3$ and the function

$$\zeta_1 = \sum_{(m,n)\neq(0,0)} (m^2 + n^2)^{-z}$$

are never zero in the same half plane. Moreover, it is believed that all the nontrivial zeros of ζ_1 and of $\zeta_2 + \zeta_3$ lie on the critical line Re $z = 1/2$.

What about their distribution?

As another generalization one may consider for example,

$$Z_k(z) = \sum_{n_1,\ldots n_k} \left(n_1^2 + \cdots + n_k^2\right)^{-z} \tag{1.8.17}$$

where the sum is taken over all integers $n_1, ..., n_k$ except when $n_1 = \cdots = n_k = 0$. The zeta function, Eq. (1.8.17), is defined with respect to a hypercubic lattice in k dimensions. Equation (1.8.1) is the special case $k = 1$ of Eq. (1.8.17). Instead of the hypercubic lattice one can take any other lattice and define a zeta function corresponding to this lattice. Or one could consider the Dirichlet series

$$L(z) = \sum_{n \in \mathcal{N}^r} P(n)^{-z} \tag{1.8.18}$$

where $P(x)$ is a polynomial in r variables $x_1, ..., x_r$ with real non-negative coefficients, the sum is taken over all positive integers $n_1, ..., n_r$ except for the singular points of $P(x)$, if any. Thus if $P(x) = x$, we have $L(z) \approx \zeta(z)$; if $P(x) = a + bx$, a, b real and > 0, then we have the Hurwitz zeta function

$$\zeta_{a,b}(z) = \sum_{n=0}^{\infty} (a + bn)^{-z}, \quad \text{Re } z > 1. \tag{1.8.19}$$

Such Dirichlet series or general zeta functions, though they do not have a functional equation in general, they have many important properties similar to those of the Riemann or Hurwitz zeta functions. For example, $\Gamma(z)\zeta_{a,b}(z)$ [with Hurwitz function $\zeta_{a,b}(z)$, Eq. (1.8.19), and the gamma function $\Gamma(z)$], has simple poles at $z = 1, 0, -1, -2, ...$ with residues rational in (a, b). The "nontrivial" zeros of the hypercubic lattice zeta function $Z_k(z)$, Eq. (1.8.17), for example, are thought to be on the critical line Re $z = 1/2$. What about their distribution?

Little is known about the zeros of such general zeta functions or of the Dirichlet L-series, even empirically. It will be interesting to see what is the distribution of their (nontrivial) zeros and if they have any thing to do with the GUE results.

The fluctuation properties of the nuclear energy levels or that of the levels of a chaotic system are quite different: they behave, in the absence of a strong magnetic field, as the eigenvalues of a matrix from the orthogonal ensemble, *cf.* Chapter 6.

1.9. Things Worth Consideration, but Not Treated in This Book

Another much studied problem is that of percolation, or a random assembly of metals and insulators, or that of normal and superconductors. Consider again a 2- or 3-dimensional lattice. Let each bond of the lattice be open with probability p and closed with probability $1-p$. Or p and $1-p$ may be considered as the respective probabilities of the bond being conducting or insulating or of it being a normal or a superonductor. The question is what is the probability that one can pass from one end to the other of a large lattice? Or, how conducting or superconducting is a large lattice? It is clear that if $p = 0$, no bond is open, and the probability of passage is zero. What is not so clear is that it remains zero for small positive values of p. Only when p increases and passes beyond a certain critical value p_c, this probability attains a nonzero value. It is known from numerical studies that the probability of passage is proportional to $(p-p_c)^{\alpha}$ for $p \geq p_c$, where the constant α, known as the critical index, is independent of the lattice and depends only on the dimension. The mathematical problem here can be characterized in its simplest form as follows. Given two fixed matrices A and B of the same order, let the matrix M_i for $i = 1, 2, ..., N$ be equal to A with probability p and equal to B with probability $1-p$. How does the product $M = M_1 M_2 \cdots M_N$ behave when N is large? For example, what is the distribution of its eigenvalues? Of course, instead of just two matrices A and B and probabilities p and $1-p$, we can take a certain number of them with their respective probabilities, but the problem is quite difficult even with two. No good analytical solution is known.

The actual problem of percolation or conduction is more complicated. Numerical simulation is performed on a ribbon of finite width, ≈ 10, for two dimensions (see Figure 1.15), or on a rod of finite cross section, $\approx$ 5×5, for three dimensions. If all the resistences of the lattice lines are given, then we can explicitly compute the currents flowing through the

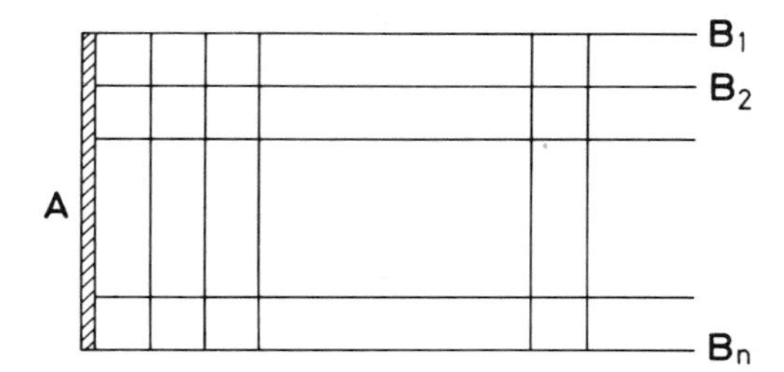

FIG. 1.15. A model for the numerical simulation of conduction through a random assembly of conductors and insulators. At each step a new layer of randomly chosen conductors and insulators is added to the right and the subsequent change in the currents through the open ends $B_1, \ldots, B_n$ is studied.

open ends B_1, B_2, ..., B_n, when an electric field is applied between A and these ends. Adding one more layer in the length of the ribbon, the currents in the ends B_1, ..., B_n change and this change can be characterized by an $n \times n$ matrix T, called the transfer matrix. This matrix T gives the new currents in terms of the old ones; it depends on the resistences of the lattice lines we have just added. As these resistences are random, say c with probability p and ∞ with probability $1-p$, the matrix T is random; its matrix elements are constructed from the added random resistences according to known laws. For a length L of the ribbon, the currents in the ends B_1, ..., B_n are determined by the product of L matrices $T_1 T_2 \cdots T_L$. Computing the spectrum of the product of L matrices for L large, $\approx 10^6$, and studying how it depends on L and on the order n, one can get a fair estimate for the critical index α.

Chaotic systems are simulated by a particle free to move inside a certain 2-dimensional domain. If the plane is Euclidean, then the shape of the domain is chosen so that the Laplace equation in the domain with Dirichlet boundary conditions has no analytic solution. This usually happens when the classical motion of the particle in the domain with elastic bouncing at the boundary has no, or almost no, periodic orbits. If the plane has a negative curvature, the boundary is usually taken to be polygonal with properly chosen angles. Sometimes one takes two quartic coupled oscillators with the potential energy $x^4 + y^4 + \lambda x^2 y^2$ as a model of a chaotic system. Why the spectrum of all these systems has to obey the predictions of the random matrix theory is not at all clear.

Still another curious instance has been lately reported (Balasz and Voros, 1989). Consider the matrix of the finite Fourier transform with matrix elements $A_{jk}(n) = n^{-1/2}\omega_n^{jk}$, $\omega_n = \exp(2\pi i/n)$. This $A(n)$ is unitary for every $n \geq 1$. Its eigenvalues are either ± 1 or $\pm i$. Let $B(n) = A^{-1}(2n)\left[A(n)\dot{+}A(n)\right]$ be the $2n \times 2n$ matrix, the product of the inverse of $A(2n)$ and the direct sum of $A(n)$ and $A(n)$. This $B(n)$ is also unitary, so that its eigenvalues lie on the unit circle. The matrix $B(n)$ is in some sense the quantum analog of the baker's transformation.

Now when n is very large, the eigenvalues of $B(n)$ and the eigenvalues of a random matrix taken from the circular unitary ensemble (see Chapters 9 and 10) have almost the same statistical properties. Instead of the above $B(n)$, we could have considered $B(n) = A^{-1}(3n)\left[A(n)\dot{+}A(2n)\right]$ or $B(n) = A^{-1}(3n)\left[A(n)\dot{+}A(n)\dot{+}A(n)\right]$, and the result would have been the same. Why this should be so is not known.

Fluctuations in the electrical conductivity of heterogeneous metal junctions in a magnetic field and in general the statistical properties of transmission through a random medium is related to the theory of random matrices. In this connection see the recent articles by Al'tshuler, Mailly, Mello, Muttalib, Pichard, Zano, and coworkers.

The hydrogen atom in an intense magnetic field has a spectrum characteristic of matrices from the Gaussian unitary ensemble; see Delande and Gay (1986).

2 / Gaussian Ensembles. The Joint Probability Density Function for the Matrix Elements

After examining the consequences of time-reversal invariance, we introduce Gaussian ensembles as a mathematical idealization. They are implied if we make the hypothesis of maximum statistical independence allowed under the symmetry constraints.

2.1. Preliminaries

In the mathematical model our systems are characterized by their Hamiltonians, which in turn are represented by Hermitian matrices. Let us look into the structure of these matrices. The low-lying energy levels (eigenvalues) are far apart and each may be described by a different set of quantum numbers. As we go to higher excitations, the levels draw closer, and because of their mutual interference most of the approximate quantum numbers lose their usefulness, for they are no longer exact. At still higher excitations the interference is so great that some quantum numbers may become entirely meaningless. However, there may be certain exact integrals of motion, such as total spin or parity, and the quantum numbers corresponding to them are conserved whatever the excitation may be. If the basis functions are chosen to be the eigenfunctions of these conserved quantities, all Hamiltonian matrices of the ensemble will reduce to the form of diagonal blocks. One block will correspond uniquely to each set of exact quantum numbers. The matrix elements lying outside these blocks will all be zero, and levels belonging to two different blocks will be statistically uncorrelated. As to the levels corresponding

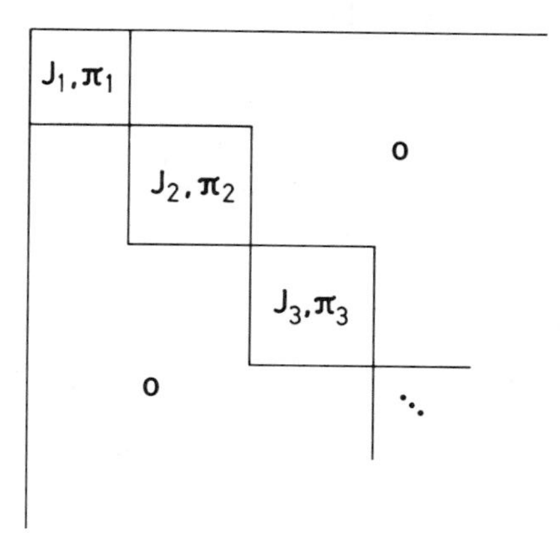

FIG. 2.1. Block diagonal structure of a Hamiltonian matrix. Each diagonal block corresponds to a set of exact symmetries or to a set of exactly conserved quantum numbers. The matrix elements connecting any two diagonal blocks are zero, whereas those inside each diagonal block are random.

to the same block, the interactions are so complex that any regularity resulting from partial diagonalization will be washed out. (See Figure 2.1.)

We shall assume that such a basis has already been chosen and restrict our attention to one of the diagonal blocks, an $(N \times N)$ Hermitian matrix in which N is a large but fixed positive integer. Because nuclear spectra contain at least hundreds of levels with the same spin and parity, we are interested in (the limit of) very large N.

With these preliminaries, the matrix elements may be supposed to be random variables and allowed the maximum statistical independence permitted under symmetry requirements. To specify precisely the correlations among various matrix elements we need a careful analysis of the consequences of time-reversal invariance.

2.2. Time-Reversal Invariance

We begin by recapitulating the basic notions of time-reversal invariance. From physical considerations, the time-reversal operator is required to be antiunitary (Wigner, 1959) and can be expressed, as any other antiunitary operator, in the form

$$T = KC, \tag{2.2.1}$$

where K is a fixed unitary operator and the operator C takes the complex conjugate of the expression following it. Thus a state under time reversal transforms to

$$\psi^R = T\psi = K\psi^*, \tag{2.2.2}$$

ψ^* being the complex conjugate of ψ. From the condition

$$(\Phi, A\psi) = \left(\psi^R, A^R\Phi^R\right)$$

for all pairs of states ψ, Φ, and (2.2.2), we deduce that under time reversal an operator A transforms to

$$A^R = KA^TK^{-1}, \tag{2.2.3}$$

where A^T is the transpose of A. A is said to be self-dual if $A^R = A$. A physical system is invariant under time reversal if its Hamiltonian is self-dual, that is, if

$$H^R = H. \tag{2.2.4}$$

When the representation of the states is transformed by a unitary transformation, $\psi \longrightarrow U\psi$, T transforms according to

$$T \longrightarrow UTU^{-1} = UTU^{\dagger} \tag{2.2.5}$$

or K transforms according to

$$K \longrightarrow UKU^T. \tag{2.2.6}$$

Because operating twice with T should leave the physical system unchanged, we have

$$T^2 = \alpha \times 1, \qquad |\alpha| = 1, \tag{2.2.7}$$

where 1 is the unit operator; or

$$T^2 = KCKC = KK^*CC = KK^* = \alpha \times 1. \tag{2.2.8}$$

But K is unitary:

$$K^* K^T = 1.$$

From these two equations we get

$$K = \alpha K^T = \alpha \left(\alpha K^T\right)^T = \alpha^2 K.$$

Therefore,

$$\alpha^2 = 1 \quad \text{or} \quad \alpha = \pm 1, \tag{2.2.9}$$

so that the unitary matrix K is either symmetric or antisymmetric. In other words, either

$$KK^* = 1 \tag{2.2.10}$$

or

$$KK^* = -1. \tag{2.2.11}$$

These alternatives correspond, respectively, to an integral or a half-odd integral total angular momentum of the system measured in units of $\hbar$ (Wigner, 1959), for the total angular momentum operator $\mathbf{J} = (J_1, J_2, J_3)$ must transform as

$$J_\ell^R = -J_\ell, \quad \ell = 1, 2, 3. \tag{2.2.12}$$

For brevity we call the two possibilities the even-spin and odd-spin case, respectively.

2.3. Gaussian Orthogonal Ensemble

Suppose now that the even-spin case holds and (2.2.10) is valid. Then a unitary operator U will exist such that (*cf.* Appendix A.3)

$$K = UU^T. \tag{2.3.1}$$

By (2.2.6) a transformation $\psi \longrightarrow U^{-1}\psi$ performed on the states ψ brings K to unity. Thus in the even-spin case the representation of states can always be chosen so that

$$K = 1. \tag{2.3.2}$$

After one such representation is found, further transformations $\psi \longrightarrow R\psi$ are allowed only with R a real orthogonal matrix so that (2.3.2) remains valid. The consequence of (2.3.2) is that self-dual matrices are symmetric. In the even-spin case every system invariant under time reversal will be associated with a real symmetric matrix H if the representation of states is suitably chosen. For even-spin systems with time-reversal invariance the Gaussian orthogonal ensemble E_{1G}, defined below, is therefore appropriate.

Definition 2.3.1. The Gaussian orthogonal ensemble E_{1G} is defined in the space T_{1G} of real symmetric matrices by two requirements:

1. The ensemble is invariant under every transformation

$$H \longrightarrow W^T H W \tag{2.3.3}$$

of T_{1G} into itself, where W is any real orthogonal matrix.

2. The various elements H_{kj}, $k \leq j$, are statistically independent.

These requirements, expressed in the form of equations, read as follows:

1. The probability $P(H)dH$ that a system of E_{1G} will belong to the volume element $dH = \Pi_{k \leq j} dH_{kj}$ is invariant under real orthogonal transformations:

$$P(H')dH' = P(H)dH, \tag{2.3.4}$$

where

$$H' = W^T H W \tag{2.3.5}$$

and

$$W^T W = W W^T = 1. \tag{2.3.6}$$

2. This probability density function $P(H)$ is a product of functions, each of which depends on at most a single variable:

$$P(H) = \prod_{k \leq j} f_{kj}\left(H_{kj}\right). \tag{2.3.7}$$

Suppose, next, that we are dealing with a system invariant under space rotations. The spin may now be even or odd. The Hamiltonian matrix H which represents the system commutes with every component of $\mathbf{J}$.

If we use the standard representation of the J matrices with J_1 and J_3 real and J_2 pure imaginary, Eq. (2.2.12) may be satisfied by the usual choice (Wigner, 1959)

$$K = e^{i\pi J_2} \tag{2.3.8}$$

for K. With this choice of K, H and K commute and H^R reduces to H^T. Thus a rotation-invariant system is represented by a real symmetric matrix H, and once again the ensemble E_{1G} is appropriate.

2.4. Gaussian Symplectic Ensemble

In this section we discuss a system to which E_{1G} does not apply, a system with odd spin, invariant under time reversal, but having no rotational symmetry. In this case, Eq. (2.2.11) holds, K cannot be diagonialized by any transformation of the form (2.2.6), and there is no integral of the motion by which the double-valuedness of the time-reversal operation can be trivially eliminated.

Every antisymmetric unitary operator can be reduced by a transformation (2.2.6) to the standard canonical form (*cf.* Appendix A.3)

$$Z = \begin{bmatrix} 0 & +1 & 0 & 0 & \cdots \\ -1 & 0 & 0 & 0 & \cdots \\ 0 & 0 & 0 & +1 & \cdots \\ 0 & 0 & -1 & 0 & \cdots \\ \cdots & \cdots & \cdots & \cdots & \cdots \\ \cdots & \cdots & \cdots & \cdots & \cdots \end{bmatrix} \equiv \begin{bmatrix} 0 & +1 \\ -1 & 0 \end{bmatrix} \dot{+} \begin{bmatrix} 0 & +1 \\ -1 & 0 \end{bmatrix} \dot{+} \cdots \tag{2.4.1}$$

which consists of (2×2) blocks

$$\begin{bmatrix} 0 & +1 \\ -1 & 0 \end{bmatrix}$$

along the leading diagonal; all other elements of Z are zero. We assume that the representation of states is chosen so that K is reduced to this form. The number of rows and columns of all matrices must now be even, for otherwise K would be singular in contradiction to (2.2.11). It is convenient to denote the order of the matrices by $2N$ instead of N.

After one such representation is chosen, for which $K = Z$, further transformations $\psi \longrightarrow B\psi$ are allowed, only with B a unitary $(2N \times 2N)$ matrix for which

$$Z = BZB^T. \tag{2.4.2}$$

Such matrices B form precisely the N-dimensional symplectic group (Weyl, 1946), usually denoted by Sp(N).

It is well known (Chevalley, 1946; Dieudonné, 1955) that the algebra of the symplectic group can be expressed most conveniently in terms of quaternions. We therefore introduce the standard quaternion notation for (2×2) matrices,

$$e_1 = \begin{bmatrix} i & 0 \\ 0 & -i \end{bmatrix}, \quad e_2 = \begin{bmatrix} 0 & 1 \\ -1 & 0 \end{bmatrix}, \quad e_3 = \begin{bmatrix} 0 & i \\ i & 0 \end{bmatrix}, \tag{2.4.3}$$

with the usual multiplication table

$$e_1^2 = e_2^2 = e_3^2 = -1, \tag{2.4.4}$$

$$e_1 e_2 = -e_2 e_1 = e_3, \quad e_2 e_3 = -e_3 e_2 = e_1, \quad e_3 e_1 = -e_1 e_3 = e_2. \tag{2.4.5}$$

Note that in (2.4.3), as well as throughout the rest of this book, i is the ordinary imaginary unit and not a quaternion unit. The matrices e_1, e_2, and e_3, together with the (2×2) unit matrix

$$1 = \begin{bmatrix} 1 & 0 \\ 0 & 1 \end{bmatrix},$$

form a complete set, and any (2×2) matrix with complex elements can be expressed linearly in terms of them with complex coefficients:

$$\begin{bmatrix} a & b \\ c & d \end{bmatrix} = \frac{1}{2}(a+d)1 - \frac{i}{2}(a-d)e_1 + \frac{1}{2}(b-c)e_2 - \frac{i}{2}(b+c)e_3. \tag{2.4.6}$$

All the $(2N \times 2N)$ matrices will be considered as cut into N^2 blocks of (2×2) and each (2×2) block expressed in terms of quaternions. In general, a $(2N \times 2N)$ matrix with complex elements thus becomes an $(N \times N)$ matrix with complex quaternion elements. In particular the matrix Z is now

$$Z = e_2 I, \tag{2.4.7}$$

where I is the $(N \times N)$ unit matrix. It is easy to verify that the rules of matrix multiplication are not changed by this partitioning.

Let us add some definitions. We call a quaternion "real" if it is of the form

$$q = q^{(0)} + \mathbf{q} \cdot \mathbf{e} \equiv q^{(0)} + q^{(1)} e_1 + q^{(2)} e_2 + q^{(3)} e_3, \tag{2.4.8}$$

with real coefficients $q^{(0)}, q^{(1)}, q^{(2)}$, and $q^{(3)}$. Thus a real quaternion does not correspond to a (2×2) matrix with real elements. Any complex quaternion has a "conjugate quaternion"

$$\bar{q} = q^{(0)} - \mathbf{q} \cdot \mathbf{e}, \tag{2.4.9}$$

which is distinct from its "complex conjugate"

$$q^* = q^{(0)*} + \mathbf{q}^* \cdot \mathbf{e}. \tag{2.4.10}$$

A quaternion with $q^* = q$ is real, one with $q^* = -q$ is pure imaginary, and one with $\bar{q} = q$ is a scalar. By applying both types of conjugation together, we obtain the "Hermitian conjugate"

$$q^\dagger = \bar{q}^* = q^{(0)*} - \mathbf{q}^* \cdot \mathbf{e}. \tag{2.4.11}$$

A quaternion with $q^\dagger = q$ is Hermitian and corresponds to the ordinary notion of a (2×2) Hermitian matrix; one with $q^\dagger = -q$ is anti-Hermitian. The conjugate (Hermitian conjugate) of a product of quaternions is the product of their conjugates (Hermitian conjugates) taken in the reverse order:

$$\overline{(q_1 q_2 \cdots q_n)} = \bar{q}_n \cdots \bar{q}_2 \bar{q}_1, \tag{2.4.12}$$

$$(q_1 q_2 \cdots q_n)^\dagger = q_n^\dagger \cdots q_2^\dagger q_1^\dagger. \tag{2.4.13}$$

Now consider a general $(2N \times 2N)$ matrix A which is to be written as an $(N \times N)$ matrix Q with quaternion elements q_{kj}; $k, j = 1, 2, ..., N$. The standard matrix operations on A are then reflected in Q in the following way.

Transposition:

$$\left(Q^T\right)_{kj} = -e_2 \bar{q}_{jk} e_2. \tag{2.4.14}$$

Hermitian conjugation:

$$\left(Q^{\dagger}\right)_{kj} = q_{jk}^{\dagger}. \tag{2.4.15}$$

Time reversal:

$$\left(Q^{R}\right)_{kj} = e_2 \left(Q^{T}\right)_{kj} e_2^{-1} = \bar{q}_{jk}. \tag{2.4.16}$$

The matrix Q^R is called the "dual" of Q. A "self-dual" matrix is one with $Q^R = Q$. That is, if

$$q_{jk} = \begin{bmatrix} a_{jk} & b_{jk} \\ c_{jk} & d_{jk} \end{bmatrix},$$

then $Q = [q_{jk}]$ is self-dual if

$$a_{jk} = d_{kj}, \quad b_{jk} = -b_{kj}, \quad \text{and} \quad c_{jk} = -c_{kj}. \tag{2.4.17}$$

The usefulness of quaternion algebra is a consequence of the simplicity of Eqs. (2.4.15) and (2.4.16). In particular, it is noteworthy that the time-reversal operator K does not appear explicitly in Eq. (2.4.16) as it did in Eq. (2.2.3). By Eqs. (2.4.15) and (2.4.16) the condition

$$Q^R = Q^{\dagger} \tag{2.4.18}$$

is necessary and sufficient for the elements of Q to be real quaternions. When (2.4.18) holds, we call Q "quaternion real."

A unitary matrix B that satisfies Eq. (2.4.2) is automatically quaternion real. In fact, it satisfies the conditions

$$B^R = B^{\dagger} = B^{-1}, \tag{2.4.19}$$

which define the symplectic group. The matrices H which represent the energy operators of physical systems are Hermitian as well as self-dual:

$$H^R = H, \quad H^{\dagger} = H, \tag{2.4.20}$$

hence are also quaternion real. From Eqs. (2.4.15) and (2.4.16) we see that the quaternion elements of a self-dual Hermitian matrix must satisfy

$$q_{jk}^{\dagger} = \bar{q}_{jk} = q_{kj} \tag{2.4.21}$$

or $q_{jk}^{(0)}$ must form a real symmetric matrix, whereas $q_{jk}^{(1)}$, $q_{jk}^{(2)}$, and $q_{jk}^{(3)}$ must form real antisymmetric matrices. Thus the number of real independent parameters that define a $(2N \times 2N)$ self-dual Hermitian matrix is

$$\frac{1}{2}N(N+1) + \frac{1}{2}N(N-1) \times 3 = N(2N-1).$$

From this notational excursion, let us come back to the point. Systems having odd-spin, invariance under time-reversal, but no rotational symmetry, must be represented by self-dual, Hermitian Hamiltonians. Therefore the Gaussian symplectic ensemble, as defined below, should be appropriate for their description.

Definition 2.4.1. The Gaussian symplectic ensemble E_{4G} is defined in the space T_{4G} of self-dual Hermitian matrices by the following properties:

1. The ensemble is invariant under every automorphism

$$H \longrightarrow W^R H W \tag{2.4.22}$$

of T_{4G} into itself, where W is any symplectic matrix.

2. Various linearly independent components of H are also statistically independent.

These requirements put in the form of equations read as follows:

1. The probability $P(H)dH$ that a system E_{4G} will belong to the volume element

$$dH = \prod_{k \le j} dH_{kj}^{(0)} \prod_{\lambda=1}^{3} \prod_{k<j} dH_{kj}^{(\lambda)} \tag{2.4.23}$$

is invariant under symplectic transformations; that is,

$$P(H')dH' = P(H)dH \tag{2.4.24}$$

if

$$H' = W^R H W, \tag{2.4.25}$$

where

$$W^R W = 1 \quad \text{or} \quad W Z W^T = Z. \tag{2.4.26}$$

2. The probability density function $P(H)$ is a product of functions each of which depends on a single variable:

$$P(H) = \prod_{k \le j} f_{kj}^{(0)}\left(H_{kj}^{(0)}\right) \prod_{\lambda=1}^{3} \prod_{k<j} f_{kj}^{(\lambda)}\left(H_{kj}^{(\lambda)}\right). \tag{2.4.27}$$

2.5. Gaussian Unitary Ensemble

Mathematically a much simpler ensemble is the Gaussian unitary ensemble E_{2G}, which applies to systems without invariance under time reversal. Such systems are easily created in principle by putting an ordinary atom or nucleus, for example, into an externally generated magnetic field. The external field is not affected by the time-reversal operation. However, for the unitary ensemble to be applicable, the splitting of levels by the magnetic field must be at least as large as the average level spacing in the absence of the magnetic field. The magnetic field must, in fact, be so strong that it will completely "mix up" the level structure that would exist in zero field; for otherwise our random hypothesis cannot be justified. This state of affairs could never occur in nuclear physics. In atomic or molecular physics a practical application of the unitary ensemble may perhaps be possible.

A system without time-reversal invariance has a Hamiltonian that may be an arbitrary Hermitian matrix not restricted to be real or self-dual. Thus we are led to the following definition.

Definition 2.5.1. The Gaussian unitary ensemble E_{2G} is defined in the space T_{2G} of Hermitian matrices by the following properties:

1. The probability $P(H)dH$ that a system of E_{2G} will belong to the volume element

$$dH = \prod_{k \le j} dH_{kj}^{(0)} \prod_{k<j} dH_{kj}^{(1)}, \tag{2.5.1}$$

where $H_{kj}^{(0)}$ and $H_{kj}^{(1)}$ are real and imaginary parts of H_{kj}, is invariant under every automorphism

$$H \longrightarrow U^{-1}HU \tag{2.5.2}$$

of T_{2G} into itself, where U is any unitary matrix.

2. Various linearly independent components of H are also statistically independent.

In mathematical language these requirements are

1.

$$P(H')\,dH' = P(H)\,dH, \tag{2.5.3}$$

if

$$H' = U^{-1}HU, \tag{2.5.4}$$

where U is any unitary matrix.

2. $P(H)$ is a product of functions, each of which depends on a single variable:

$$P(H) = \prod_{k\leq j} f_{kj}^{(0)}\left(H_{kj}^{(0)}\right)\prod_{k<j} f_{kj}^{(1)}\left(H_{kj}^{(1)}\right). \tag{2.5.5}$$

2.6. Joint Probability Density Function for Matrix Elements

We now come to the question of the extent to which we are still free to specify the joint probability density function $P(H)$. It will be seen that the two postulates of invariance and statistical independence elaborated above fix uniquely the functional form of $P(H)$.

The postulate of invariance restricts $P(H)$ to depend only on a finite number of traces of the powers of H. We state this fact as a lemma (Weyl, 1946).

Lemma 2.6.1. *All the invariants of an* $(N \times N)$ *matrix* H *under nonsingular similarity transformations* A,

$$H \longrightarrow H' = AHA^{-1},$$

can be expressed in terms of the traces of the first N *powers of* H.

Actually the trace of the jth power of H is the sum of the jth powers of its eigenvalues λ_k, $k = 1, 2, ..., N$, of H,

$$\text{tr}\ H^j = \sum_{k=1}^{N} \lambda_k^j \equiv p_j, \qquad \text{say},$$

and it is a well-known fact that any symmetric function of the λ_k can be expressed in terms of the first N of the p_j; see, for example, Macdonald (1979) or Mehta (1989).

The postulate of statistical independence excludes everything except the traces of the first two powers, and these, too, may occur only in an exponential. To see this we will need the following lemma.

Lemma 2.6.2. *If three continuous and differentiable functions $f_k(x)$, $k = 1, 2, 3$, satisfy the equation*

$$f_1(xy) = f_2(x) + f_3(y), \tag{2.6.1}$$

then they are necessarily of the form $a \ln x + b_k$, $(k = 1, 2, 3)$, with $b_1 = b_2 + b_3$.

Proof. Differentiating Eq. (2.6.1) with respect to x, we have

$$f_1'(xy) = \frac{1}{y} f_2'(x),$$

which, on integration with respect to y, gives

$$\frac{1}{x} f_1(xy) = f_2'(x) \ln y + \frac{1}{x} g(x), \tag{2.6.2}$$

where $g(x)$ is still arbitrary. Substituting $f_1(xy)$ from Eq. (2.6.2) into Eq. (2.6.1),

$$x f_2'(x) \ln y + g(x) - f_2(x) = f_3(y). \tag{2.6.3}$$

Therefore the left-hand side of Eq. (2.6.3) must be independent of x; this is possible only if

$$x f_2'(x) = a \quad \text{and} \quad g(x) - f_2(x) = b_3,$$

that is, only if

$$f_2(x) = a \ln x + b_2 = g(x) - b_3,$$

where a, b_2, and b_3 are arbitrary constants.

Now Eq. (2.6.3) gives

$$f_3(y) = a \ln y + b_3$$

and finally Eq. (2.6.1) gives

$$f_1(xy) = a \ln (xy) + (b_2 + b_3)\,. \qquad \text{Q.E.D.}$$

Let us now examine the consequences of the statistical independence of the various components of H. Consider the particular transformation

$$H = U^{-1} H' U, \tag{2.6.4}$$

where

$$U = \begin{bmatrix} \cos\theta & \sin\theta & 0 & \dots & 0 \\ -\sin\theta & \cos\theta & 0 & \dots & 0 \\ 0 & 0 & 1 & \dots & 0 \\ \dots & \dots & \dots & \dots & \dots \\ 0 & 0 & 0 & \dots & 1 \end{bmatrix} \tag{2.6.5}$$

or, in quaternion notation (provided N is even),

$$U = \begin{bmatrix} \cos\theta + e_2 \sin\theta & 0 & \dots & 0 & \dots & 0 \\ 0 & 1 & \dots & 0 & \dots & 0 \\ \dots & \dots & \dots & \dots & \dots & \dots \\ 0 & 0 & \dots & 0 & \dots & 1 \end{bmatrix}. \tag{2.6.6}$$

This U is, at the same time, orthogonal, symplectic, and unitary.

Differentiation of Eq. (2.6.4) with respect to θ gives

$$\frac{\partial H}{\partial \theta} = \frac{\partial U^T}{\partial \theta} H' U + U^T H' \frac{\partial U}{\partial \theta} = \frac{\partial U^T}{\partial \theta} U H + H U^T \frac{\partial U}{\partial \theta}, \tag{2.6.7}$$

and by substituting for $U, U^T, \partial U/\partial\theta$, and $\partial U^T/\partial\theta$ from Eq. (2.6.5) or Eq. (2.6.6) we get

$$\frac{\partial H}{\partial \theta} = AH + HA^T, \tag{2.6.8}$$

where

$$A = \frac{\partial U^T}{\partial \theta} U = \begin{bmatrix} 0 & -1 & 0 & 0 \\ 1 & 0 & 0 & 0 \\ 0 & 0 & 0 & 0 \\ \cdots & \cdots & \cdots & \cdots \\ 0 & 0 & 0 & 0 \end{bmatrix} \tag{2.6.9}$$

or, in quaternion notation, A is diagonal:

$$A = \begin{bmatrix} -e_2 & 0 & \ldots & 0 \\ 0 & 0 & \ldots & 0 \\ \cdots & \cdots & \cdots & \cdots \\ 0 & 0 & \ldots & 0 \end{bmatrix}. \tag{2.6.10}$$

If the probability density function

$$P(H) = \prod_{(\alpha)} \prod_{j \le k} f_{kj}^{(\alpha)} \left(H_{kj}^{(\alpha)} \right) \tag{2.6.11}$$

is invariant under the transformation U, its derivative with respect to θ must vanish; that is,

$$\sum \frac{1}{f_{kj}^{(\alpha)}} \frac{\partial f_{kj}^{(\alpha)}}{\partial H_{kj}^{(\alpha)}} \frac{\partial H_{kj}^{(\alpha)}}{\partial \theta} = 0. \tag{2.6.12}$$

Let us write this equation explicitly, say, for the unitary case. Equations (2.6.8) and (2.6.12) give

$$\begin{aligned} &\left[\left(-\frac{1}{f_{11}^{(0)}} \frac{\partial f_{11}^{(0)}}{\partial H_{11}^{(0)}} + \frac{1}{f_{22}^{(0)}} \frac{\partial f_{22}^{(0)}}{\partial H_{22}^{(0)}} \right) \left(2H_{12}^{(0)} \right) \right. \\ &\quad \left. + \frac{1}{f_{12}^{(0)}} \frac{\partial f_{12}^{(0)}}{\partial H_{12}^{(0)}} \left(H_{11}^{(0)} - H_{22}^{(0)} \right) \right] \\ &\quad + \sum_{k=3}^{N} \left(-\frac{1}{f_{1k}^{(0)}} \frac{\partial f_{1k}^{(0)}}{\partial H_{1k}^{(0)}} H_{2k}^{(0)} + \frac{1}{f_{2k}^{(0)}} \frac{\partial f_{1k}^{(0)}}{\partial H_{2k}^{(0)}} H_{1k}^{(0)} \right) \\ &\quad + \sum_{k=3}^{N} \left(-\frac{1}{f_{1k}^{(1)}} \frac{\partial f_{1k}^{(1)}}{\partial H_{1k}^{(1)}} H_{2k}^{(1)} + \frac{1}{f_{2k}^{(1)}} \frac{\partial f_{2k}^{(1)}}{\partial H_{2k}^{(1)}} H_{1k}^{(1)} \right) = 0. \end{aligned} \tag{2.6.13}$$

The braces at the left-hand side of this equation depend on mutually exclusive sets of variables and their sum is zero. Therefore, each must be a constant; for example,

$$-\frac{H_{2k}^{(0)}}{f_{1k}^{(0)}}\frac{\partial f_{1k}^{(0)}}{\partial H_{1k}^{(0)}}+\frac{H_{1k}^{(0)}}{f_{2k}^{(0)}}\frac{\partial f_{2k}^{(0)}}{\partial H_{2k}^{(0)}}=C_k^{(0)}. \tag{2.6.14}$$

On dividing both side of Eq. (2.6.14) by $H_{1k}^{(0)}H_{2k}^{(0)}$ and applying the Lemma 2.6.2, we conclude that the constant $C_k^{(0)}$ must be zero, that is,

$$\frac{1}{H_{1k}^{(0)}}\frac{1}{f_{1k}^{(0)}}\frac{\partial f_{1k}^{(0)}}{\partial H_{1k}^{(0)}}=\frac{1}{H_{2k}^{(0)}}\frac{1}{f_{2k}^{(0)}}\frac{\partial f_{2k}^{(0)}}{\partial H_{2k}^{(0)}}=\text{constant}=-2a,\quad\text{say}, \tag{2.6.15}$$

which on integration gives

$$f_{1k}^{(0)}\left(H_{1k}^{(0)}\right)=\exp\left[-a\left(H_{1k}^{(0)}\right)^2\right]. \tag{2.6.16}$$

In the other two cases we also derive a similar equation. Now because the off-diagonal elements come on as squares in the exponential and all invariants are expressible in terms of the traces of powers of H, the function $P(H)$ is an exponential that contains traces of at most the second power of H.

Because $P(H)$ is required to be invariant under more general transformations than we have here considered, one might think that the form of $P(H)$ is further restricted. This, however, is not so, for

$$\begin{aligned}P(H)&=\exp\left(-a\ \mathrm{tr}\ H^2+b\ \mathrm{tr}\ H+c\right)\\&=e^c\prod_j\exp\left(bH_{jj}^{(0)}\right)\prod_{k\le j}\exp\left[-a\left(H_{kj}^{(0)}\right)^2\right]\\&\quad\times\prod_\lambda\prod_{k<j}\exp\left[-a\left(H_{kj}^{(\lambda)}\right)^2\right]\end{aligned} \tag{2.6.17}$$

is already a product of functions, each of which depends on a separate variable. Moreover, because we require $P(H)$ to be normalizable and real, a must be real and positive and b and c must be real.

Therefore we have proved the following theorem (Porter and Rosenzweig, 1960a).

Theorem 2.6.3. *In all the above three cases the form of $P(H)$ is automatically restricted to*

$$P(H) = \exp\left(-a \text{ tr } H^2 + b \text{ tr } H + c\right), \tag{2.6.18}$$

where a is real and positive and b and c are real.

In the foregoing discussion we have emphasized the postulate of statistical independence of various components of H even at the risk of frequent repetitions. This statistical independence is important in restricting $P(H)$ to the simple form (2.6.18), and hence makes the subsequent analytical work tractable. However, it lacks a clear physical motivation and therefore looks somewhat artificial.

The main objection to the assumption of statistical independence, leading to Eq. (2.6.18), is that all values of $H_{kj}^{(\lambda)}$ are not equally weighted and therefore do not correspond to all "interactions" being "equally probable." By a formal change Dyson (1962) has defined his "circular ensembles," which are esthetically more satisfactory and equally easy to work with. We shall come to them in Chapters 9 to 11. They give equivalent results as we will see in Chapter 10. On the other hand, Rosenzweig (1963), has emphasized the "fixed strength" ensemble. Others (Leff, 1963; Fox and Kahn, 1964) have arbitrarily tried the so-called "generalized" ensembles related to classical orthogonal polynomials other than the Hermite polynomials. A brief review of these topics is given later in Chapter 19.

2.7. Another Gaussian Ensemble of Hermitian Matrices

The ensembles so far considered were characterized by two requirements: (i) the probability $P(H)dH$ that a system belongs to the volume element dH is such that $P(H)$ is invariant under $H \to U^{-1}HU$, where U is any matrix which is either real orthogonal, symplectic, or unitary according to the symmetry of the system; and (ii) various linearly independent components of H are also statistically independent.

If for our system the time reversal invariance is only weakly violated, then the appropriate ensemble will be almost an orthogonal or symplectic ensemble slightly mixed with the unitary ensemble. Keeping the hypothesis (ii) that various linearly independent parts of H are also statistically independent, we should now take

$$P(H) \propto \exp\left[-\mathrm{tr}\left(H_1^2/c_1 + H_2^2/c_2\right)\right] \tag{2.7.1}$$

where $H = H_1 + H_2$, H_1 and H_2 are Hermitian, H_1 is symmetric (self-dual) and H_2 is anti-symmetric (anti-self-dual). If $c_2 = 0$, then $H_2 = 0$ with probability 1, and we have the orthogonal (symplectic) ensemble; if $c_2 = c_1$, then we have the unitary ensemble. For a small violation of the time reversal invariance, $c_2 \ll c_1$. Since it does not increase the mathematical difficulties and the analytical solution is as elegant, we will treat in Chapter 14 the general case where c_1 and c_2 are arbitrary real positive numbers.

Note that under real orthogonal transformations the traces of powers of H_1 and H_2 (i.e., of real and imaginary parts of H) are invariant and so is the probability density $P(H)$ of Eq. (2.7.1). However, under unitary transformations the real and imaginary parts of H mix up and the above $P(H)$ is no longer invariant unless $c_1 = c_2$.

2.8. Antisymmetric Hermitian Matrices

Though physically not relevant, the mathematical analysis of a Gaussian ensemble of antisymmetric (or that of anti-self-dual quaternion) Hermitian matrices is equally elegant. As above, the probability is again

$$P(H)dH, \quad dH = \prod_{j<k} dH_{jk}, \quad \text{and} \quad P(H) \propto \exp\left(-a\ \mathrm{tr}H^2\right).$$

Summary of Chapter 2

The probability density $P(H)$ of a random matrix H is proportional to $\exp(-a\ \mathrm{tr}\ H^2 + b\ \mathrm{tr}\ H + c)$ with certain constants a, b, c in the following three cases:

(1) If H is a Hermitian symmetric random matrix, its elements H_{jk} with $j \geq k$ are statistically independent, and $P(H)$ is invariant under all

real orthogonal transformations of H. The resulting ensemble is named as Gaussian orthogonal.

(2) If H is a Hermitian random matrix, its diagonal elements H_{jj} and the real and imaginary parts of its off-diagonal elements H_{jk} for $j > k$ are statistically independent, and $P(H)$ is invariant under all unitary transformations of H. The resulting ensemble is named as Gaussian unitary.

(3) If H is a Hermitian self-dual random matrix, its diagonal elements H_{jj} and the four quaternionic components of its off-diagonal elements H_{jk} with $j > k$ are statistically independent, and $P(H)$ is invariant under all symplectic transformations of H. The resulting ensemble is named as Gaussian symplectic.

Moreover,

(4) For a Hermitian antisymmetric random matrix H, it is not unreasonable to take the elements H_{jk} with $j > k$ as Gaussian variables with the same variance.

(5) Similarly, for a Hermitian random matrix H, with $P(H)$ not invariant under unitary transformations of H, it is not unreasonable to take its symmetric and antisymmetric parts to have the probability densities prescribed under cases (1) and (4) above.

Invariance of $P(H)$ under orthogonal, unitary or symplectic transformations of H is required by physical considerations and depend on whether the system described by the Hamiltonian H has or does not have certain symmetries like time-reversal or rotational symmetry. The statistical independence of the various real parameters entering H is assumed for simplicity.

3 / Gaussian Ensembles. The Joint Probability Density Function for the Eigenvalues

In this chapter we will derive the joint probability density function for the eigenvalues of H implied by the Gaussian densities for its matrix elements. Finally an argument based on information theory is given. This argument rationalizes any of the Gaussian probability densities for the matrix elements; it even allows one to define an ensemble having a preassigned eigenvalue density.

3.1. Orthogonal Ensemble

The joint probability density function (abbreviated j.p.d.f. later in the chapter) for the eigenvalues $\theta_1, \theta_2, ..., \theta_N$ can be obtained from Eq. (2.6.18) by expressing the various components of H in terms of the N eigenvalues θ_j and other mutually independent variables p_μ, say, which together with the θ_j form a complete set. In an $(N \times N)$ real symmetric matrix the number of independent real parameters which determine all H_{kj} is $N(N+1)/2$. We may take these as H_{kj} with $k \leq j$. The number of extra parameters p_μ needed is therefore

$$\ell = \frac{1}{2}N(N+1) - N = \frac{1}{2}N(N-1). \tag{3.1.1}$$

Because

$$\text{tr } H^2 = \sum_1^N \theta_j^2, \qquad \text{tr } H = \sum_1^N \theta_j, \tag{3.1.2}$$

the probability density that the N roots and $N(N-1)/2$ parameters will occur around the values $\theta_1, ..., \theta_N$ and $p_1, p_2, ..., p_\ell$ is, according to

Eq. (2.6.18)

$$P(\theta_1, ..., \theta_N; p_1, ..., p_\ell) = \exp\left(-a\sum_1^N \theta_j^2 + b\sum_1^N \theta_j + c\right) J(\theta, p), \tag{3.1.3}$$

where J is the Jacobian

$$J(\theta, p) = \left|\frac{\partial(H_{11}, H_{12}, ..., H_{NN})}{\partial(\theta_1, ..., \theta_N, p_1, ..., p_\ell)}\right|. \tag{3.1.4}$$

Hence, the j.p.d.f. of the eigenvalues θ_j can be obtained by integrating Eq. (3.1.3) over the parameters $p_1, ..., p_\ell$. It is usually possible to choose these parameters so that the Jacobian (3.1.4) becomes a product of a function f of the θ_j and a function g of the p_μ. If this is the case, the integration provides the required j.p.d.f. as a product of the exponential in Eq. (3.1.3), the function f of the θ_j, and a constant. The constant can then be absorbed in c in the exponential.

To define the parameters p_μ (Wigner, 1962) we recall that any real symmetric matrix H can be diagonalized by a real orthogonal matrix (*cf.* Appendix A.3):

$$H = U\Theta U^{-1} \tag{3.1.5}$$

$$= U\Theta U^T, \tag{3.1.5$'$}$$

where Θ is the diagonal matrix with diagonal elements $\theta_1, \theta_2, ..., \theta_N$ arranged in some order, say, $\theta_1 \le \theta_2 \le \cdots \le \theta_N$, and U is a real orthogonal matrix

$$UU^T = U^T U = 1, \tag{3.1.6}$$

whose columns are the normalized eigenvectors of H. These eigenvectors are, or may be chosen to be, mutually orthogonal. To define U completely we must in some way fix the phases of the eigenvectors, for instance by requiring that the first nonvanishing component be positive. Thus U depends on $N(N-1)/2$ real parameters and may be chosen to be U_{kj}, $k > j$. If H has multiple eigenvalues, further conditions are needed to fix U completely. It is not necessary to specify them, for they apply only in regions of lower dimensionality which are irrelevant to the probability density function. At any rate, enough appropriate conditions are imposed on U so that it is uniquely characterized by the $N(N-1)/2$ parameters

p_μ. Once this is done, the matrix H, which completely determines the Θ and the U subject to the preceding conditions, also determines the θ_j and the p_μ uniquely. Conversely, the θ_j and p_μ completely determine the U and Θ, and hence by Eq. (3.1.5) all the matrix elements of H.

Differentiating Eq. (3.1.6), we get

$$\frac{\partial U^T}{\partial p_\mu} U + U^T \frac{\partial U}{\partial p_\mu} = 0, \tag{3.1.7}$$

and because the two terms in Eq. (3.1.7) are the Hermitian conjugates of each other,

$$S^{(\mu)} = U^T \frac{\partial U}{\partial p_\mu} = -\frac{\partial U^T}{\partial p_\mu} U \tag{3.1.8}$$

is an antisymmetric matrix.

Also from Eq. (3.1.5) we have

$$\frac{\partial H}{\partial p_\mu} = \frac{\partial U}{\partial p_\mu} \Theta U^T + U \Theta \frac{\partial U^T}{\partial p_\mu}. \tag{3.1.9}$$

On multiplying Eq. (3.1.9) by U^T on the left and by U on the right, we get

$$U^T \frac{\partial H}{\partial p_\mu} U = S^{(\mu)} \Theta - \Theta S^{(\mu)}. \tag{3.1.10}$$

In terms of its components, Eq. (3.1.10) reads

$$\sum_{j,k} \frac{\partial H_{jk}}{\partial p_\mu} U_{j\alpha} U_{k\beta} = S_{\alpha\beta}^{(\mu)} \left(\theta_\beta - \theta_\alpha\right). \tag{3.1.11}$$

In a similar way, by differentiating Eq. (3.1.5) with respect to θ_γ,

$$\sum_{j,k} \frac{\partial H_{jk}}{\partial \theta_\gamma} U_{j\alpha} U_{k\beta} = \frac{\partial \Theta_{\alpha\beta}}{\partial \theta_\gamma} = \delta_{\alpha\beta} \delta_{\alpha\gamma}. \tag{3.1.12}$$

The matrix of the Jacobian in Eq. (3.1.4) can be written in partitioned form as

$$[J(\theta, p)] = \begin{bmatrix} \dfrac{\partial H_{jj}}{\partial \theta_\gamma} & \dfrac{\partial H_{jk}}{\partial \theta_\gamma} \\ \dfrac{\partial H_{jj}}{\partial p_\mu} & \dfrac{\partial H_{jk}}{\partial p_\mu} \end{bmatrix}. \tag{3.1.13}$$

The two columns in Eq. (3.1.13) correspond to N and $N(N-1)/2$ actual columns; $1 \le j < k \le N$. The two rows in Eq. (3.1.13) correspond again to N and $N(N-1)/2$ actual rows: $\gamma = 1, 2, ..., N$; $\mu = 1, 2, ..., N(N-1)/2$. If we multiply the $[J]$ in Eq. (3.1.13) on the right by the $N(N+1)/2 \times N(N+1)/2$ matrix written in partitioned form as

$$[V] = \begin{bmatrix} (U_{j\alpha} U_{j\beta}) \\ (2U_{j\alpha} U_{k\beta}) \end{bmatrix}, \tag{3.1.14}$$

in which the two rows correspond to N and $N(N-1)/2$ actual rows, $1 \le j < k \le N$, and the column corresponds to $N(N+1)/2$ actual columns, $1 \le \alpha \le \beta \le N$, we get by using Eqs. (3.1.11) and (3.1.12)

$$[J]\,[V] = \begin{bmatrix} \delta_{\alpha\beta}\delta_{\alpha\gamma} \\ S^{(\mu)}_{\alpha\beta}\,(\theta_\beta - \theta_\alpha) \end{bmatrix}. \tag{3.1.15}$$

The two rows on the right-hand side correspond to N and $N(N-1)/2$ actual rows and the column corresponds to $N(N+1)/2$ actual columns. Taking the determinant on both sides of Eq. (3.1.15), we have

$$J\,(\theta, p)\,\det V = \prod_{\alpha<\beta} (\theta_\beta - \theta_\alpha) \det \begin{bmatrix} \delta_{\alpha\beta}\delta_{\alpha\gamma} \\ S^{\mu}_{\alpha\beta} \end{bmatrix}$$

or

$$J\,(\theta, p) = \prod_{\alpha<\beta} |\theta_\beta - \theta_\alpha|\, f\,(p)\,, \tag{3.1.16}$$

where $f(p)$ is independent of the θ_j and depends only on the parameters p_μ.

By inserting this result into Eq. (3.1.3) and integrating over the variables p_μ we get the j.p.d.f. for the eigenvalues of the matrices of an orthogonal ensemble

$$P\,(\theta_1, ..., \theta_N) = \exp\left[-\sum_1^N \left(a\theta_j^2 - b\theta_j - c\right)\right] \prod_{j<k} |\theta_k - \theta_j|\,, \tag{3.1.17}$$

where c is some new constant. Moreover, if we shift the origin of the θ to $b/2a$ and change the energy scale everywhere by a constant factor

$\sqrt{2a}$, we may replace θ_j with $\left(1/\sqrt{2a}\right) x_j + b/2a$. By this formal change (3.1.17) takes the simpler form

$$P_{N1}(x_1, ..., x_N) = C_{N1} \exp\left(-\frac{1}{2}\sum_1^N x_j^2\right) \prod_{j<k} |x_j - x_k|, \qquad (3.1.18)$$

where C_{N1} is a constant. (Subscript 1 is to remind of the power of the product of differences.)

3.2. Symplectic Ensemble

As the analysis is almost identical in all three invariant cases, we have presented the details for one particular ensemble, the orthogonal one. Here and in Section 3.3 and Section 3.4 we indicate briefly the modifications necessary to arrive at the required j.p.d.f. in the other cases.

Corresponding to the result that a real symmetric matrix can be diagonalized by a real orthogonal matrix, we have the following:

Theorem 3.2.1. *Given a quaternion-real, self-dual matrix H, there exists a symplectic matrix U such that*

$$H = U\Theta U^{-1} = U\Theta U^R, \qquad (3.2.1)$$

where Θ is diagonal, real, and scalar (cf. Appendix A.3).

The fact that Θ is scalar means that it consists of N blocks of the form

$$\begin{bmatrix} \theta_j & 0 \\ 0 & \theta_j \end{bmatrix} \qquad (3.2.2)$$

along the main diagonal. Thus, the eigenvalues of H consist of N equal pairs. The Hamiltonian of any system that is invariant under time reversal, has odd spin, and has no rotational symmetry satisfies the conditions of Theorem 3.2.1. All energy levels of such a system will be doubly degenerate. This is Kramer's degeneracy (Kramer, 1930), and Theorem 3.2.1 shows how it appears naturally in the quaternion language.

Apart from the N eigenvalues θ_j, the number of real independent parameters p_μ needed to characterize an $N \times N$ quaternion-real, self-dual matrix H is

$$\ell = 4 \times \frac{1}{2} N(N-1) = 2N(N-1). \tag{3.2.3}$$

Equations (3.1.2) and (3.1.3) are replaced, respectively, by

$$\text{tr } H^2 = 2\sum_{j=1}^{N} \theta_j^2, \qquad \text{tr } H = 2\sum_{j=1}^{N} \theta_j, \tag{3.2.4}$$

and

$$P(\theta_1, ..., \theta_N; p_1, ..., p_\ell) = \exp\left[-\sum_{j=1}^{N} \left(2a\theta_j^2 - 2b\theta_j - c\right)\right] J(\theta, p), \tag{3.2.5}$$

where $J(\theta, p)$ is now given by

$$J(\theta, p) = \left| \frac{\partial\left(H_{11}^{(0)}, ..., H_{NN}^{(0)}, H_{12}^{(0)}, ..., H_{12}^{(3)}, ..., H_{N-1,N}^{(0)}, ..., H_{N-1,N}^{(3)}\right)}{\partial\left(\theta_1, ..., \theta_N, p_1, ..., p_{2N(N-1)}\right)} \right|. \tag{3.2.6}$$

Equation (3.1.5) is replaced by Eq. (3.2.1); Eqs. (3.1.6)–(3.1.10) are valid if U^T is replaced by U^R. Note that these equations are now in the quaternion language, and we need to separate the four quaternion parts of the modified Eq. (3.2.1). For this we let

$$H_{jk} = H_{jk}^{(0)} + H_{jk}^{(1)} e_1 + H_{jk}^{(2)} e_2 + H_{jk}^{(3)} e_3, \tag{3.2.7}$$

$$S_{\alpha\beta}^{(\mu)} = S_{\alpha\beta}^{(0\mu)} + S_{\alpha\beta}^{(1\mu)} e_1 + S_{\alpha\beta}^{(2\mu)} e_2 + S_{\alpha\beta}^{(3\mu)} e_3, \tag{3.2.8}$$

and write Eq. (3.1.10) and the equation corresponding to Eq. (3.1.12) in the form of partitioned matrices:

$$\begin{bmatrix} \dfrac{\partial H_{jj}^{(0)}}{\partial \theta_\gamma} & \dfrac{\partial H_{jk}^{(0)}}{\partial \theta_\gamma} & \dfrac{\partial H_{jk}^{(1)}}{\partial \theta_\gamma} & \cdots & \dfrac{\partial H_{jk}^{(3)}}{\partial \theta_\gamma} \\ \dfrac{\partial H_{jj}^{(0)}}{\partial p_\mu} & \dfrac{\partial H_{jk}^{(0)}}{\partial p_\mu} & \dfrac{\partial H_{jk}^{(1)}}{\partial p_\mu} & \cdots & \dfrac{\partial H_{jk}^{(3)}}{\partial p_\mu} \end{bmatrix} \begin{bmatrix} v & w \\ A^{(0)} & B^{(0)} \\ \cdots & \cdots \\ A^{(3)} & B^{(3)} \end{bmatrix}$$

$$= \begin{bmatrix} \rho_{\gamma,\alpha} & \sigma^{(0)}_{\gamma,\alpha\beta} & \cdots & \sigma^{(3)}_{\gamma,\alpha\beta} \\ \varepsilon^{(\mu)}_{\alpha} & S^{(0\mu)}_{\alpha\beta}(\theta_\beta - \theta_\alpha) & \dots & S^{(3\mu)}_{\alpha\beta}(\theta_\beta - \theta_\alpha) \end{bmatrix} \tag{3.2.9}$$

$$1 \le j < k \le N, \quad 1 \le \alpha < \beta \le N, \quad 1 \le \gamma \le N, \quad 1 \le \mu \le 2N(N-1),$$

where the matrices $\partial H^{(0)}_{jj}/\partial\theta_\gamma$, v, and ρ are $N \times N$; the matrices $\partial H^{(\lambda)}_{jk}/\partial\theta_\gamma$ and $\sigma^{(\lambda)}_{\gamma,\alpha\beta}$, with $\lambda = 0, 1, 2, 3$, are $N \times N(N-1)/2$; the $A^{(\lambda)}$ are all $N(N-1)/2 \times N$; the $\partial H^{(0)}_{jj}/\partial p_\mu$ and the $\varepsilon^{(\mu)}_\alpha$ are $2N(N-1) \times N$; the w is $N \times 2N(N-1)$; the $\partial H^{(\lambda)}_{jk}/\partial p_\mu$ and the $S^{(\lambda\mu)}_{\alpha\beta}$ are $2N(N-1) \times N(N-1)/2$; and the matrices $B^{(\lambda)}$ are $N(N-1)/2 \times 2N(N-1)$. The matrices ρ and the σ appear as we separate the result of differentiation of Eq. (3.2.1) with respect to θ_γ into quaternion components. Because Θ is diagonal and scalar, the $\sigma^{(\lambda)}$ are all zero matrices. Moreover, the matrix ρ does not depend on θ_γ, for Θ depends linearly on the θ_γ. The computation of the matrices $v, w, A^{(\lambda)}$, and $B^{(\lambda)}$ is straightforward, but we do not require them. All we need is to note that they are formed of the various components of U, and hence do not depend on θ_γ.

Now we take the determinant on both sides of Eq. (3.2.9). The determinant of the first matrix on the left is the Jacobian (3.2.6). Because the $\sigma^{(\lambda)}$ are all zero, the determinant of the right-hand side breaks into a product of two determinants:

$$\det\left[\rho_{\gamma,\alpha}\right] \det\left[S^{(\lambda\mu)}_{\alpha\beta}(\theta_\beta - \theta_\alpha)\right], \tag{3.2.10}$$

the first one being independent of the θ_γ, whereas the second is

$$\prod_{\alpha<\beta} (\theta_\beta - \theta_\alpha)^4 \det\left[S^{(\lambda\mu)}_{\alpha\beta}\right]. \tag{3.2.11}$$

Thus

$$J(\theta, p) = \prod_{\alpha<\beta} (\theta_\beta - \theta_\alpha)^4 f(p), \tag{3.2.12}$$

which corresponds to Eq. (3.1.16).

By inserting Eq. (3.2.12) into Eq. (3.2.5) and integrating over the parameters, we obtain the j.p.d.f.

$$P\left(\theta_1, ..., \theta_N\right) = \exp\left(-2a\sum_{j=1}^{N}\theta_j^2 + 2b\sum_{j=1}^{N}\theta_j + c\right)\prod_{j<k}\left(\theta_j - \theta_k\right)^4. \tag{3.2.13}$$

As before, we may shift the origin to make $b = 0$ and change the scale of energy to make $a = 1$. Thus, the j.p.d.f. for the eigenvalues of matrices in the symplectic ensemble in its simple form is

$$P_{N4}\left(x_1, ..., x_N\right) = C_{N4}\exp\left(-2\sum_{j=1}^{N}x_j^2\right)\prod_{j<k}\left(x_j - x_k\right)^4, \tag{3.2.14}$$

where C_{N4} is a constant. (Subscript 4 to remind again of the power of the product of differences.)

3.3. Unitary Ensemble

In addition to the real eigenvalues, the number of real independent parameters p_μ needed to specify an arbitrary Hermitian matrix H completely is $N(N-1)$. Equations (3.1.2) and (3.1.3) remain unchanged, but Eq. (3.1.4) is replaced by

$$J(\theta, p) = \frac{\partial\left(H_{11}^{(0)}, ..., H_{NN}^{(0)}, H_{12}^{(0)}, H_{12}^{(1)}, ..., H_{N-1,N}^{(0)}, H_{N-1,N}^{(1)}\right)}{\partial\left(\theta_1, ..., \theta_N, p_1, ..., p_{N(N-1)}\right)}, \tag{3.3.1}$$

where $H_{jk}^{(0)}$ and $H_{jk}^{(1)}$ are the real and imaginary parts of H_{jk}. Equations (3.1.5) to (3.1.10) are valid if U^T replaced by $U^\dagger$. Instead of Eqs. (3.1.11) and (3.1.12) we now have

$$\sum_{j,k}\frac{\partial H_{jk}}{\partial p_\mu}U_{j\alpha}^{*}U_{k\beta} = S_{\alpha\beta}^{(\mu)}\left(\theta_\beta - \theta_\alpha\right), \tag{3.3.2}$$

$$\sum_{j,k}\frac{\partial H_{jk}}{\partial p_\gamma}U_{j\alpha}^{*}U_{k\beta} = \frac{\partial\Theta_{\alpha\beta}}{\partial\theta_\gamma} = \delta_{\alpha\beta}\delta_{\alpha\gamma}. \tag{3.3.3}$$

By separating the real and imaginary parts we may write these equations in partitioned matrix notation as

$$\begin{bmatrix} \dfrac{\partial H_{jj}^{(0)}}{\partial \theta_\gamma} & \dfrac{\partial H_{jk}^{(0)}}{\partial \theta_\gamma} & \dfrac{\partial H_{jk}^{(1)}}{\partial \theta_\gamma} \\ \dfrac{\partial H_{jj}^{(0)}}{\partial p_\mu} & \dfrac{\partial H_{jk}^{(0)}}{\partial p_\mu} & \dfrac{\partial H_{jk}^{(1)}}{\partial p_\mu} \end{bmatrix} \begin{bmatrix} v & w \\ A^{(0)} & B^{(0)} \\ A^{(1)} & B^{(1)} \end{bmatrix}$$

$$= \begin{bmatrix} \rho_{\gamma,\alpha} & \sigma_{\gamma,\alpha\beta}^{(0)} & \sigma_{\gamma,\alpha\beta}^{(1)} \\ \varepsilon_\alpha^{(\mu)} & S_{\alpha\beta}^{(0\mu)}(\theta_\beta - \theta_\alpha) & S_{\alpha\beta}^{(1\mu)}(\theta_\beta - \theta_\alpha) \end{bmatrix}, \quad (3.3.4)$$

$$1 \le j < k \le N, \quad 1 \le \alpha < \beta \le N,$$

$$1 \le \mu \le N(N-1), \quad 1 \le \gamma \le N.$$

where $S_{\alpha\beta}^{(0\mu)}$ and $S_{\alpha\beta}^{(1\mu)}$ are the real and imaginary parts of $S_{\alpha\beta}^{(\mu)}$. The matrices $\partial H_{jj}^{(0)}/\partial\theta_\gamma, v$, and ρ are $N \times N$; the $\partial H_{jk}^{(\lambda)}/\partial\theta_\gamma$ and the $\sigma_{\gamma,\alpha\beta}^{(\lambda)}$ are $N \times N(N-1)/2$; the $A^{(\lambda)}$ are $N(N-1)/2 \times N$; the $\partial H_{jk}^{(\lambda)}/\partial p_\mu$ and $S_{\alpha\beta}^{(\lambda\mu)}$ are $N(N-1) \times N(N-1)/2$; the $B^{(\lambda)}$ are $N(N-1)/2 \times N(N-1)$; the $\partial H_{jj}^{(0)}/\partial p_\mu$ and the $\varepsilon_\alpha^{(\mu)}$ are $N(N-1) \times N$; and the matrix w is $N \times N(N-1)$. To compute $v, w, A^{(\lambda)}, \rho, \varepsilon, \sigma^{(\lambda)}$, etc., is again straightforward, but we do not need them explicitly. What we want to emphasize is that they are either constructed from the components of U or arise from the differentiation of Θ with respect to θ_j and consequently are all independent of the eigenvalues θ_j. Similarly, $S^{(\mu)}$ is independent of θ_j. One more bit of information we need is that $\sigma^{(0)}$ and $\sigma^{(1)}$ are zero matrices, which can easily be verified.

Thus, by taking the determinants on both sides of Eq. (3.3.4) and removing the factors $(\theta_\beta - \theta_\alpha)$ we have

$$J(\theta, p) = \prod_{\alpha<\beta} (\theta_\beta - \theta_\alpha)^2 f(p), \quad (3.3.5)$$

where $f(p)$ is some function of the p_μ.

By inserting Eq. (3.3.5) into Eq. (3.1.3) and integrating over the parameters p_μ we get the j.p.d.f. for the eigenvalues of matrices in the unitary ensemble

$$P(\theta_1, ..., \theta_N) = \exp\left(-a\sum_1^N \theta_j^2 + b\sum_1^N \theta_j + c\right)\prod_{j<k}(\theta_j - \theta_k)^2, \quad (3.3.6)$$

and, as before, by a proper choice of the origin and the scale of energy we have

$$P_{N2}(x_1, ..., x_N) = C_{N2}\exp\left(-\sum_1^N x_j^2\right)\prod_{j<k}(x_j - x_k)^2. \quad (3.3.7)$$

We record Eqs. (3.1.18), (3.2.14), and (3.3.7) as a theorem.

Theorem 3.3.1. *The joint probability density function for the eigenvalues of matrices from a Gaussian orthogonal, Gaussian symplectic, or Gaussian unitary ensemble is given by*

$$P_{N\beta}(x_1, ..., x_N) = C_{N\beta}\exp\left(-\frac{1}{2}\beta\sum_1^N x_j^2\right)\prod_{j<k}|x_j - x_k|^\beta, \quad (3.3.8)$$

where $\beta = 1$ if the ensemble is orthogonal, $\beta = 4$ if it is symplectic, and $\beta = 2$ if it is unitary. The constant $C_{N\beta}$ is chosen in such a way that the $P_{N\beta}$ is normalized to unity:

$$\int_{-\infty}^{\infty} \cdots \int_{-\infty}^{\infty} P_{N\beta}(x_1, ..., x_N)\, dx_1 \cdots dx_N = 1. \quad (3.3.9)$$

According to Selberg the normalization constant $C_{N\beta}$ is given by (see Chapter 17)

$$C_{N\beta}^{-1} = (2\pi)^{N/2}\beta^{-N/2-\beta N(N-1)/4}\left[\Gamma(1+\beta/2)\right]^{-N}\prod_{j=1}^N \Gamma(1+\beta j/2). \quad (3.3.10)$$

For the physically interesting cases $\beta = 1, 2$, and 4 we will recalculate this value in a different way later (see Sections 5.2, 6.4 and 7.2).

It is possible to understand the different powers β that appear in Eq. (3.3.8) by a simple mathematical argument based on counting dimensions. The dimension of the space T_{1G} is $N(N+1)/2$, whereas the dimension of the subspace T'_{1G}, composed of the matrices in T_{1G} with two equal eigenvalues, is $N(N+1)/2-2$. Because of the single restriction, the equality of two eigenvalues, the dimension should normally have decreased by one; as it is decreased by two, it indicates a factor in Eq. (3.3.8) linear in $(x_j - x_k)$. Similarly, when $\beta = 2$, the dimension of T_{2G} is N^2, whereas that of T'_{2G} is N^2-3. When $\beta = 4$, the dimension of T_{4G} is $N(2N-1)$, whereas that T'_{4G} is $N(2N-1)-5$ (see Appendix A.4).

3.4. Ensemble of Antisymmetric Hermitian Matrices

The eigenvalues and eigenvectors of anti-symmetric Hermitian matrices come in pairs; if θ is an eigenvalue with the eigenvector V_θ, then $-\theta$ is an eigenvalue with the eigenvector V_θ^*. The vectors V_θ and V_θ^* can be normalized, and if $\theta \neq 0$ they are orthogonal. Thus if $V_\theta = \xi^{(\theta)} + i\eta^{(\theta)}$, $\xi^{(\theta)}$ and $\eta^{(\theta)}$ are real, then

$$(V_\theta, V_\theta^*) = \sum_j \left(\xi_j^{(\theta)} + i\eta_j^{(\theta)}\right)^2 = 0, \tag{3.4.1}$$

$$(V_\theta, V_\theta) = \sum_j \left|\xi_j^{(\theta)} + i\eta_j^{(\theta)}\right|^2 = 2. \tag{3.4.2}$$

These two equations are equivalent to

$$\sum_j \left(\xi_j^{(\theta)}\right)^2 = \sum_j \left(\eta_j^{(\theta)}\right)^2 = 1, \quad \sum_j \xi_j^{(\theta)} \eta_j^{(\theta)} = 0. \tag{3.4.3}$$

Therefore, if $\theta \neq 0$, the real and imaginary parts of V_θ have the same length and are orthogonal to each other. Moreover, the real and imaginary parts of V_θ are each orthogonal to the real and imaginary parts of V_λ if $\theta \neq \lambda$. If the matrices are of odd order, $2N+1$, then zero is an additional eigenvalue with an essentially real eigenvector V_0 orthogonal to all the ξ and η. As before, it is not necessary to consider the case in which two or more eigenvalues coincide.

With this information on the matrix diagonalizing H, we can derive the j.p.d.f. for the eigenvalues. Denoting the positive eigenvalues of H by $\theta_1, ..., \theta_n$, one has

$$\mathrm{tr}\ H^2 = \sum_{j=1}^{n} 2\theta_j^2.$$

The Jacobian will again be a product of the differences of all pairs of eigenvalues and of a function independent of them. Thus the j.p.d.f. for the eigenvalues will be proportional to

$$\prod_{1 \le j < k \le n} \left(\theta_j^2 - \theta_k^2\right)^2 \exp\left(-2\sum_{j=1}^{n} \theta_j^2\right), \tag{3.4.4}$$

if the order N of H is even, $N = 2n$, and

$$\prod_{j=1}^{n} \theta_j^2 \prod_{1 \le j < k \le n} \left(\theta_j^2 - \theta_k^2\right)^2 \exp\left(-2\sum_{j=1}^{n} \theta_j^2\right), \tag{3.4.5}$$

if $N = 2n + 1$ is odd. The constants of normalization will be calculated in chapter 13.

3.5. Another Gaussian Ensemble of Hermitian Matrices

To derive the j.p.d.f. for the eigenvalues is a little tricky in this case. This is so because the matrix element probability densities depend on the eigenvalues and the angular variables characterizing the eigenvectors; and one has to integrate over these angular variables. When either $c_1 = 0$ or $c_2 = 0$ or $c_1 = c_2$, the matrix element probability densities depend only on the eigenvalues. Also the Jacobian separates into a product of two functions, one involving only the eigenvalues and the other only the eigenvectors; therefore the integral over the eigenvectors, giving only a constant need not be calculated. For arbitrary c_1 and c_2 this simplification is not there. In view of these difficulties we will come back to this question in Chapter 14, when we are better prepared with the method of integration over alternate variables (Section 6.5), with quaternion determinants (Section 6.2) and the integral over the unitary group

$$\int \exp\left(\operatorname{tr}\left(A - U^{\dagger}BU\right)^{2}\right) dU$$

$$= \text{const} \times \det\left(\exp\left(a_j - b_k\right)^2\right) \prod_{j<k} \left[\left(a_j - a_k\right)\left(b_j - b_k\right)\right]^{-1}, \tag{3.5.1}$$

valid for arbitrary Hermitian matrices A and B having eigenvalues $a_1, ..., a_n$ and $b_1, ..., b_n$, respectively (see Appendix A.5).

3.6. Random Matrices and Information Theory

A reasonable specification of the probability density $P(H)$ for the random matrix H can be supplied from a different point of view, which is more satisfactory in some ways. Let us define the amount of information $\mathcal{I}(P(H))$ carried by the probability density $P(H)$. For discrete events $1, ..., m$, with probabilities $p_1, ..., p_m$, additivity and continuity of the information fixes it uniquely, apart from a constant multiplicative factor, to be (Shanon 1948; Khinchin 1957)

$$\mathcal{I} = -\sum_j p_j \ln p_j. \tag{3.6.1}$$

For continuous variables entering H it is reasonable to write

$$\mathcal{I}(P(H)) = -\int dH\ P(H) \ln P(H), \tag{3.6.2}$$

where dH is given by Eq. (2.5.1). One may now adopt the point of view that H should be as random as possible and compatible with the constraints it must satisfy. In other words, among all possible probabilities $P(H)$ of a matrix H constrained to satisfy some given properties, we must choose the one which minimizes the information $\mathcal{I}(P(H))$; $P(H)$ must not carry more information than what is required by the constraints. The constraints may, for example, be the fixed expectation values k_i of some functions $f_i(H)$,

$$\langle f_i(H) \rangle \equiv \int dH \; P(H) \; f_i(H) = k_i. \tag{3.6.3}$$

To minimize the information in Eq. (3.6.2) subject to Eq. (3.6.3) one may use Lagrange multipliers. This gives us for an arbitrary variation $\delta P(H)$ of $P(H)$,

$$\int dH \; \delta P(H) \left(1 + \ln P(H) - \sum_i \lambda_i f_i(H) \right) = 0 \tag{3.6.4}$$

or

$$P(H) \propto \exp \left(\sum_i \lambda_i f_i(H) \right), \tag{3.6.5}$$

and the Lagrange multipliers λ_i are then determined from Eqs. (3.6.3) and (3.6.5).

For example, requiring H Hermitian, $H = H_1 + iH_2$, H_1, H_2 real, and

$$\left\langle \text{tr } H_1^2 \right\rangle = k_1, \quad \left\langle \text{tr } H_2^2 \right\rangle = k_2,$$

will give us Eq. (2.7.1) for $P(H)$.

Another example is to require the level density to be a given function $\sigma(x)$

$$\langle \text{tr } \delta(H - x) \rangle = \sigma(x). \tag{3.6.6}$$

The Dirac delta function $\delta(H - x)$ is defined through the diagonalization of H, $\delta(H - x) = U^\dagger \delta(\theta - x) U$, if $H = U^\dagger \theta U$, U unitary, θ [or $\delta(\theta - x)$] diagonal real with diagonal elements θ_i [or $\delta(\theta_i - x)$]. This gives then

$$P(H) \propto \exp \left(\int dx \; \lambda(x) \; \text{tr } \delta(H - x) \right) = \exp[\text{tr } \lambda(H)] = \det\left[\mu(H)\right]; \tag{3.6.7}$$

the Lagrange multipliers $\lambda(x) \equiv \ln \mu(x)$ are then determined by Eq. (3.6.6). Thus one may determine a $P(H)$ giving a preassigned level density. For more details see Balian (1968).

It is important to note that if $P(H)$ depends only on the traces of powers of H, then the joint probability density of the eigenvalues will contain the factor $\prod |\theta_j - \theta_k|^\beta$, $\beta = 1$, 2, or 4; coming from the Jacobian of the elements of H with respect to its eigenvalues. And the local fluctuation properties are mostly governed by this factor.

Summary of Chapter 3

For the random Hermitian matrices H considered in Chapter 2, the joint probability density of its real eigenvalues x_1, x_2, ..., x_N is derived. It turns out to be proportional to

$$\prod_{1\le j<k\le N} |x_j - x_k|^\beta \exp\left(-\beta \sum_{j=1}^{N} x_j^2/2\right), \tag{3.3.8}$$

where $\beta = 1,\ 2$, or 4 according as the ensemble is Gaussian orthogonal, Gaussian unitary, or Gaussian symplectic.

For the Gaussian ensemble of Hermitian antisymmetric random matrices the joint probability density of the eigenvalues $\pm x_1$, ..., $\pm x_m$, with $x_j \ge 0$, is proportional to

$$(x_1 \cdots x_m)^\gamma \prod_{1\le j<k\le m} \left(x_j^2 - x_k^2\right)^2 \exp\left(-\sum_{j=1}^{m} x_j^2\right),$$

where $\gamma = 0$ if the order N of the matrices is even, $N = 2m$, and $\gamma = 2$ if this order is odd, $N = 2m + 1$. In the $N = 2m + 1$, odd case, zero is always an eigenvalue.

The noninvariant ensemble of Hermitian random matrices being more complicated is taken up later in Chapter 14.

An additional justification of our choices of the ensembles is supplied from an argument of the information theory.

4 / Gaussian Ensembles. Level Density

In this short chapter we reproduce a statistical mechanical argument of Wigner to "derive" the level density for the Gaussian ensembles. The joint probability density of the eigenvalues is written as the Boltzmann factor for a gas of charged particles interacting via a two dimensional Coulomb force. The equilibrium density of this Coulomb gas is such as to make the potential energy a minimum, and this density is identified with the level density of the corresponding Gaussian ensembles. That the eigenvalue density so deduced is correct, will be seen later when in Chapters 5 to 7 and 14 we will compute it for each of the four Gaussian ensembles for any finite $N \times N$ matrices and take the limit as $N \longrightarrow \infty$.

Another argument, again essentially due to Wigner, is given to show that the same level density, a "semicircle," holds for Hermitian matrices with elements having an average value zero and a common mean square value.

4.1. The Partition Function

Consider a gas of N point charges with positions $x_1, x_2, ..., x_N$ free to move on the infinite straight line $-\infty < x < \infty$. Suppose that the potential energy of the gas is given by

$$W = \frac{1}{2} \sum_i x_i^2 - \sum_{i<j} \ln \, |x_i - x_j| \,. \tag{4.1.1}$$

The first term in W represents a harmonic potential that attracts each charge independently toward the point $x = 0$; the second term represents an electrostatic repulsion between each pair of charges. The logarithmic function comes in if we assume the universe to be two-dimensional. Let this charged gas be in thermodynamical equilibrium at a temperature T, so that the probability density of the positions of the N charges is given by

$$P(x_1, ..., x_N) = C \exp(-W/kT), \tag{4.1.2}$$

where k is the Boltzmann constant. We immediately recognize that Eq. (4.1.2) is identical to Eq. (3.3.8) provided β is related to the temperature by

$$\beta = (kT)^{-1}. \tag{4.1.3}$$

This system of point charges in thermodynamical equilibrium is called the Coulomb gas model, corresponding to the Gaussian ensembles.

Following Dyson (1962), we can define various expressions that relate to our energy-level series in complete analogy with the classical notions of entropy, specific heat, and the like. These expressions, when computed from the observed experimental data and compared with the theoretical predictions, provide a nice method of checking the theory.

In classical mechanics the joint probability density in the velocity space is a product of exponentials

$$\prod_j \exp\left(-C_j v_j^2\right)$$

with constant C_j, and its contribution to the thermodynamic quantities of the model are easily calculated. We simply discard these trivial terms. The nontrivial contributions arise from the partition function

$$\psi_N(\beta) = \int_{-\infty}^{\infty} \cdots \int_{-\infty}^{\infty} e^{-\beta W} dx_1 \cdots dx_N \tag{4.1.4}$$

and its derivatives with respect to β. Therefore it is important to have an analytical expression for $\psi_N(\beta)$. Fortunately, this can be deduced from an integral evaluated by Selberg (see Chapter 17).

Theorem 4.1.1. *For any positive integer N and real or complex β we have, identically,*

$$\psi_N(\beta) = (2\pi)^{N/2} \beta^{-N/2-\beta N(N-1)/4} \left[\Gamma(1+\beta/2)\right]^{-N} \prod_{j=1}^{N} \Gamma(1+\beta j/2). \tag{4.1.5}$$

Let us note the fact that the energy W given by (4.1.1) is bounded from below. More precisely,

$$W \geq W_0 = \frac{1}{4} N(N-1)(1+\ln 2) - \frac{1}{2} \sum_{j=1}^{N} j \ln j, \tag{4.1.6}$$

and this minimum is attained when the positions of the charges coincide with the zeros of the Hermite polynomial $H_N(x)$ (*cf.* Appendix A.6).

Once the partition function is known, other thermodynamic quantities such as free energy, entropy, and specific heat can be calculated by elementary differentiation. Because all the known properties are identical to those of the circular ensembles, studied at length in Chapter 11, we do not insist on this point here.

4.2. The Asymptotic Formula for the Level Density. Gaussian Ensembles

Since the expression (3.3.8) for $P(x_1, ..., x_N)$, the probability that the eigenvalues will lie in unit intervals around $x_1, x_2, ..., x_N$, is valid for all values of x_i, the density of levels

$$\sigma_N(x) = N \int_{-\infty}^{\infty} \cdots \int_{-\infty}^{\infty} P(x, x_2, ..., x_N)\, dx_2 \cdots dx_N \tag{4.2.1}$$

can be calculated for any N by actual integration (Mehta and Gaudin, 1960). The details of this tedious calculation are not given here, since an expression for $\sigma_N(x)$, derived by a different method, appears in Chapters 5, 6, and 7.

However, if one is interested in the limit of large N, as we certainly are, these complications can be avoided by assuming that the corresponding Coulomb gas is a classical fluid with a continuous macroscopic density. More precisely, this amounts to the following two assumptions:

(1) The potential energy W given by (4.1.1) can be approximated by the functional

$$W(\sigma) = \frac{1}{2}\int_{-\infty}^{\infty} dx\, x^2\sigma(x) - \frac{1}{2}\int_{-\infty}^{\infty} dx\, dy\, \sigma(x)\, \sigma(y)\, \ln |x-y|. \quad (4.2.2)$$

(2) The level density $\sigma(x)$ will be such as to minimize the expression (4.2.2), consistent with the requirements

$$\int_{-\infty}^{\infty} dx\, \sigma(x) = N \quad (4.2.3)$$

and

$$\sigma(x) \geq 0. \quad (4.2.4)$$

The first integral in Eq. (4.2.2) reproduces the first sum in Eq. (4.1.1) accurately in the limit of large N. The same is not true of the second integral, for it neglects the two-level correlations, which may be expected to extend over a few neighboring levels; however, because the total number of levels is large their effect may be expected to be small. The factor $\frac{1}{2}$ in the second term of Eq. (4.2.2) comes from the condition $i < j$ in Eq. (4.1.1).

The problem of finding the stationary points of the functional $W(\sigma)$, Eq. (4.2.2), with the restriction Eq. (4.2.3) leads us to the integral equation

$$-\frac{1}{2}x^2 + \int_{-\infty}^{\infty} dy\, \sigma(y)\, \ln |x-y| = C, \quad (4.2.5)$$

where C is a Lagrange constant. Actually, Eq. (4.2.5) has to hold only for those values of x for which $\sigma(x) > 0$. One cannot add a negative increment to $\sigma(x)$ where $\sigma(x) = 0$, and therefore the functional differentiation is not valid; hence Eq. (4.2.5) cannot be derived for such values of x. It is not difficult to solve Eq. (4.2.5) (Mushkhelishvili, 1953). This will not be done here, but the solution will be given and then verified.

Differentiation of Eq. (4.2.5) with respect to x eliminates C. Before carrying it out, we must replace the integral with

$$\lim_{\varepsilon \longrightarrow 0} \left(\int_{-\infty}^{x-\varepsilon} dy + \int_{x+\varepsilon}^{\infty} dy \right) \sigma(y) \ln |x - y| . \tag{4.2.6}$$

When Eq. (4.2.6) is differentiated with respect to x, the terms arising from the differentiation of the limits drop out and only the derivative of $\ln |x - y|$ remains. The integral becomes a principal value integral and Eq. (4.2.5) becomes

$$P \int_{-\infty}^{\infty} \frac{\sigma(y)}{x - y} dy = x. \tag{4.2.7}$$

Conversely, if Eq. (4.2.7) is satisfied by some $\sigma(y)$ and this σ is an even function, then it will satisfy Eq. (4.2.5) also. We try

$$\begin{aligned} \sigma(y) &= C \left(A^2 - y^2\right)^{1/2}, \qquad |y| < A, \\ &= 0, \qquad\qquad\qquad\qquad |y| > A. \end{aligned} \tag{4.2.8}$$

Elementary integration gives

$$\begin{aligned} \int & \frac{\left(A^2 - y^2\right)^{1/2}}{x - y} dy \\ &= x \sin^{-1}\left(\frac{y}{A}\right) - \left(A^2 - y^2\right)^{1/2} \\ &\quad + \left(A^2 - x^2\right)^{1/2} \ln \left(\frac{A\,(x - y) - x\left(A^2 - y^2\right)^{1/2} - y\left(A^2 - x^2\right)^{1/2}}{A\,(x - y) - x\left(A^2 - y^2\right)^{1/2} + y\left(A^2 - x^2\right)^{1/2}} \right) \end{aligned} \tag{4.2.9}$$

Taking the principal value of Eq. (4.2.9) between the limits $(-A, A)$, we find that only the first term gives a nonzero contribution, which is πx. Hence, Eq. (4.2.7) gives

$$C = 1/\pi, \tag{4.2.10}$$

and Eq. (4.2.3) gives

$$\frac{1}{\pi}\frac{\pi}{2}A^2 = N. \tag{4.2.11}$$

Thus,

$$\sigma(x) = \begin{cases} \frac{1}{\pi}\left(2N - x^2\right)^{1/2}, & |x| < (2N)^{1/2}, \\ 0, & |x| > (2N)^{1/2}. \end{cases} \tag{4.2.12}$$

This is the so-called "semicircle law" first derived by Wigner.

Actually the two-level correlation function can be calculated (*cf.* Sections 5.2, 6.4, and 7.2) and the above intuitive arguments put to test. Instead, we shall derive an exact expression for the level density valid for any N. The limit $N \longrightarrow \infty$ can then be taken (*cf.* Appendix A.9) to obtain the "semicircle law."

We have noted in Section 4.1 that without any approximation whatever the energy W attains its minimum value when the points x_1, $x_2, ..., x_N$ are the zeros of the Nth order Hermite polynominal. The postulate of classical statistical mechanics then implies that in the limit of very large N the level density is the same as the density of zeros of the Nth order Hermite polynominal. This later problem has been investigated by many authors, and we may conveniently refer to the relevant mathematical literature (Szegö, 1959).

4.3. The Asymptotic Formula for the Level Density. Other Ensembles

Numerical evidence shows, as we said in Chapter 1, that the local statistical properties of the Gaussian ensembles are shared by a much wider class of matrices. In particular, the eigenvalue density follows the "semicircle law." Wigner (1955, 1957a) first considered bordered matrices, i.e., real symmetric matrices H with elements

$$\begin{aligned} H_{jk} &= \pm h, \qquad \text{if} \quad |j-k| \le m, \\ &= 0, \qquad \text{if} \quad |j-k| > m. \end{aligned} \tag{4.3.1}$$

Except for the symmetry of H, the signs of H_{jk} are random. He then calculates the moments of the level density and derives an integral equation for it. The calculations are long. The final result is that in the limit that $h^2/m \to 0$ and the order of the matrices is infinite, the eigenvalue density is a "semicircle." Here we present still another heuristic argument in its support, again essentially due to Wigner.

Consider a matrix H with elements H_{ij} all having an average value zero and a mean square value V^2. Let the order N be large enough so that the density of its eigenvalues may be taken to be a continuous function. Let this function be $\sigma(\varepsilon, V^2)$, so that the number of eigenvalues lying between ε and $\varepsilon + \delta\varepsilon$ is given by $\sigma(\varepsilon, V^2)d\varepsilon$. If we change the matrix elements by small quantities δH_{ij} such that the δH_{ij} themselves all have the average value zero and a mean square value v^2, the change in a particular eigenvalue at ε_i can be calculated by the second order perturbation theory

$$Z(\varepsilon, V^2) = \delta H_{ii} + \sum_{j \neq i} \frac{|\delta H_{ij}|^2}{\varepsilon_i - \varepsilon_j} + \cdots . \tag{4.3.2}$$

The δH_{ii} do not produce, on the average, any change in ε_i. The eigenvalues ε_j which lie nearest to ε_i give the largest contribution to Eq. (4.3.2) with an absolute value $v^2/\bar{s}$ where $\bar{s}$ is the mean spacing at ε_i. But as there are eigenvalues on both sides of ε_i, the two contributions arising from the two nearest eigenvalues nearly cancel out, leaving quantities of a higher order in v^2. The sum in Eq. (4.3.2) can therefore be approximated by

$$Z(\varepsilon, V^2) \approx v^2 \int \frac{\sigma(\varepsilon', V^2)}{\varepsilon - \varepsilon'} d\varepsilon', \tag{4.3.3}$$

where the integral in Eq. (4.3.3) is a principal value integral and

$$V^2 = \left\langle |H_{ij}|^2 \right\rangle, \quad v^2 = \left\langle |\delta H_{ij}|^2 \right\rangle . \tag{4.3.4}$$

The ensemble averages being indicated by $\langle \ \rangle$. Let us calculate the change in the number of eigenvalues lying in an interval $(\varepsilon, \varepsilon + \delta\varepsilon)$. This can be done in two ways; one gives, as is obvious from the way of writing,

$$\sigma(\varepsilon, V^2) Z(\varepsilon, V^2) - \sigma(\varepsilon + \delta\varepsilon, V^2) Z(\varepsilon + \delta\varepsilon, V^2) \approx -\frac{\partial(\sigma Z)}{\partial\varepsilon} \delta\varepsilon, \tag{4.3.5}$$

while the other gives in a similar way

$$v^2 \frac{\partial \sigma}{\partial V^2}. \tag{4.3.6}$$

If all the matrix elements H_{ij} are multiplied by a constant c, the values ε_i are also multiplied by c, while V^2 is multiplied by c^2. Hence,

$$\sigma(c\varepsilon, c^2V^2)cd\varepsilon = \sigma(\varepsilon, V^2)d\varepsilon. \tag{4.3.7}$$

Setting $cV = 1$ the last equation gives

$$\sigma(\varepsilon, V^2) = \frac{1}{V}\sigma(\varepsilon/V, 1),$$

which could have been inferred by dimensional arguments. Putting

$$Z(\varepsilon, V^2) = \frac{v^2}{V} Z_1(\varepsilon/V), \quad \sigma(\varepsilon, V^2) = \frac{1}{V}\sigma_1(\varepsilon/V) \tag{4.3.8}$$

in Eqs. (4.3.3), (4.3.5), and (4.3.6), we obtain

$$\frac{\partial(Z_1\sigma_1)}{\partial x} = \frac{1}{2}\frac{\partial(x\sigma_1)}{\partial x}, \quad x = \varepsilon/V, \tag{4.3.9}$$

$$Z_1(x) = P\int \frac{\sigma_1(x')}{x - x'}dx'. \tag{4.3.10}$$

When $x = 0$, by symmetry requirement $Z_1 = 0$; therefore, Eq. (4.3.9) gives, on integration,

$$Z_1(x) = x/2. \tag{4.3.11}$$

Finally we have the boundary condition

$$\int \sigma(\varepsilon, V^2)d\varepsilon = \int \sigma_1(x)dx = N. \tag{4.3.12}$$

Equations (4.3.10), (4.3.11), and (4.3.12) together are equivalent to the integral equation (4.2.7) together with Eq. (4.2.3). The solution, as there, is the semicircle law (4.2.12):

$$\sigma(\varepsilon, V^2) = \begin{cases} \dfrac{1}{2\pi V^2}(2NV^2 - \varepsilon^2)^{1/2}, & \varepsilon^2 < 2NV^2, \\ 0, & \varepsilon^2 > 2NV^2. \end{cases} \tag{4.3.13}$$

Olson and Uppulury and later Wigner extended these considerations to include a still wider class of matrices to have the "semicircle law" as their eigenvalue density.

For the two level correlation function or the spacing distribution no such argument has yet been found.

Summary of Chapter 4

As a consequence of Selberg's integral one has the partition function

$$\psi_N(\beta) = \int_{-\infty}^{\infty} \cdots \int_{-\infty}^{\infty} e^{-\beta W} dx_1 \cdots dx_N \tag{4.1.4}$$

$$= (2\pi)^{N/2} \beta^{-N/2 - \beta N(N-1)/4} [\Gamma(1+\beta/2)]^{-N} \prod_{j=1}^{N} \Gamma(1+\beta j/2), \tag{4.1.5}$$

where

$$W = \frac{1}{2}\sum_{j=1}^{N} x_j^2 - \sum_{1 \le j < k \le N} \ln|x_j - x_k|. \tag{4.1.1}$$

For a large class of random matrices the asymptotic density of eigenvalues is the "semicircle"

$$\sigma(x) = \begin{cases} \dfrac{1}{\pi}(2N - x^2)^{1/2}, & |x| < (2N)^{1/2}, \\ 0, & |x| > (2N)^{1/2}. \end{cases} \tag{4.2.12}$$

5 / Gaussian Unitary Ensemble

In Chapter 2 it was suggested that if the zero and the unit of the energy scale are properly chosen, then the statistical properties of the fluctuations of energy levels of a sufficiently complicated system will behave as if they were the eigenvalues of a matrix taken from one of the Gaussian ensembles. In Chapter 3 we derived the joint probability density function for the eigenvalues of matrices from orthogonal, symplectic, or unitary ensembles. To examine the behavior of a few eigenvalues, one has to integrate out the others, and this we will undertake now.

In this chapter we take up the Gaussian unitary ensemble, which is the simplest from the mathematical point of view, and derive expressions for the probability of a few eigenvalues to lie in certain specified intervals. For other ensembles the same thing will be undertaken in later chapters.

In Section 5.1 we define correlation, cluster, and spacing probability functions. These definitions are general and apply to any of the ensembles. In Section 5.2 we derive the correlation and cluster functions by a method which will be generalized in later chapters to other ensembles. Section 5.3 explains Gaudin's way of using Fredholm theory of integral equations to write the probability $E_2(0; s)$, that an interval of length s contains none of the eigenvalues, as an infinite product. In Section 5.4 we derive an expression for $E_2(n; s)$, the probability that an interval of length s contains exactly n eigenvalues. Finally, Section 5.5 contains a few remarks about the convergence and numerical computation of $E_2(n; s)$ as well as some other equalities.

The two-level cluster function is given by Eq. (5.2.20), its Fourier transform by Eq. (5.2.22); the probability $E_2(n; s)$ for an interval of length s to contain exactly n levels is given by Eq. (5.4.30) while the nearest neighbor spacing density $p_2(0; s)$ is given by Eq. (5.4.32) or Eq.

(5.4.33). These various functions are tabulated in Appendices A.14 and A.15.

5.1. Generalities

The joint probability density function of the eigenvalues (or levels) of a matrix from one of the Gaussian invariant ensembles is [*cf.* Theorem 3.3.1, Eq. (3.3.8)]

$$P_{N\beta}(x_1, ..., x_N) = \text{const} \times \exp\left(-\frac{\beta}{2}\sum_{j=1}^{N} x_j^2\right) \prod_{1\le j<k\le N} |x_j - x_k|^\beta, \quad -\infty < x_i < \infty, \tag{5.1.1}$$

where $\beta = 1, 2,$ or 4 according as the ensemble is orthogonal, unitary or symplectic. The considerations of this section are general, they apply to all the three ensembles. From Section 5.2 onwards we will concentrate our attention to the simplest case $\beta = 2$. For simplicity of notation the index β will sometimes be suppressed; for example, we will write P_N instead of $P_{N\beta}$.

5.1.1. ABOUT CORRELATION FUNCTIONS

The n-point correlation function is defined by (Dyson, 1962)

$$R_n(x_1, ..., x_n) = \frac{N!}{(N-n)!}\int_{-\infty}^{\infty}\cdots\int_{-\infty}^{\infty} P_N(x_1, ..., x_N)\, dx_{n+1}\cdots dx_N, \tag{5.1.2}$$

which is the probability density of finding a level (regardless of labelling) around each of the points x_1, x_2, ..., x_n, the positions of the remaining levels being unobserved. In particular, $R_1(x)$ will give the overall level density. Each function R_n for $n > 1$ contains terms of various kinds describing the grouping of n levels into various subgroups or clusters. For practical purposes it is convenient to work with the n-level cluster function (or the cumulant) defined by

$$T_n(x_1, ..., x_n) = \sum_G (-1)^{n-m}(m-1)! \prod_{j=1}^{m} R_{G_j}(x_k, \text{ with } k \text{ in } G_j). \tag{5.1.3}$$

Here G stands for any division of the indices $(1, 2, ..., n)$ into m subgroups $(G_1, G_2, ..., G_m)$. For example,

$$\begin{aligned} T_1(x) &= R_1(x), \\ T_2(x_1, x_2) &= -R_2(x_1, x_2) + R_1(x_1)R_1(x_2), \\ T_3(x_1, x_2, x_3) &= R_3(x_1, x_2, x_3) - R_1(x_1)R_2(x_2, x_3) - \cdots - \cdots \\ &\quad + 2R_1(x_1)R_1(x_2)R_1(x_3), \end{aligned}$$

$$\begin{aligned} &T_4(x_1, x_2, x_3, x_4) \\ &\quad = -R_4(x_1, x_2, x_3, x_4) + [R_1(x_1)R_3(x_2, x_3, x_4) + \cdots + \cdots + \cdots] \\ &\quad + [R_2(x_1, x_2)R_2(x_3, x_4) + \cdots + \cdots] \\ &\quad - 2\,[R_2(x_1, x_2)R_1(x_3)R_1(x_4) + \cdots + \cdots + \cdots + \cdots + \cdots] \\ &\quad + 6R_1(x_1)R_1(x_2)R_1(x_3)R_1(x_4)R_1(x_5)R_1(x_6), \end{aligned}$$

where in the last equation the first bracket contains four terms, the second contains three terms, and the third contains six terms. Equation (5.1.3) is a finite sum of products of the R functions, the first term in the sum being $(-1)^{n-1}R_n(x_1, x_2, ..., x_n)$ and the last term being $(n-1)!R_1(x_1)\cdots R_1(x_n)$.

We would be particularly pleased if these functions R_n and T_n turn out to be functions only of the differences $|x_i - x_j|$. Unfortunately, this is not true in general. Even the level density R_1 will turn out to be a semicircle rather than a constant (*cf.* Section 5.2). However, as long as we remain in the region of maximum (constant) density, we will see that the functions R_n and T_n satisfy this requirement. Outside the central region of maximum density one can always approximate the overall level density to be a constant in a small region ("small" meaning that the number of levels n in this interval is such that $1 \ll n \ll N$). And one believes that the statistical properties of these levels are independent of where this interval is chosen. However, if the interval is not chosen symmetrically about the origin, the mathematics is more difficult.

It was this unsatisfactory feature of the Gaussian ensembles that led Dyson to define the circular ensembles discussed in Chapters 9, 10, and 11.

The inverse of (5.1.3) (*cf.* Appendix A.7) is

$$R_n(x_1, ..., x_n) = \sum_G (-1)^{n-m} \prod_{j=1}^{m} T_{G_j}\,(x_k, \text{ with } k \text{ in } G_j)\,. \qquad (5.1.4)$$

Thus each set of functions R_n and T_n is easily determined in terms of the other. The advantage of the cluster functions is that they have the property of vanishing when any one (or several) of the separations $|x_i - x_j|$ becomes large in comparison to the local mean level spacing. The function T_n describes the correlation properties of a single cluster of n levels, isolated from the more trivial effects of lower order correlations.

Of special interest for comparison with experiments are those features of the statistical model that tend to definite limits as $N \to \infty$. The cluster functions are convenient also from this point of view. While taking the limit $N \to \infty$, we must measure the energies in units of the mean level spacing α and introduce the variables

$$y_j = x_j/\alpha. \qquad (5.1.5)$$

The y_j then form a statistical model for an infinite series of energy levels with mean spacing $\alpha = 1$. The cluster functions

$$Y_n\,(y_1, y_2, ..., y_n) = \lim_{N \to \infty} \alpha^n T_n\,(x_1, x_2, ..., x_n) \qquad (5.1.6)$$

are well defined and finite everywhere. In particular,

$$Y_1(y) = 1,$$

whereas $Y_2(y_1, y_2)$ defines the shape of the neutralizing charge cloud induced by each particle around itself when the model is interpreted as a classical Coulomb gas (see Section 5.2).

We will calculate all correlation and cluster functions. The R_n will be exhibited as a certain determinant and T_n as a sum over $(n-1)!$ permutations. The *two-level form factor* or the Fourier transform of Y_2

$$b(k) = \int_{-\infty}^{\infty} Y_2(r) \exp(2\pi i k r) dr, \quad r = |y_1 - y_2|\,, \qquad (5.1.7)$$

is of special interest, as many important properties of the level distribution, such as mean square values, depend only on it. The Fourier transform of Y_n, or the *n-level form factor*

$$\int Y_n(y_1, ..., y_n) \exp\left(2\pi i \sum_{j=1}^{n} k_j y_j\right) dy_1 \cdots dy_n$$
$$= \delta(k_1 + ... + k_n) b(k_1, ..., k_n), \tag{5.1.8}$$

will be given as a single integral. The presence of δ reflects the fact that $Y_n(y_1, ..., y_n)$ depends only on the differences of the y_j.

5.1.2. ABOUT LEVEL SPACINGS

The probability density for several consecutive levels inside an interval is defined by

$$A_\beta(\theta; x_1, ..., x_n) = \frac{N!}{n!(N-n)!} \underset{\text{out}}{\int \cdots \int} P_{N\beta}(x_1, ..., x_N)\, dx_{n+1} \cdots dx_N, \tag{5.1.9}$$

where the subscript "out" means that integrations over $x_{n+1}, ..., x_N$ are taken outside the interval $(-\theta, \theta)$, $|x_j| \geq \theta$, $j = n+1, ..., N$, while the remaining variables are inside that interval, $|x_j| \leq \theta$, $j = 1, ..., n$.

The quantities of interest are again those which remain finite in the limit $N \to \infty$. We put

$$t = \theta/\alpha, \tag{5.1.10}$$

and define

$$B_\beta(t; y_1, ..., y_n) = \lim_{N\to\infty} \alpha^n A_\beta(\theta; x_1, ..., x_n). \tag{5.1.11}$$

This is the probability density that in a series of eigenvalues with mean spacing unity an interval of length $2t$ contains exactly n levels at positions around the points $y_1, ..., y_n$. To get the probability that a randomly chosen interval of length $s = 2t$ contains exactly n levels one has to integrate $B_\beta(t; y_1, ..., y_n)$ over $y_1, ..., y_n$ from $-t$ to t,

$$E_\beta(n; s) = \int_{-t}^{t} \cdots \int_{-t}^{t} B_\beta(t; y_1, ..., y_n)\, dy_1 \cdots dy_n. \tag{5.1.12}$$

If we put $y_1 = -t$, and integrate over $y_2, y_3, ..., y_n$ from $-t$ to t, we get

$$F_\beta(0; s) = B_\beta(t; -t), \tag{5.1.13a}$$

$$F_\beta(n-1; s) = n \int_{-t}^{t} \cdots \int_{-t}^{t} B_\beta(t; -t, y_2, ..., y_n) dy_2 \cdots dy_n, \quad n > 1. \tag{5.1.13b}$$

If we put $y_1 = -t$, $y_2 = t$ and integrate over $y_3, ..., y_n$ from $-t$ to t, we get

$$p_\beta(0; s) = 2B_\beta(t; -t, t), \tag{5.1.14a}$$

$$p_\beta(n-2; s) = n(n-1) \int_{-t}^{t} \cdots \int_{-t}^{t} B_\beta\,(t; -t, t, y_3, ...y_n)\, dy_3 \cdots dy_n, \quad n > 2. \tag{5.1.14b}$$

Omitting the subscript β for the moment, $E(n; s)$ is the probability that an interval of length s $(= 2t)$ chosen at random contains exactly n levels. If we choose a level at random and measure a distance s from it, say to the right, then the probability that this distance contains exactly n levels, not counting the level at the beginning, is $F(n; s)$. Lastly, let us choose a level at random, denote it by A_0, move to the right, say, and denote the successive levels by A_1, A_2, A_3, ... Then $p(n; s)ds$ is the probability that the distance between levels A_0 and A_{n+1} lies between s and $s + ds$. Let us also write

$$I(n) = \int_0^\infty E(n; s)\, ds. \tag{5.1.15}$$

The functions $E(n; s)$, $F(n; s)$, and $p(n; s)$ are of course related (*cf.* Appendix A.8),

$$-\frac{d}{ds} E(0; s) = F(0; s), \quad -\frac{d}{ds} F(0; s) = p(0; s), \tag{5.1.16a}$$

$$-\frac{d}{ds} E(n; s) = F(n; s) - F(n-1; s), \quad n \geq 1, \tag{5.1.16b}$$

and

$$-\frac{d}{ds} F(n;s) = p(n;s) - p(n-1;s), \quad n \geq 1, \tag{5.1.16c}$$

or

$$F(n;s) = -\frac{d}{ds} \sum_{j=0}^{n} E(j;s), \quad n \geq 0, \tag{5.1.17}$$

and

$$\begin{aligned} p(n;s) &= -\frac{d}{ds} \sum_{j=0}^{n} F(j;s), \\ &= \frac{d^2}{ds^2} \sum_{j=0}^{n} (n-j+1) E(j;s), \quad n \geq 0. \end{aligned} \tag{5.1.18}$$

The functions $E(n;s)$ and $F(n;s)$ satisfy a certain number of constraints. Evidently, they are non-negative, and

$$\sum_{n=0}^{\infty} E(n;s) = \sum_{n=0}^{\infty} F(n;s) = 1, \tag{5.1.19}$$

for any s, since the number of levels in the inteval s is either 0, or 1, or 2, or ... . As we have chosen the average level density to be unity, we have

$$\sum_{n=0}^{\infty} n\, E(n;s) = s. \tag{5.1.20}$$

If the levels never overlap, then

$$E(n;0) = F(n;0) = \delta_{n0} \equiv \begin{cases} 1, & \text{if } n = 0, \\ 0, & \text{if } n > 0. \end{cases} \tag{5.1.21}$$

Moreover, the $F(n;s)$ and $p(n;s)$ are probabilities, so that the right-hand sides of Eqs. (5.1.17) and (5.1.18) are non-negative for n and s.

Also $p(n;s)$ are normalized and $E(n;s)$ and $F(n;s)$ decrease fast to zero as $s\to\infty$, so that

$$\begin{aligned}\int_0^s p(n;x)dx &= 1-\sum_{j=0}^{n} F(j;s)\\ &= 1+\frac{d}{ds}\sum_{j=0}^{n}(n-j+1)E(j;s)\end{aligned}\tag{5.1.22}$$

increases from 0 to 1 as s varies from 0 to ∞. Thus,

$$\sum_{j=0}^{n} F(j;s) \quad \text{and} \quad \sum_{j=0}^{n}(n-j+1)E(j;s) \tag{5.1.23}$$

are nonincreasing functions of s.

The numbers $I(n)$ are also not completely arbitrary. For example, consider the mean square scatter of s_n, the distance between the levels A_0 and A_n, separated by $n-1$ other levels,

$$\left\langle (s_n-\langle s_n\rangle)^2\right\rangle = \langle s_n^2\rangle - \langle s_n\rangle^2 \geq 0. \tag{5.1.24}$$

Integrating by parts and using Eqs. (5.1.16), (5.1.17), and (5.1.18), we get

$$\langle s_n^2\rangle \equiv \int_0^\infty s^2 p(n-1;s)\,ds = 2\sum_{j=0}^{n-1}(n-j)\,I(j), \tag{5.1.25}$$

$$\langle s_n\rangle \equiv \int_0^\infty s\,p(n-1;s)\,ds = n, \tag{5.1.26}$$

where we have dropped the integrated end terms. From Eqs. (5.1.24)–(5.1.26) one gets the inequalities

$$2\sum_{j=0}^{n-1}(n-j)I(j) \geq n^2, \quad n\geq 1. \tag{5.1.27}$$

In particular, $I(0)\geq 1/2$. Also $I(0)\geq I(n)/2$ (see Section 16.4).

For illustration purposes, let us consider 2 or 3 examples.

Example 1. The positions of the various levels are independent random variables with no correlation whatsoever. This case is known as the Poisson distribution. For all $n \geq 0$, one has

$$E(n;s) = F(n;s) = p(n;s) = s^n e^{-n}/n! \ , \tag{5.1.28}$$

$$I(n) = 1. \tag{5.1.29}$$

Example 2. Levels are regularly spaced, i.e., they are located at integer coordinates. In this case

$$E(n;s) = \begin{cases} 1 - |s-n|, & \text{if } |s-n| \leq 1, \\ 0, & \text{if } |s-n| \geq 1, \end{cases} \tag{5.1.30}$$

$$F(n;s) = \begin{cases} 1, & \text{if } n < s < n+1, \\ 0, & \text{otherwise,} \end{cases} \tag{5.1.31}$$

$$p(n;s) = \delta(n+1-s), \tag{5.1.32}$$

$$I(0) = 1/2, \quad I(n) = 1, \quad n \geq 1. \tag{5.1.33}$$

The δ in Eq. (5.1.32) above is the Dirac delta function.

Example 3. Levels occur regularly at intervals a, b and c apart with $a + b + c = 3$, so that the average density is one. In this case

$$\begin{aligned} I(0) &= \frac{1}{2}\, I(3) = \frac{1}{6}\, (a^2 + b^2 + c^2), \\ I(1) &= I(2) = \frac{1}{3}\, (bc + ca + ab), \\ I(n) &= I(n-3), \quad n \geq 4. \end{aligned} \tag{5.1.34}$$

If we take $a \le b < a + b \le c$, for example, $a = b = 1/4$, $c = 5/2$, then $I(0) > 1$, $I(1) < 1$, $I(3) > 2$, etc. This example shows that $I(n)$ are not necessarily monotone.

To get the probability that $n - 1$ consecutive spacings have values $s_1, ..., s_{n-1}$ it is sufficient first to order the y_j as

$$-t \le y_1 \le \cdots \le y_n \le t, \tag{5.1.35}$$

in Eq. (5.1.11) and then to substitute

$$t = -y_1 = \frac{1}{2} \sum_{j=1}^{n-1} s_j, \tag{5.1.36}$$

$$y_i = \sum_{j=1}^{i-1} s_j - \frac{1}{2} \sum_{j=1}^{n-1} s_j, \quad i = 2, ..., n; \tag{5.1.37}$$

(thus $y_n = t$); in the expression for $B_\beta\,(t; y_1, ..., y_n)$.

In particular, let $B_\beta(t; y) = \mathcal{B}_\beta(x_1, x_2)$ and $B_\beta(t; -t, t, y) = \mathcal{P}_\beta(x_1, x_2)$ with $x_1 = t + y$, $x_2 = t - y$. Then $\mathcal{B}_\beta(x_1, x_2)$ is the probability that no eigenvalues lie for a distance x_1 on one side and x_2 on the other side of a given eigenvalue. Similarly, $\mathcal{P}_\beta(x_1, x_2)$ is the joint probability density function for two adjacent spacings x_1 and x_2; the distances are measured in units of mean spacing. Also $\mathcal{P}_\beta(x_1, x_2) = \partial^2 \mathcal{B}_\beta(x_1, x_2)/(\partial x_1 \partial x_2)$.

Explicit expressions will be derived for $B_\beta\,(t; y_1, ..., y_n)$, and $E_\beta(n; s)$, for $\beta = 2$ in this chapter, and for $\beta = 1$ and 4 in later chapters.

5.1.3. SPACING DISTRIBUTION

The functions E_β, F_β, and p_β are of particular interest for small values of n. Thus the probability that the distance between a randomly chosen pair of consecutive eigenvalues lies between s and $s + ds$ is $p_\beta(0; s)ds$; $p_\beta(0; s)$ is known as the probability density for the spacings between the nearest neighbors. The probability that this distance (between the nearest neighbors) be less than or equal to s is

$$\Psi_\beta(s) = \int_0^s p_\beta(0; s) ds = 1 - F_\beta(0; s). \tag{5.1.38}$$

This $\Psi_\beta(s)$ is known as the distribution function of the spacings (between the nearest neighbors). It increases from 0 to 1 as s varies from 0 to ∞ [*cf.* Eq. (5.1.22) and what follows].

5.1.4. CORRELATIONS AND SPACINGS

From the interpretation of $Y_2(s)$, $s = |y_1 - y_2|$, and $p(n; s)$, Eqs. (5.1.6) and (5.1.14), as probability densities one has for all s,

$$Y_2(s) = \sum_{n=0}^{\infty} p(n; s) \tag{5.1.39}$$

The E_β and $R_{n\beta}$ or $T_{n\beta}$ are also related. From this relation and the expression for $R_{n\beta}$ or $T_{n\beta}$, one can derive an expression for $E_\beta(n; s)$. See Appendix A.7.

5.2. The n-Point Correlation Function

From now on in the rest of this chapter we will consider the Gaussian unitary ensemble corresponding to $\beta = 2$ in Eq. (5.1.1).

For the calculation of $R_{n2}(x_1, ..., x_n)$ it is convenient to use the following.

Theorem 5.2.1. *Let $J_N = [J_{ij}]$ be an $N \times N$ Hermitian matrix such that*

$$\begin{aligned}
&(i)\ J_{ij} = f(x_i, x_j), \text{ i.e., } J_{ij} \text{ depends only on } x_i \text{ and } x_j,\\
&(ii)\ \int f(x, x)\,d\mu(x) = c, \qquad\qquad (5.2.1)\\
&(iii)\ \int f(x, y) f(y, z)\,d\mu(y) = f(x, z),
\end{aligned}$$

where $d\mu$ is a suitable measure and c is a constant. Then

$$\int \det J_N d\mu(x_N) = (c - N + 1) \det J_{N-1}, \tag{5.2.2}$$

where J_{N-1} is the $(N-1) \times (N-1)$ matrix obtained from J_N by removing the row and the column containing x_N.

Proof. From the definition of the determinant

$$\begin{aligned}\det J_N &= \sum_{(\pi)} \varepsilon(\pi) \prod_1^{\ell} (J_{\xi\eta} J_{\eta\zeta} \cdots J_{\theta\xi}) \\ &= \sum_{(\pi)} \varepsilon(\pi) \prod_1^{\ell} [f(x_\xi, x_\eta) f(x_\eta, x_\zeta) \cdots f(x_\theta, x_\xi)], \qquad (5.2.3)\end{aligned}$$

where the permutation π consists of ℓ cycles of the form $(\xi \to \eta \to \zeta \to \cdots \to \theta \to \xi)$. Now the index N occurs somewhere. There are two possibilities: (i) it forms a cycle by itself, and $f(x_N, x_N)$ gives on integration the constant c; the remaining factor is by definition $\det J_{N-1}$; (ii) it occurs in a longer cycle and integration on x_N reduces the length of this cycle by one. This can happen in $(N-1)$ ways since the index N can be inserted between any two indices of the cyclic sequence $1, 2, \ldots, N-1$. Also the resulting permutation over $N-1$ indices has an opposite sign. The remaining expression is by definition $\det J_{N-1}$, and the contribution from this possibility is $-(N-1)\det J_{N-1}$. Adding the two contributions we get the result.

It remains now to write the integrand P_{N2} in Eq. (5.1.2) as a determinant having the above properties, and for this we will introduce orthogonal polynomials.

The product of the differences

$$\Delta(x) = \prod_{i<j} (x_j - x_i)$$

is the Vandermonde determinant, its jth row being

$$x_1^{j-1}, x_2^{j-1}, \ldots, x_N^{j-1}, \qquad (5.2.4)$$

j varying from 1 to N. Multiplying the jth row by $2^{j-1}$and adding to it an appropriate linear combination of the other rows with lower powers of the variables, we may replace this jth row by

$$H_{j-1}(x_1), H_{j-1}(x_2), \ldots, H_{j-1}(x_N), \qquad (5.2.5)$$

where $H_j(x)$ is the Hermite polynomial of order j,

$$H_j(x) = \exp(x^2)(-d/dx)^j \exp(-x^2) = j! \sum_{i=0}^{[j/2]} (-1)^i \frac{(2x)^{j-2i}}{i!(j-2i)!}.$$

Also for later convenience, we multiply the jth row by a normalization factor $[2^{j-1}(j-1)!\sqrt{\pi}]^{-1/2}$and the kth column by $\exp(-x_k^2/2)$, so that

$$\exp\left(-\frac{1}{2}\sum_1^N x_j^2\right) \Delta(x) = \text{const} \times \det M, \tag{5.2.6}$$

$$M_{jk} = \varphi_{j-1}(x_k), \qquad j, k = 1, 2, ..., N, \tag{5.2.7}$$

where

$$\begin{aligned} \varphi_j(x) &= \left(2^j j! \sqrt{\pi}\right)^{-1/2} \exp(-x^2/2) H_j(x) \\ &= \left(2^j j! \sqrt{\pi}\right)^{-1/2} \exp\left(x^2/2\right) \left(-d/dx\right)^j \exp\left(-x^2\right), \end{aligned} \tag{5.2.8}$$

are the "oscillator wave functions," orthogonal over $(-\infty, \infty)$,

$$\begin{aligned} &\int_{-\infty}^{\infty} \varphi_j(x)\varphi_k(x)dx \\ &\quad = \left(2^{j+k} j! k! \pi\right)^{-1/2} \int_{-\infty}^{\infty} H_j(x) H_k(x) \exp(-x^2) dx \\ &\quad = \delta_{jk} = \begin{cases} 1, & \text{if} \quad j = k, \\ 0, & \text{otherwise.} \end{cases} \end{aligned} \tag{5.2.9}$$

Therefore [for the constant, see the remark after Eq. (5.2.13)],

$$P_{N2}(x_1, ..., x_N) = \frac{1}{N!}\det(M^T M) = \frac{1}{N!} Q_{N2}(x_1, ..., x_N), \tag{5.2.10}$$

with

$$Q_{N2}(x_1, ..., x_N) = \det\left[K_N(x_i, x_j)\right]_{i,j=1,...,N}, \tag{5.2.11}$$

$$K_N(x, y) = \sum_{k=0}^{N-1} \varphi_k(x)\varphi_k(y). \tag{5.2.12}$$

From the orthonormality of the $\varphi_j(x)$, Eq. (5.2.9), it is straightforward to verify that

$$\begin{aligned}\int_{-\infty}^{\infty} K_N(x,x)dx &= \sum_{k=0}^{N-1}\int_{-\infty}^{\infty}\varphi_k^2(x)dx = \sum_{k=0}^{N-1} 1 = N,\\ \int_{-\infty}^{\infty} K_N(x,y)K_N(y,z)dy &= \sum_{j,k=0}^{N-1}\varphi_j(x)\varphi_k(z)\int_{-\infty}^{\infty}\varphi_j(y)\varphi_k(y)dy\\ &= \sum_{j,k=0}^{N-1}\varphi_j(x)\varphi_k(z)\delta_{jk} = K_N(x,z)\end{aligned} \tag{5.2.13}$$

The Q_{N2} therefore satisfies the conditions of theorem Section 5.2.1, and we can integrate over any number of variables. Integrating over all of them fixes the constant in Eq. (5.2.10) to be $(N!)^{-1}$. Integrating over $N-n$ variables $x_{n+1},...,x_N$, we get

$$R_n = \det\left[K_N(x_i,x_j)\right]_{i,j=1,\dots,n}, \tag{5.2.14}$$

with $K_N(x,y)$ given by Eq. (5.2.12).

We may expand the determinant in Eq. (5.2.14),

$$R_n = \sum_P (-1)^{n-m}\prod_1^m K_N(x_a,x_b)K_N(x_b,x_c)\cdots K_N(x_d,x_a).$$

where the permutation P is a product of m exclusive cycles of lengths $h_1,...,h_m$ of the form $(a\to b\to c\to\cdots\to d\to a)$, $\sum_1^m h_j = n$. Taking h_j as the number of indices in G_j, and comparing with Eq. (5.1.4), we get

$$T_n(x_1,...,x_n) = \sum_P K_N(x_1,x_2)K_N(x_2,x_3)\cdots K_N(x_n,x_1), \tag{5.2.15}$$

where the sum is over all the $(n-1)!$ distinct cyclic permutations of the indices in $(1,...,n)$.

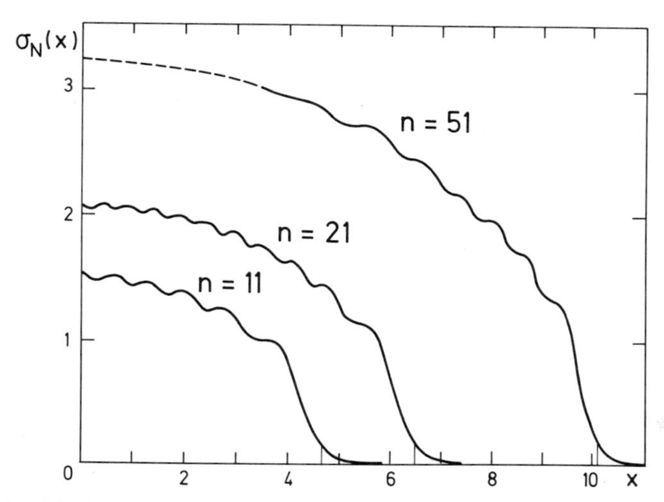

FIG. 5.1. The level density $\sigma_N(x)$, Eq. (5.2.16), for $N = 11$, 21, and 51. The oscillations are noticeable even for $N = 51$. The "semicircle," Eq. (5.2.17), ends at points marked at $\sqrt{22} \approx 4.7$, $\sqrt{42} \approx 6.5$, and $\sqrt{102} \approx 10.1$. Reprinted with permission from E. P. Wigner, "Distribution laws for the roots of a random Hermitian matrix" (1962), in *Statistical theories of spectra: fluctuations*, ed. C. E. Porter, Academic Press, New York, (1965).

Putting $n = 1$ in Eq. (5.2.14), we get the level density

$$\sigma_N(x) = K_N(x, x) = \sum_{j=0}^{N-1} \varphi_j^2(x). \tag{5.2.16}$$

As $N \to \infty$, this goes to the semi-circle law (*cf.* Appendix A.9)

$$\sigma_N(x) \to \sigma(x) = \begin{cases} \dfrac{1}{\pi}(2N - x^2)^{1/2}, & |x| < (2N)^{1/2}, \\ 0, & |x| > (2N)^{1/2}. \end{cases} \tag{5.2.17}$$

The mean spacing at the origin is thus

$$\alpha = 1/\sigma(0) = \pi/(2N)^{1/2}. \tag{5.2.18}$$

Figure 5.1 shows $\sigma_N(x)$ for a few values of N.

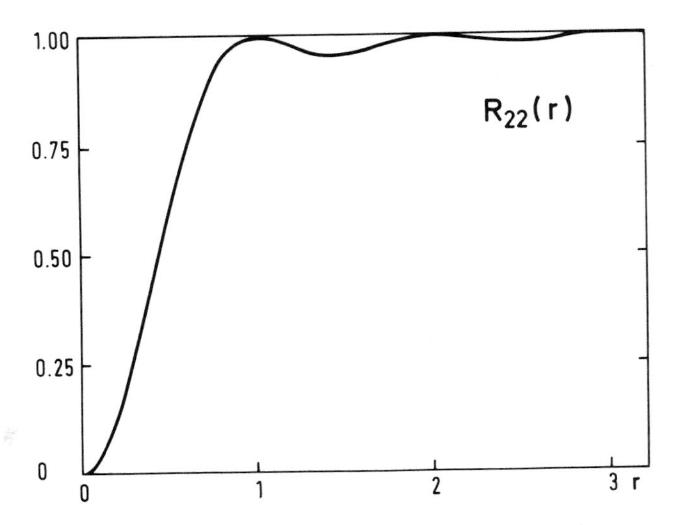

FIG. 5.2. Two-point correlation function, $1 - [\sin(\pi r)/(\pi r)]^2$, for the unitary ensemble (*cf.* Section 5.2 end).

Putting $n = 2$ in (5.2.15) we get the two-level cluster function

$$T_2(x_1, x_2) = [K_N(x_1, x_2)]^2 = \left(\sum_{j=0}^{N-1} \varphi_j(x_1)\varphi_j(x_2)\right)^2. \tag{5.2.19}$$

Taking the limit as $N \to \infty$, this equation, the definitions (5.1.6) and (5.2.18) (*cf.* Appendix A.10) give

$$\begin{aligned} Y_2(x_1, x_2) &= \lim_{N \to \infty} \left(\frac{\pi}{(2N)^{1/2}} \sum_{j=0}^{N-1} \varphi_j(x_1)\varphi_j(x_2)\right)^2 \\ &= [s(r)]^2 \equiv \left(\frac{\sin(\pi r)}{\pi r}\right)^2, \end{aligned} \tag{5.2.20}$$

with

$$r = |y_1 - y_2|, \quad y_1 = (2N)^{1/2} x_1/\pi, \quad y_2 = (2N)^{1/2} x_2/\pi.$$

Figure 5.2 shows the limiting two level correlation function $1 - [\sin(\pi r)/(\pi r)]^2$. Note the oscillations around integer values of r.

In the limit $N \to \infty$, the n-level cluster function is

$$Y_n(y_1, ..., y_n) = \sum_P s(r_{12})s(r_{23}) \cdots s(r_{n1}), \tag{5.2.21}$$

the sum being taken over the $(n-1)!$ distinct cyclic permutations of the indices $(1, 2, ..., n)$, and $r_{ij} = |y_i - y_j| = (2N)^{1/2} |x_i - x_j| / \pi$.

The two-level form factor is (*cf.* Appendix A.11)

$$\begin{aligned} b(k) &= \int_{-\infty}^{\infty} Y_2(r)\exp(2\pi ikr)dr \\ &= \begin{cases} 1 - |k|, & |k| \le 1, \\ 0, & |k| \ge 1. \end{cases} \end{aligned} \tag{5.2.22}$$

The n-level form-factor or the Fourier transform of Y_n is

$$\begin{aligned} &\int Y_n(y_1, ..., y_n)\exp\left(2\pi i \sum_{j=1}^{n} k_j y_j\right) dy_1 \cdots dy_n \\ &= \delta(k_1 + ...k_n) \int_{-\infty}^{\infty} dk \sum_P f_2(k) f_2(k + k_1) \cdots f_2(k + k_1 + \cdots + k_{n-1}), \end{aligned} \tag{5.2.23}$$

with

$$f_2(k) = \begin{cases} 1, & \text{if} \quad |k| < 1/2, \\ 0, & \text{if} \quad |k| > 1/2. \end{cases} \tag{5.2.24}$$

It must be noted that the three and higher order correlation functions can all be expressed in terms of the level density and the two-level function $K_N(x, y)$, for every finite N, as is evident from Eq. (5.2.14). This is a particular feature of the unitary ensemble.

5.3. Level Spacings

In this section we will express $A_2(\theta)$, the probability that the interval $(-\theta, \theta)$ does not contain any of the points $x_1, ..., x_N$, as a Fredholm determinant of a certain integral equation. In the limit of large N the solutions of this integral equation are the so called spheroidal functions satisfying a second order differential equation. The limit of $A_2(\theta)$ will thus be expressed as a fast converging infinite product.

To start with, let us calculate the simplest level spacing function

$$A_2(\theta) = \underset{\text{out}}{\int \cdots \int} P_{N2}\left(x_1, ..., x_N\right) dx_1 \cdots dx_N, \tag{5.3.1}$$

case $n = 0$, of Eq. (5.1.9). Substituting from Eq. (5.2.6) or Eq. (5.2.10) for P_{N2}, we get

$$\begin{aligned} A_2(\theta) &= \frac{1}{N!} \underset{\text{out}}{\int \cdots \int} (\det\ M)^2\, dx_1 \cdots dx_N \\ &= \frac{1}{N!} \underset{\text{out}}{\int \cdots \int} \{\det\left[\varphi_{j-1}(x_k)\right]\}^2 dx_1 \cdots dx_N. \end{aligned} \tag{5.3.2}$$

At this point we apply Gram's result (*cf.* Appendix A.12) to get

$$A_2(\theta) = \det\ G \tag{5.3.3}$$

where the elements of the matrix G are given by

$$g_{jk} = \int_{\text{out}} \varphi_{j-1}(x)\varphi_{k-1}(x)dx = \delta_{jk} - \int_{-\theta}^{\theta} \varphi_{j-1}(x)\varphi_{k-1}(x)dx. \tag{5.3.4}$$

To diagonalize G, consider the integral equation

$$\lambda\psi(x) = \int_{-\theta}^{\theta} K(x,y)\psi(y)dy, \tag{5.3.5}$$

$$K(x,y) \equiv K_N(x,y) = \sum_{i=0}^{N-1} \varphi_i(x)\varphi_i(y). \tag{5.2.12}$$

As the kernel $K(x, y)$ is a sum of separable ones, the eigenfunction $\psi(x)$ is necessarily of the form

$$\psi(x) = \sum_{i=0}^{N-1} c_i\varphi_i(x). \tag{5.3.6}$$

Substituting Eq. (5.3.6) in Eq. (5.3.5) and remembering that $\varphi_i(x)$ for $i = 0, 1, ..., N-1$, are linearly independent functions, we obtain the system of linear equations

$$\lambda c_i = \sum_{j=0}^{N-1} c_j \int_{-\theta}^{\theta} \varphi_i(y)\varphi_j(y)dy, \quad i = 0, 1, ..., N-1. \tag{5.3.7}$$

This system will have a nonzero solution if and only if λ satisfies the following equation

$$\det\left[\lambda\delta_{ij} - \int_{-\theta}^{\theta} \varphi_i(y)\varphi_j(y)dy\right]_{i,j=0,\ldots,N-1} = 0. \tag{5.3.8}$$

This algebraic equation in λ has N real roots; let them be $\lambda_0, ..., \lambda_{N-1}$, so that

$$\det\left[\lambda\delta_{ij} - \int_{-\theta}^{\theta} \varphi_i(y)\varphi_j(y)dy\right] \equiv \prod_{i=0}^{N-1} (\lambda - \lambda_i)\,. \tag{5.3.9}$$

Comparing with Eqs. (5.3.3)–(5.3.4), we see that

$$A_2(\theta) = \prod_{i=0}^{N-1} (1 - \lambda_i)\,, \tag{5.3.10}$$

where λ_i are the eigenvalues of the integral equation (5.3.5).

As we are interested in large N, we take the limit $N \to \infty$. In this limit the quantity which is finite is not $K_N(x, y)$, but

$$Q_N(\xi, \eta) = \left(\pi t/(2N)^{1/2}\right) K_N(x, y), \tag{5.3.11}$$

with

$$\pi t = (2N)^{1/2}\theta, \quad \pi t\xi = (2N)^{1/2}x, \quad \pi t\eta = (2N)^{1/2}y, \tag{5.3.12}$$

and (*cf.* Appendix A.10)

$$\lim Q_N(\xi, \eta) = Q(\xi, \eta) = \frac{\sin(\xi - \eta)\pi t}{(\xi - \eta)\pi}. \tag{5.3.13}$$

The limiting integral equation is then

$$\lambda f(\xi) = \int_{-1}^{1} Q(\xi, \eta) f(\eta) d\eta, \tag{5.3.14}$$

where

$$f(\xi) = \psi\left(\pi t \xi / (2N)^{1/2}\right),$$

and

$$2t = 2\theta(2N)^{1/2}/\pi = \text{spacing}/(\text{mean spacing at the origin}). \tag{5.3.15}$$

As $Q(\xi, \eta) = Q(-\xi, -\eta)$, and the interval of integration in Eq. (5.3.14) is symmetric about the origin, if $f(\xi)$ is a solution of the integral equation (5.3.14) with the eigenvalue λ, then so is $f(-\xi)$ with the same eigenvalue. Hence every solution of Eq. (5.3.14) is (or can be chosen to be) either even or odd. Consequently, instead of $\sin(\xi - \eta)\pi t/[(\xi - \eta)\pi]$, one can as well study the kernel $\sin(\xi + \eta)\pi t/[(\xi + \eta)\pi]$.

Actually, from the orthogonality of the $\varphi_j(x)$, $g_{ij} = 0$ whenever $i + j$ is odd; the matrix G has a checker board structure, every alternate element being zero; det G is a product of two determinants, one containing only even functions $\varphi_{2i}(x)$, and the other only odd functions $\varphi_{2i+1}(x)$; and $A_2(\theta)$ in Eq. (5.3.10) is a product of two factors, one containing the λ_i corresponding to the even solutions of Eq. (5.3.5) and the other containing those corresponding to the odd solutions.

The kernel $Q(x, y)$ is the square of another symmetric kernel $(t/2)^{1/2} \exp(\pi i x y t)$,

$$\int_{-1}^{1} \left((t/2)^{1/2} e^{\pi i x z t}\right) \left((t/2)^{1/2} e^{\pi i z y t}\right)^* dz = \frac{\sin(x - y)\pi t}{(x - y)\pi}. \tag{5.3.16}$$

Therefore, Eq. (5.3.14) may be replaced by the integral equation

$$\mu f(x) = \int_{-1}^{1} \exp(\pi i x y t) f(y) dy, \tag{5.3.17}$$

with

$$\lambda = \frac{1}{2} t \left|\mu\right|^2. \tag{5.3.18}$$

Taking the limit $N \to \infty$, Eqs. (5.3.10) and (5.3.18) then give

$$B_2(t) = \prod_{i=0}^{\infty} \left(1 - \frac{1}{2} t \left|\mu_i\right|^2\right), \tag{5.3.19}$$

where μ_i are determined from Eq. (5.3.17).

The integral Eq. (5.3.17) can be written as a pair of equations corresponding to even and odd solutions

$$\mu_{2j} f_{2j}(x) = 2 \int_0^1 \cos(\pi x y t) f_{2j}(y) dy, \tag{5.3.20}$$

$$\mu_{2j+1} f_{2j+1}(x) = 2i \int_0^1 \sin(\pi x y t) f_{2j+1}(y) dy. \tag{5.3.21}$$

The eigenvalues corresponding to the even solutions are real and those corresponding to odd solutions are pure imaginary.

A careful examination of Eq. (5.3.17) shows that its solutions are the spheroidal functions (Robin, 1959) that depend on the parameter t. These functions are defined as the solutions of the differential equation

$$(L - \ell) f(x) \equiv \left((x^2 - 1) \frac{d^2}{dx^2} + 2x \frac{d}{dx} + \pi^2 t^2 x^2 - \ell \right) f(x) = 0, \tag{5.3.22}$$

which are regular at the points $x = \pm 1$. In fact, it is easy to verify that the self-adjoint operator L commutes with the kernel $\exp(\pi i x y t)$ defined over the interval $(-1,1)$; that is,

$$\int_{-1}^{1} \exp(\pi i x y t) L(y) f(y) dy = L(x) \int_{-1}^{1} \exp(\pi i x y t) f(y) dy, \tag{5.3.23}$$

provided

$$(1 - x^2) f(x) = 0 = (1 - x^2) f'(x), \qquad x \to \pm 1. \tag{5.3.24}$$

Equation (5.3.24) implies that $f(x)$ is regular at $x = \pm 1$. Hence Eqs. (5.3.17) and (5.3.22) both have the same set of eigenfunctions. Once the

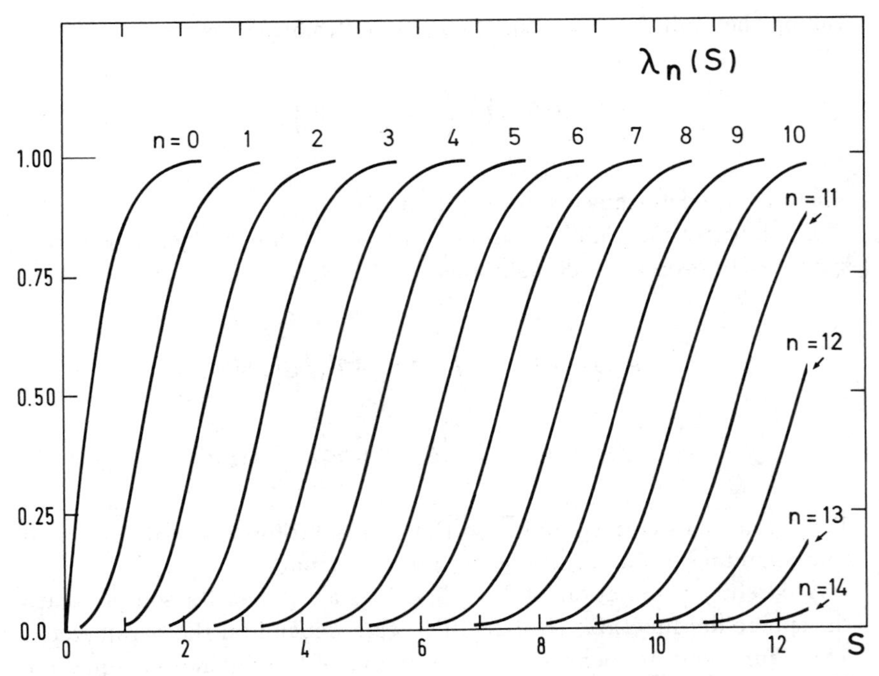

FIG. 5.3. The eigenvalues λ_i; Eq. (5.3.14). Reprinted with permission from D. V. S. Jain, M. L. Mehta, and des Cloizeaux, J. The probabilities for several consecutive eigen-values of a random matrix, *Indian J. Pure Appl. Math.*, 3, 329–351 (1972).

eigenfunctions are known, the corresponding eigenvalues can be computed from Eq. (5.3.17). For example, for even solutions put $x = 0$, to get

$$\mu_{2i} = [f_{2j}(0)]^{-1} \int_{-1}^{1} f_{2j}(y) dy, \tag{5.3.25}$$

while for the odd solutions, differentiate first with respect to x and then put $x = 0$,

$$\mu_{2j+1} = i\pi t \left(f'_{2j+1}(0) \right)^{-1} \int_{-1}^{1} y f_{2j+1}(y) dy. \tag{5.3.26}$$

The spheroidal functions $f_j(x)$, $j = 0, 1, 2, ...$ form a complete set of functions integrable over (−1,1). They are therefore the complete set of solutions of the integral equation (5.3.17), or that of Eq. (5.3.14). We therefore have

$$E_2(0;s) = B_2(t) = \lim_{N \to \infty} A_2\left(\theta(2N)^{1/2}/\pi\right) = \prod_{i=0}^{\infty}\left(1 - \frac{1}{2}t\,|\mu_i|^2\right), \tag{5.3.27}$$

$s = 2t$, where μ_i are given by Equations (5.3.25) and (5.3.26), the functions $f_i(x)$ being the spheroidal functions, solutions of the differential equation (5.3.22).

Figure 5.3 shows the eigenvalues λ_i for some values of i and s.

5.4. Several Consecutive Spacings

Next we take the case of several consecutive spacings, integral (5.1.9), with $n > 0$. From Eqs. (5.1.1) and (5.2.6) or (5.2.10) one has

$$\begin{aligned} P_{N2}(x_1, ..., x_N) &= \frac{1}{N!}\det\left[MM^T\right] \\ &= \frac{1}{N!}\det\left[\sum_{k=1}^{N}\varphi_i(x_k)\varphi_j(x_k)\right]_{i,j=0,1,...,N-1}, \end{aligned} \tag{5.4.1}$$

where the matrix M is given by Eq. (5.2.7). We want to calculate

$$A_2(\theta; x_1, ..., x_n) = \frac{N!}{n!(N-n)!}\underset{\text{out}}{\int \cdots \int} P_{N2}(x_1, ..., x_N)dx_{n+1}\cdots dx_N. \tag{5.4.2}$$

The subscript "out" means that the variables $x_1, ..., x_N$ belong to the intervals

$$\begin{aligned} |x_j| &\le \theta, && \text{if} \quad 1 \le j \le n \\ |x_j| &\ge \theta, && \text{if} \quad n+1 \le j \le N \end{aligned} \tag{5.4.3}$$

In Eq. (5.4.1) the row with index i can be considered as a sum of N rows with elements $\varphi_i(x_k)\varphi_j(x_k)$, $k = 1, ..., N$. Expressing each row as such sums, we write the determinant in Eq. (5.4.1) as a sum of determinants. If a variable occurs in two or more rows, these rows are proportional and therefore the corresponding determinant is zero. The nonzero determinants are those in which all the indices k are different.

$$P_{N2}(x_1, ..., x_N) = \frac{1}{N!} \sum_{(k_i)} \det \left[\varphi_i(x_{k_i}) \varphi_j(x_{k_i}) \right]_{i,j=0,...,N-1}. \qquad (5.4.4)$$

The indices $k_0, ..., k_{N-1}$ are obtained by a permutation of $1, ..., N$ and in Eq. (5.4.4) the summation is over all permutations. As each variable $x_1, ..., x_N$ occurs in one and only one row of any of the determinants of Eq. (5.4.4), one may easily integrate over as many variables as one likes. After integrating over $x_{n+1}, ..., x_N$ we expand each determinant in the Laplace manner (see for example, Mehta, 1989) according to the n rows containing the variables $x_1, ..., x_n$. Introducing the matrix G with elements

$$g_{ij} = \delta_{ij} - \int_{-\theta}^{\theta} \varphi_i(x) \varphi_j(x) dx, \qquad (5.3.4)$$

we obtain the result

$$A_2(\theta; x_1, ..., x_n) = \frac{1}{n!} \sum_{(i,j)} \left(\det \left[\varphi_{i_k}(x_k) \varphi_{j_\ell}(x_k) \right]_{k,\ell=1,...,n} \right) \times G'(i_1, ..., i_n; j_1, ..., j_n). \qquad (5.4.5)$$

The indices $i_1, ..., i_n$ are chosen from $0, ..., N-1$ as are also the indices $j_1, ..., j_n$. The indices $i_1, ..., i_n$ are not ordered, while the indices $j_1, ..., j_n$ are ordered $j_1 < \cdots < j_n$. The summation in Eq. (5.4.5) is extended over all possible choices of indices satisfying the above conditions. The cofactor $G'(i_1, ... i_n; j_1, ..., j_n)$ is, apart from a sign, the determinant of the $(N-n) \times (N-n)$ matrix obtained from G by omitting the rows $i_1, ..., i_n$ and the columns $j_1, ..., j_n$. The sign is plus or minus according to whether $\sum_{k=1}^{n} (i_k + j_k)$ is even or odd. Therefore (*cf.* Mehta, 1989, Section 3.9)

$$G'(i_1, ..., i_n; j_1, ..., j_n) = \det G \det \left(\left[\left(G^{-1} \right)_{ji} \right]_{i=i_1,...,i_n; j=j_1,...j_n} \right). \qquad (5.4.6)$$

If we integrate Eq. (5.4.5) over $x_1, ..., x_n$ in the interval $(-\theta, \theta)$, we get

$$\begin{aligned} C_2(n;\theta) &\equiv \int_{-\theta}^{\theta} \cdots \int_{-\theta}^{\theta} A_2(\theta; x_1, ..., x_n) dx_1 \cdots dx_n \\ &= \sum_{(i,j)} \det\gamma(i;j) G'(i;j) \end{aligned} \tag{5.4.7}$$

where $\gamma(i;j) \equiv \gamma(i_1, ..., i_n; j_1, ..., j_n)$ is the $n \times n$ matrix formed from the rows $i_1, ..., i_n$ and columns $j_1, ..., j_n$ of the matrix γ with elements

$$\gamma_{pq} = \int_{-\theta}^{\theta} \varphi_p(x) \varphi_q(x) dx \tag{5.4.8}$$

and as above

$$G'(i;j) = \det G \det \left[G^{-1}(j;i) \right]. \tag{5.4.6}$$

The summation in Eq. (5.4.7) is over all possible choices of indices with $0 \le i_1 < i_2 < \cdots < i_n \le N-1$; $0 \le j_1 < j_2 < \cdots < j_n \le N-1$. (The ordering of the $i_1, ..., i_n$ removes the factor $n!$ between Eqs. (5.4.5) and (5.4.7).) Hence,

$$C_2(n;\theta) = \det G \sum_{(i,j)} \det \gamma(i;j) \det G^{-1}(j;i) \tag{5.4.9}$$

which in the limit $N \to \infty$, $\theta \to 0$, $s = 2\theta(2N)^{1/2}/\pi$ finite, goes to $E_2(n;s)$,

$$E_2(n;s) = \lim \alpha^n C_2(n;\theta). \tag{5.4.10}$$

Recalling the diagonalization of G from Section 5.3 above, Eq. (5.4.5) reads, for $n = 0$,

$$A_2(\theta) = \det G = \prod_{i=0}^{N-1} (1 - \lambda_i), \tag{5.4.11}$$

where the λ_i, $i = 0, ..., N-1$, are the eigenvalues of the matrix $[\gamma]$ defined by Eq. (5.4.8), i.e.,

$$\sum_{j=0}^{N-1} \gamma_{ij} h_{jk} = h_{ik} \lambda_k. \tag{5.4.12}$$

Equivalently, the λ_i are the eigenvalues of the integral equation (5.3.5),

$$\int_{-\theta}^{\theta} K(x,y)\psi_i(y)dy = \lambda_i \psi_i(x) \tag{5.4.13}$$

with the kernel $K(x,y)$, Eq. (5.2.12),

$$K(x,y) = \sum_{i=0}^{N-1} \varphi_i(x)\varphi_i(y), \tag{5.4.14}$$

and the eigenfunctions, Eq. (5.3.6),

$$\psi_i(x) = \sum_{j=0}^{N-1} h_{ji}\varphi_j(x). \tag{5.4.15}$$

The normalization of the $\psi_i(x)$ depends on that of the eigenvectors of γ. As γ is real and symmetric we may choose its eigenvectors to be real, orthogonal, and normalized to unity (*cf.* Mehta, 1989), so that $h = [h_{ij}]$ is a real orthogonal matrix:

$$\sum_{j=0}^{N-1} h_{ij}h_{kj} = \sum_{j=0}^{N-1} h_{ji}h_{jk} = \delta_{ik}. \tag{5.4.16}$$

From Eqs. (5.4.14)–(5.4.16) we deduce

$$K(x,y) = \sum_{i=0}^{N-1} \psi_i(x)\psi_i(y). \tag{5.4.17}$$

Since K is real and symmetric, its eigenfunctions $\psi_i(x)$ are orthogonal

$$\int_{-\theta}^{\theta} \psi_i(x)\psi_j(x)dx = 0, \quad i \neq j. \tag{5.4.18}$$

However, their normalization is not unity. From Eqs. (5.4.13), (5.4.17), and (5.4.18) one gets

$$\int_{-\theta}^{\theta} \psi_i^2(x)dx = \lambda_i. \tag{5.4.19}$$

To take the limits $N \to \infty$, let us put as before

$$\alpha = \pi/(2N)^{1/2}, \quad \theta = \alpha t, \quad x = \alpha t \xi, \quad y = \alpha t \eta, \tag{5.4.20}$$

and keep ξ, η, t finite. Then

$$K(x, y) \to (\alpha t)^{-1} Q(\xi, \eta), \tag{5.4.21}$$

$$\psi_i(x) \to a_i f_i(\xi), \tag{5.4.22}$$

where $Q(\xi, \eta)$ is given by Eq. (5.3.13) and $f_i(\xi)$ are the spheroidal functions, depending on the parameter t, solutions of the integral equation (5.3.14) or of Eq. (5.3.17). If we normalize the spheroidal functions as

$$\int_{-1}^{1} f_i(\xi) f_j(\xi) d\xi = \delta_{ij}, \tag{5.4.23}$$

then the constants a_i are given by

$$\lambda_i = \int_{-\theta}^{\theta} \psi_i^2(x) dx = a_i^2 \alpha t \int_{-1}^{1} f_i^2(\xi) d\xi,$$

or

$$a_i = (\lambda_i / \alpha t)^{1/2}. \tag{5.4.24}$$

After this revision of Section 5.3, let us come back to Eqs. (5.4.5) and (5.4.6). The matrix h diagonalizes G, and hence also G^{-1}, while it transforms the functions $\varphi_i(x)$ into $\psi_i(x)$. Using Eqs. (5.4.12), (5.4.15), and (5.4.16) we can write Eqs. (5.4.5) and (5.4.6) as

$$A_2(\theta; x_1, \ldots x_n) = \frac{1}{n!} \sum_{(i;j)} \det \left[\psi_{i_k}(x_k) \psi_{j_\ell}(x_k) \right]_{k,\ell=1,\ldots,n}$$
$$\times \det [1 - \Lambda] \det \left[(1 - \Lambda)_{ji}^{-1} \right]_{j=j_1,\ldots,j_n; i=i_1,\ldots,i_n}, \tag{5.4.25}$$

where Λ is the diagonal matrix with diagonal elements λ_i. As $[1-\Lambda]$ is diagonal, non-vanishing terms result if and only if the indices i can be obtained by a permutation P of the indices j :

$$j_1 < ... < j_n, \quad i_k = j_{Pk}, \quad k = 1, ..., n. \tag{5.4.26}$$

Thus Eq. (5.4.25) can be written as

$$A_2(\theta; x_1, ..., x_n) = \frac{1}{n!} \prod_{\rho=0}^{N-1} (1-\lambda_\rho) \sum_{(j)} (1-\lambda_{j_1})^{-1} \cdots (1-\lambda_{j_n})^{-1} \\ \times \sum_P \varepsilon_P \det [\psi_{j_{Pk}}(x_k) \psi_{j_\ell}(x_k)]_{k,\ell=1,...,n} . \tag{5.4.27}$$

where P a permutation of the indices $(1, 2, ..., n)$ and ε_P its sign. Eq. (5.4.27) coincides with

$$A_2(\theta; x_1, ..., x_n) = \frac{1}{n!} \prod_{\rho=0}^{N-1} (1-\lambda_\rho) \sum_{(j)} (1-\lambda_{j_1})^{-1} \cdots (1-\lambda_{j_n})^{-1} \\ \times (\det [\psi_{j_\ell}(x_k)])^2 . \tag{5.4.28}$$

When $N \to \infty$, $\theta \to 0$ while t and y_j given by Eqs. (5.1.5), (5.1.10), (5.2.18) are finite, we get from Eqs. (5.1.9), (5.1.11), (5.4.28), (5.4.22), and (5.4.24)

$$B_2(t; y_1, ..., y_n) = \frac{t^{-n}}{n!} \prod_\rho (1-\lambda_\rho) \sum_{(j)} \frac{\lambda_{j_1}}{1-\lambda_{j_1}} \cdots \frac{\lambda_{j_n}}{1-\lambda_{j_n}} \\ \times \left[\det \left([f_{j_\ell}(y_k/t)]_{\ell,k=1,...,n} \right)\right]^2 . \tag{5.4.29}$$

Let us recall that the eigenvalues λ_i and the functions $f_i(\xi)$ depend on t as a parameter and that the indices $0 \le j_1 < \cdots < j_n$ are integers.

Due to the orthonormality of the $f_i(\xi)$ we may easily integrate over $y_1, ..., y_n$ in Eq. (5.4.29). Actually, for $j = j_1, ..., j_n$; $k = 1, ..., n$;

$$(\det [f_j(x_k)])^2 = \sum_{P,Q} \varepsilon_P \varepsilon_Q f_{P1}(x_1) \cdots f_{Pn}(x_n) f_{Q1}(x_1) \cdots f_{Qn}(x_n),$$

where P and Q are permutations of the indices $j_1, ..., j_n$ and ε_P, ε_Q their signs. On integration one gets 1 when $P = Q$ and 0 when $P \neq Q$. Hence,

$$\begin{aligned} E_2(n; s) &= \int_{-t}^{t} \cdots \int_{-t}^{t} B_2(t; y_1, ..., y_n) dy_1 \cdots dy_n \\ &= t^{-n} \prod_\rho (1 - \lambda_\rho) \sum_{(j)} \frac{\lambda_{j_1}}{1 - \lambda_{j_1}} \cdots \frac{\lambda_{j_n}}{1 - \lambda_{j_n}}. \end{aligned} \tag{5.4.30}$$

In the same way, from Eq. (5.1.14), (5.4.29), and (5.4.23), we get

$$\begin{aligned} p_2(n-2; s) = \frac{4}{s^2} \prod_\rho (1 - \lambda_\rho) \sum_{(j)} \frac{\lambda_{j_1}}{1 - \lambda_{j_1}} \cdots \frac{\lambda_{j_n}}{1 - \lambda_{j_n}} \\ \times \sum_{\ell, m=1}^{n} f_{j_\ell}(1) f_{j_m}(-1) [f_{j_\ell}(1) f_{j_m}(-1) - f_{j_\ell}(-1) f_{j_m}(1)]. \end{aligned} \tag{5.4.31}$$

On the other hand, to obtain the probability density of $(n-1)$ consecutive spacings $s_1, ..., s_{n-1}$ one has only to make the substitutions (5.1.36), (5.1.37) in the expression (5.4.29) as explained at the end of Section 5.1.

The case $n = 2$ is of special interest. Putting $s = 2t$, we get the probability for a single spacing s as

$$\begin{aligned} p_2(0; s) &= 2\ B_2(s/2; -s/2, s/2) \\ &= \frac{16}{s^2} \prod_\rho (1 - \lambda_\rho) \sum_j \frac{\lambda_{2j}}{1 - \lambda_{2j}} f_{2j}^2(1) \sum_k \frac{\lambda_{2k+1}}{1 - \lambda_{2k+1}} f_{2k+1}^2(1). \end{aligned} \tag{5.4.32}$$

Using Eqs. (5.1.18) and (5.4.30) we may also write

$$p_2(0; s) = \frac{d^2}{ds^2} E_2(0; s) = \frac{d^2}{ds^2} \prod_\rho (1 - \lambda_\rho). \tag{5.4.33}$$

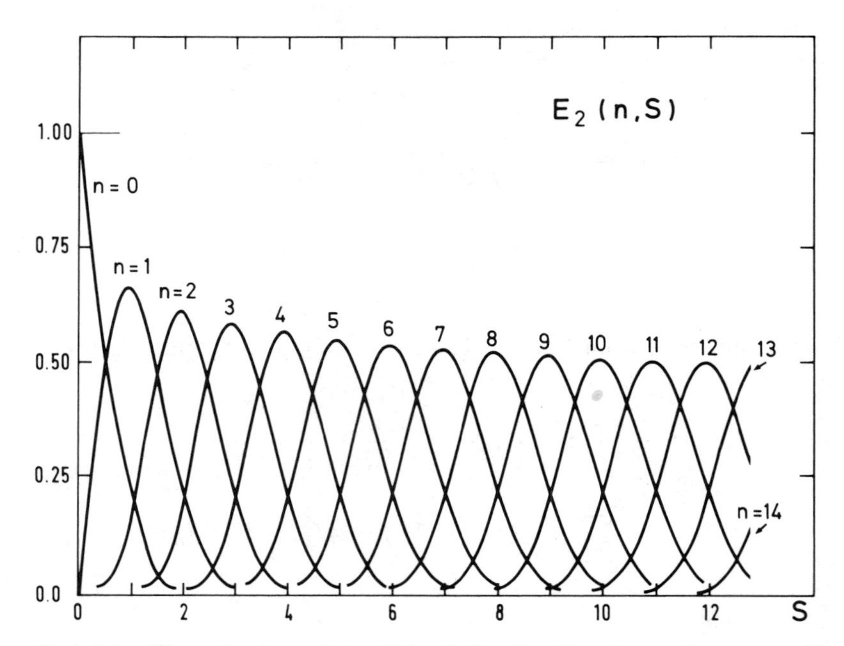

FIG. 5.4. The n-level spacings $E_2(n, s)$ for the Gaussian unitary ensemble. Reprinted with permission from D. V. S. Jain, M. L. Mehta, and des Cloizeaux, J. The probabilities for several consecutive eigen-values of a random matrix, *Indian J. Pure Appl. Math.*, 3, 329–351 (1972).

Figure 5.4 gives a graphical representation of $E_2(n; s)$ for small values of n. For comparison the corresponding probabilities

$$E_0(n; s) = s^n \exp(-s)/n! \tag{5.4.34}$$

for a set of independent random levels, Poisson process, are drawn in Figure 5.5, and the probabilities $E_\infty(n; s)$ corresponding to equally spaced levels in Figure 5.6. From these figures one sees that the set of eigenvalues of a matrix from the Gaussian unitary ensemble is more or less equally spaced and each individual peak is quite isolated, loosing its height and gaining in width only slowly as n increases. Figures 5.7 and 5.8 are the contour maps for two consecutive spacings.

For the empirical probability density of the nearest neighbor spacings of the zeros of the Riemann zeta function on the critical line see Figures 1.12 to 1.14 in Chapter 1. Figures 5.9 and 5.10 represent those for

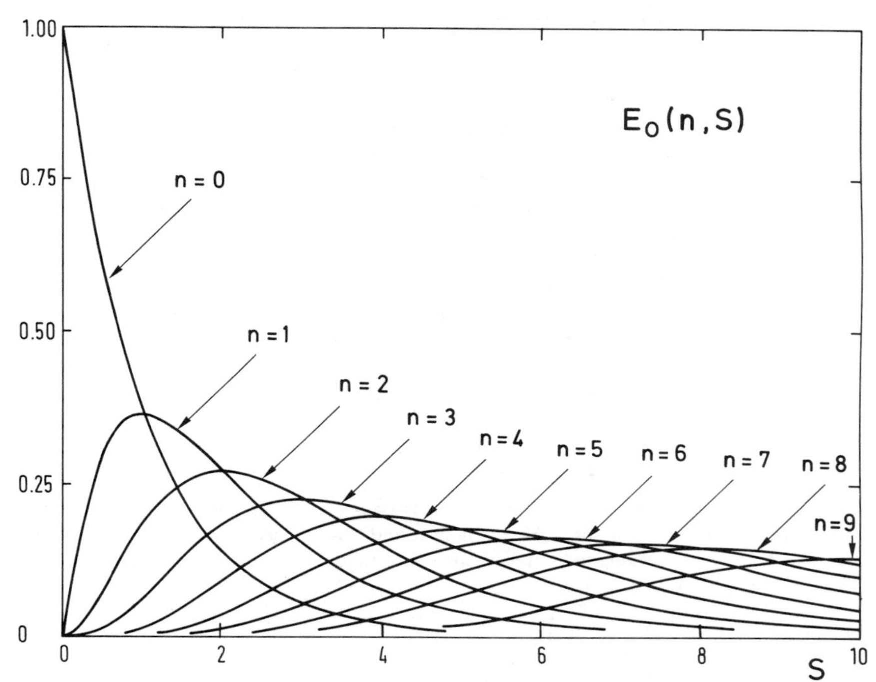

FIG. 5.5. The n-level spacings $E_0(n; s)$ for the Poisson process. Reprinted with permission from D. V. S. Jain, M. L. Mehta, and des Cloizeaux, J. The probabilities for several consecutive eigen-values of a random matrix, *Indian J. Pure Appl. Math.*, 3, 329–351 (1972).

the next nearest neighbour spacings in comparison to $p_2(1; s)$ for the Gaussian unitary ensemble.

5.5. Some Remarks

A few remarks about the analysis of the previous two sections are in order.

5.5.1. We have put aside the question of convergence. In fact, for fixed ξ and η, $Q_N(\xi, \eta)$ tends to $Q(\xi, \eta)$ uniformly (Goursat, 1956) with respect to ξ and η in any finite interval $|\xi|, |\eta| \leq 1$. Hence, the Fredholm

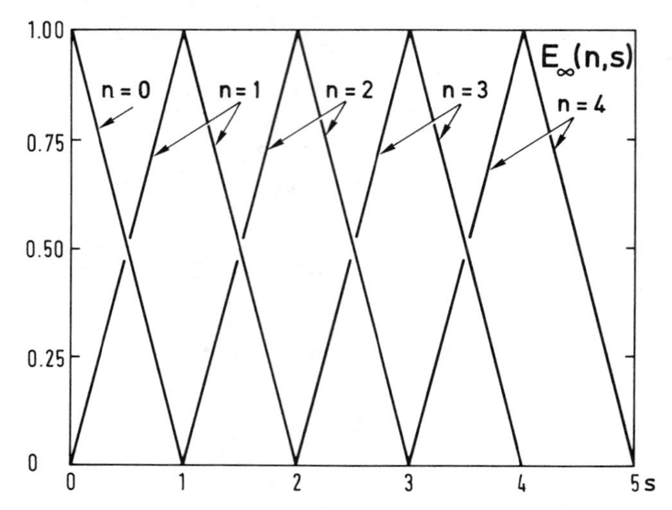

FIG. 5.6. The same as figure 5.5, but for equally spaced levels. Reprinted with permission from D. V. S. Jain, M. L. Mehta, and des Cloizeaux, J. The probabilities for several consecutive eigen-values of a random matrix, *Indian J. Pure Appl. Math.*, 3, 329–351 (1972).

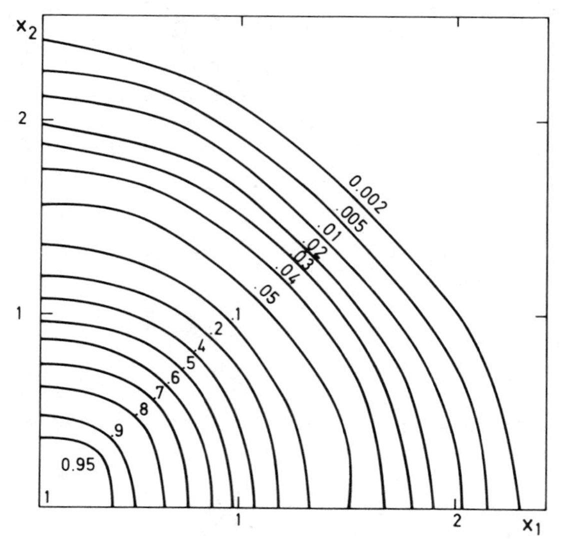

FIG. 5.7. Contour map of the probability $\mathcal{B}_2(x_1, x_2)$ that no eigenvalues (of a random matrix chosen from the Guassian unitary ensemble) lie for a distance x_1 on one side and x_2 on the other side of a given eigenvalue, the distances being measured in units of the mean spacing.

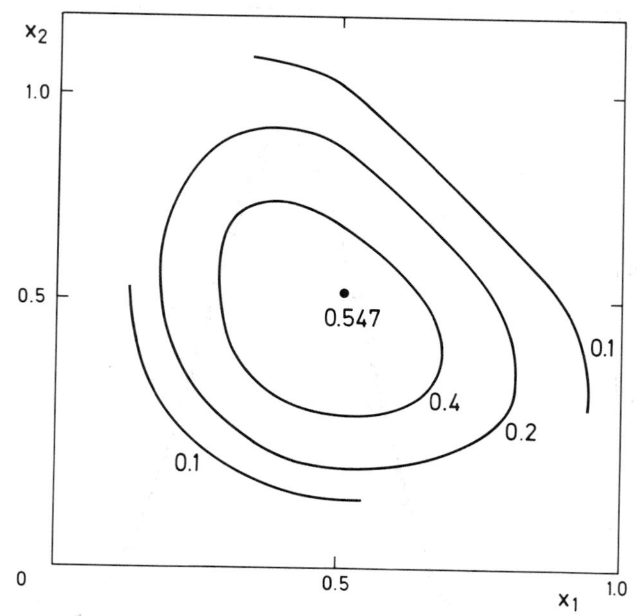

FIG. 5.8. Contour map of $B_2(t; -t, t, y)$ as a function of $x_1 = t + y$, $x_2 = t - y$, or of the function $\mathcal{P}_2(x_1, x_2) = \partial^2 \mathcal{B}_2(x_1, x_2)/\partial x_1 \partial x_2$, the joint probability density function for the two adjacent spacings x_1 and x_2 measured in units of the mean spacing.

determinant of the kernel $Q_N(\xi, \eta)$ converges (Goursat, 1956) to the Fredholm determinant of the limiting kernel $Q(\xi, \eta)$; that is,

$$\lim A_2(\theta) = B_2(t) = E_2(0; s), \tag{5.5.1}$$

and in general

$$\lim A_2(\theta; x_1, ..., x_n) = B_2(t; y_1, ..., y_n). \tag{5.5.2}$$

5.5.2. For $t = 0$ the spheroidal functions $f_i(\xi)$ are proportional to the

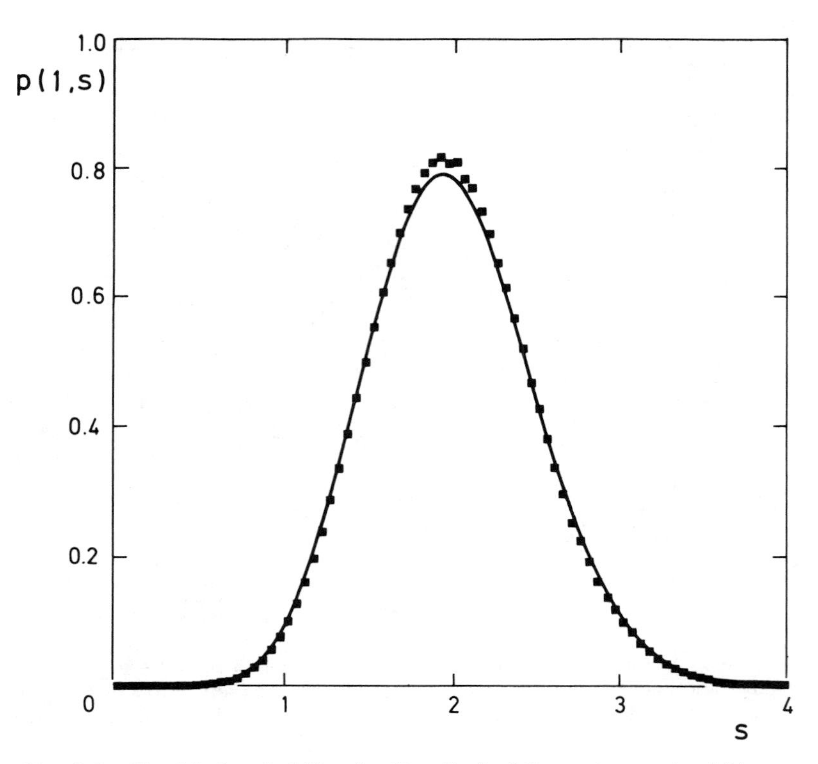

FIG. 5.9. Empirical probability density $p(1; s)$ of the next nearest neighbor spacings for the zeros $\frac{1}{2}+i\gamma_n$ of the Riemann zeta function for $10^{12} < n < 10^{12}+10^5$ compared with that for the Gaussian unitary ensemble. From Odlyzko (1989). Copyright ©, 1989, American Telephone and Telegraph Company reprinted with permission.

Legendre polynomials

$$f_i(\xi) = \left(\frac{2i+1}{2}\right)^{1/2} P_i(\xi) \quad (t=0), \tag{5.5.3}$$

and for small t, they can be expanded in terms of them (Stratton et al., 1956; Robin 1959):

$$f_i(x) = \sum_j d_j(t,i) P_j(x). \tag{5.5.4}$$

Because of parity only those j occur in the summation for which $j-i$ is an even integer. For example, to the smallest order we get from Eqs.

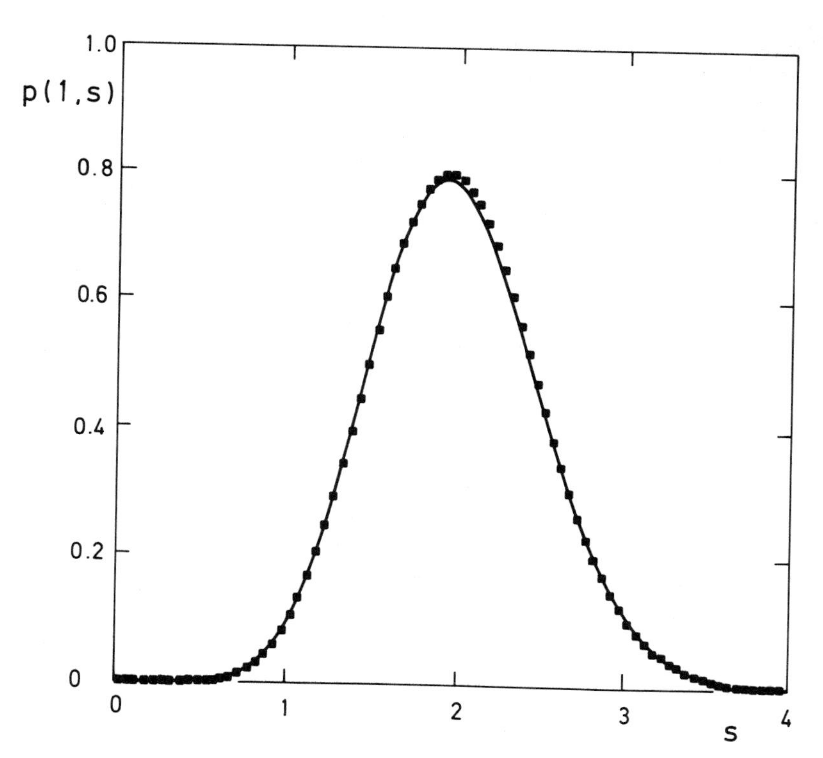

FIG. 5.10. Same as Figure 5.9, but for the 79 million zeros around $n \approx 10^{20}$. From Odlyzko (1989). Copyright ©, 1989, American Telephone and Telegraph Company reprinted with permission.

(5.3.25) and (5.3.26)

$$\mu_{2j} = \int_{-1}^{1} P_{2j}(y)dy/P_{2j}(0) = 2\delta_{j0}, \tag{5.5.5}$$

and

$$\mu_{2j+1} = i\pi t \int_{-1}^{1} yP_{2j+1}(y)dy/P'_{2j+1}(0) = \frac{2}{3} i\pi t\delta_{j0}. \tag{5.5.6}$$

A few terms in such power series expansions are given in Appendix A.13.

5.5.3. Extensive numerical tables of expansion coefficients $d_j(t,i)$, Eq. (5.5.4), are available (Stratton et al., 1956) or can be computed (Van

Buren, 1976). Using them one can calculate $f_i(x)$, μ_i, λ_i, and so on. From Eqs. (5.3.25), (5.3.26), we get, e.g.,

$$\mu_{2k} = 2d_0(t,2k)\left(\sum_j (-1)^j \frac{1\times 3\times\cdots\times(2j-1)}{2\times 4\times\cdots\times(2j)} d_{2j}(t,2k)\right)^{-1},$$

and (5.5.7)

$$\mu_{2k+1} = \frac{2}{3} i\pi t d_1(t,2k+1) \times \left(\sum_j (-1)^j \frac{3\times 5\times\cdots\times(2j+1)}{2\times 4\times\cdots\times(2j)} d_{2j+1}(t,2k+1)\right)^{-1}. \tag{5.5.8}$$

Let us note that for $p_2(0;s)$, expression (5.4.32) is better than (5.4.33) for the following reason. If a function is known only numerically, its derivative is known with a lesser precision. Equation (5.4.33) involves two numerical differentiations, hence its precision is considerably less than that of Eq. (5.4.32), which involves no numerical differentiation.

5.5.4. Let us write

$$D_+ \equiv D_+(t) = \prod_{j=0}^{\infty} (1-\lambda_{2j}), \tag{5.5.9}$$

$$D_- \equiv D_-(t) = \prod_{j=0}^{\infty} (1-\lambda_{2j+1}), \tag{5.5.10}$$

so that they are the products over the eigenvalues corresponding respectively to the even and odd solutions of the integral equation (5.3.14). In other words $D_\pm(t)$ is the Fredholm determinant of the kernel

$$Q_\pm(\xi,\eta) = \frac{1}{2}\left(\frac{\sin(\xi-\eta)\pi t}{(\xi-\eta)\pi} \pm \frac{\sin(\xi+\eta)\pi t}{(\xi+\eta)\pi}\right). \tag{5.5.11}$$

over the interval $(-1,1)$. Computing the logarithmic derivatives of $D_\pm(t)$ (see Appendix A.16), one has

$$\frac{d}{dt}\ln D_+(t) = -\frac{2}{t}\sum_{i=0}^{\infty} \frac{\lambda_{2i}}{1-\lambda_{2i}} f_{2i}^2(1), \tag{5.5.12}$$

and

$$\frac{d}{dt}\ln D_-(t) = -\frac{2}{t}\sum_{i=0}^{\infty} \frac{\lambda_{2i+1}}{1-\lambda_{2i+1}} f_{2i+1}^2(1). \tag{5.5.13}$$

The equality of the expressions (5.4.32) and (5.4.33) is then expressed as

$$\frac{d^2}{dt^2}(D_+D_-) = 4\left(\frac{dD_+}{dt}\right)\left(\frac{dD_-}{dt}\right). \tag{5.5.14}$$

A direct proof of relation (5.5.14), due to Gaudin, is given in Appendix A.16.

We will encounter $D_+(t)$ and $D_-(t)$ later in connection with (Gaussian or circular) orthogonal and symplectic ensembles. Following Dyson, Eq. (5.5.14) will help us in Chapter 12 to compute a few terms in the asymptotic series of $E_\beta(0;s)$, $\beta = 1, 2$, or 4, valid for large s.

5.5.5. In Section 5.2 we saw that the level density is a semi-circle. And still all our considerations about correlations and spacings were restricted to the central part of the spectrum, i.e., all the eigenvalues were supposed to lie near the origin where the level density is flat and equals $(2N)^{1/2}/\pi$. This was done only for simplicity. Actually the p-point correlation function for any finite p is stationary under translations over the spectrum provided we measure distances in terms of the local mean spacing. For example, Eqs. (5.3.11) and (5.3.12) can be replaced by

$$Q_N(\xi,\eta) = \pi t K_N(x,y)(2N-a^2)^{-1/2}$$

and

$$\pi t = \theta(2N-a^2)^{1/2}, \quad \pi t\xi = x(2N-a^2)^{1/2}, \quad \pi t\eta = y(2N-a^2)^{1/2},$$

without changing Eq. (5.3.13); the points x and y are now near the point a and not near the origin. Similarly, the 2-point cluster function for levels not near the origin is again given by (*cf.* Eq. (5.2.20))

$$\begin{aligned} Y_2(x_1,x_2) &= \lim_{N\to\infty} \frac{\pi^2}{(2N-x_1^2)^{1/2}(2N-x_2^2)^{1/2}}\left(\sum_{j=0}^{N-1}\varphi_j(x_1)\varphi_j(x_2)\right)^2 \\ &= \left(\frac{\sin(\pi r)}{\pi r}\right)^2, \end{aligned}$$

with $r = |y_1 - y_2|$, $\pi y_1 = (2N - x_1^2)^{1/2} x_1$, $\pi y_2 = (2N - x_2^2)^{1/2} x_2$, with $y_1 - y_2$ finite. The stationarity property of the p-point correlation function then follows from the fact that it can be expressed in terms of the 2-point functions. Thus, all the local fluctuation properties of the ensemble are stationary.

Expressing all distances in terms of the local mean spacing for comparison with the theory has usually been called "unfolding."

Another property of some conceptual importance is the so called "ergodicity." The basic ergodic theorem, familiar in the statistical mechanics and in the theory of random processes, states that the time average of a physical quantity equals its ensemble average for "almost all" members of the ensemble in some reasonable limit. For random matrices we have an ordering of the eigenvalues instead of time, the spectrum is discrete rather than continuous, and some physical quantities, such as the level density, may not be stationary. By the ergodic theorem for random matrices one means that in the limit of large matrices the spectral average of a fluctuation measure equals in "almost all" cases its ensemble average. For a detailed discussion and a proof of this theorem, see Pandey (1979).

5.5.6. If we set

$$\mathcal{F}_2(z,t) = \prod_i (1 - z\lambda_i) = \prod_i \left(1 - \frac{1}{2} zt\, |\mu_i|^2\right), \tag{5.5.15}$$

then

$$E_2(n;s) = \frac{1}{n!} \left(-\frac{\partial}{\partial z}\right)^n \mathcal{F}_2(z, s/2)\Big|_{z=1}. \tag{5.5.16}$$

This equation can be derived from the relation between $E(n;s)$ and the n-level correlation or cluster functions. See Appendix A.7.

5.5.7. From very different considerations, Jimbo et al. (1980) derived a complicated differential equation for the Fredholm determinant related to $E_2(0.s)$. Let

$$\sigma(z) = z \frac{\partial}{\partial z} \ln\, \mathcal{F}_2(z,t), \tag{5.5.17}$$

with $\mathcal{F}_2(z,t)$ given by Eq. (5.5.15). Then $\sigma(z)$ satisfies a Painlevé equation of the fifth kind

$$\left(z\frac{d^2\sigma}{dz^2}\right)^2 = -4\left(z\frac{d\sigma}{dz} - \sigma - 1\right)\left(z\frac{d\sigma}{dz} - \sigma + z^2\left(\frac{d\sigma}{dz}\right)^2\right). \tag{5.5.18}$$

For numerical computations, it may be much faster to use this differential equation with proper initial conditions.

They also derived the following relation between $D_+(t)$, $D_-(t)$ and their product $D(t) = D_+(t)D_-(t)$, (Eqs. (5.5.9) and (5.5.10):

$$2\ln D_\pm(t) = \ln D(t) \mp \int_0^t d\tau \left(-\frac{d^2}{d\tau^2} \ln D(\tau) \right)^{1/2} . \tag{5.5.19}$$

We will not need these and so we do not have to copy their proofs here.

5.5.8. From the explicit expressions for, say, the correlation functions or the level spacings, one can compute them and represent them as graphs. However, the question of the probable (or mean square) thickness of such curves was not asked. The reason lies in the difficulty of giving a precise definition of this "thickness" and then in estimating it. Other similar questions arise when one tries to compare the empirical or experimental data with the predictions of a theory. We will come back to it in Chapter 16.

5.5.9. For every integer m, let the sides of a fixed $2m$-gon be labeled s_1, s_2, ..., s_{2m} consecutively around its boundary. Let $\varepsilon_g(m)$ be the number of ways of putting these sides into m pairs, each side belonging to one and only one pair, and such that if the sides in each pair are identified we obtain an orientable surface of genus g. (An orientable surface of genus g is topologically equivalent to a sphere with g handles.) The generating function

$$C(m, N) = \sum_{g=0}^{\infty} \varepsilon_g(m) N^{m+1-2g} \tag{5.5.20}$$

was studied by Harer and Zagier (1986). They showed from combinatorial arguments that

$$C(m, N) = 2^m \left\langle \text{tr } A^{2m} \right\rangle \equiv 2^m \int \text{tr } A^{2m} e^{-\text{tr } A^2} dA \div \int e^{-\text{tr } A^2} dA \tag{5.5.21}$$

where A is an $N \times N$ Hermitian matrix, $A_{jk} = A_{jk}^{(0)} + iA_{jk}^{(1)} = A_{kj}^{(0)} - iA_{kj}^{(1)}$, $1 \le j \le k \le N$, $A_{kj}^{(0)}$, $A_{kj}^{(1)}$ real and

$$dA = \prod_{j \le k} dA_{jk}^{(0)} \prod_{j<k} dA_{jk}^{(1)}. \tag{2.5.1}$$

Here and in what follows, all the integrals are taken from $-\infty$ to $+\infty$.

Taking the eigenvalues of A as new variables, one has

$$C(m,N) = 2^m \left\langle \sum_{i=1}^{N} x_i^{2m} \right\rangle = 2^m N \left\langle x_1^{2m} \right\rangle, \tag{5.5.22}$$

with the notation

$$\langle f(x) \rangle = \int f(x)\Phi(x)dx \div \int \Phi(x)dx, \tag{5.5.23}$$

$$\begin{aligned} \Phi(x) &\equiv \exp\left(-\sum_{i=1}^{N} x_i^2\right) \prod_{1\le i<k\le N} (x_i - x_j)^2, \\ dx &\equiv dx_1 ... dx_N. \end{aligned} \tag{5.5.24}$$

From Eqs. (5.1.2), (5.2.14), and (5.2.16) we can therefore write

$$\begin{aligned} C(m,N) &= 2^m N \int x^{2m} \sum_{j=0}^{N-1} \varphi_j^2(x)dx \div \int \sum_{j=0}^{N-1} \varphi_j^2(x)dx \\ &= 2^m \int x^{2m} \left\{ N\varphi_N^2(x) - \sqrt{N(N+1)}\varphi_{N-1}(x)\varphi_{N+1}(x) \right\} dx, \end{aligned} \tag{5.5.25}$$

where we have used the orthonormality of the functions $\varphi_j(x)$, Eq. (5.2.9), and the Christoffel–Darboux formula (*cf.* Bateman, 1953)

$$\sum_{j=0}^{N-1} \varphi_j^2(x) = N\varphi_N^2(x) - \sqrt{N(N+1)}\varphi_{N-1}(x)\varphi_{N+1}(x). \tag{5.5.26}$$

Now using the relations

$$(2x)^k = \sum_{j=0}^{[k/2]} \frac{k!}{j!(k-2j)!} H_{k-2j}(x), \tag{5.5.27}$$

and (Bateman, 1954)

$$\int e^{-x^2} H_j(x) H_k(x) H_\ell(x) dx$$
$$= \begin{cases} \frac{j!k!\ell!}{(s-j)!(s-k)!(s-\ell)!} 2^s \sqrt{\pi}, & \text{if } j+k+\ell = 2s \text{ is even}, \\ 0, & \text{otherwise}, \end{cases} \tag{5.5.28}$$

we can express $C(m, N)$ as a finite sum

$$\begin{aligned} C(m,N) &= \sum_{j=0}^{m} \frac{(2m)!N!2^{-j}}{j!(m-j)!(m-j+1)!(N-m+j)!} \\ &\quad \times (N(m-j+1) - (N+1)(m-j)) \\ &= \frac{(2m)!}{2^m m!} \sum_{j=0}^{m} \binom{m}{j} \binom{N}{m-j+1} 2^{m-j} \\ &= \frac{(2m)!}{2^m m!} c(m,N), \end{aligned} \tag{5.5.29}$$

with

$$c(m,N) = \sum_{j=0}^{m} \binom{m}{j} \binom{N}{m-j+1} 2^{m-j} = \sum_{j=0}^{m} \binom{m}{j} \binom{N}{j+1} 2^{j}. \tag{5.5.30}$$

Actually, the summation over j can be formally extended to all integers since the binomial coefficient is zero outside the allowed range. There is a nice recurrence relation for $c(m, N)$, which can be derived as follows. From

$$\begin{aligned} c(m,N) - c(m,N-1) &= \sum_j \binom{m}{j} \left(\binom{N}{j+1} - \binom{N-1}{j+1} \right) 2^j \\ &= \sum_j \binom{m}{j} \binom{N-1}{j} 2^j, \end{aligned} \tag{5.5.31}$$

and

$$c(m-1,N) - c(m-1,N-1) = \sum_j \binom{m-1}{j} \binom{N-1}{j} 2^j \tag{5.5.32}$$

one has by subtraction

$$
\begin{aligned}
c(m,N) &- c(m,N-1) - (c(m-1,N) - c(m-1,N-1)) \\
&= \sum_j \left(\binom{m}{j} - \binom{m-1}{j} \right) \binom{N-1}{j} 2^j \\
&= \sum_j \binom{m-1}{j-1} \binom{N-1}{j} 2^j = \sum_j \binom{m-1}{j} \binom{N-1}{j+1} 2^{j+1} \\
&= 2c(m-1,N-1) \qquad (5.5.33)
\end{aligned}
$$

or

$$
c(m,N) = c(m,N-1) + c(m-1,N) + c(m-1,N-1). \qquad (5.5.34)
$$

This relation is symmetric in m and N. The initial values

$$
c(m,1) = m, \quad c(0,N) = N, \qquad (5.5.35)
$$

computed easily from Eq. (5.5.30), are not symmetric. Equation (5.5.30) or Eqs. (5.5.34) and (5.5.35) determine completely the $c(m,N)$. From Eqs. (5.5.34) and (5.5.35) one can also derive the generating function

$$
1 + 2\sum_{m=0}^{\infty} c(m,N) x^{m+1} = \left(\frac{1+x}{1-x} \right)^N \qquad (5.5.36)
$$

and by expanding the binomials $(1+x)^N(1-x)^{-N}$ one can get still another form for the $c(m,N)$:

$$
c(m,N) = \frac{1}{2} \sum_{j_1+j_2=m+1} \binom{N}{j_1} \binom{N+j_2-1}{j_2}. \qquad (5.5.37)
$$

5.5.10. Instead of the $2m$-gon of Section 5.5.9 above, consider now a compact surface of genus 0 with k boundary components, and divide the ith boundary component in n_i edges. Let $f_g(n_1, ..., n_k)$ be the number of ways of identifying these edges in pairs to obtain a closed orientable connected surface of genus g. Clearly $f_g(n_1, ..., n_k)$ is symmetric in the variables, $f_g(n_1, ..., n_k) = 0$ unless $n_1 + ... + n_k$ is even, and $f_g(2m) = \varepsilon_g(m)$.

An equation similar to Eq. (5.5.21) seems to hold here,

$$\sum_{g=0}^{\infty} N^{m+1-2g} f_g(n_1, ..., n_k) = 2^m \langle \text{tr } A^{n_1} ... \text{tr } A^{n_k} \rangle, \quad n_1 + ... + n_k = 2m. \tag{5.5.38}$$

One can easily compute the average

$$\left\langle (\text{tr } A)^{j_1} \left(\text{tr } A^2\right)^{j_2} \right\rangle \tag{5.5.39}$$

for any non-negative integers j_1, j_2 by partial differentiation with respect to a and b at $a = 1$, $b = 0$ of

$$\begin{aligned} \int \exp\left(-a \text{ tr } A^2 - 2b \text{ tr } A\right) &= e^{Nb^2/a} \int \exp\left(-a \text{ tr}(A + b/a)^2\right) dA \\ &= \text{const} \times \exp\left(Nb^2/a\right) a^{-N^2/2}. \end{aligned} \tag{5.5.40}$$

Also, the average $\left\langle x_1^{j_1} ... x_k^{j_k} \right\rangle$ with $j_1 + ... + j_k = 2m$ can be expressed as a linear combination of similar averages with $j_1 + ... + j_k = 2m - 2$, by an argument of Aomoto (see Chapter 17, Section 17.8). A general formula for such averages is not known.

Summary of Chapter 5

For the Gaussian unitary ensemble with the joint probability density of the eigenvalues

$$P_{N2}(x_1, ..., x_N) \propto \exp\left(-\sum_{j=1}^{N} x_j^2\right) \prod_{1 \le j < k \le N} (x_j - x_k)^2, \tag{5.1.1}$$

the asymptotic two level cluster function is

$$Y_2(r) = \left(\frac{\sin(\pi r)}{\pi r}\right)^2, \tag{5.2.20}$$

and its Fourier transform is

$$b(k) = \begin{cases} 1 - |k|, & |k| \le 1, \\ 0, & |k| \ge 1. \end{cases} \tag{5.2.22}$$

The n-level cluster function is

$$Y_n(y_1, ..., y_n) = \sum_P \left(\frac{\sin(\pi r_{12})}{\pi r_{12}}\right)\left(\frac{\sin(\pi r_{23})}{\pi r_{23}}\right)\cdots\left(\frac{\sin(\pi r_{n1})}{\pi r_{n1}}\right) \tag{5.2.21}$$

where $r_{ij} = |y_i - y_j|$, and the sum is taken over all $(n-1)!$ cyclic permutation of the indices $(1, 2, ..., n)$.

The probability $E_2(n; s)$ that a randomly chosen interval of length s contains exactly n levels is given by the formulas

$$E_2(0; s) = \prod_{i=0}^{\infty} (1 - \lambda_i), \tag{5.3.27}$$

and for $n > 0$,

$$E_2(n; s) = E_2(0; s) \sum_{0 \le j_1 < j_2 < \cdots < j_n} \frac{\lambda_{j_1}}{1 - \lambda_{j_1}} \cdots \frac{\lambda_{j_n}}{1 - \lambda_{j_n}}, \tag{5.4.30}$$

where $\lambda_i = s |\mu_i|^2 / 4$, and μ_i and $f_i(x)$ are the eigenvalues and eigenfunctions of the integral equation

$$\mu f(x) = \int_{-1}^{1} \exp(i\pi x y s/2) f(y) dy. \tag{5.3.17}$$

The eigenfuctions $f_i(x)$ are known as the prolate spheroidal functions.

6 / Gaussian Orthogonal Ensemble

The Gaussian unitary ensemble studied in chapter 5 is the simplest from the mathematical point of view. However, for physical applications the most important one is the Gaussian orthogonal ensemble (GOE). The eigenvalues of a real symmetric matrix chosen at random from the GOE are real and have the joint probability distribution function

$$P_{N1}(x_1, ..., x_N) = \text{const} \times \exp\left(-\frac{1}{2}\sum_{i=1}^{N} x_i^2\right) \prod_{1\leq i<j\leq N} |x_i - x_j| . \quad (6.1.1)$$

As we will be dealing with the GOE throughout this chapter, we will sometimes omit the subscript 1.

It was suggested in Chapter 2 that if the energy scale is properly chosen, then the fluctuation properties of the series of points $x_1, ..., x_N$ should provide a good model for the eigenvalues of a complicated system having the time reversal invariance and rotational symmetry.

The main objective of this chapter is to derive expressions for the n-level correlation function R_n, cluster function T_n, the form factor Y_n, and the spacing distributions $E(n; s)$, $B(n; y_1, ..., y_n)$, and $p(n; s)$ corresponding to Eqs. (5.2.14), (5.2.15), (5.2.21), (5.4.29), and (5.4.30) of the preceding chapter.

6.1. Generalities

The first serious difficulty with the function (6.1.1) is its unfavorable symmetry caused by the presence of the absolute value sign. This will be taken care of by the method of integration over alternate variables and the introduction of matrices whose elements are no longer ordinary complex numbers, but quaternions. Once this is understood, one can quietly follow step by step all the manipulations of Chapter 5 and arrive at the corresponding expressions for all the quantities pertaining to the GOE. The quaternion matrices form the subject of a chapter in "Matrix Theory" of Mehta (1989), and so we will be brief in discussing them. The method of integration over alternate variables to modify the symmetry of the integrand will be explained in greater detail.

Except for the introduction of quaternion matrices in Sections 6.2 and 6.3 and the method of integration over alternate variables in Section 6.5, we will follow the plan of Chapter 5. Thus Section 6.4 deals with correlation functions, Sections 6.5, 6.6, and 6.7 with the spacing probabilities and Section 6.8 with bounds of the spacing distribution. The two-level cluster function is given by Eq. (6.4.14), its Fourier transform by Eq. (6.4.17); the probability $E_1(n; s)$ that an interval of length s contains exactly n levels is given by Eqs. (6.5.19), (6.6.12), and (6.7.21), while the nearest neighbor spacing probability density $p_1(0; s)$ is given by Eq. (6.6.18) or (6.6.19). These various functions are shown graphically on figures 6.1 and 6.3 and are tabulated in Appendices A.14 and A.15.

As in Chapter 5, we can write the integrand in (6.1.1) as a determinant containing "oscillator wave functions" [*cf.* Eqs. (5.2.6)–(5.2.8)].

$$\exp\left(-\frac{1}{2}\sum_{i=1}^{N} x_i^2\right)\prod_{j<i}(x_i - x_j) = \text{const} \times \det\left[\varphi_{i-1}(x_j)\right]_{i,j=1,\ldots,N}, \tag{6.1.2}$$

where

$$\begin{aligned}\varphi_j(x) &= \left(2^j j!\sqrt{\pi}\right)^{-1/2}\exp(-x^2/2)H_j(x)\\ &= \left(2^j j!\sqrt{\pi}\right)^{-1/2}\exp(x^2/2)(-d/dx)^j\exp(-x^2),\end{aligned} \tag{6.1.3}$$

are the "oscillator wave functions" orthogonal over $(-\infty,\infty)$. From the recurrence relations for the Hermite polynomials $H_j(x)$,

$$H_{j+1}(x) = 2xH_j(x) - 2jH_{j-1}(x), \quad H_j'(x) = 2jH_{j-1}(x), \tag{6.1.4}$$

where a prime denotes the derivative, one deduces

$$\sqrt{2}\,\varphi_j'(x) = \sqrt{j}\,\varphi_{j-1}(x) - \sqrt{j+1}\,\varphi_{j+1}(x), \tag{6.1.5}$$

i.e., $\varphi_{j+1}(x)$ is a linear combination of $\varphi_j'(x)$ and $\varphi_{j-1}(x)$. Therefore one can, for example, replace everywhere $\varphi_{2j+1}(x)$ by $\varphi_{2j}'(x)$ in the determinant (6.1.2). Thus for N even, we have

$$\exp\left(-\frac{1}{2}\sum_{i=1}^{N} x_i^2\right)\prod_{j<i}(x_i - x_j)$$
$$= c\det\left[\varphi_{2j-2}(x_i) \qquad \varphi_{2j-2}'(x_i)\right]_{\substack{i=1,\ldots,N \\ j=1,\ldots,N/2,}} , \tag{6.1.6}$$

where c is a constant. If N is odd, there is an extra column $\varphi_{N-1}(x_i)$. For $N = 2m+1$, we may replace $\varphi_{2j}(x)$ by $\varphi_{2j+1}'(x)$ for $j = 0, 1, ..., m-1$, using the last column $\varphi_{2m}(x)$ to eliminate the extra terms,

$$\exp\left(-\frac{1}{2}\sum_{i=1}^{N} x_i^2\right)\prod_{j<i}(x_i - x_j)$$
$$= c\det\left[\varphi_{2j-1}'(x_i) \qquad \varphi_{2j-1}(x_i) \qquad \varphi_{2m}(x_i)\right]$$
$$i = 1, ..., N, \quad j = 1, ..., (N-1)/2, \tag{6.1.7}$$

where c is again a constant, not necessarily the same.

In Eq. (6.1.6) the $(2j-1)$th column is $\varphi_{2j-2}(x_i)$ and the $2j$th column is $\varphi_{2j-2}'(x_i)$, i standing for the ith component of the column. In Eq. (6.1.7) the $(2j-1)$th column is $\varphi_{2j-1}'(x_i)$, the $2j$th column is $\varphi_{2j-1}(x_i)$, while the last, i.e., the Nth or the $(2m+1)$th column is $\varphi_{2m}(x_i)$. Such a convenient self-explanatory notation will often be used in what follows.

6.2. Quaternion Matrices

A quaternion is a linear combination of four units 1, e_1, e_2, and e_3 :

$$q = q^{(0)} + q^{(1)}e_1 + q^{(2)}e_2 + q^{(3)}e_3. \tag{6.2.1}$$

The four basic units satisfy the multiplication laws

$$1 \times 1 = 1;$$
$$1 \times e_j = e_j \times 1 = e_j, \quad j = 1, 2, 3; \tag{6.2.2}$$
$$e_1^2 = e_2^2 = e_3^2 = e_1 e_2 e_3 = -1;$$

and the multiplication is associative. We may think of these basic units as 2×2 matrices, Eq. (2.4.3):

$$\begin{aligned} 1 &\to \begin{bmatrix} 1 & 0 \\ 0 & 1 \end{bmatrix}, \quad e_1 \to \begin{bmatrix} i & 0 \\ 0 & -i \end{bmatrix}, \\ e_2 &\to \begin{bmatrix} 0 & 1 \\ -1 & 0 \end{bmatrix}, \quad e_3 \to \begin{bmatrix} 0 & i \\ i & 0 \end{bmatrix}. \end{aligned} \tag{6.2.3}$$

The coefficients $q^{(i)}$ are usually taken to be real numbers, but we will allow them to be complex. Thus an $N \times N$ matrix Q with quaternion elements may be thought of as a $2N \times 2N$ ordinary matrix $C(Q)$ obtained by the replacements (6.2.3). Conversely, any $2N \times 2N$ matrix with ordinary complex numbers as elements can be thought of as cut into N^2 2×2 blocks and each block expressed as a linear combination of the basic four 2×2 matrices (6.2.3) with complex coefficients. Thus an $N \times N$ quaternion matrix Q is equivalent to an ordinary $2N \times 2N$ matrix $C(Q)$.

One may define the Hermitian conjugate and the dual of a quaternion matrix in the usual way (*cf.* Section 2.4). The scalar part of q is by definition $q^{(0)}$ [or $\frac{1}{2}\mathrm{tr}\ C(q)$, if thought as a 2×2 matrix; see Eq. (2.4.6)]. The determinant of $Q = [q_{jk}]$ will be defined as the complex number

$$\begin{aligned} \det Q &= \sum_P \varepsilon_P \left(q_{j_1 j_2} q_{j_2 j_3} \cdots q_{j_r j_1}\right)^{(0)} \left(q_{k_1 k_2} q_{k_2 k_3} \cdots q_{k_s k_1}\right)^{(0)} \cdots \\ &= \sum_P (-1)^{N-\ell} \prod_1^{\ell} \left(q_{ab} q_{bc} \cdots q_{da}\right)^{(0)}, \end{aligned} \tag{6.2.4}$$

where P is any permutation of the indices $(1, 2, ..., N)$ consisting of ℓ exclusive cycles of the form

$$(a \to b \to c \to \cdots \to d \to a), \tag{6.2.5}$$

$\varepsilon_P = (-1)^{N-\ell}$ is the sign of P, and the superscript (0) means the scalar part (or half trace, in terms of 2×2 matrices). If Q is a self-dual matrix (*cf.* Section 2.4), then $(q_{ab} q_{bc} \cdots q_{da})$ and $(q_{ad} \cdots q_{cb} q_{ba})$ are duals of each other and their sum is a scalar. Thus, if an appropriate convention is made as to the order of the cycles in Eq. (6.2.4) for a permutation and all other permutations obtained from it by reversing one or more of its cycles, then the superscript (0) may be dropped when Q is self-dual.

If we denote as above, by $C(Q)$ the ordinary $2N \times 2N$ matrix corresponding to the $N \times N$ self-dual quaternion matrix Q, and $Z = C(e_2 I)$, I the $N \times N$ unit matrix, [see Eq. (2.4.1)], then $ZC(Q)$ is antisymmetric [*cf.* Eq. (2.4.16)] and its Pfaffian is equal to the determinant of Q,

$$\det Q = \text{Pf}\,[ZC(Q)] = \{\det\ [C(Q)]\}^{1/2}. \tag{6.2.6}$$

It is well-known that, for an antisymmetric matrix of even order, the determinant is the square of the Pfaffian (see, e.g., Mehta, 1989). As $\det Z = 1$, the second relation above is immediate. For a proof of the first relation, see for example, Mehta (1989), Chapter 8.

Corresponding to Theorem 5.2.1, we have the

Theorem 6.2.1. *Let $Q_N = [f(x_i, x_j)]$ be an $N \times N$ self-dual quaternion matrix such that*

(*i*) *the (i,j) element $f(x_i, x_j)$ depends only on x_i and x_j;*

(*ii*) $\displaystyle\int f(x,x)d\mu(x) = c;$

(*iii*) $$\int f(x,y)f(y,z)d\mu(y) = f(x,z) + g(x,z); \tag{6.2.7}$$

(*iv*) $$g(x,y) = \lambda f(x,y) - f(x,y)\lambda; \tag{6.2.8}$$

where $d\mu(x)$ is a suitable measure, c is a constant scalar and λ is a constant quaternion. Then

$$\int \det\, Q_N\ d\mu(x_N) = (c - N + 1)\det\, Q_{N-1}, \tag{6.2.9}$$

where Q_{N-1} is the $(N-1) \times (N-1)$ quaternion matrix obtained from Q_N by removing the row and column containing x_N.

The proof is almost the same as in Section 5.2, except that one now has quaternion matrices and determinants. The extra terms due to $g(x,y)$ in

Eq. (6.2.7) finally disappear on summation over cycles in the definition (6.2.4) of quaternion determinants. For details, see Mehta (1989).

To take advantage of Theorem 6.2.1, one has to express $P_{N1}(x_1, ..., x_N)$, Eq. (6.1.1), as the determinant of an $N \times N$ self-dual quaternion matrix. This is what we will do in the next section.

6.3. The Probability Density Function as a Quaternion Determinant

In view of the relations (6.2.6) it will be sufficient to form an ordinary $2N \times 2N$ matrix $C(Q)$ such that $ZC(Q)$ is antisymmetric (or Q is self-dual) and det $C(Q)$ is the square of the function (6.1.1). Instead of explaining the method of construction (*cf.* Mehta, 1989; see also Appendix A.17), we will explicitly give $C(Q)$ and verify that it satisfies the above requirements.

Let

$$\varepsilon(x) = \frac{1}{2}\mathrm{sgn}(x) = \begin{cases} 1/2, & x > 0, \\ 0\ , & x = 0, \\ -1/2, & x < 0, \end{cases} \tag{6.3.1}$$

$$S_N(x,y) = \sum_{j=0}^{N-1} \varphi_j(x)\varphi_j(y) + \left(\frac{N}{2}\right)^{1/2} \varphi_{N-1}(x) \int_{-\infty}^{\infty} \varepsilon(y-t)\varphi_N(t)dt, \tag{6.3.2}$$

$$DS_N(x,y) = -\frac{d}{dy}S_N(x,y), \tag{6.3.3}$$

$$IS_N(x,y) = \int_{-\infty}^{\infty} \varepsilon(x-t)S_N(t,y)dt, \tag{6.3.4}$$

$$\alpha(x) = \begin{cases} \varphi_{2m}(x) \Big/ \displaystyle\int_{-\infty}^{\infty} \varphi_{2m}(t)dt, & \text{if} \quad N = 2m+1 \quad \text{odd}, \\ 0, & \text{if} \quad N = 2m \quad \text{even}, \end{cases} \tag{6.3.5}$$

$$u(x) = \int_0^x \alpha(t)dt \tag{6.3.6}$$

$$JS_N(x,y) = IS_N(x,y) - \varepsilon(x-y) + u(x) - u(y), \tag{6.3.7}$$

$$\begin{aligned}\sigma_{N1}(x,y) &= \begin{bmatrix} S_N + \alpha & DS_N \\ JS_N & S_N^T + \alpha^T \end{bmatrix} \\ &\equiv \begin{bmatrix} S_N(x,y) + \alpha(x) & DS_N(x,y) \\ JS_N(x,y) & S_N(y,x) + \alpha(y) \end{bmatrix},\end{aligned} \tag{6.3.8}$$

note the interchange of x and y in the lower right corner, and

$$Q_{N1} = \left[\sigma_{N1}(x_j, x_k)\right]_{j,k=1,\ldots,N}. \tag{6.3.9}$$

Now we verify that Q_{N1} is self-dual, $\det[C(Q_{N1})]$ is proportional to the square of $P_{N1}(x_1, ..., x_N)$, Eq. (6.1.1), and it satisfies all the conditions of Theorem 6.2.1.

6.3.1. From Eq. (6.1.5) for $i \geq 0$, denoting the derivative by a prime,

$$\begin{aligned}-\varphi_{2i}'(x)\varphi_{2i}(t) &= \left(\sqrt{i+1/2}\;\varphi_{2i+1}(x) - \sqrt{i}\;\varphi_{2i-1}(x)\right)\varphi_{2i}(t) \\ &= \varphi_{2i+1}(x)\left(\varphi_{2i+1}'(t) + \sqrt{i+1}\;\varphi_{2i+2}(t)\right) \\ &\quad - \sqrt{i}\;\varphi_{2i-1}(x)\varphi_{2i}(t) \\ &= \varphi_{2i+1}(x)\varphi_{2i+1}'(t) + \sqrt{i+1}\;\varphi_{2i+1}(x)\varphi_{2i+2}(t) \\ &\quad - \sqrt{i}\;\varphi_{2i-1}(x)\varphi_{2i}(t).\end{aligned} \tag{6.3.10}$$

Summing this for $i = 0$ to $i = m-1$, we get

$$\begin{aligned}&-\sum_{i=0}^{m-1}\varphi_{2i}'(x)\varphi_{2i}(y) \\ &\quad = \sum_{i=0}^{m-1}\varphi_{2i+1}(x)\varphi_{2i+1}'(y) + \sqrt{m}\;\varphi_{2m-1}(x)\varphi_{2m}(y),\end{aligned} \tag{6.3.11}$$

Or interchanging x and y and using once more Eq. (6.1.5),

$$\begin{aligned}&-\sum_{i=0}^{m-1}\varphi_{2i+1}'(x)\varphi_{2i+1}(y) \\ &\quad = \sum_{i=0}^{m}\varphi_{2i}(x)\varphi_{2i}'(y) + \sqrt{m+1/2}\;\varphi_{2m}(x)\varphi_{2m+1}(y).\end{aligned} \tag{6.3.12}$$

Hence, if $N = 2m$ is even, we can write

$$S_N(x,y) = \sum_{i=0}^{m-1}\left(\varphi_{2i}(x)\varphi_{2i}(y) - \varphi'_{2i}(x)\int_0^y \varphi_{2i}(t)dt\right), \qquad (6.3.13)$$

$$DS_N(x,y) = \sum_{i=0}^{m-1}\left(-\varphi_{2i}(x)\varphi'_{2i}(y) + \varphi'_{2i}(x)\varphi_{2i}(y)\right), \qquad (6.3.14)$$

and

$$IS_N(x,y) = \sum_{i=0}^{m-1}\left(\int_0^x \varphi_{2i}(t)dt\varphi_{2i}(y) - \varphi_{2i}(x)\int_0^y \varphi_{2i}(t)dt\right). \qquad (6.3.15)$$

And if $N = 2m+1$ is odd, we can write

$$S_N(x,y) = \sum_{i=0}^{m-1}\left(\varphi_{2i+1}(x)\varphi_{2i+1}(y) - \varphi'_{2i+1}(x)\int_{-\infty}^y \varphi_{2i+1}(t)dt\right), \qquad (6.3.16)$$

$$DS_N(x,y) = \sum_{i=0}^{m-1}\left(-\varphi_{2i+1}(x)\varphi'_{2i+1}(y) + \varphi'_{2i+1}(x)\varphi_{2i+1}(y)\right), \qquad (6.3.17)$$

and

$$IS_N(x,y) = \sum_{i=0}^{m-1}\left(\int_{-\infty}^x \varphi_{2i+1}(t)dt\varphi_{2i+1}(y) - \varphi_{2i+1}(x)\int_{-\infty}^y \varphi_{2i+1}(t)dt\right). \qquad (6.3.18)$$

Therefore in any case DS_N, IS_N, and JS_N change sign under the exchange of x and y,

$$DS_N(x,y) = -DS_N(y,x), \qquad (6.3.19)$$

$$IS_N(x,y) = -IS_N(y,x), \qquad (6.3.20)$$

$$JS_N(x,y) = -JS_N(y,x). \qquad (6.3.21)$$

From Eqs. (6.3.19)–(6.3.21) and the definition of the dual, Eq. (2.4.17), it is straightforward to verify that Q_{N1} is self-dual. To see that $\det[C(Q_{N1})]$ is proportional to the square of $P_{N1}(x_1, ..., x_N)$, Eq. (6.1.1), we proceed as follows.

6.3.2. (i) Case $N = 2m$ is even.

In this case the $2N \times 2N$ matrix

$$\begin{bmatrix} S_N(x_j, x_k) & DS_N(x_j, x_k) \\ IS_N(x_j, x_k) & S_N(x_k, x_j) \end{bmatrix} \tag{6.3.22}$$

can be written as a product of two matrices

$$\begin{bmatrix} \varphi_{2i}(x_j) & \varphi'_{2i}(x_j) \\ \displaystyle\int_0^{x_j} \varphi_{2i}(t)dt & \varphi_{2i(x_j)} \end{bmatrix} \tag{6.3.23}$$

and

$$\begin{bmatrix} \varphi_{2i}(x_k) & -\varphi'_{2i}(x_k) \\ \displaystyle -\int_0^{x_k} \varphi_{2i}(t)dt & \varphi_{2i}(x_k) \end{bmatrix} \tag{6.3.24}$$

of orders $2N \times N$ and $N \times 2N$, respectively. Therefore the rank of the matrix (6.3.22) is N; i.e., its N rows

$$[IS_N(x_j, x_k) \quad S_N(x_k, x_j)] \tag{6.3.25}$$

are linear combinations of the N rows

$$[S_N(x_j, x_k) \quad DS_N(x_j, x_k)] \,. \tag{6.3.26}$$

So we may subtract the rows (6.3.25) from the corresponding rows of the matrix

$$C(Q_{N1}) = \begin{bmatrix} S_N(x_j, x_k) & DS_N(x_j, x_k) \\ JS_N(x_j, x_k) & S_N(x_k, x_j) \end{bmatrix} \tag{6.3.27}$$

without changing its determinant,

$$\begin{aligned} \det[C(Q_{N1})] &= \det \begin{bmatrix} S_N(x_j, x_k) & DS_N(x_j, x_k) \\ -\varepsilon(x_j - x_k) & 0 \end{bmatrix} \\ &= \pm \det[DS_N(x_j, x_k)] \det[\varepsilon(x_j - x_k)] \,. \end{aligned} \tag{6.3.28}$$

Now for the first determinant

$$[DS_N(x_j, x_k)] = \begin{bmatrix}\varphi_{2i}(x_j) & \varphi'_{2i}(x_j)\end{bmatrix} \begin{bmatrix} -\varphi'_{2i}(x_k) \\ \varphi_{2i}(x_k) \end{bmatrix}, \tag{6.3.29}$$

gives with Eq. (6.1.6),

$$\begin{aligned}\det\,[DS_N(x_j, x_k)] &= \left(\det\begin{bmatrix}\varphi_{2i}(x_j) & \varphi'_{2i}(x_j)\end{bmatrix}\right)^2 \\ &= \text{const} \times [P_{N1}(x_1, ..., x_N)]^2. \end{aligned} \tag{6.3.30}$$

In the second determinant if we interchange any two variables x, two rows and two columns of $[\text{sgn}(x_j - x_k)]$ are interchanged, so that $\det\,[\text{sgn}(x_j - x_k)]$ is independent of the values of the x_j, except where $x_j = x_k$. Let $x_1 > x_2 > \cdots > x_N$, so that $[\text{sgn}(x_j - x_k)]$ is an anti-symmetric matrix with all elements above the main diagonal equal to 1. The determinant of this last matrix is easily seen to be 1.

(ii) Case $N = 2m + 1$, odd.

Instead of (6.3.22)–(6.3.24) now consider

$$\begin{bmatrix} S_N(x_j, x_k) + \alpha(x_j) & DS_N(x_j, x_k) + \delta\varphi_{2m}(x_j)\varphi_{2m}(x_k) \\ IS_N(x_j, x_k) + \left(\sqrt{\delta}\int_{-\infty}^{\infty} \varphi_{2m}(t)dt\right)^{-2} & S_N(x_k, x_j) + \alpha(x_k) \end{bmatrix}, \tag{6.3.31}$$

with δ a small constant, as the product of two matrices

$$\begin{bmatrix} \varphi_{2i+1}(x_j) & \varphi'_{2i+1}(x_j) & \sqrt{\delta}\varphi_{2m}(x_j) \\ \int_{-\infty}^{x_j} \varphi_{2i+1}(t) & \varphi_{2i+1}(x_j) & \left(\sqrt{\delta}\int_{-\infty}^{\infty} \varphi_{2m}(t)dt\right)^{-1} \end{bmatrix}, \tag{6.3.32}$$

and

$$\begin{bmatrix} \varphi_{2i+1}(x_k) & -\varphi'_{2i+1}(x_k) \\ -\int_{-\infty}^{x_j} \varphi_{2i+1}(t)dt & \varphi_{2i+1}(x_k) \\ \left(\sqrt{\delta}\int_{-\infty}^{\infty} \varphi_{2m}(t)dt\right)^{-1} & \sqrt{\delta}\varphi_{2m}(x_k) \end{bmatrix}, \tag{6.3.33}$$

of orders $2N \times N$ and $N \times 2N$, respectively. The rank of Eq. (6.3.31) is again N and by the same argument we can subtract its last N rows from those of

$$C_\delta(Q_{N1}) = \begin{bmatrix} S_N(x_j, x_k) + \alpha(x_j) & DS_N(x_j, x_k) + \delta\varphi_{2m}(x_j)\varphi_{2m}(x_k) \\ JS_N(x_j x_k) & S_N(x_k, x_j) + \alpha(x_k) \end{bmatrix}, \tag{6.3.34}$$

without changing the determinant of this later. Thus

$$\begin{aligned} &\det\left[C_\delta(Q_{N1})\right] \\ &\quad = \det\left[DS_N(x_j, x_k) + \delta\varphi_{2m}(x_j)\varphi_{2m}(x_k)\right] \\ &\quad \times \det\left[-\varepsilon(x_j - x_k) + u(x_j) - u(x_k) - \delta^{-1}\left(\int_{-\infty}^{\infty} \varphi_{2m}(t)dt\right)^{-2}\right]. \end{aligned} \tag{6.3.35}$$

The first factor is

$$\begin{aligned} &\left(\det\left[\varphi_{2i+1}(x_j) \quad \varphi'_{2i+1}(x_j) \quad \sqrt{\delta}\varphi_{2m}(x_j)\right]_{j=1,2,\ldots,2m+1}^{i=0,1,\ldots,m-1}\right)^2 \\ &\quad = \delta \times (\text{const}) \times \left[P_{N1}(x_1, \ldots, x_N)\right]^2, \end{aligned} \tag{6.3.36}$$

from Eq. (6.1.7). The second factor in Eq. (6.3.35) is calculated by bordering it with extra rows and columns to be (*cf.* Appendix A.19)

$$\pm\delta^{-1}\left(\int_{-\infty}^{\infty} \varphi_{2m}(t)dt\right)^{-2} = \delta^{-1} \times (\text{const}). \tag{6.3.37}$$

Finally take the limit $\delta \to 0$.

6.3.3. From Eqs. (6.3.13), (6.3.16), and the orthonormality of the $\varphi_j(x)$, it is not difficult to verify that

$$\int_{-\infty}^{\infty} S_N(x, x)dx = 2m, \quad N = 2m \quad \text{or} \quad N = 2m + 1. \tag{6.3.38}$$

so that

$$\int_{-\infty}^{\infty} \sigma_{N1}(x,x)dx = N. \tag{6.3.39}$$

Also introducing the notation

$$f * g = \int_{-\infty}^{\infty} f(x,y)g(y,z)dy, \tag{6.3.40}$$

it is not difficult from Eqs. (6.3.13)–(6.3.18), from the parity of $\varphi_j(x)$ and the orthonormality of the $\varphi_j(x)$, to verify that

$$S_N * S_N = S_N, \quad \alpha * S_N = 0 = S_N * \alpha, \quad \alpha * \alpha = \alpha, \tag{6.3.41}$$

$$S_N * DS_N = DS_N, \quad \alpha * DS_N = 0, \tag{6.3.42}$$

$$IS_N * S_N = IS_N = \varepsilon * S_N, \quad IS_N * \alpha = 0, \tag{6.3.43}$$

$$\varepsilon * \alpha = u = u * \alpha, \quad u^T * \alpha = 0, \tag{6.3.44}$$

$$DS_N * IS_N = S_N = DS_N * \varepsilon, \tag{6.3.45}$$

$$D * u = 0 = D * u^T. \tag{6.3.46}$$

From which by exchanging the variables [with $f^T(x,y) = f(y,x)$],

$$DS_N * S_N^T = DS_N, \quad S_N^T * IS_N = IS_N = S_N^T * \varepsilon, \quad \text{etc.} \tag{6.3.47}$$

Collecting these we get

$$\sigma_{N1} * \sigma_{N1} = \begin{bmatrix} S_N + \alpha & 2DS_N \\ 0 & S_N^T + \alpha^T \end{bmatrix} = \sigma_{N1} + \lambda\sigma_{N1} - \sigma_{N1}\lambda, \tag{6.3.48}$$

with

$$\lambda = \frac{1}{2}\begin{bmatrix} 1 & 0 \\ 0 & -1 \end{bmatrix}. \tag{6.3.49}$$

Thus, the $N \times N$ quaternion matrix Q_{N1}, Eq. (6.3.9), satisfies all the conditions of Theorem 6.2.1.

6.3.4. Finally, integrating over all the variables and using Theorem 6.2.1, one can fix the constants. The conclusion is

$$P_{N1}(x_1, ..., x_N) = \frac{1}{N!}\det\left[\sigma_{N1}(x_j, x_k)\right]_{j,k=1,...,N}. \tag{6.3.50}$$

6.4. The Correlation and Cluster Functions

From Theorem 6.2.1 and Eq. (6.3.50) one immediately has the n-level correlation function

$$\begin{aligned} R_n(x_1, ..., x_n) &= \frac{N!}{(N-n)!} \int \cdots \int P_N(x_1, ..., x_N) dx_{n+1} \cdots dx_N \\ &= \det \left[\sigma_{N1}(x_j, x_k)\right]_{j,k=1,\ldots,n}. \end{aligned} \tag{6.4.1}$$

And as in Section 5.2, the n-level cluster function is

$$T_n(x_1, ..., x_n) = \sum_P \sigma_{N1}(x_1, x_2)\sigma_{N1}(x_2, x_3) \cdots \sigma_{N1}(x_n, x_1), \tag{6.4.2}$$

the sum being taken over all $(n-1)!$ distinct cyclic permutations of the indices $(1, 2, ..., n)$.

Setting $n = 1$, we get the level density

$$R_1(x) = S_N(x, x) + \alpha(x). \tag{6.4.3}$$

In the limit $N \to \infty$, this goes to the "semicircle law" (*cf.* Appendix A.9),

$$R_1(x) = \begin{cases} \dfrac{1}{\pi}(2N - x^2)^{1/2}, & |x| < (2N)^{1/2}, \\ 0, & |x| > (2N)^{1/2}. \end{cases} \tag{6.4.4}$$

The mean spacing at the origin is thus

$$\alpha = 1/R_1(0) = \pi/(2N)^{1/2}. \tag{6.4.5}$$

Setting $n = 2$, we get the two-level cluster function

$$T_2(x, y) = \sigma_{N1}(x, y)\sigma_{N1}(y, x). \tag{6.4.6}$$

In the limit $N \to \infty$ (*cf.* Appendix A.10),

$$\lim\, (\pi/(2N))^{1/2} S_N(x, y) = s(r), \tag{6.4.7}$$

$$\lim\, [\pi/(2N)]\mathrm{sgn}(x-y) D S_N(x, y) = \frac{d}{dr} s(r) \equiv Ds(r), \tag{6.4.8}$$

$$\lim\, \mathrm{sgn}(x-y) I S_N(x, y) = -\int_0^r s(t)dt \equiv Is(r), \tag{6.4.9}$$

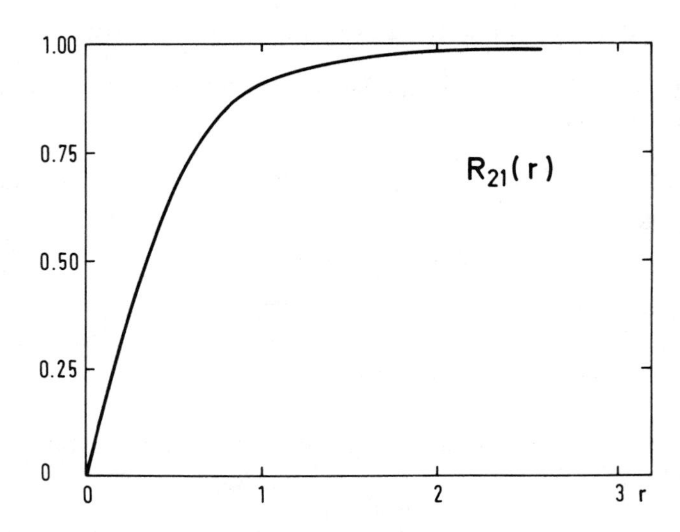

FIG. 6.1. Two-level correlation function for the orthogonal ensemble.

with

$$x = \pi\xi/(2N)^{1/2}, \quad y = \pi\eta/(2N)^{1/2}, \quad r = |\xi - \eta|, \tag{6.4.10}$$

and

$$s(r) = \sin(\pi r)/(\pi r). \tag{6.4.11}$$

Hence

$$Y_2(r) = \lim(\pi/2N)T_2(x, y) = \sigma_1(r)\sigma_1(-r), \tag{6.4.12}$$

with

$$\sigma_1(r) = \begin{bmatrix} s(r) & Ds(r) \\ Is(r) - 1/2 & s(r) \end{bmatrix}, \tag{6.4.13}$$

or

$$\begin{aligned} Y_2(r) &= \left(\frac{1}{2} - \int_0^r s(t)dt\right)\left(\frac{d}{dr}s(r)\right) + [s(r)]^2 \\ &= \left(\int_r^\infty s(t)dt\right)\left(\frac{d}{dr}s(r)\right) + [s(r)]^2. \end{aligned} \tag{6.4.14}$$

Figure 6.1 shows the limiting two-level correlation function. Note that the oscillations are almost imperceptible in contrast to Figure 5.2.

The behavior of $Y_2(r)$ for small and large r is given by

$$Y_2(r) = 1 - \frac{\pi^2 r}{6} + \frac{\pi^4 r^3}{60} - \frac{\pi^4 r^4}{135} + \cdots \tag{6.4.15}$$

and

$$Y_2(r) = \frac{1}{\pi^2 r^2} - \frac{1+\cos^2 \pi r}{\pi^4 r^4} + \cdots, \tag{6.4.16}$$

respectively. The two-level form factor is (*cf.* Appendix A.11)

$$\begin{aligned} b(k) &= \int_{-\infty}^{\infty} Y_2(r) \exp(2\pi i k r) dr \\ &= \begin{cases} 1 - 2|k| + |k| \ln(1 + 2|k|), & |k| \le 1, \\ -1 + |k| \ln\left(\dfrac{2|k|+1}{2|k|-1}\right), & |k| \ge 1. \end{cases} \end{aligned} \tag{6.4.17}$$

This has the behavior

$$b(k) = 1 - 2|k| + 2k^2 + \cdots, \tag{6.4.18}$$

$$b(k) = \frac{1}{12k^2} + \frac{1}{80k^4} + \cdots, \tag{6.4.19}$$

for small and large k, respectively.

In the limit $N \to \infty$, the n-level cluster function is

$$Y_n(\xi_1, ..., \xi_n) = \sum_P \sigma_1(r_{12}) \sigma_1(r_{23}) \cdots \sigma_1(r_{n1}), \tag{6.4.20}$$

the sum being taken over the $(n-1)!$ distinct cyclic permutations of the indices $(1, 2, ..., n)$, and $r_{ij} = |\xi_i - \xi_j| = (2N)^{1/2} |x_i - x_j| / \pi$. The

n-level form-factor or the Fourier transform of Y_n is

$$\int Y_n(\xi_1, ..., \xi_n) \exp\left(2\pi i \sum_{j=1}^{n} k_j \xi_j\right) d\xi_1 \cdots d\xi_n$$
$$= \delta(k_1 + \cdots + k_n).$$

$$\times \int_{-\infty}^{\infty} dk \sum_P f_1(k) f_1(k + k_1) \cdots f_1(k + k_1 + \cdots + k_{n-1}), \tag{6.4.21}$$

with

$$f_1(k) = \begin{bmatrix} f_2(k) & k f_2(k) \\ (f_2(k) - 1)/k & f_2(k) \end{bmatrix}, \quad f_2(k) = \begin{cases} 1, & |k| < 1/2, \\ 0, & |k| > 1/2. \end{cases} \tag{6.4.22}$$

The sum in Eq. (6.4.21) is over all $(n-1)!$ permutations of the indices $(1, ..., n-1)$.

6.5. Level Spacings. Integration over Alternate Variables

To begin with, we calculate the level-spacing function

$$A_1(\theta) = \int \cdots \int_{\text{out}} P_{N1}(x_1, ..., x_N) dx_1 \cdots dx_N, \tag{6.5.1}$$

case $n = 0$ of Eq. (5.1.9). To take care of the absolute value sign in Eq. (6.1.1), we order the variables as $-\infty < x_1 \leq x_2 \leq \cdots \leq x_N < \infty$, and multiply the result by $N!$. Substituting from Eq. (6.1.6) we can write

$$A_1(\theta) = cN! \int_{R(\text{out})} \cdots \int \det \begin{bmatrix} \varphi_{2j-2}(x_i) \\ \varphi'_{2j-2}(x_i) \end{bmatrix} dx_1 \cdots dx_N, \tag{6.5.2}$$

where the region of integration $R(\text{out})$ is $-\infty < x_1 \leq \cdots \leq x_N < \infty$, and $|x_j| > \theta$, $j = 1, ..., N$. If we integrate over x_1 we replace the first column by the column $[F_j(x_2)]$, $j = 0, ..., [N/2] - 1$, where the functions $F_j(x)$ are defined by

$$F_{2j}(x) = \int_{-\infty}^{x} u(y)\varphi_{2j}(y)dy, \quad F_{2j+1}(x) = \int_{-\infty}^{x} u(y)\varphi'_{2j}(y)dy, \tag{6.5.3}$$

with

$$u(x) = \begin{cases} 0, & |x| \le \theta, \\ 1, & |x| > \theta. \end{cases} \tag{6.5.4}$$

Now x_2 occurs in two columns and we cannot integrate over it. However, integrating over x_3 we replace the third column by the column $[F_j(x_4) - F_j(x_2)]$. But we have already a column $[F_j(x_2)]$, so we may drop the terms with a negative sign. Thus, at each integration over x_3, x_5, ..., the corresponding column is replaced by a column of an F_j. In case N is odd, the last column is replaced by a column of pure numbers $F_j(\infty)$.

Thus an integration over the alternate variables $x_1, x_3, x_5, ...$, gives

$$A_1(\theta) = cN! \int_{R(\text{out})} \cdots \int \det \begin{bmatrix} F_{2j}(x_{2i}) & \varphi_{2j}(x_{2i}) \\ F_{2j+1}(x_{2i}) & \varphi'_{2j}(x_{2i}) \end{bmatrix} dx_2 dx_4 \cdots dx_{2m}, \tag{6.5.5}$$

where either $N = 2m$ or $N = 2m+1$. In the latter case, there is an extra column $[F_j(\infty)]$ at the right end. To avoid minor complications we take in the rest of the chapter N even, $N = 2m$.

The integrand in Eq. (6.5.5) is now symmetric in the remaining variables, therefore one can integrate over them independently and divide by $(N/2)!$; the result is a determinant (*cf.* Appendix A.18),

$$\begin{aligned} A_1(\theta) &= cN![(N/2)!]^{-1} \int_{\text{out}} \cdots \int \det \begin{bmatrix} F_{2j}(x_{2i}) & \varphi_{2j}(x_{2i}) \\ F_{2j+1}(x_{2i}) & \varphi'_{2j}(x_{2i}) \end{bmatrix} \\ &\quad \times dx_2 dx_4 \cdots dx_{2m} \\ &= cN!(-2)^{N/2} \det [\bar{g}_{ij}]_{i,j=0,1,\ldots,m-1}, \end{aligned} \tag{6.5.6}$$

where the region of integration is now the entire real line outside the interval $(-\theta, \theta)$ for each of the variables $x_2, x_4, ..., x_{2m}$, and

$$\begin{aligned}\bar{g}_{ij} &= -\frac{1}{2}\int_{-\infty}^{\infty}\left(F_{2i}(x)\varphi_{2j}'(x) - F_{2j+1}(x)\varphi_{2i}(x)\right)u(x)dx \\ &= \delta_{ij} - \int_{-\theta}^{\theta}\varphi_{2i}(x)\varphi_{2j}(x)dx. \end{aligned} \tag{6.5.7}$$

We can fix the constant by observing that $A_1(0) = 1$, so that

$$c(-2)^{N/2}N! = 1. \tag{6.5.8}$$

$$A_1(\theta) = \det \bar{G} = \det\left[\bar{g}_{ij}\right], \tag{6.5.9}$$

with the elements $\bar{g}_{ij}$ of $\bar{G}$ given by Eq. (6.5.7).

From this point on, the analysis follows Section 5.3. The diagonalization of $\bar{G}$ and the passage to the limit $m \to \infty$, $\theta \to 0$, proceed exactly as in Sections 5.3–5.4 except that we now need only the even part in x of $K(x, y)$, and only the even functions $\psi_{2i}(x)$ and $f_{2i}(\xi)$. Relations corresponding to Eqs. (5.4.8) and (5.4.11)–(5.4.24) are

$$\bar{\gamma}_{ij} = \int_{-\theta}^{\theta}\varphi_{2i}(x)\varphi_{2j}(x)dx, \tag{6.5.10}$$

$$A_1(\theta) = \prod_{i=0}^{m-1}\left(1 - \lambda_{2i}\right), \tag{6.5.11}$$

$$\sum_{j=0}^{m-1}\bar{\gamma}_{ij}\;\bar{h}_{jk} = \bar{h}_{ik}\lambda_{2k}, \tag{6.5.12}$$

$$\int_{-\theta}^{\theta}\bar{K}(x, y)\psi_{2i}(y)dy = \lambda_{2i}\psi_{2i}(x), \tag{6.5.13}$$

$$\bar{K}(x, y) = \sum_{i=0}^{m-1}\varphi_{2i}(x)\varphi_{2i}(y) = \sum_{i=0}^{m-1}\psi_{2i}(x)\psi_{2i}(y), \tag{6.5.14}$$

$$\psi_{2i}(x) = \sum_{j=0}^{m-1}\bar{h}_{ji}\;\varphi_{2j}(x), \tag{6.5.15}$$

$$\sum_{j=0}^{m-1} \bar{h}_{ij}\,\bar{h}_{kj} = \sum_{j=0}^{m-1} \bar{h}_{ji}\,\bar{h}_{jk} = \delta_{ik}, \tag{6.5.16}$$

$$\alpha = \pi/(2N)^{1/2}, \quad \theta = \alpha t, \quad x = \alpha t\xi, \quad y = \alpha t\eta, \quad s = 2t, \tag{6.5.17}$$

$$\bar{K}(x, y) = (\alpha t)^{-1}\,\bar{Q}(\xi, \eta), \tag{6.5.18}$$

$$E_1(0; s) = B_1(t) = \lim A_1(\theta) = \prod_i (1 - \lambda_{2i}) = \prod_i \left(1 - \frac{1}{2} t\mu_{2i}^2\right), \tag{6.5.19}$$

$$\bar{Q}(\xi, \eta) = \frac{1}{2}[Q(\xi, \eta) + Q(\xi, -\eta)]. \tag{6.5.20}$$

The symbols $Q(\xi, \eta)$, $\psi_i(x)$, $f_i(\xi)$, and λ_i have the same meaning as in Section 5.4. The μ_{2i} are the eigenvalues of the integral equation (5.3.20)

$$\mu_{2i} f(\xi) = 2\int_0^1 \cos(\pi\xi\eta t) f(\eta) d\eta. \tag{5.3.20}$$

Equations (5.4.22) and (5.4.24) may be completed by

$$\psi_{2i}^{(j)}(\xi) \to (\alpha t)^{-j-1/2} \lambda_{2i}^{1/2} f_{2i}^{(j)}(\xi), \tag{6.5.21}$$

where the superscript j denotes the jth derivative.

Note that the constant in Eq. (6.1.1) is fixed by Eqs. (6.1.6), (6.5.2), and (6.5.8),

$$P_{N1}(x_1, \ldots, x_{2m}) = \frac{(-2)^{-m}}{(2m)!} \det \begin{bmatrix} \varphi_{2i}(x_j) \\ \varphi_{2i}'(x_j) \end{bmatrix}_{i=0,\ldots,m-1;\, j=1,\ldots,2m;}$$

or written in full,

$$P_{N1}(x_1, \ldots, x_{2m}) = \frac{(-2)^{-m}}{(2m)!} \det \begin{bmatrix} \varphi_0(x_1) & \varphi_0(x_2) & \ldots & \varphi_0(x_{2m}) \\ \varphi_0'(x_1) & \varphi_0'(x_2) & \ldots & \varphi_0'(x_{2m}) \\ \ldots & \ldots & \ldots & \ldots \\ \varphi_{2m-2}(x_1) & \varphi_{2m-2}(x_2) & \ldots & \varphi_{2m-2}(x_{2m}) \\ \varphi_{2m-2}'(x_1) & \varphi_{2m-2}'(x_2) & \ldots & \varphi_{2m-2}'(x_{2m}) \end{bmatrix}. \tag{6.5.22}$$

6.6. Several Consecutive Spacings: $n = 2r$

Next we take the case of several consecutive spacings, the integral

$$A_1(\theta; x_1, ..., x_n) = \frac{N!}{n!(N-n)!} \int \cdots \int_{\text{out}} P_{N1}(x_1, ..., x_N) dx_{n+1} \cdots dx_N, \tag{6.6.1}$$

with $n > 0$. In writing the expression (6.1.6) we took the ordering $x_1 \leq x_2 \leq \cdots \leq x_{2m}$. However, the same expression is valid also when $-\theta \leq x_1 \leq \cdots \leq x_{2r} \leq \theta$, $x_{2r+1} \leq x_{2r+2} \leq \cdots \leq x_{2m}$, and $|x_j| \geq \theta$ for $j = 2r+1, 2r+2, ..., 2m$; in other words, none or some of the $x_{2r+1}, ..., x_{2m}$ are less than $-\theta$ and others are greater than θ. This pertains to the fact that the determinant in Eq. (6.1.6) does not change sign when the columns containing the variables x_{2r+1},...,x_{2m} are passed one by one over the $2r$ columns containing the variables x_1,...,x_{2r}. Let us therefore take $n = 2r$, $r \geq 1$. We expand the determinant in Eq. (6.1.6) by the first $2r$ columns in the Laplace manner and integrate every term so obtained over $x_{2r+1}, ..., x_{2m}$ outside the interval $(-\theta, \theta)$, while $-\theta \leq x_1 \leq \cdots \leq x_{2r} \leq \theta$, using the method of integration over alternate variables (Section 6.5). Let us remark that the $2r \times 2r$ determinants formed from the first $2r$ columns of (6.1.6) not all will have a nonzero coefficient. Only those containinrg r rows of functions φ and another r rows of functions φ' will survive. This is so because

$$\int \cdots \int_{(\text{out}) y \leq x} (\varphi_{2i}(y)\varphi_{2j}(x) - \varphi_{2i}(x)\varphi_{2j}(y))\, dxdy = 0,$$

$$\int \cdots \int_{(\text{out}) y \leq x} \left(\varphi'_{2i}(y)\varphi'_{2j}(x) - \varphi'_{2i}(x)\varphi'_{2j}(y)\right) dxdy = 0. \tag{6.6.2}$$

We also have

$$\int \cdots \int_{(\text{out}) y \leq x} \left(\varphi_{2i}(y)\varphi'_{2j}(x) - \varphi_{2i}(x)\varphi'_{2j}(y)\right) dxdy = 2\, \bar{g}_{ij}, \tag{6.6.3}$$

the $\bar{g}_{ij}$ given by Eq. (6.5.7). The result is

$$A_1(\theta; x_1, ..., x_{2r}) = (-2)^{-r} \sum_{(i,j)} \det \left(\begin{bmatrix} \varphi_{2i_k}(x_\ell) \\ \varphi'_{2j_k}(x_\ell) \end{bmatrix}_{1 \le \ell \le 2r}^{1 \le k \le r} \right)$$

$$\times \bar{G}'(i_1, ..., i_r; j_1, ..., j_r) \tag{6.6.4}$$

where the summation is extended over all possible choices of the indices $i_1 < \cdots < i_r$, $j_1 < \cdots < j_r$ from $0, ..., m-1$. The $\bar{G}'(i_1, ..., i_r; j_1, ..., j_r)$ is apart from a sign the $(m-r) \times (m-r)$ determinant obtained from $\bar{G}$ by omitting the rows $i_1, ..., i_r$ and the columns $j_1, ..., j_r$. The sign is plus or minus according to whether $\sum_{k=1}^{r} (i_k + j_k)$ is even or odd. Thus, $\bar{G}'(i_1, ..., i_r; j_1, ..., j_r)$ is equal to (*cf.* Mehta, 1989, Section 3.9) the determinant of $\bar{G}$ multiplied by an $r \times r$ determinant from the elements of the inverse of $\bar{G}$,

$$\bar{G}'(i_1, ..., i_r; j_1, ..., j_r) = \det\left[\bar{G}\right] \det\left(\left[(\bar{G}^{-1})_{ji}\right]_{i=i_1,...,i_r}^{j=j_1,...,j_r}\right). \tag{6.6.5}$$

Diagonalizing $\bar{G}$ as explained above in Section 6.5, and using Eqs. (6.5.9), (6.5.11), and (6.5.15),

$$A_1(\theta; x_1, ..., x_{2r}) = (-2)^{-r} \prod_{\rho=0}^{m-1} (1 - \lambda_{2\rho})$$

$$\times \sum_{(i)} (1 - \lambda_{2i_1})^{-1} \cdots (1 - \lambda_{2i_r})^{-1}$$

$$\times \det \left(\begin{bmatrix} \psi_{2i_k}(x_j) \\ \psi'_{2i_k}(x_j) \end{bmatrix}_{1 \le j \le 2r}^{1 \le k \le r} \right). \tag{6.6.6}$$

With

$$\alpha = \pi/(2\sqrt{m}), \quad \theta = \alpha t, \quad x_j = \alpha y_j, \tag{6.6.7}$$

we get in the limit $m \to \infty$,

$$\begin{aligned} B_1(t; y_1, ..., y_{2r}) &= \lim \alpha^{2r} A_1(\theta; x_1, ..., x_{2r}) \\ &= (-2)^{-r} t^{-2r} \prod_\rho (1 - \lambda_{2\rho}) \\ &\times \sum_{(i)} \frac{\lambda_{2i_1}}{1 - \lambda_{2i_1}} \cdots \frac{\lambda_{2i_r}}{1 - \lambda_{2i_r}} \\ &\times \det \left(\begin{bmatrix} f_{2i_k}(y_j/t) \\ f'_{2i_k}(y_j/t) \end{bmatrix}_{1 \le j \le 2r}^{1 \le k \le r} \right). \end{aligned} \tag{6.6.8}$$

The variables y_j satisfy of course the inequalities

$$-t \le y_1 \le \cdots \le y_{2r} \le t, \tag{6.6.9}$$

the λ_{2i} and the $f_{2i}(\xi)$ depend on t as a parameter, and the indices $0 \le i_1 < \cdots < i_r$ are non-negative integers.

To get $E_1(2r; s)$, Eq. (5.1.12), from Eq. (6.6.8) one may again use the method of integration over alternate variables (Section 6.5 and Appendix A.18). It is easy to see that

$$\begin{aligned} \int \cdots \int_{-1 \le y \le x \le 1} dy dx \left(f_{2i}(y) f_{2j}(x) - f_{2i}(x) f_{2j}(y) \right) &= 0 \\ \int \cdots \int_{-1 \le y \le x \le 1} dy dx \left(f'_{2i}(y) f'_{2j}(x) - f'_{2i}(x) f'_{2j}(y) \right) &= 0, \end{aligned} \tag{6.6.10}$$

and that

$$\begin{aligned} &\int \cdots \int_{-1 \le y \le x \le 1} dy dx \left(f_{2i}(y) f'_{2j}(x) - f_{2i}(x) f'_{2j}(y) \right) \\ &\quad = -2 \left(\delta_{ij} - f_{2j}(1) \int_{-1}^{1} f_{2i}(y) dy \right). \end{aligned} \tag{6.6.11}$$

Therefore, the method of integration over alternate variables gives us

$$E_1(2r;s) = \prod_\rho (1-\lambda_{2\rho}) \sum_{i_1<\ldots<i_r} \frac{\lambda_{2i_1}}{1-\lambda_{2i_1}} \cdots \frac{\lambda_{2i_r}}{1-\lambda_{2i_r}} \times \det\left[\delta_{ij} - f_{2i}(1)\int_{-1}^{1} f_{2j}(y)dy\right]_{i,j=i_1,i_2,\ldots,i_r}, \tag{6.6.12}$$

or (*cf.* Appendix A.19)

$$E_1(2r;s) = \prod_\rho (1-\lambda_{2\rho}) \sum_{i_1<\cdots<i_r} \left\{ \frac{\lambda_{2i_1}}{1-\lambda_{2i_1}} \cdots \frac{\lambda_{2i_r}}{1-\lambda_{2i_r}} \times \left(1 - \sum_{j=1}^{r} f_{2i_j}(1)\int_{-1}^{1} f_{2i_j}(y)dy\right)\right\}. \tag{6.6.13}$$

As we obtained Eq. (6.6.4) by integrating the expression (6.5.22) so we obtain from Eq. (6.6.8) the following expression for $p_1(2r-2;s)$:

$$p_1(2r-2;s) = -\frac{2}{s^2}\prod_\rho (1-\lambda_{2\rho}) \times \sum_{i_1<\cdots<i_r}\left(\frac{\lambda_{2i_1}}{1-\lambda_{2i_1}} \cdots \frac{\lambda_{2i_r}}{1-\lambda_{2i_r}} \sum_{j,k} a_{jk}b_{jk}\right), \tag{6.6.14}$$

where the summation over j, k in the above equation is over the indices $i_1, \ldots, i_r$, while

$$a_{jk} = \det\begin{bmatrix} f_{2j}(-1) & f_{2j}(1) \\ f_{2k}'(-1) & f_{2k}'(1) \end{bmatrix} = 2f_{2j}(1)f_{2k}'(1), \tag{6.6.15}$$

and b_{jk} is the cofactor of the element (j,k) in

$$\det\left[\delta_{ij} - f_{2j}(1)\int_{-1}^{1} f_{2i}(y)dy\right]_{i,j=i_1,\ldots,i_r}. \tag{6.6.16}$$

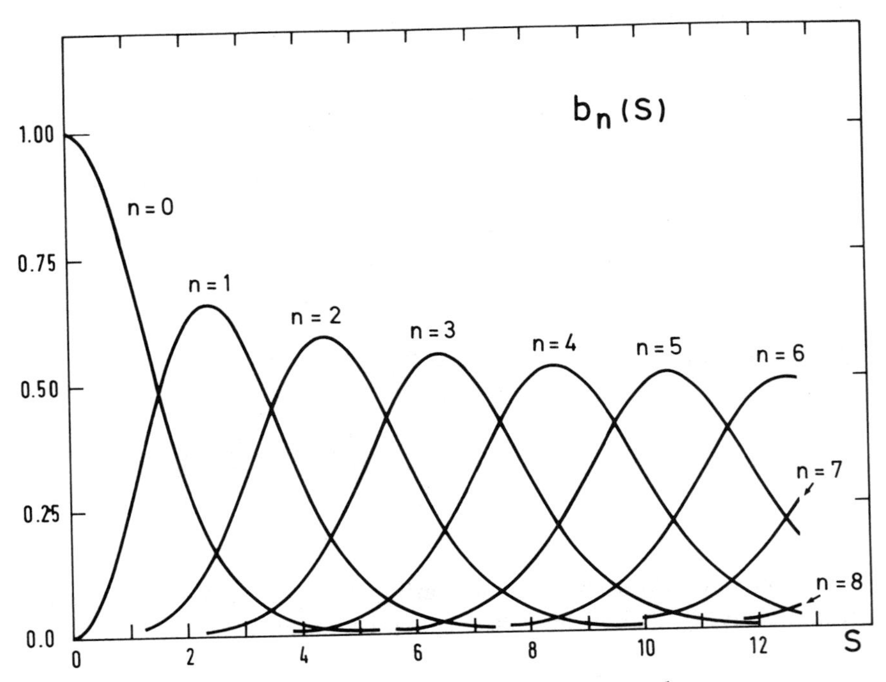

FIG. 6.2. The functions $b_n(s) = f_{2n}(1)\int_{-1}^{1} f_{2n}(x)\,dx / \int_{-1}^{1} f_{2n}^2(x)\,dx$, where $f_n(x)$ is the prolate spheroidal function.

That is to say (*cf.* Appendix A.19)

$$b_{jk} = f_{2j}(1)\int_{-1}^{1} f_{2k}(y)dy + \delta_{jk}\left(1 - \sum_{\ell} f_{2\ell}(1)\int_{-1}^{1} f_{2\ell}(y)dy\right), \tag{6.6.17}$$

where the indix ℓ takes all the values $i_1, ..., i_r$.

A case of special interest is $r = 1$,

$$p_1(0;s) = -\frac{4}{s^2}\prod_{\rho}(1-\lambda_{2\rho})\sum_i \frac{\lambda_{2i}}{1-\lambda_{2i}} f_{2i}(1)f'_{2i}(1), \tag{6.6.18}$$

which from Eqs. (5.1.18) and (6.5.19) can also be written as

$$p_1(0;s) = \frac{d^2}{ds^2}\prod_{\rho}(1-\lambda_{2\rho}). \tag{6.6.19}$$

As we remarked in Section 5.5.3, for numerical computations Eq. (6.6.18) is better suited than Eq. (6.6.19).

Figure 6.2 shows the functions $b_j(s) = f_{2j}(1) \int_{-1}^{1} f_{2j}(x)dx$, for small values of j and s.

6.7. Several Consecutive Spacings: $n = 2r - 1$

The case $n = 2r - 1$ is more complicated for the following reason. If we order the variables as $-\infty < x_{2r} \leq x_{2r+1} \leq \cdots \leq x_{2m} < \infty$; $|x_j| \geq \theta$, $j = 2r, 2r+1, ..., 2m$; then the determinant in Eq. (6.5.22) changes sign each time a variable x_j, $2r \leq j \leq 2m$, passes over the excluded interval $(-\theta, \theta)$. To overcome this difficulty one divides the integral into two parts according to whether an odd or an even number of variables are greater than θ;

$$A_1(\theta; x_1, ..., x_{2r-1}) = \mathcal{O}(\theta; x_1, ..., x_{2r-1}) + \mathcal{E}(\theta; x_1, ..., x_{2r-1}), \quad (6.7.1)$$

with

$$\mathcal{O}(\theta; x_1, ..., x_{2r-1}) = \sum_{k=0}^{m-r} \int \cdots \int P_{N1}(x_1, ..., x_{2m}) dx_{2r} dx_{2r+1} \cdots dx_{2m}, \quad (6.7.2)$$

where the integration limits are

$$\begin{aligned} x_j &< -\theta, \qquad 2r \leq j \leq 2r + 2k - 1, \\ x_j &> \theta, \qquad 2r + 2k \leq j \leq 2m; \end{aligned} \quad (6.7.3)$$

and $-\mathcal{E}(\theta; x_1, ..., x_{2r-1})$ is the same sum of integrals (6.7.2) but with the integration limits

$$\begin{aligned} x_j &< -\theta, \qquad 2r \leq j \leq 2r + 2k, \\ x_j &> \theta, \qquad 2r + 2k + 1 \leq j \leq 2m. \end{aligned} \quad (6.7.4)$$

If for some k, the upper limit of variation of j is less than its lower limit, the corresponding line in Eq. (6.7.3) or Eq. (6.7.4) is ignored.

In $\mathcal{O}$ we integrate over alternate variables x_{2r+1}, x_{2r+3}, ..., x_{2m-1}; introducing functions $u(x)$, $F_j(x)$, Eqs. (6.5.3), (6.5.4), and

$$\varepsilon(x) = \begin{cases} 1, & \text{if } x > \theta, \\ 0, & \text{if } x < \theta, \end{cases} \tag{6.7.5}$$

so that

$$\mathcal{O}(\theta; x_1, ..., x_{2r-1}) = (-2)^{-m} \int \cdots \int \prod_{j=r}^{m} u(x_{2j}) dx_{2j}.$$
$$\times \det \begin{bmatrix} 0 & \varepsilon(x_{2j}) & 0 \\ \varphi_{2i}(x_k) & F_{2i}(x_{2j}) & \varphi_{2i}(x_{2j}) \\ \varphi'_{2i}(x_k) & F_{2i+1}(x_{2j}) & \varphi'_{2i}(x_{2j}) \end{bmatrix}, \tag{6.7.6}$$

$$i = 0, 1, ..., m-1; \quad k = 1, 2, ..., 2r-1; \quad j = r, r+1, ..., m. \tag{6.7.7}$$

In Eq. (6.7.6) one has $-\theta \le x_1 \le x_2 \le \cdots \le x_{2r-1} \le \theta$, and integrations are carried over the domain $-\infty < x_{2r} \le x_{2r+1} \le \cdots \le x_{2m} < \infty$.

Now one can drop the ordering and integrate independently over the remaining variables. One expands the determinant according to its first $(2r-1)$ columns and uses the orthonormality of the functions $\varphi_i(x)$.

In $\mathcal{E}$ we integrate over the other alternate variables, x_{2m}, x_{2m-2}, ..., x_{2r}, introducing similar functions

$$F'_{2i}(x) = \int_x^\infty u(t)\varphi_{2i}(t)dt, \quad F'_{2i+1}(x) = \int_x^\infty u(t)\varphi'_{2i}(t)dt, \tag{6.7.8}$$

and

$$\varepsilon'(x) = \begin{cases} 1, & \text{if } x < -\theta, \\ 0, & \text{if } x > -\theta; \end{cases} \tag{6.7.9}$$

change variables $x_1, ..., x_{2r-1}$ to their negatives and compare with $\mathcal{O}$:

$$\mathcal{E}(\theta; x_1, ..., x_{2r-1}) = \mathcal{O}(\theta; -x_{2r-1}, ..., -x_1). \tag{6.7.10}$$

For clarity of the exposition we take in greater detail the case $n = 1$. The general case is similar, though more cumbersome.

6.7.1. Case $n = 1$

Equation (6.7.6) gives now (see Appendix A.18),

$$\mathcal{O}(\theta; x) = (-2)^{-m} \int \cdots \int \prod_{j=1}^{m} u(x_{2j}) dx_{2j} \times \det \begin{bmatrix} 0 & \varepsilon(x_{2j}) & 0 \\ \varphi_{2i}(x) & F_{2i}(x_{2j}) & \varphi_{2i}(x_{2j}) \\ \varphi'_{2i}(x) & F_{2i+1}(x_{2j}) & \varphi'_{2i}(x_{2j}) \end{bmatrix} \tag{6.7.11}$$

$$= (-2)^{-m} \left(\det \begin{bmatrix} 0 & \rho_{2j+1}(\theta) \\ \varphi_{2i}(x) & -2\bar{g}_{ij} \end{bmatrix} + \det \begin{bmatrix} 0 & -\rho_{2j}(\theta) \\ \varphi'_{2i}(x) & -2\bar{g}_{ij} \end{bmatrix} \right), \tag{6.7.12}$$

the range of integration in Eq. (6.7.11) being $x_2 \leq x_4 \leq \cdots \leq x_{2m}$,

$$\rho_{2j}(\theta) = \int_{-\infty}^{\infty} \varepsilon(x) u(x) \varphi_{2j}(x) dx = \int_{\theta}^{\infty} \varphi_{2j}(x) dx, \tag{6.7.13}$$

$$\rho_{2j+1}(\theta) = \int_{-\infty}^{\infty} \varepsilon(x) u(x) \varphi'_{2j}(x) dx = -\varphi_{2j}(\theta), \tag{6.7.14}$$

and $\bar{g}_{ij}$ given by Eq. (6.5.7). From Eqs. (6.7.1), (6.7.10), and (6.7.12), one has

$$A_1(\theta; x) = -\det \begin{bmatrix} 0 & \varphi_{2j}(\theta) \\ \varphi_{2i}(x) & \bar{g}_{ij} \end{bmatrix}_{i,j=0,1,\ldots,m-1} \tag{6.7.15}$$

$$= \sum_{i,j=0}^{m-1} \varphi_{2i}(x) \varphi_{2j}(\theta) \, \bar{G}(i; j). \tag{6.7.16}$$

Following the diagonalization of $\bar{G}$ as in Section 6.5 above, we make the replacements

$$\begin{aligned} \varphi_{2i}(\theta) &\to \psi_{2i}(t) \to (\alpha t)^{-1/2} \lambda_{2i}^{1/2} f_{2i}(1), \\ \varphi_{2i}(x) &\to \psi_{2i}(y) \to (\alpha t)^{-1/2} \lambda_{2i}^{1/2} f_{2i}(y/t) \end{aligned} \tag{6.7.17}$$

to get

$$\begin{aligned} B_1(t;y) &= \lim \alpha A_1(\theta;x) \\ &= -\frac{1}{t}\det\begin{bmatrix} 0 & \sqrt{\lambda_{2j}}f_{2j}(1) \\ \sqrt{\lambda_{2i}}f_{2i}(y/t) & (1-\lambda_{2i})\,\delta_{ij} \end{bmatrix} \\ &= \frac{1}{t}\prod_\rho (1-\lambda_{2\rho})\sum_j \frac{\lambda_{2j}}{1-\lambda_{2j}} f_{2j}(1)f_{2j}(y/t). \qquad (6.7.18) \end{aligned}$$

6.7.2. Case $n = 2r - 1$

Equation (6.7.18) is now replaced by

$$\begin{aligned} B_1(t;y_1,\ldots,y_{2r-1}) &= \lim \alpha^{2r-1}A_1(\theta;x_1,\ldots,x_{2r-1}) \\ &= (-2)^{1-r}t^{-2r+1}\prod_\rho(1-\lambda_{2\rho})\sum_{(i)}\frac{\lambda_{2i_1}}{1-\lambda_{2i_1}}\cdots\frac{\lambda_{2i_r}}{1-\lambda_{2i_r}} \\ &\quad \times\sum_{(j)} f_{2j}(1)\det M_j(i_1,\ldots,i_r) \qquad (6.7.19) \end{aligned}$$

where the $(2r-1)\times(2r-1)$ matrix $M_j(i_1,\ldots,i_r)$ depends on the variables $y_1, \ldots, y_{2r-1}$ as

$$M_j(i_1,\ldots,i_r) =$$

$$\begin{bmatrix} f_{2j}\left(\frac{y_1}{t}\right) & f_{2i_1}\left(\frac{y_1}{t}\right) & f'_{2i_1}\left(\frac{y_1}{t}\right) & \ldots & f_{2i_{r-1}}\left(\frac{y_1}{t}\right) & f'_{2i_{r-1}}\left(\frac{y_1}{t}\right) \\ \ldots & \ldots & \ldots & \ldots & \ldots & \ldots \\ f_{2j}\left(\frac{y_{2r-1}}{t}\right) & f_{2i_1}\left(\frac{y_{2r-1}}{t}\right) & f'_{2i_1}\left(\frac{y_{2r-1}}{t}\right) & \ldots & f_{2i_{r-1}}\left(\frac{y_{2r-1}}{t}\right) & f'_{2i_{r-1}}\left(\frac{y_{2r-1}}{t}\right) \end{bmatrix},$$

(6.7.20)

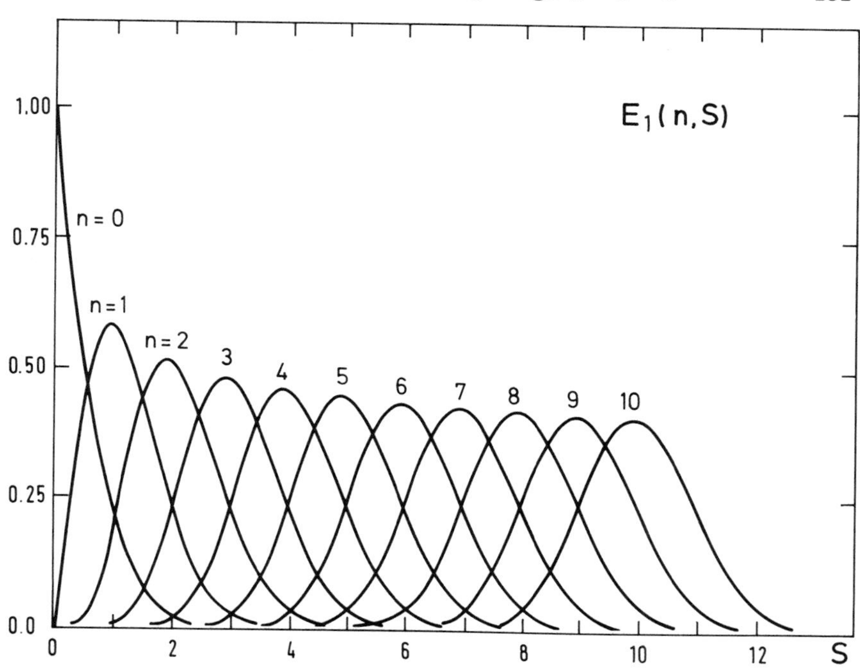

FIG. 6.3. The n-level spacings $E_1(n,s)$ for the Gaussian orthogonal ensemble. Reprinted with permission from D. V. S. Jain, M. L. Mehta, and des Cloizeaux, J. The probabilities for several consecutive eigen-values of a ransom matrix, *Indian J. Pure Appl. Math.*, 3, 329–351 (1972).

j is one of the indices $i_1, ..., i_r$ and the summation in Eq. (6.7.19) is over all possible choices of non-negative integers $0 \leq i_1 < \cdots < i_r$ and of j among these integers. The variables y_j are supposed to be ordered $-t \leq y_1 \leq \cdots \leq y_{2r-1} \leq t$.

Integration over alternate variables $y_1, y_3, ..., y_{2r-1}$ and then independently over the rest from $-t$ to t, using the orthonormality of the functions f_j gives

$$E_1(2r-1;s) = \prod_\rho (1-\lambda_{2\rho}) \sum_{(i)} \frac{\lambda_{2i_1}}{1-\lambda_{2i_1}} \cdots \frac{\lambda_{2i_r}}{1-\lambda_{2i_r}} \times \left(\sum_{j=1}^{r} f_{2i_j}(1) \int_{-1}^{1} f_{2i_j}(y)dy \right). \qquad (6.7.21)$$

The expression for $p_1(2r-3;s)$ is obtained by putting $y_1=-t$, $y_{2r-1}=t$ and integrating over the other variables. We get

$$p_1(2r-3;s) = -\frac{2}{s^2}\prod_\rho (1-\lambda_{2\rho})\sum_{(i)} \frac{\lambda_{2i_1}}{1-\lambda_{2i_1}} \cdots \frac{\lambda_{2i_r}}{1-\lambda_{2i_r}} \times \sum_{(j)} f_{2j}(1) a_{j_1 j_2 j_3} b_{j_2 j_3; j_1 j}. \tag{6.7.22}$$

By definition

$$a_{j_1 j_2 j_3} = 2f'_{2j_1}(1)\left(f_{2j_2}(1)\int_{-1}^{1} f_{2j_3}(t)dt - f_{2j_3}(1)\int_{-1}^{1} f_{2j_2}(t)dt\right), \tag{6.7.23}$$

and $b_{j_2 j_3; j_1 j}$ is, apart from a sign, the determinant obtained from Eq. (6.6.16) by omitting the rows corresponding to the values j_2, j_3 and the columns corresponding to the values j_1, j. The sign can be fixed, as always, by bringing this minor to the leading position in the upper left-hand corner. The indices j_1, j_2, j_3 and j are chosen from $i_1, ..., i_r$, and the summation in Eq. (6.7.22) above is over all such choices and then over all choices of the non-negative integers $0 \le i_1 < i_2 < \cdots < i_r$.

Figure 6.3 shows $E_1(n;s)$ for a few values of n. Figures 6.4 and 6.5 are the contour maps for two consecutive spacings. As in Figure 5.4, we notice that the set of eigenvalues of a matrix from the Gaussian orthogonal ensemble is more or less equally spaced, though not as good, since here the individual peaks are less high and a little wider than for the Gaussian unitary case.

6.8. Bounds for the Distribution Function of the Spacings

Equations (6.5.9) and (6.5.7) can also be written as

$$A_1(\theta) = \det\left[2\int_\theta^\infty \varphi_{2i}(x)\varphi_{2j}(x)dx\right]_{i,j=0,1,\ldots,m-1}. \tag{6.8.1}$$

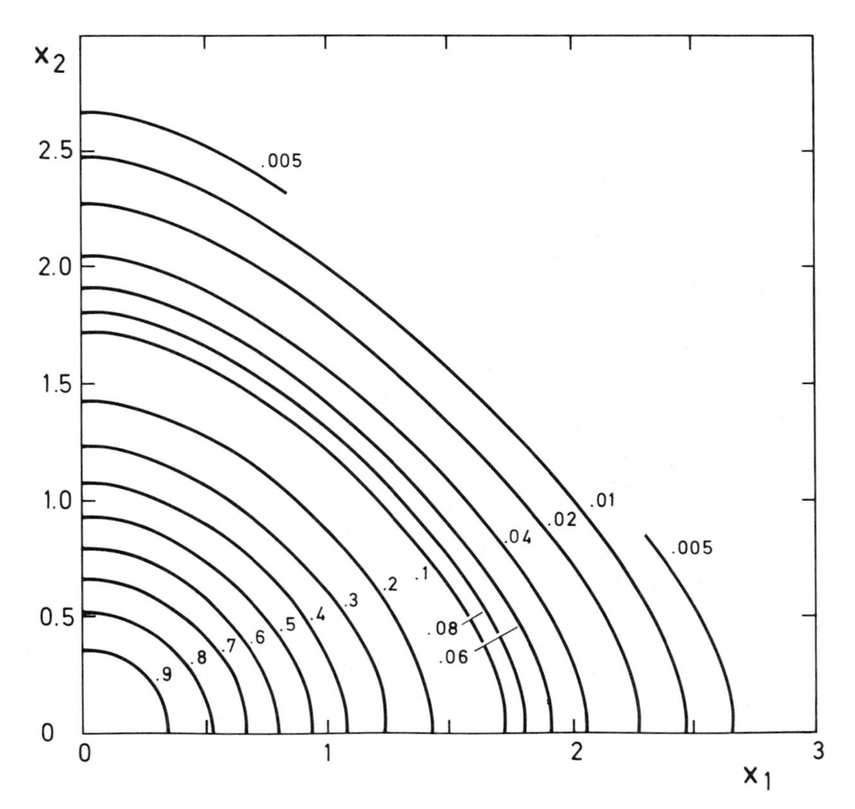

FIG. 6.4. The contour map of $\mathcal{B}_1(x, y)$; the same as Figure 5.7 but this time the random matrix is chosen from the Gaussian orthogonal ensemble.

We then apply Gram's result (*cf.* Appendix A.12) to write

$$\det\left[\int_\theta^\infty \varphi_{2i}(x)\varphi_{2j}(x)dx\right]$$

$$= \frac{1}{m!}\int_\theta^\infty \cdots \int_\theta^\infty \left(\det\left[\varphi_{2i-2}(x_j)\right]_{i,j=1,\ldots,m}\right)^2 dx_1 \cdots dx_m. \tag{6.8.2}$$

In Section 5.2 we wrote $\exp\left(-\frac{1}{2}\sum_i x_i^2\right)\prod_{i<j}(x_i - x_j)$ as a determinant containing oscillator functions $\varphi_j(x)$. Following step by step the same

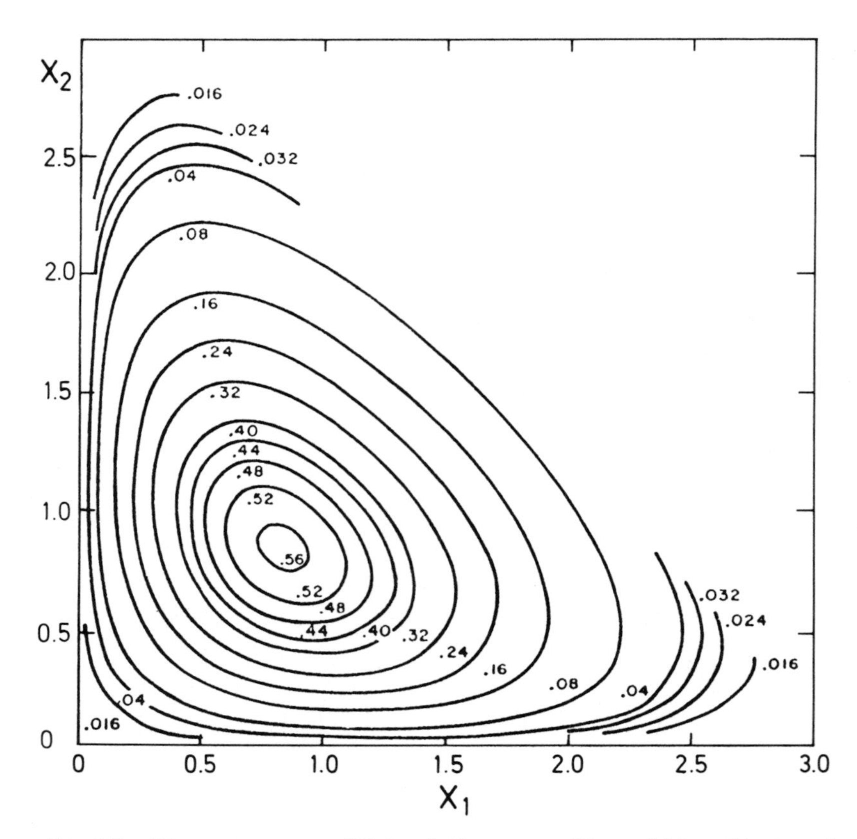

FIG. 6.5. The contour map of $\mathcal{P}_1(x, y)$; the same as Figure 5.8 but this time the random matrix is chosen from the Gaussian orthogonal ensemble.

procedure but in the reverse direction we can convert the integrand of Eq. (6.8.2) back to the form

$$\exp\left(-\sum_{i=1}^{m} x_i^2\right) \prod_{1 \le i < j \le m} \left(x_i^2 - x_j^2\right)^2, \tag{6.8.3}$$

except for some multiplicative factors. Since Eq. (6.8.2) contains only even functions $\varphi_j(x)$, we have in Eq. (6.8.3) the product of differences of the x_j^2 instead of those of the x_j. Thus,

$$A_1(\theta) = \text{const} \times \int_\theta^\infty \cdots \int_\theta^\infty dx_1 \cdots dx_m \exp\left(-\sum_{i=1}^m x_i^2\right) \times \prod_{1 \le i < j \le m} \left(x_i^2 - x_j^2\right)^2. \tag{6.8.4}$$

Differentiation with respect to θ gives

$$\frac{dA_1(\theta)}{d\theta} = \text{const} \times e^{-\theta^2} \int_\theta^\infty \cdots \int_\theta^\infty \exp\left(-\sum_{i=1}^{m-1} x_i^2\right) \cdot \times \prod_{1 \le i < j \le m-1} \left(x_i^2 - x_j^2\right)^2 \prod_{i=1}^{m-1} \left(x_i^2 - \theta^2\right)^2 dx_i. \tag{6.8.5}$$

Introducing new variables y_i defined by

$$x_i^2 = y_i^2 + \theta^2, \tag{6.8.6}$$

one may write

$$\frac{dA_1(\theta)}{d\theta} = -I_m(\theta) \exp(-m\theta^2), \tag{6.8.7}$$

with

$$I_m(\theta) = \text{const} \times \int_0^\infty \cdots \int_0^\infty \exp\left(-\sum_{i=1}^{m-1} y_i^2\right) \cdot \times \prod_{1 \le i < j \le m-1} \left(y_i^2 - y_j^2\right)^2 \prod_{i=1}^{m-1} y_i^5 \left(y_i^2 + \theta^2\right)^{-1/2} dy_i. \tag{6.8.8}$$

Applying Gram's result once again (*cf.* Appendix A.12) we write $I_m(\theta)$ as a determinant

$$I_m(\theta) = \text{const} \times \det\left[\eta_{i+j}\right]_{i,j=1,\ldots,m-1}, \tag{6.8.9}$$

where

$$\eta_j(\theta) = 2 \int_0^\infty e^{-y^2} y^{2j+1} (y^2 + \theta^2)^{-1/2} dy. \tag{6.8.10}$$

We may expand $\eta_j(\theta)$ and hence $I_m(\theta)$ in a power series in θ (*cf.* Appendix A.20),

$$\eta_j(\theta) = \Gamma\left(j+\frac{1}{2}\right) - \frac{1}{2}\theta^2\Gamma\left(j-\frac{1}{2}\right) + \frac{3}{8}\theta^2\Gamma\left(j-\frac{3}{2}\right) - \cdots, \tag{6.8.11}$$

$$I_m(\theta) = I_m(0)\left(1 - \frac{1}{3}(m-1)\theta^2 + \frac{1}{30}(m-1)(7m+1)\theta^4 + \cdots\right). \tag{6.8.12}$$

Finally we may take the limit $m \to \infty$, $2\theta\sqrt{m} = \pi t$ finite,

$$\frac{dE_1(0;2t)}{2dt} = \lim \frac{\sqrt{m}}{\pi}\frac{dA_1(\theta)}{d\theta} = -I(t)\exp\left(-(\pi t/2)^2\right). \tag{6.8.13}$$

As $dE_1(0;s)/ds = -1$, at $s = 0$ (*cf.* Section 5.1.3), we get from Eq. (6.8.12) the power series expansion of $I(t)$,

$$I(t) = \lim \frac{\sqrt{m}}{\pi} I_m(\theta) = 1 - \frac{1}{3}\left(\frac{\pi t}{2}\right)^2 + \frac{7}{30}\left(\frac{\pi t}{2}\right)^4 + \cdots. \tag{6.8.14}$$

The form Eq. (6.8.9) expressing $I_m(\theta)$ as a determinant is convenient for calculations, as in arriving at Eqs. (6.8.12) and (6.8.14), whereas the integral form (6.8.8) is useful to find bounds for $I(t)$.

It is easy to prove (*cf.* Appendix A.21) that for all positive values of y_i

$$1 - \frac{1}{2}\sum_{i=1}^{m}\frac{\theta^2}{y_i^2} \le \prod_{i=1}^{m}\left(y_i\left(y_i^2+\theta^2\right)^{-1/2}\right) \le 1. \tag{6.8.15}$$

The expansion (6.8.12) in the limit $m \to \infty$ gives then the inequalities

$$1 - \frac{1}{3}\left(\frac{\pi t}{2}\right)^2 \le I(t) \le 1. \tag{6.8.16}$$

Hence we get rigorous lower and upper bounds for the distribution function $\Psi(s)$, $(s = 2t)$, of the spacings (*cf.* Section 5.1.3)

$$\Psi_L(s) \le \Psi(s) \le \Psi_U(s), \tag{6.8.17}$$

where

$$\Psi_L(s) = 1 - \exp\left(-\frac{\pi^2 s^2}{16}\right), \tag{6.8.18}$$

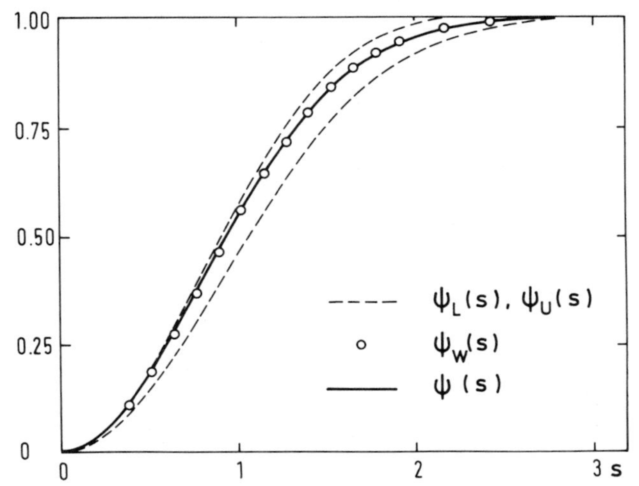

FIG. 6.6. The distribution function of the spacings $\Psi(s)$, the lower and upper bounds $\Psi_L(s)$, $\Psi_U(s)$ and the Wigner surmise $\Psi_W(s)$. Reprinted with permission from Elsevier Science Publishers, Gaudin, M. Sur la loi limite de l'éspacement des valeurs propres d'une matrice aléatoire, *Nucl. Phys.*, 25, 447–458 (1961).

and

$$\Psi_U(s) = 1 - \left(1 - \frac{1}{3}\frac{\pi^2 s^2}{16}\right) \exp\left(-\frac{\pi^2 s^2}{16}\right). \tag{6.8.19}$$

Because the differences $\Psi - \Psi_L$ and $\Psi_U - \Psi$ are every where non-negative, the difference between unity and the approximate mean values $\langle s \rangle_L$ and $\langle s \rangle_U$ obtained by substituting Ψ_L and Ψ_U for Ψ in

$$\langle s \rangle = \int_0^\infty s p_1(0; s) ds = \int_0^\infty [1 - \Psi(s)] ds \tag{6.8.20}$$

[*cf.* Eqs. (5.1.18) and (5.1.38)], provides a good estimation of the accuracy of the corresponding approximations to Ψ. One has

$$\begin{aligned} \langle s \rangle_L - 1 &= \frac{2}{\sqrt{\pi}} - 1 \cong 0.1284, \\ 1 - \langle s \rangle_U &= 1 - \frac{5}{3\sqrt{\pi}} \cong 0.0597. \end{aligned} \tag{6.8.21}$$

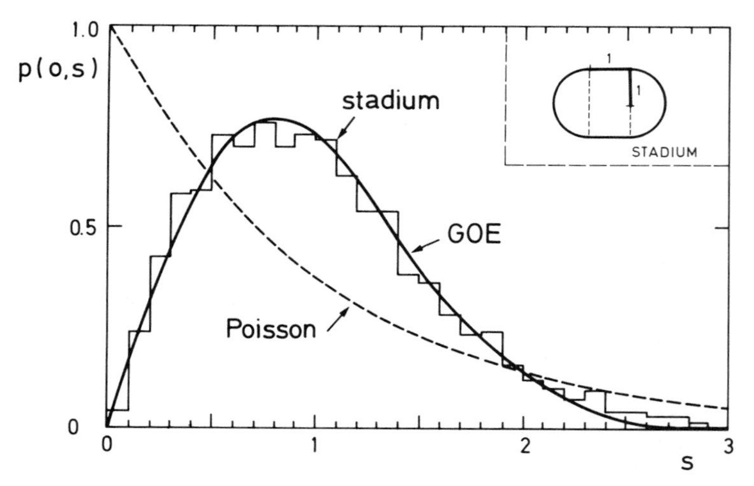

FIG. 6.7. Empirical probability density of the nearest neighbor spacings of the possible energies of a particle free to move on the stadium consisting of a rectangle of size 1×2 with semicircular caps of radius 1, depicted in the right upper corner. The stadium can be defined by the inequalities $|y| \leq 1$, and either $|x| \leq 1/2$ or $(x \pm 1/2)^2 + y^2 \leq 1$. The solid curve represents Eq. (6.5.19) corresponding to the Gaussian orthogonal ensemble (GOE), while the dashed curve is for the Poisson process corresponding to no correlations. Supplied by O. Bohigas, from Bohigas et al. (1984c).

For visual comparision, Figure 6.6 is a plot of the functions Ψ_L, Ψ, Ψ_U and the Wigner surmise

$$\Psi_W(s) = 1 - \exp\left(-\frac{\pi s^2}{4}\right). \tag{6.8.22}$$

It is a surprise that Wigner surmise is so close to the real distribution.

Figures 1.3, 1.4, 1.6, 1.7, and 1.8, Chapter 1, represent histograms of the nearest neighbor spacings of the nuclear and atomic levels and Figures 6.7 and 6.8 represent those for chaotic systems. Finally, Figure 6.9 corresponds to the ultrasonic resonance frequencies of an aluminium block.

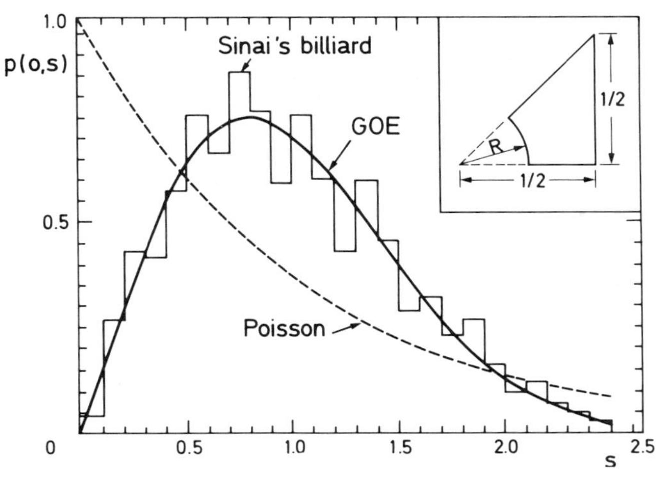

FIG. 6.8. Same as Figure 6.7 but when the particle moves on Sinai's billiard table consisting of 1/8 of a square cut by a circular arc, depicted in the right upper corner. One may define it by the inequalities $y \geq 0$, $x \geq y$, $x \leq 1$ and $x^2 + y^2 \geq r$. Only 1/8 of the square is taken so that all obvious symmetries of the square are disposed of. Supplied by O. Bohigas, from Bohigas et al. (1984b).

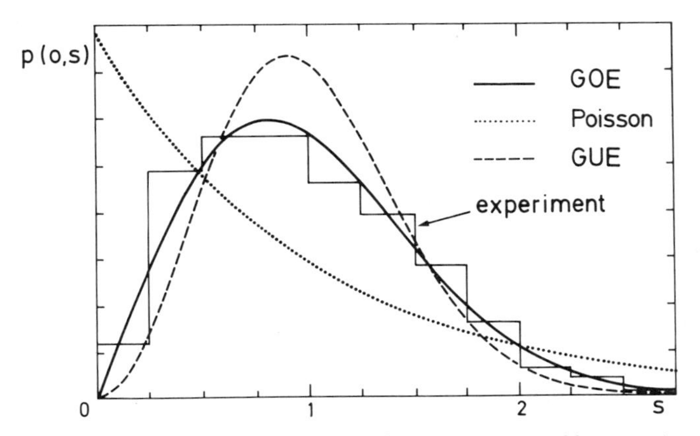

FIG. 6.9. Empirical probability density of the nearest neighbor spacings for the ultrasonic frequencies of an aluminium block. The three curves correspond respectively to the Poisson process with no correlations, to the Gaussian orthogonal ensemble (GOE) and to the Gaussian unitary ensemble (GUE). From Weaver (1989). Reprinted with permission from American Institute of Physics, Weaver, R. L. Spectral statistics in elastodynamics, *J. Acoust. Soc. Amer.*, 85, 1005–1013 (1989).

Summary of Chapter 6

For the Gaussian orthogonal ensemble with the joint probability density of the eigenvalues

$$P_{N1}(x_1, ..., x_N) \propto \exp\left(-\sum_{j=1}^{N} x_j^2/2\right) \prod_{1\le j<k\le N} |x_j - x_k|, \qquad (6.1.1)$$

the asymptotic two level cluster function is

$$Y_2(r) = \left(\frac{\sin(\pi r)}{\pi r}\right)^2 + \left[\int_r^\infty \left(\frac{\sin(\pi t)}{\pi t}\right) dt\right] \left[\frac{d}{dr}\left(\frac{\sin(\pi r)}{\pi r}\right)\right]. \qquad (6.4.14)$$

The n-level cluster function is a little complicated; see Eq. (6.4.20).

The Fourier transform of $Y_2(r)$ is

$$b(k) = \begin{cases} 1 - 2|k| + |k| \ln(1 + 2|k|), & |k| \le 1, \\ -1 + |k| \ln\left(\dfrac{2|k|+1}{2|k|-1}\right), & |k| \ge 1. \end{cases} \qquad (6.4.17)$$

The probability that a randomly chosen interval of length s contains exactly n levels is given by the formulas

$$E_1(0; s) = \prod_{i=0}^{\infty} (1 - \lambda_{2i}), \qquad (6.5.19)$$

and for $r > 0$,

$$E_1(2r; s) = E_1(0; s) \sum_{0\le j_1<j_2<\cdots<j_r} \prod_{i=1}^{r} \left(\frac{\lambda_{j_i}}{1 - \lambda_{j_i}}\right) \times [1 - (b_{j_1} + \cdots + b_{j_r})], \qquad (6.6.13)$$

$$E_1(2r-1; s) = E_1(0; s) \sum_{0\le j_1<j_2<\cdots<j_r} \prod_{i=1}^{r} \left(\frac{\lambda_{j_i}}{1 - \lambda_{j_i}}\right) \times (b_{j_1} + \cdots + b_{j_r}), \qquad (6.7.21)$$

where

$$b_j = f_{2j}(1) \int_{-1}^{1} f_{2j}(x)dx \Big/ \int_{-1}^{1} f_{2j}^2(x)dx,$$

$\lambda_j = s\,|\mu_j|^2/4$, and μ_j and $f_j(x)$ are the eigenvalues and the eigenfunctions of the integral equation

$$\mu f(x) = \int_{-1}^{1} \exp(i\pi xys/2) f(y)dy. \tag{5.3.17}$$

or μ_{2j} and $f_{2j}(x)$ are the eigenvalues and the eigenfunctions of the integral equation

$$\mu f(x) = 2\int_{0}^{1} \cos(\pi xys/2) f(y)dy. \tag{5.3.20}$$

7 / Gaussian Symplectic Ensemble

In this chapter we will consider briefly the Gaussian symplectic ensemble. The joint probability density function of the eigenvalues, Eq. (3.2.14), or (5.1.1) with $\beta = 4$, will be expressed as an $N \times N$ quaternion self-dual matrix satisfying the requirements of Theorem 6.2.1. The correlation and cluster functions will then follow from that theorem. In particular, the two-level cluster function is given by Eq. (7.2.6) and its Fourier transform by Eq. (7.2.9). The question of level spacings is more complicated and will not receive any detailed attention here. In fact, in Chapter 10 we will show the equivalence of the Gaussian and circular ensembles and a profound relation between orthogonal and symplectic ensembles. This will allow us to deduce the level spacings for the symplectic ensemble from that of the orthogonal ensemble.

7.1. A Quaternion Determinant

To take advantage of Theorem 6.2.1, we will give explicitly an $N \times N$ quaternion self-dual matrix $Q_{N4} = [f(x_i, x_j]$, or an ordinary $2N \times 2N$ matrix $C(Q_{N4})$ with $ZC(Q_{N4})$ antisymmetric, and verify that

$$\text{(i)} \quad \det\ [C(Q_{N4})] = \text{const} \times [P_{N4}(x_1, ..., x_N)]^2, \tag{7.1.1}$$

$$\text{(ii)} \quad (i, j) \text{ element } f(x_i, x_j) \text{ depends only on } x_i \text{ and } x_j, \tag{7.1.2}$$

$$\text{(iii)} \quad \int_{-\infty}^{\infty} f(x, x) dx = N, \tag{7.1.3}$$

and

$$\text{(iv)} \quad \int_{-\infty}^{\infty} f(x,y)f(y,z)dy = f(x,z). \tag{7.1.4}$$

7.1.1. Let $S_N(x,y)$, $DS_N(x,y)$ and $IS_N(x,y)$ be defined as in Section 6.3, Eqs. (6.3.2)–(6.3.4). Now let

$$\sigma_{N4}(x,y) = \frac{1}{\sqrt{2}} \begin{bmatrix} S_{2N+1}(x\sqrt{2}, y\sqrt{2}) & DS_{2N+1}(x\sqrt{2}, y\sqrt{2}) \\ IS_{2N+1}(x\sqrt{2}, y\sqrt{2}) & S_{2N+1}(y\sqrt{2}, x\sqrt{2}) \end{bmatrix} \tag{7.1.5}$$

be a quaternion in its 2×2 matrix form (*cf.* Section 2.4). Note that, as in Eq. (6.3.7), x and y are interchanged in the lower right corner. Let

$$Q_{N4} = [\sigma_{N4}(x_i, x_j)]_{i,j=1,\ldots,N} \,. \tag{7.1.6}$$

7.1.2. From the antisymmetry under the exchange $x \longleftrightarrow y$ of DS_{2N+1} and IS_{2N+1}, *cf.* Eqs. (6.3.17) and (6.3.18), and the definition of the dual, Eq. (2.4.17), it is straightforward to verify that Q_{N4} is self-dual.

7.1.3. The $2N \times 2N$ matrix $C(Q_{N4})$ is the product of two $2N \times 2N$ matrices

$$2^{-1/4} \begin{bmatrix} \varphi_{2i+1}(x_j\sqrt{2}) & \varphi_{2i+1}'(x_j\sqrt{2}) \\ \int_{-\infty}^{x_j\sqrt{2}} \varphi_{2i+1}(t)dt & \varphi_{2i+1}(x_j\sqrt{2}) \end{bmatrix} \tag{7.1.7}$$

and

$$2^{-1/4} \begin{bmatrix} \varphi_{2i+1}(x_k\sqrt{2}) & -\varphi_{2i+1}'(x_k\sqrt{2}) \\ -\int_{-\infty}^{x_k\sqrt{2}} \varphi_{2i+1}(t)dt & \varphi_{2i+1}(x_k\sqrt{2}) \end{bmatrix}. \tag{7.1.8}$$

Hence $\det[C(Q_{N4})]$ is the product of the determinants of these two matrices, and these determinants are visibly equal. Moreover, from Eq. (6.1.5) we can replace $\varphi_{2i+1}(x\sqrt{2})$ by a linear combination of $\varphi_{2i}'(x\sqrt{2})$

and $\varphi_{2i-1}(x\sqrt{2})$ and eliminate the $\varphi_{2i-1}(x\sqrt{2})$ using a row with a lower index i. The result is

$$\begin{aligned}\det\left[C(Q_{N4})\right] &= \text{const} \times \left(\det\begin{bmatrix}\varphi_{2i+1}(x_k\sqrt{2}) & \varphi'_{2i+1}(x_k\sqrt{2})\\ \varphi_{2i}(x_k\sqrt{2}) & \varphi'_{2i}(x_k\sqrt{2})\end{bmatrix}\right)^2\\ &= \text{const} \times \left(\det\left[\,\varphi_p(x_j\sqrt{2}) \quad \varphi'_p(x_j\sqrt{2})\,\right]\right)^2,\end{aligned}$$

$$i = 0, 1, ..., N-1; \quad j = 1, ..., N; \quad p = 0, 1, ..., 2N-1. \tag{7.1.9}$$

Now replace φ_p and φ'_p,

$$\varphi_p(x\sqrt{2}) = \left(2^p p!\sqrt{\pi}\right)^{-1/2} \exp(-x^2) H_p(x\sqrt{2}), \tag{7.1.10}$$

$$\varphi'_p(x\sqrt{2}) = \left(2^p p!\sqrt{\pi}\right)^{-1/2} \exp(-x^2)\left(H'_p(x\sqrt{2}) - 2xH_p(x\sqrt{2})\right), \tag{7.1.11}$$

take the factorials and the exponential factors out and eliminate the $-2xH_p(x\sqrt{2})$ terms using a previous column. Finally one can substitute the expression of the Hermite polynomials $H_p(x\sqrt{2})$ and following the procedure of Section 5.2 in the reverse direction keep only the highest power in each matrix element. The result is

$$\det\left[C(Q_{N4})\right] = \text{const} \times \exp\left(-2\sum_{i=1}^{N} x_i^2\right)\left(\det\left[\,x_j^p \quad p x_j^{p-1}\,\right]\right)^2. \tag{7.1.12}$$

The determinant in the last line above is the confluent alternant (*cf.* Mehta, 1989, Section 7.1)

$$\det\left[\,x_j^p \quad p x_j^{p-1}\,\right] = \prod_{j<k} (x_j - x_k)^4. \tag{7.1.13}$$

Thus Equation (7.1.1) is verified.

7.1.4. The fact that $\sigma_{N4}(x_i, x_j)$ depends only on x_i and x_j is evident.

7.1.5. From Eq. (6.3.38) one sees that

$$\int_{-\infty}^{\infty} \sigma_{N4}(x,x)dx = \frac{1}{\sqrt{2}} \int_{-\infty}^{\infty} S_{2N+1}(x\sqrt{2}, x\sqrt{2})dx = N. \tag{7.1.14}$$

We would like to remind the reader that throughout this book a quaternion is often denoted by its 2×2 matrix representation, *cf.* Section 2.4. For example, in the equation above, $\sigma_{N4}(x,x)$ is a scalar quaternion (the off diagonal elements are zero and the two diagonal elements are each) equal to the function $S_{2N+1}(x\sqrt{2}, x\sqrt{2})/\sqrt{2}$. Similarly, the sum of the product on the right hand side of Eq. (7.2.3) is a scalar quaternion giving the expression of $T_n(x_1, ..., x_n)$.

With the notation of Section 6.3, Eqs. (6.3.40)–(6.3.47), one verifies without much trouble that

$$\int_{-\infty}^{\infty} \sigma_{N4}(x,y)\sigma_{N4}(y,z)dy = \sigma_{N4}(x,z). \tag{7.1.15}$$

Thus the quaternion matrix Q_{N4} satisfies all the requirements announced at the beginning of this section.

7.2. Correlation and Cluster Functions

From Eq. (7.1.6) and Theorem 6.2.1 integrating over all the variables one can fix the constants,

$$P_{N4}(x_1, ..., x_N) = \frac{1}{N!} \det Q_{N4} = \frac{1}{N!} \det \left[\sigma_{N4}(x_i, x_j)\right]_{i,j=1,...,N}. \tag{7.2.1}$$

By the same Theorem 6.2.1 the n-level correlation function is

$$\begin{aligned} R_n(x_1, ..., x_n) &= \frac{N!}{(N-n)!} \int \cdots \int P_{N4}(x_1, ..., x_N) dx_{n+1} \cdots dx_N \\ &= \det \left[\sigma_{N4}(x_i, x_j)\right]_{i,j=1,...,n}. \end{aligned} \tag{7.2.2}$$

And as in Section 5.2, the n-level cluster function is

$$T_n(x_1, ..., x_n) = \sum_P \sigma_{N4}(x_1, x_2)\sigma_{N4}(x_2, x_3) \cdots \sigma_{N4}(x_n, x_1), \tag{7.2.3}$$

the sum being taken over all $(n-1)!$ distinct cyclic permutations of the indices $(1, 2, ..., n)$.

Setting $n = 1$, we get the level density that in the limit $N \to \infty$ is given again by Eq. (5.2.17) or Eq. (6.4.4). Setting $n = 2$, we get the two-level cluster function

$$\begin{aligned} T_2(x, y) &= \sigma_{N4}(x, y)\sigma_{N4}(y, x) \\ &= \frac{1}{2}\big[S_{2N+1}(x\sqrt{2}, y\sqrt{2})S_{2N+1}(y\sqrt{2}, x\sqrt{2}) \\ &\quad - DS_{2N+1}(x\sqrt{2}, y\sqrt{2})IS_{2N+1}(x\sqrt{2}, y\sqrt{2})\big]. \end{aligned} \tag{7.2.4}$$

In the limit $N \to \infty$,

$$\begin{aligned} Y_2(r) &= \lim \frac{\pi}{2N} T_2(x, y) = \sigma_4(r)\sigma_4(-r), \\ \sigma_4(r) &= \begin{bmatrix} s(2r) & Ds(2r) \\ Is(2r) & s(2r) \end{bmatrix}, \end{aligned} \tag{7.2.5}$$

with the functions s, Ds, and Is given by Eqs. (6.4.7)–(6.4.9), (6.4.11). Or

$$Y_2(r) = \left(\frac{\sin(2\pi r)}{2\pi r}\right)^2 - \frac{d}{dr}\left(\frac{\sin(2\pi r)}{2\pi r}\right)\int_0^r \left(\frac{\sin(2\pi t)}{2\pi t}\right) dt. \tag{7.2.6}$$

The behavior of $Y_2(r)$ for small and large r is given by

$$Y_2(r) = 1 - \frac{(2\pi r)^4}{135} + \cdots, \tag{7.2.7}$$

and

$$Y_2(r) = -\frac{\pi}{2}\frac{\cos(2\pi r)}{2\pi r} + \frac{1 + (\pi/2)\sin(2\pi r)}{(2\pi r)^2} - \frac{1 + \cos^2(2\pi r)}{(2\pi r)^4} + ..., \tag{7.2.8}$$

respectively. The two-level form factor is (*cf.* Appendix A.11)

$$\begin{aligned} b(k) &= \int_{-\infty}^{\infty} Y_2(r)\exp(2\pi i k r)dr \\ &= \begin{cases} 1 - \frac{1}{2}|k| + \frac{1}{4}|k|\ln(1 - |k|), & |k| < 2, \\ 0, & |k| > 2. \end{cases} \end{aligned} \tag{7.2.9}$$

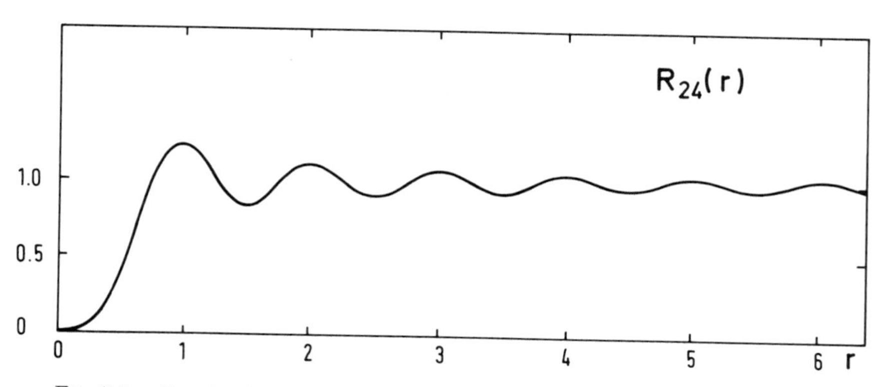

FIG. 7.1. Two-level correlation function for the symplectic ensemble.

In the limit $N \to \infty$, the n-level cluster function is given by Eq. (5.2.21) or (6.4.20), where $s(r)$ or $\sigma_1(r)$ is now replaced by $\sigma_4(r)$ of Eq. (7.2.5). The n-level form factor is given by Eq. (5.2.23) or (6.4.21), where $f_2(k)$ or $f_1(k)$ is now replaced by

$$f_4(k) = \frac{1}{2} f_2(k) \begin{bmatrix} 1 & k \\ k^{-1} & 1 \end{bmatrix}, \tag{7.2.10}$$

with $f_2(k)$ given by Eq. (5.2.24),

$$f_2(k) = \begin{cases} 1, & |k| < 1/2, \\ 0, & |k| > 1/2. \end{cases} \tag{7.2.11}$$

Figure 7.1 shows the limiting two-level correlation function. Note that the oscillations persist much longer than for the Gaussian unitary ensemble, Figure 5.2.

7.3. Level Spacings

The joint probability density function for the eigenvalues of a matrix taken from the Gaussian symplectic ensemble can be written by Eqs. (7.2.1), (7.1.5), and (7.1.9) as

$$P_{N4}(x_1, ..., x_N) = \text{const} \times \det \left[\varphi_i(x_j\sqrt{2}) \quad \varphi_i'(x_j\sqrt{2}) \right], \\ i = 0, 1, ..., 2N-1; \quad j = 1, 2, ..., N. \tag{7.3.1}$$

Note that each variable occurs in two columns, we do not need to order the variables and one can integrate over any number of them by using Pfaffians; *cf.* Appendix A.18. Moreover, since our integration limits are $(-\infty, -\theta)$ and (θ, ∞), the excluded interval $(-\theta, \theta)$ situated symmetrically about the origin, we have

$$\int_{\text{out}} \varphi_{2i}(x\sqrt{2})\varphi_{2j}'(x\sqrt{2})dx = \int_{\text{out}} \varphi_{2i+1}(x\sqrt{2})\varphi_{2j+1}'(x\sqrt{2})dx = 0. \tag{7.3.2}$$

Therefore the Pfaffians reduce to determinants. For example, the probability that the interval $(-\theta, \theta)$ contains none of the eigenvalues is

$$\begin{aligned} A_4(\theta) &= \underset{\text{out}}{\int \cdots \int} P_{N4}(x_1, ..., x_N) dx_1 \cdots dx_N \\ &= \text{const} \times \det \left[B_{ij}\right]_{i,j=0,1,\ldots,N-1}, \end{aligned} \tag{7.3.3}$$

with

$$B_{ij} = \int_{\text{out}} \left(\varphi_{2i}(x\sqrt{2})\varphi_{2j+1}'(x\sqrt{2}) - \varphi_{2i}'(x\sqrt{2})\varphi_{2j+1}(x\sqrt{2}) \right) dx, \tag{7.3.4}$$

where the subscript "out" means, as usual, that we integrate outside the interval $(-\theta, \theta)$.

Similarly, the probability that the interval $(-\theta, \theta)$ contains exactly n eigenvalues can be written as

$$\text{const} \times \sum_{(i;j)} B(i;j) \times \det \begin{bmatrix} \varphi_{2i_p}(x_k\sqrt{2}) & \varphi_{2i_p}'(x_k\sqrt{2}) \\ \varphi_{2j_p+1}(x_k\sqrt{2}) & \varphi_{2j_p+1}'(x_k\sqrt{2}) \end{bmatrix}_{p,k=1,\ldots,n}, \tag{7.3.5}$$

where the sum is taken over all choices of the indices satisfying $0 \le i_1 < \cdots < i_n \le N-1$; $0 \le j_1 < \cdots < j_n \le N-1$; and $B(i;j)$ is the $(N-n) \times (N-n)$ minor of B obtained by removing the rows $i_1, ..., i_n$ and the columns $j_1, ..., j_n$. The x_k lie in the interval $(-\theta, \theta)$, $|x_k| \le \theta$.

However, taking the limit $N \to \infty$ is difficult, since the element B_{ij} is a sum of two terms of different nature, one involving the integration of the product of two even functions φ, and the other of two odd functions φ.

Another expression for $E_4(n; s)$ is given, for example, by Eq. (A.7.21), Appendix A.7, where λ_i are the eigenvalues of the integral equation (A.7.23), with $\phi(x, y)$ replaced now by $\sigma_4(x-y)$, σ_4 given by Eq. (7.2.5).

Instead of either of these we will deduce these level spacing probabilities from the equivalence of Gaussian and circular ensembles and a profound relation between orthogonal and symplectic ensembles, proved in Chapter 10 (*cf.* Sections 10.6 and 10.7).

Summary of Chapter 7

For the Gaussian symplectic ensemble with the joint probability density of the eigenvalues

$$P_{N4}(x_1, ..., x_N) \propto \exp\left(-2\sum_{j=1}^{N} x_j^2\right) \prod_{1 \le j < k \le N} (x_j - x_k)^4,$$

the asymptotic two-level cluster function is

$$Y_2(r) = \left(\frac{\sin(2\pi r)}{2\pi r}\right)^2 - \frac{d}{dr}\left(\frac{\sin(2\pi r)}{2\pi r}\right) \int_0^r \left(\frac{\sin(2\pi t)}{2\pi t}\right) dt, \tag{7.2.6}$$

with the Fourier transform

$$b(k) = \begin{cases} 1 - \frac{1}{2}\,|k| + \frac{1}{4}\,|k| \ln\left|1 - |k|\right|, & |k| \le 2, \\ 0, & |k| \ge 2. \end{cases} \tag{7.2.9}$$

The probability $E_4(n; s)$ that a randomly chosen interval of length s contains exactly n levels will be given in Chapter 10.

8 / Gaussian Ensembles: Brownian Motion Model

8.1. Stationary Ensembles

In Chapter 4 we exploited the idea that the probability $P(x_1, ..., x_N)$, Eq. (3.3.8), for the eigenvalues of a random matrix to lie in unit intervals around the points $x_1, ..., x_N$,

$$P(x_1, ..., x_N) = C_{N\beta} e^{-\beta W}, \tag{8.1.1}$$

$$W = \frac{1}{2} \sum_1^N x_j^2 - \sum_{i<j} \ell n |x_i - x_j|, \tag{8.1.2}$$

is identical with the probability density of the positions of N unit charges free to move on the infinite straight line $-\infty < x < \infty$ under the influence of forces derived from the potential energy (8.1.2), according to the laws of classical mechanics, in a state of thermodynamical equilibrium at a temperature given by

$$kT = \beta^{-1}. \tag{8.1.3}$$

This system of point charges in thermodynamical equilibrium is called the stationary Coulomb gas model or simply the Coulomb gas model, which corresponds to the Gaussian ensembles.

8.2. Nonstationary Ensembles

In this chapter we present an idea of Dyson, generalizing the notion of a matrix ensemble in such a way that the Coulomb gas model acquires meaning not only as a static model in timeless thermodynamical equilibrium but as a dynamical system that may be in an arbitrary nonequilibrium state changing with time. The word "time" in this chapter always refers to a fictitious time which is a property of the mathematical model and has nothing to do with real physical time.

When we try to interpret Coulomb gas as a dynamical system, we naturally consider it first as an ordinary conservative system in which the charges move as Newtonian particles and exchange energy with one another only through the electric forces arising from the potential (8.1.2). We then have to give meaning to the velocity of each particle and to regulate the behavior of the random matrix H in such a way that the eigenvalues have the normal Newtonian Property of inertia. No reasonable way of doing this has yet been found. Perhaps there is no such way.

After considerable and fruitless efforts to develop a Newtonian theory of ensembles, Dyson (1962b) discovered that the correct procedure is quite different and much simpler. The x_j should be interpreted as positions of particles in Brownian motion (Chandrasekhar, 1943; Uhlenbeck and Ornstein, 1930; Wang and Uhlenbeck, 1945). This means that the particles have no well-defined velocities nor do they possess inertia. Instead, they feel frictional forces resisting their motion. The gas is not a conservative system, for it is constantly exchanging energy with its surroundings through these frictional forces. The potential (8.1.2) still operates on the particles in the following way. The particle at x_j experiences an external electric force

$$E(x_j) = -\frac{\partial W}{\partial x_j} = -x_j + \sum_{\substack{i \\ (i \neq j)}} \frac{1}{x_j - x_i} \tag{8.2.1}$$

in addition to the local frictional force and the constantly fluctuating force giving rise to the Brownian motion.

The equation of motion of the Brownian particle at x_j may be written as

$$\frac{d^2 x_j}{dt^2} = -f\frac{dx_j}{dt} + E(x_j) + A(t) \tag{8.2.2}$$

where f is the friction coefficient and $A(t)$ is a rapidly fluctuating force. For $A(t)$ we postulate the usual properties (Uhlenbeck and Ornstein, 1930)

$$\langle A(t_1) A(t_2) \cdots A(t_{2n+1}) \rangle = 0, \tag{8.2.3}$$

$$\langle A(t_1) A(t_2) \cdots A(t_{2n}) \rangle = \sum_{\text{pairs}} \langle A(t_i) A(t_j) \rangle \langle A(t_k) A(t_\ell) \rangle \cdots, \tag{8.2.4}$$

and

$$\langle A(t_1) A(t_2) \rangle = \frac{2kT}{f} \delta(t_1 - t_2), \tag{8.2.5}$$

where the summation in Eq. (8.2.4) extends over all distinct ways in which the $2n$ indices can be divided into n pairs.

There is nothing new in the integration of the Langevin equation (8.2.2). After a long enough time for the effect of the initial velocities to become negligible, let $x_1, x_2, ..., x_N$ be the positions of the particles at time t. At a later time $t + \delta t$ let these positions be changed to $x_1 + \delta x_1, x_2 + \delta x_2, ..., x_N + \delta x_N$. The δx_j, $j = 1, 2, ..., N$, will in general be different for every member of the ensemble. They are random variables. Using Eqs. (8.2.3) to (8.2.5) we find that to the first order in the small quantities

$$f \langle \delta x_j \rangle = E(x_j) \ \delta t, \tag{8.2.6}$$

$$f \left\langle (\delta x_j)^2 \right\rangle = 2kT \ \delta t, \tag{8.2.7}$$

and all other ensemble averages, for example, $\langle \delta x_j \delta x_\ell \rangle, \left\langle (\delta x_j)^2 \delta x_\ell \right\rangle,$ $\left\langle (\delta x_j)^3 \right\rangle$, are of a higher order in δt.

An alternative description of the Brownian motion is obtained by deriving the Fokker-Planck or Smoluchowski equation. Let $P(x_1, x_2, ..., x_N; t)$ be the time-dependent joint probability density that the particles will be at the positions x_j at time t. Assuming that the future evolution of the system is completely determined by its present state, with no reference to its past (that is, the process is a Markov process), we obtain

$$\begin{aligned} P(x_1, ..., x_N; t + \delta t) &= \int \cdots \int P(x_1 - \delta x_1, ..., x_N - \delta x_N; t) \\ &\times \psi(x_1 - \delta x_1, ..., x_N - \delta x_N; \delta x_1, ..., \delta x_N; \delta t)\, d(\delta x_1) \cdots d(\delta x_N), \end{aligned} \tag{8.2.8}$$

where ψ under the integral sign is the probability that the positions of the particles will change from $x_1 - \delta x_1, ..., x_N - \delta x_N$ to $x_1, ..., x_N$ in a time interval δt. Expanding both sides of Eq. (8.2.8) in a power series of δx_j, δt, using Eqs. (8.2.6) and (8.2.7), and going to the limit $\delta t \longrightarrow 0$, we get (Uhlenbeck and Ornstein, 1930)

$$f\,\frac{\partial P}{\partial t} = \sum_{j=1}^{N} \left\{ kT\,\frac{\partial^2 P}{\partial x_j^2} - \frac{\partial}{\partial x_j}\left(E(x_j)\,P\right) \right\}. \tag{8.2.9}$$

Equation (8.2.9) describes the evolution of the Coulomb gas with time. If we start from an arbitrary initial probability density P at time $t = t_0$, a unique solution of Eq. (8.2.9) will exist for all $t \geq t_0$. Any solution of this sort we call a *time-dependent Coulomb gas model.*

Equation (8.2.9) implies in turn Eqs. (8.2.6) and (8.2.7). To see this we multiply both sides of Eq. (8.2.9) by x_j and integrate over all x_i. Making the usual assumptions that $P(x_1, ..., x_N; t)$, as well as its derivatives, vanish quite fast on the boundary, we get on partial integration

$$f\frac{d}{dt}\langle x_j \rangle = \langle E(x_j) \rangle, \tag{8.2.10}$$

where

$$\langle F \rangle = \int F(x_1, ..., x_N)\, P(x_1, ..., x_N; t)\, dx_1 \cdots dx_N$$

is the ensemble average of F. Starting at the positions $x_1, ..., x_N$ and executing the motion for a small time interval δt, we find that Eq. (8.2.10) is the same as Eq. (8.2.6). Similarly, by multiplying by x_j^2 and integrating Eq. (8.2.9) we have

$$f\frac{d}{dt}\langle x_j^2 \rangle = 2kT + 2\langle x_j E(x_j) \rangle,$$

which together with $\left\langle (\delta x_j)^2 \right\rangle = \langle x_j^2 \rangle - \langle x_j \rangle^2$ yields Eq. (8.2.7).

Thus, the descriptions of the motion by Eqs. (8.2.6) and (8.2.7), and by Eq. (8.2.9) are equivalent. Also there exists a unique solution to (8.2.9) that is independent of time, and this time independent solution is given by Eqs. (8.1.1) and (8.1.2).

A Brownian motion model can also be constructed for the matrix H, of which x_j are the eigenvalues. The independent real parameters H_{ii}, $1 \le i \le N$; $H_{ij}^{(\lambda)}$; $1 \le i < j \le N$, $0 \le \lambda \le \beta - 1$, which determine all the matrix elements of H, are $p = N + N(N-1)\beta/2$ in number. Let us denote them by H_μ, where μ is a single index that runs from 1 to p and replaces the three indices i, j and λ. Suppose that the parameters H_μ have the values $H_1, H_2, ..., H_p$ at time t and $H_1 + \delta H_1, ..., H_p + \delta H_p$ at a later time $t + \delta t$. Brownian motion of H is defined by requiring that each δH_μ be a random variable with the ensemble averages

$$f \langle \delta H_\mu \rangle = -H_\mu \delta t, \tag{8.2.11}$$

$$f \left\langle (\delta H_\mu)^2 \right\rangle = g_\mu kT \; \delta t, \tag{8.2.12}$$

where

$$g_\mu = g_{ij}^{(\lambda)} = 1 + \delta_{ij} = \begin{cases} 2, & \text{if } \; i = j, \\ 1, & \text{if } \; i \neq j. \end{cases} \tag{8.2.13}$$

All other averages are of a higher order in δt. This is a Brownian motion of the simplest type, the various components H_μ being completely uncoupled and each being subject to a fixed simple harmonic force. The Smoluchowski equation which corresponds to Eq. (8.2.11) and Eq. (8.2.12) is

$$f \frac{\partial P}{\partial t} = \sum_\mu \left(\frac{1}{2} g_\mu kT \frac{\partial^2 P}{\partial H_\mu^2} + \frac{\partial}{\partial H_\mu} (H_\mu P) \right), \tag{8.2.14}$$

where $P(H_1, ..., H_p; t)$ is the time dependent joint probability density of the H_μ. The solution of Eq. (8.2.14) which corresponds to a given initial condition $H = H'$ at $t = 0$, is known explicitely (Uhlenbeck and Ornstein, 1930).

$$P(H, t) = C \left(1 - q^2\right)^{-p/2} \exp\left(- \frac{\operatorname{tr}(H - qH')^2}{2kT(1 - q^2)} \right), \tag{8.2.15}$$

where

$$q = \exp(-t/f). \tag{8.2.16}$$

The solution shows that the Brownian process is invariant under symmetry preserving unitary transformations of the matrix H; in fact, the awkward looking factor g_μ in (8.2.12) is put in to ensure this invariance. When $t \to \infty$, $q \to 0$, and the probability density (8.2.15) tends to the stationary form,

$$P(H_1, ..., H_p) = \text{const} \times \exp\left(-\text{tr}H^2/2kT,\right) \tag{8.2.17}$$

which is the unique time independent solution of Eq. (8.2.14). Note that with the relation Eq. (8.1.3) between β and the temperature kT, Eq. (8.2.17) is essentially the same as Eq. (2.6.18)

We are now in a position to state the main result of this chapter.

Theorem 8.2.1. *When the matrix H executes a Brownian motion according to the simple harmonic law* (8.2.11), (8.2.12), *starting from any initial conditions, its eigenvalues $x_1, x_2, ..., x_N$ execute a Brownian motion that obeys the Equations of motion* (8.2.6) *and* (8.2.7) *of the time dependent Coulomb gas.*

To prove the theorem we need only show that Eqs. (8.2.6) and (8.2.7) follow from Eqs. (8.2.11) and (8.2.12). Suppose, then, that Eqs. (8.2.11) and (8.2.12) hold. We have seen that the process described by Eqs. (8.2.11) and (8.2.12) is independent of the representation of H. Therefore, we may choose the representation so that H is diagonal at time t. The instantaneous values of H_μ at time t are then

$$H_{jj}^{(0)} = x_j, \quad j = 1, 2, ..., N, \tag{8.2.18}$$

and all other components are zero. At a later time $t + \delta t$ the matrix $H + \delta H$ is no longer diagonal and its eigenvalues $x_j + \delta x_j$ must be calculated by perturbation theory. We have to the second order in δH

$$\delta x_j = \delta H_{jj}^{(0)} + \sum_{\substack{i \\ i \neq j}} \sum_{\lambda=0}^{\beta-1} \frac{(\delta H_{ij}^{(\lambda)})^2}{x_j - x_i}. \tag{8.2.19}$$

Higher terms in the perturbation series will not contribute to the first order in δt. When we take the ensemble average on each side of Eq. (8.2.19) and use Eqs. (8.2.11), (8.2.18), (8.2.12), (8.1.3), and (8.2.1), the result is Eq. (8.2.6). When we take the ensemble average of $(\delta x_j)^2$, only the first term on the right side of Eq. (8.2.19) contributes to the order δt, and this term gives Eq. (8.2.7) by virtue of Eqs. (8.2.12) and (8.2.13). The theorem is thus proved.

When the limit $t \to \infty$ is taken, Theorem 8.2.1 reduces to Theorem 3.3.1. This new proof of Theorem 3.3.1 is in some respects more illuminating. It shows how the repulsive Coulomb potential (8.1.2), pushing apart each pair of eigenvalues, arises directly from the perturbation formula (8.2.19). It has long been known that perturbations generally split levels that are degenerate in an unperturbed system. We now see that this splitting effect of perturbations is quantitatively identical with the repulsive force of the Coulomb gas model.

Theorem 8.2.1 is a much stronger statement than Theorem 3.3.1. It shows that the electric force (8.2.1), acting on the eigenvalues x_j, has a concrete meaning for any matrix H whatever, not only for an ensemble of matrices in stationary thermal equilibrium. The force $E(x_j)$ is precisely proportional to the mean rate of drift of x_j which occurs when the matrix H is subjected to a random perturbation.

8.3. Some Ensemble Averages

We now describe a general property of the time-dependent Coulomb gas model which may be used to calculate a few ensemble averages. Dyson observed that if $G = G(x_1, ..., x_N)$ is any function of the positions of the charges, not depending explicitly on time, then the time variation of $\langle G \rangle$, the ensemble average of G, is governed by the equation

$$f \frac{d}{dt} \langle G \rangle = -\sum_j \left\langle \frac{\partial W}{\partial x_j} \frac{\partial G}{\partial x_j} \right\rangle + kT \sum_j \left\langle \frac{\partial^2 G}{\partial x_j^2} \right\rangle . \tag{8.3.1}$$

This equation is obtained by multiplying Eq. (8.2.9) throughout by G and partial integrations; W is given by Eq. (8.1.2).

As a first example, choose

$$R = \sum_j x_j^2 \tag{8.3.2}$$

for G so that

$$\frac{\partial W}{\partial x_j}\frac{\partial R}{\partial x_j} = -2 \sum_{\substack{i \\ (i \neq j)}} \frac{x_j}{x_j - x_i} + 2x_j^2,$$

$$\frac{\partial^2 R}{\partial x_j^2} = 2,$$

and Eq. (8.3.1) becomes

$$\begin{aligned} f\frac{\partial \langle R \rangle}{\partial t} &= - 2\langle R \rangle + N(N-1) + 2kTN \\ &= 2\left(R_\infty - \langle R \rangle\right), \end{aligned} \tag{8.3.3}$$

with

$$R_\infty = \frac{1}{2}N(N-1) + kTN. \tag{8.3.4}$$

The solution of Eq. (8.3.3) is

$$\langle R \rangle = R_0 q^2 + R_\infty \left(1 - q^2\right), \tag{8.3.5}$$

where q is given by Eq. (8.2.16) and R_0 is the value of $\langle R \rangle$ at $t = 0$. Equation (8.3.5) shows that the ensemble average $\langle R \rangle$ approaches its equilibrium value R_∞ with exponential speed as $t \longrightarrow \infty$.

Next take $G = W$ in Eq. (8.3.1), so that

$$\begin{aligned} \left(\frac{\partial W}{\partial x_j}\right)^2 = x_j^2 &+ \sum_{\substack{i \\ (i \neq j)}} \left[\left(\frac{1}{x_j - x_i}\right)^2 - \frac{2x_j}{x_j - x_i}\right] \\ &+ \sum_{\substack{i,\ell \\ (i,j,\ell \text{ all different})}} \left[(x_j - x_i)(x_j - x_\ell)\right]^{-1}, \end{aligned} \tag{8.3.6}$$

and

$$\frac{\partial^2 W}{\partial x_j^2} = 1 + \sum_{\substack{i \\ (i \neq j)}} (x_j - x_i)^{-2}. \tag{8.3.7}$$

On performing a summation over j the last term in Eq. (8.3.6) drops out [*cf.* Appendix A.5, paragraph preceding Eq. (A.5.14)], whereas the second term in the first bracket gives $-N(N-1)$. Substituting in Eq. (8.3.1) and simplifying, we get

$$\begin{aligned} f\frac{\partial \langle W \rangle}{\partial t} = &(kT - 1) \sum_{\substack{i,j \\ (i \neq j)}} \left\langle (x_j - x_i)^{-2} \right\rangle \\ &+ \left(N^2 - N + NkT\right) - \sum_j \langle x_j^2 \rangle . \end{aligned} \tag{8.3.8}$$

For the stationary Coulomb gas at temperature kT the left side of Eq. (8.3.8) vanishes and Eq. (8.3.4) may be used on the right. Thus, we find a "virial theorem" for the stationary gas:

$$\sum_{\substack{i,j \\ (i \neq j)}} \left\langle (x_j - x_i)^{-2} \right\rangle = \frac{N(N-1)}{2(1-kT)}. \tag{8.3.9}$$

The probability density of eigenvalues becomes proportional to $|x_j - x_i|^\beta$, when two eigenvalues x_i, x_j come close together. The ensemble average of $(x_j - x_i)^{-2}$ is therefore defined only for $\beta > 1$ and Eqs. (8.3.8) and (8.3.9) hold only for $kT < 1$.

An especially interesting case $\beta = 1$, requires a passage to the limit in (8.3.8). As $kT \longrightarrow 1$, we have for any fixed value of Δ

$$\begin{aligned} \lim\ (kT - 1) \int_{-\Delta}^{\Delta} |y|^{\beta - 2}\, dy &= \lim\ (kT - 1)(\beta - 1)^{-1}\, 2\Delta^{\beta - 1} \\ &= -2. \end{aligned} \tag{8.3.10}$$

We obtain the correct limit in (8.3.8) if we replace

$$(kT - 1)\,(x_j - x_i)^{-2}$$

with

$$-2\,(x_j - x_i)^{-1}\,\delta\,(x_j - x_i)\,, \tag{8.3.11}$$

which has a well-defined meaning as an ensemble average when $kT = 1$, for the probability density then contains a factor $|x_j - x_i|$; Eq. (8.3.8) thus becomes in the limit

$$f\frac{d\,\langle W\rangle}{dt} = -2 \sum_{\substack{i,j\\(i\neq j)}} \left\langle |x_j - x_i|^{-1}\,\delta\,(x_j - x_i)\right\rangle + N^2 - \sum_j \langle x_j^2\rangle\,, \qquad kT = 1. \tag{8.3.12}$$

The corresponding "virial theorem" is

$$\sum_{\substack{i,j\\(i\neq j)}} \left\langle (x_j - x_i)^{-1}\,\delta\,(x_j - x_i)\right\rangle = \frac{1}{4}N\,(N-1)\,, \qquad kT = 1 \tag{8.3.13}$$

for the stationary gas.

Summary of Chapter 8

Suppose that all the real parameters entering the definition of a Hermitian matrix H execute Brownian motion with the averages

$$\langle \delta H_{ij}\rangle \;= -H_{ij}\delta t, \tag{8.2.11}$$

$$\left\langle (\delta H_{ij})^2\right\rangle \;= (1 + \delta_{ij})\delta t/\beta. \tag{8.2.12}$$

Then starting at whatever initial values, after a long enough time the matrix elements H_{ij} reach an equilibrium distribution given by

$$P(H) \propto \exp(-\beta \text{ tr } H^2/2). \tag{8.2.17}$$

Some ensemble averages can also be computed, such as

$$\sum_{\substack{i,j \\ (i \neq j)}} \left\langle (x_j - x_i)^{-2} \right\rangle = \frac{N(N-1)}{2(1-kT)}, \qquad \beta > 1, \tag{8.3.9}$$

or

$$\sum_{\substack{i,j \\ (i \neq j)}} \left\langle (x_j - x_i)^{-1} \delta(x_j - x_i) \right\rangle = \frac{1}{4} N(N-1), \qquad \beta = 1. \tag{8.3.13}$$

9 / Circular Ensembles

In the preceding chapters we presented a detailed study of the Gaussian ensembles. We pointed out at the end of Section 2.6 that the requirements of invariance and the statistical independence of various independent components seem to overrestrict the possible choices. In particular, the various values of the matrix elements are not equally weighted. Rosenzweig has tried to answer this question in part with a consideration of the "fixed-strength ensemble" in which the joint probability density function for the matrix elements is taken to be proportional to the Dirac delta function $\delta\left(\text{tr } H^2 - r^2\right)$, where r is a fixed number; see Chapter 19. However, a uniform probability density cannot be defined on the infinite real line.

Because of this unsatisfactory feature Dyson (1962a-I) introduced his circular ensembles, as follows.

Suppose that the system is characterized not by its Hamiltonian H but by a unitary matrix S, whose elements give the transition probabilities between the various states. The matrix S is unitary; its eigenvalues are therefore of the form $e^{i\theta_j}$, where the angles θ_j are real and can be taken to lie between 0 and 2π. The matrix S is a function of the Hamiltonian H of the system. This functional dependence need not be specified. All that is needed is that for small ranges of variation, the θ_j be linear functions of the eigenvalues x_j of H. To help the reader's imagination he may think of a relation such as

$$S = e^{i\tau H} \quad \text{or} \quad S = \frac{1 - i\tau H}{1 + i\tau H}. \tag{9.0.1}$$

However, such a definite relation between S and H cannot be correct except in a limited range of energy. We will deliberately leave this relation vague because we are going to restrict the order of our matrices to $N \times N$, where N is very large but finite. And this cannot represent, say, a nucleus, for the real nucleus has an infinite number of energy levels. Like the Gaussian ensembles, the circular ensembles are gross mutilations of the overall actual situation. The most we can expect of such models is that in any energy region that is small compared to the total excitation energy the statistical distribution of the levels will be correctly reproduced. With no further apologies we make the following fundamental assumption:

The statistical behavior of n consecutive levels of an actual system, whenever n is small compared with the total number of levels, is the same as that of n consecutive angles $\theta_1, \theta_2, ..., \theta_n$, where n is small compared with N.

9.1. The Orthogonal Ensemble

According to the analysis of Chapter 2, a system having time-reversal invariance and rotational symmetry or having time-reversal invariance and integral spin will be characterized by a symmetric S. Following Dyson, we define the orthogonal circular ensemble E_{1c} of symmetric unitary matrices S by assigning the probabilities in the following way.

Every symmetric unitary S (*cf.* Appendix A.3) can be written as

$$S = U^T U, \tag{9.1.1}$$

where U is unitary. Define a small neighborhood of S by

$$S + dS = U^T (1 + idM) U, \tag{9.1.2}$$

where dM is a real symmetric matrix with elements dM_{ij} and the elements dM_{ij} for $i \leq j$ vary independently in some small intervals of length $d\mu_{ij}$. The "volume" of this neighborhood is defined by

$$\mu_1(dS) = \prod_{i \leq j} d\mu_{ij}. \tag{9.1.3}$$

The ensemble E_{1c} is defined by the statement: *The probability that a matrix from the ensemble* E_{1c} *lies between* S and $S + dS$ *is proportional to*

$$P(S)dS = \frac{1}{V_1} \mu_1(dS), \tag{9.1.4}$$

where V_1 is the total volume of the space T_{1c} of unitary symmetric matrices of order $N \times N$ and therefore, a normalization constant.

For this definition to have a meaning one must be sure that $\mu_1(dS)$ does not depend on the choice of U in Eq. (9.1.1). The fact that it is so can be easily verified. Let

$$S = U^T U = V^T V, \tag{9.1.5}$$

where both U and V are unitary. The matrix

$$R = VU^{-1} \tag{9.1.6}$$

is unitary and also satisfies

$$R^T R = \left(U^T\right)^{-1} V^T V U^{-1} = \left(U^T\right)^{-1} U^T U U^{-1} = 1 \tag{9.1.7}$$

Therefore, R is real and orthogonal. Let

$$\mu_1^{'}(dS) = \prod_{i \le j} d\mu_{ij}^{'} \tag{9.1.8}$$

be the volume derived from V as $\mu_1(dS)$ was derived from U. We now have

$$S + dS = V^T(1 + idM')V \tag{9.1.9}$$

with

$$dM' = R\, dM\, R^{-1}. \tag{9.1.10}$$

To prove that $\mu_1(dS) = \mu_1'(dS)$ we need to show that the Jacobian

$$J = \det \left[\frac{\partial \left(dM_{ij}^{'} \right)}{\partial \left(dM_{ij} \right)} \right] \tag{9.1.11}$$

has absolute value unity when dM, dM' are real symmetric matrices related by Eq. (9.1.10). A proof of this is given in Appendix A.22. Thus the volume $\mu_1(dS)$ is unique. Incidentally, we have established that for a fixed S the unitary matrix U in Eq. (9.1.1) is undetermined precisely to the extent of a transformation

$$U \longrightarrow RU, \tag{9.1.12}$$

where R is an arbitrary real orthogonal matrix.

The motivation for the choice of the ensemble E_{1c} will be made clearer by the following theorem.

Theorem 9.1.1. *The orthogonal ensemble E_{1c} is uniquely defined in the space of unitary symmetric matrices of order $N \times N$ by the property of being invariant under every automorphism*

$$S \longrightarrow W^T SW \tag{9.1.13}$$

of T_{1c} into itself where W is any $N \times N$ unitary matrix.

Theorem 9.1.1 comprises two statements: (a) *that E_{1c} is invariant under the automorphisms* (9.1.13) *and* (b) *that it is unique.* To prove (a) we suppose that a neighborhood $S + dS$ of S is transformed into a neighborhood $S' + dS'$ of S' by the automorphism (9.1.13). Equations (9.1.1) and (9.1.2) then hold and

$$S' = W^T SW = V^T V, \quad V = UW \tag{9.1.14}$$

$$S' + dS' = V^T(1 + idM)V. \tag{9.1.15}$$

The volumes $\mu_1(dS)$ and $\mu_1(dS')$ are then identical by definition as the same dM is occuring in Eqs. (9.1.2) and (9.1.15). This proof of (a) is trivial, for we could choose a convenient unitary matrix V in Eq. (9.1.14) and it was already shown that the value of $\mu_1(dS)$ does not depend on the choice of V. To prove (b) let E_1' be any ensemble invariant under Eq. (9.1.13). The probability density $P'(S)dS$ associated with E_1' will define a certain volume $\mu_1'(dS)$ of the neighborhood of S in the space T_{1c}. The ratio

$$\varphi(S) = \frac{\mu_1'(dS)}{\mu_1(dS)} \tag{9.1.16}$$

is a function of S defined on T_{1c} and invariant under the transformations (9.1.13). If $S = U^T U$, we choose $W = U^{-1}$ in Eq. (9.1.13) so that S is transformed to unity and therefore $\varphi(S)$ is a constant. Thus the probability densities in E_{1c} and E'_{1c} are proportional. Also they are both normalized to unity and are therefore identical.

9.2. Symplectic Ensemble

Next we consider systems with half-integral spin and time-reversal invariance but no rotational symmetry. In this section we use the quaternion notation developed in Chapter 2. The systems are described by self-dual unitary quaternion matrices (*cf.* Chapter 2)

$$S^R \equiv -ZS^T Z = S, \quad S^\dagger = S^{-1}. \tag{9.2.1}$$

Once again we have to assign a probability that a matrix chosen randomly from the space T_{4c} of self-dual unitary quaternion matrices of order $N \times N$ will lie between S and $S + dS$. This is done as follows.

Every matrix S in T_{4c} can be written as

$$S = U^R U, \tag{9.2.2}$$

where U is unitary. To see that this is possible, observe that in the ordinary language without quaternions SZ is an antisymmetric unitary matrix and can be reduced to the canonical form

$$SZ = VZV^T, \tag{9.2.3}$$

where V is unitary. Choosing $U = (ZV)^T$ then gives Eq. (9.2.2). (See also Appendix A.3.) For a given S, the unitary matrix U in Eq. (9.2.2) is precisely undetermined to the extent of a transformation

$$U \longrightarrow BU, \tag{9.2.4}$$

where B is an arbitrary symplectic matrix. For a proof it is sufficient to observe that the dual of a product of matrices is the product of their duals taken in the reverse order; and a symplectic matrix is, by definition, one that satisfies

$$B^R B = BB^R = 1. \tag{9.2.5}$$

A small neighborhood of S in T_{4c} is defined by

$$S + dS = U^R(1 + idM)U \tag{9.2.6}$$

where dM is a self-dual quaternion real matrix with elements

$$dM_{ij} = dM_{ij}^{(0)} + \sum_{\alpha=1}^{3} dM_{ij}^{(\alpha)} e_\alpha. \tag{9.2.7}$$

The real coefficients $dM_{ij}^{(\alpha)}$ satisfy

$$dM_{ij}^{(0)} = dM_{ji}^{(0)}, \quad dM_{ij}^{(\alpha)} = -dM_{ji}^{(\alpha)}, \quad \alpha = 1, 2, 3. \tag{9.2.8}$$

There are $N(2N-1)$ independent real variables $dM_{ij}^{(\alpha)}$ and they are allowed to vary over some small intervals of lengths $d\mu_{ij}^{(\alpha)}$. The neighborhood of S, thus defined, is assigned a volume

$$\mu_4(dS) = \prod_{i<j} \prod_{\alpha=1}^{3} d\mu_{ij}^{(\alpha)} \prod_{i \le j} d\mu_{ij}^{(0)}. \tag{9.2.9}$$

In terms of this volume the symplectic ensemble E_{4c} is defined in exactly the same way as E_{1c} was defined in terms of the volume (9.1.3). The statistical weight of the neighborhood dS in T_{4c} is

$$P(S)dS = \frac{1}{V_4} \mu_4(dS), \tag{9.2.10}$$

where V_4 is the total volume of the space T_{4c} of self-dual unitary quaternion matrices of order $N \times N$.

We can now repeat almost without change the arguments in Section 9.1. We must first prove that the volume $\mu_4(dS)$ is independent of the choice of U in Eq. (9.2.2). This involves showing that the Jacobian J has absolute value unity, where

$$J = \det \left[\frac{\partial \left(dM_{ij}^{\prime(\alpha)} \right)}{\partial \left(dM_{ij}^{(\alpha)} \right)} \right], \tag{9.2.11}$$

$$\mathrm{d}M' = B\, dM B^{-1}, \tag{9.2.12}$$

and B is symplectic. As before Appendix A.22 contains a proof of this.

The analog of Theorem 9.1.1 is the following.

Theorem 9.2.1. *The symplectic ensemble E_{4c} is uniquely defined in the space T_{4c} of self-dual unitary quaternion matrices of order $N \times N$ by the property of being invariant under every automorphism*

$$S \longrightarrow W^R S W \tag{9.2.13}$$

of T_{4c} into itself, where W is any $N \times N$ unitary quaternion matrix.

Theorem 9.2.1 can be proved by following word for word the proof of Theorem 9.1.1, the operation of transposition being replaced by that of taking the dual.

9.3. Unitary Ensemble

A system without time-reversal symmetry is associated with an arbitrary unitary matrix S not restricted to be symmetric or self-dual. A neighborhood of S in the space T_{2c} of all unitary $N \times N$ matrices is defined by

$$S + dS = U(1 + idM)V, \tag{9.3.1}$$

where U and V are any unitary matrices that satisfy the equation $S = UV$ and dM is an infinitesimal Hermitian matrix with elements $dM_{ij} = dM_{ij}^{(0)} + idM_{ij}^{(1)}$. The real components $dM_{ij}^{(0)}$, $dM_{ij}^{(1)}$ are N^2 in number and are allowed to vary independently over small intervals of lengths $d\mu_{ij}^{(0)}$, $d\mu_{ij}^{(1)}$. The volume $\mu_2(dS)$ is defined by the equation

$$\mu_2(dS) = \prod_{i \le j} d\mu_{ij}^{(0)} \prod_{i<j} d\mu_{ij}^{(1)} \tag{9.3.2}$$

and is independent of the choice of U and V. The ensemble E_{2c} gives to each neighborhood dS the statistical weight

$$P(S)dS = \frac{1}{V_2} \mu_2(dS), \tag{9.3.3}$$

where V_2 is the total volume of the space T_{2c}.

The invariance property of E_{2c}, analogous to Theorems 9.1.1 and 9.2.1, is stated in Theorem 9.3.1.

Theorem 9.3.1. *The unitary ensemble E_{2c} is uniquely defined in the space T_{2c} of all $N \times N$ unitary matrices by the property of being invariant under every automorphism*

$$S \longrightarrow USV \tag{9.3.4}$$

of T_{2c} into itself, where U and V are any two $N \times N$ unitary matrices.

This theorem merely expresses the well-known result that $\mu_2(dS)$ is the invariant group-measure of the N-dimensional unitary group $U(N)$.

9.4. The Joint Probability Density Function for the Eigenvalues

We give below a few lemmas which will be used subsequently. For their proofs see Appendix A.3.

Lemma 9.4.1.

Let S be any unitary symmetric $N \times N$ matrix. Then there exists a real orthogonal matrix R which diagonalizes S; that is

$$S = R^{-1}ER, \tag{9.4.1}$$

where E is diagonal. The diagonal elements of E are complex numbers $e^{i\theta_j}$ lying on the unit circle.

Lemma 9.4.2.

Let S be a unitary self-dual quaternion matrix of order $N \times N$. Then there exists a symplectic matrix B such that

$$S = B^{-1}EB, \tag{9.4.2}$$

where E is diagonal and scalar (cf. Section 2.4). The diagonal elements of E are N complex numbers $e^{i\theta_j}$ on the unit circle; each is repeated twice.

Lemma 9.4.3.

Let S be a unitary matrix. Then there exists a unitary matrix U such that

$$S = U^{-1}EU, \tag{9.4.3}$$

where E is diagonal. The diagonal elements of E are N complex numbers $e^{i\theta_j}$ on the unit circle.

Though this result is well known, we stated it here for completeness.

We are now in a position to prove the main result of this chapter.

Theorem 9.4.4.

In the ensemble $E_{\beta c}$ the probability of finding the eigenvalues $e^{i\phi_j}$ of S with an angle in each of the intervals $(\theta_j, \theta_j + d\theta_j)$; $j = 1, ..., N$ is given by

$$P_{N\beta}(\theta_1, ..., \theta_N)\, d\theta_1 \cdots d\theta_N, \tag{9.4.4}$$

where

$$P_{N\beta}(\theta_1, ..., \theta_N) = C'_{N\beta} \prod_{1 \le \ell < j \le N} \left| e^{i\theta_\ell} - e^{i\theta_j} \right|^\beta . \tag{9.4.5}$$

Here $\beta = 1$ for orthogonal, $\beta = 4$ for symplectic, and $\beta = 2$ for unitary circular ensembles. The constant $C'_{N\beta}$ is fixed by normalization.

Proof 1. Let $\beta = 1$. By Lemma 9.4.1 every S in T_{1c} can be diagonalized in the form

$$S = R^{-1}ER, \tag{9.4.6}$$

with R orthogonal. We now wish to express the volume $\mu_1(dS)$ in terms of the volumes $\mu(dE)$ and $\mu(dR)$, defined for the neighborhoods of the matrices E and R, respectively. A small neighborhood of E is given by

$$dE = iE\, d\theta, \tag{9.4.7}$$

where $d\theta$ means a diagonal matrix with elements $d\theta_1, ..., d\theta_N$. To find the neighborhood of R we differentiate

$$RR^T = 1, \tag{9.4.8}$$

thus getting

$$R(dR)^T + (dR)R^T = 0 \tag{9.4.9}$$

showing that the infinitesimal matrix

$$dA \equiv (dR)R^T = -R(dR)^T \tag{9.4.10}$$

is a real antisymmetric matrix with elements dA_{ij}. The volumes $\mu(dE)$ and $\mu(dR)$ are given by

$$\mu(dE) = \prod_{j=1}^{N} d\theta_j, \tag{9.4.11}$$

$$\mu(dR) = \prod_{1 \le \ell < j \le N} dA_{\ell j}. \tag{9.4.12}$$

The volume $\mu(dS)$ is defined by Eq. (9.1.3), where dM is given by Eq. (9.1.2) and U is any unitary matrix satisfying Eq. (9.1.1). Differentiating Eq. (9.4.6) and using Eqs. (9.1.1), (9.1.2), (9.4.7), (9.4.8), and (9.4.10), we obtain

$$iRU^T dMUR^{-1} = -dA\ E + iE\ d\theta + E\ dA, \tag{9.4.13}$$

which is the relation between $dM, d\theta$, and dA. Since E is a diagonal unitary matrix, it has a square root F which is also diagonal with elements $\pm e^{i\theta_j/2}$. There is an ambiguity in the sign of each element, but it does not matter how these signs are chosen. A convenient choice for U satisfying Eq. (9.1.1) is then

$$U = FR \tag{9.4.14}$$

by virtue of Eq. (9.4.6). With this choice of U, Eq. (9.4.13) reduces to

$$iFdMF = -dAF^2 + iF^2 d\theta + F^2 dA, \tag{9.4.15}$$

or

$$dM = d\theta + i\left(F^{-1}dAF - FdAF^{-1}\right). \tag{9.4.16}$$

Equation (9.4.16) gives $dM_{\ell j}$ in terms of the quantities $d\theta_j$, $dA_{\ell j}$, and θ_j for each pair of indices ℓ, j; namely

$$dM_{jj} = d\theta_j, \tag{9.4.17}$$

$$dM_{\ell j} = 2\,\sin\left[\frac{1}{2}\left(\theta_\ell - \theta_j\right)\right] dA_{\ell j}, \quad \ell \neq j. \tag{9.4.18}$$

Assembling the definitions (9.1.3), (9.4.11), and (9.4.12), we deduce from (9.4.17) and (9.4.18)

$$\begin{aligned} \mu(dS) &= \prod_{\ell<j}\left|2\,\sin\left(\frac{\theta_\ell-\theta_j}{2}\right)\right|\mu(dE)\mu(dR) \\ &= \prod_{\ell<j}\left|e^{i\theta_\ell} - e^{i\theta_j}\right|\mu(dE)\mu(dR). \end{aligned} \tag{9.4.19}$$

Now keep the angles $\theta_1, \ldots, \theta_N$ fixed and integrate Eq. (9.4.19) with respect to the parameters $dA_{\ell j}$ over the entire allowed range. This will give Eq. (9.4.4) with P_{N1} given by Eq. (9.4.5). Thus, the theorem is proved for the orthogonal case.

Proof 2. Next let $\beta = 4$. The matrix S is now diagonalized with the help of a symplectic matrix

$$S = B^{-1}EB \tag{9.4.20}$$

(Lemma 9.4.2). The infinitesimal matrix

$$dA = dB\ B^R \tag{9.4.21}$$

is quaternion real and anti-Hermitian. The components of dA are $dA_{\ell j}^{(\alpha)}$, which are real. They are antisymmetric in ℓ, j for $\alpha = 0$ and symmetric in ℓ, j for $\alpha = 1, 2$, and 3. The volume $\mu(dB)$ is now given by

$$\mu(dB) = \prod_{\ell,j,\alpha} dA_{\ell j}^{(\alpha)}. \tag{9.4.22}$$

The volume $\mu(dS)$ is given by Eq. (9.2.9) with dM given by Eq. (9.2.6). The matrix dM is Hermitian and quaternion real. The algebra leading up to Eq. (9.4.16) goes exactly as before. Equation (9.4.17) still holds, the diagonal elements dM_{jj} being real scalar quaternions with only one independent component. Equation (9.4.18) now holds separately for each of the four quaternion components $\alpha = 0, 1, 2, 3$. The equation analogous to Eq. (9.4.19) is

$$\mu(dS) = \left(\prod_{\ell<j} \left| e^{i\theta_\ell} - e^{i\theta_j} \right|^4 \right) \mu(dE) C, \tag{9.4.23}$$

where C does not depend on the θ_j. The power 4 in Eq. (9.4.23) arises from the fact that every nondiagonal element $dM_{\ell j}$, $\ell < j$, gives according to Eq. (9.4.18) four factors corresponding to the four components $\alpha = 0, 1, 2$ and 3. Note also that the C in Eq. (9.4.23) is not equal to $\mu(dB)$, for the diagonal components $dA_{jj}^{(\alpha)}$ with $\alpha = 1, 2, 3$ do not occur in Eq. (9.4.17), whereas $\mu(dB)$ contains their product. For our purposes it is sufficient to know that C does not depent on the θ_j.

The rest of the proof proceeds as in the case $\beta = 1$.

Proof 3. Lastly let $\beta = 2$. In this case (9.4.6) holds with a unitary R. The infinitesimal matrix dA is now anti-Hermitian, and the diagonal elements dA_{jj} are pure imaginary. The real part $dA_{\ell j}^{(0)}$ of the nondiagonal elements $dA_{\ell j}$ is antisymmetric in ℓ, j, whereas the imaginary part $dA_{\ell j}^{(1)}$ is symmetric in ℓ, j. Equation (9.4.17) holds in this case as well; Eq. (9.4.18) holds separately for the real and imaginary components $dM_{\ell j}^{(0)}$ and $dM_{\ell j}^{(1)}$ of the nondiagonal elements $dM_{\ell j}$. The equation analogous to Eq. (9.4.19) is therefore

$$\mu(dS) = \left\{ \prod_{\ell<j} \left| e^{i\theta_\ell} - e^{i\theta_j} \right|^2 \right\} \mu(dE) C, \tag{9.4.24}$$

where C does not depend on the θ_j. As in the case $\beta = 4$, this is sufficient for our purposes.

The theorem is thus established for all three cases.

As in the Gaussian ensembles, it is possible to account for the different powers of the product of differences appearing in Eqs. (9.4.19), (9.4.23), and (9.4.24) by a dimensional argument. The dimension of the space T_{1c} of all symmetric unitary matrices is $N(N+1)/2$, whereas the dimension of the subspace T'_{1c} composed of all symmetric unitary matrices with two equal eigenvalues is $N(N+1)/2 - 2$ (see Appendix A.4). Because of the single restriction, the equality of two eigenvalues, the dimension should normally have decreased by one; as it is decreased by 2, it indicates a factor in Eq. (9.4.19) linear in $|\exp(i\theta_j) - \exp(i\theta_k)|$. Similarly, for unitary matrices the dimension decreases from N^2 to $N^2 - 3$ on equating two eigenvalues, and for quaternion self-dual unitary matrices it decreases from $N(2N-1)$ to $N(2N-1) - 5$. See the end of Section 3.3 and Appendix A.4.

Summary of Chapter 9

Three ensembles of unitary matrices S are considered; (i) that of unitary symmetric matrices S invariant under the transformations $S \to W^T SW$, known as the orthogonal (circular) ensemble; (ii) that of unitary matrices S invariant under the transformations $S \to USV$, known as the unitary (circular) ensemble; and (iii) that of self-dual unitary matrices S invariant under the transformations $S \to W^R SW$, known as the symplectic (circular) ensemble. Here W, U and V are any unitary matrices, W^T is the transpose of W, and W^R is the dual of W.

The joint probability density of the eigenvalues $\exp(i\theta_j)$, $j = 1, ..., N$, is found to be

$$P(\theta_1, ..., \theta_N) \propto \prod_{1 \le j < k \le N} |\exp(i\theta_j) - \exp(i\theta_k)|^\beta , \tag{9.4.5}$$

where $\beta = 1$ for the orthogonal, $\beta = 2$ for the unitary and $\beta = 4$ for the symplectic (circular) ensemble.

10 / Circular Ensembles (Continued)

The joint probability density function for the eigenvalues of a matrix taken from one of the three circular ensembles (orthogonal, unitary or symplectic) was derived in Chapter 9. We will now evaluate their correlation and cluster functions. For this purpose we will again use the theory of quaternion matrices. In the limit of large N, these will be seen to be identical to those for the corresponding Gaussian ensembles.

It will be shown that alternate eigenvalues of a matrix from the orthogonal ensemble behave as if they were the eigenvalues of a matrix from the symplectic ensemble. This will allow us in Section 10.7 to deduce the spacing probabilities for the symplectic ensemble from those of the orthogonal ensemble.

As for the Gaussian ensembles, a Brownian motion model can be constructed for the circular ensembles as well. Finally, for historical reasons, we will give an old method of integration for the circular orthogonal ensemble.

10.1. Unitary Ensemble. Correlation and Cluster Functions

Systems with no time reversal invariance may be characterized by unitary random matrices. The joint probability density function for the eigenvalue angles of such matrices taken from the unitary ensemble (*cf.* Chapter 9, Theorem 9.4.4) is

$$P_{N2}(\theta_1, ..., \theta_N) = \text{const} \times \prod_{j<k} \left| e^{i\theta_j} - e^{i\theta_k} \right|^2 . \qquad (10.1.1)$$

Calculating averages with the probability density P_{N2} is mathematically the simplest. One may write

$$\prod_{j<k} \left| e^{i\theta_j} - e^{i\theta_k} \right|^2 = \prod_{j<k} \left(e^{i\theta_j} - e^{i\theta_k} \right) \prod_{j<k} \left(e^{-i\theta_j} - e^{-i\theta_k} \right)$$

$$= \det \left[e^{ip\theta_j} \right] \det \left[e^{-ip\theta_j} \right], \tag{10.1.2}$$

$$p = \frac{1}{2}(1 - N),\ \frac{1}{2}(3 - N),\ \ldots,\ \frac{1}{2}(N - 3),\ \frac{1}{2}(N - 1); \tag{10.1.3}$$

$$j, k = 1,\ 2,\ \ldots,\ N. \tag{10.1.4}$$

Or

$$\prod_{j<k} \left| e^{i\theta_j} - e^{i\theta_k} \right|^2 = (2\pi)^N \det \left[S_N(\theta_j - \theta_k) \right]_{j,k=1,\ldots,N}, \tag{10.1.5}$$

with

$$S_N(\theta) = \frac{1}{2\pi} \sum_p e^{ip\theta} = \frac{1}{2\pi} \frac{\sin(N\theta/2)}{\sin(\theta/2)}, \tag{10.1.6}$$

the sum on p being over the values as in Eq. (10.1.3).

Now

$$\int_0^{2\pi} S_N(0) d\theta = N, \tag{10.1.7}$$

and

$$\int_0^{2\pi} S_N(\theta_j - \theta_k) S_N(\theta_k - \theta_\ell) d\theta_k = S_N(\theta_j - \theta_\ell), \tag{10.1.8}$$

so that all the conditions of Theorem 5.2.1 are satisfied.

Integrating over all the variables, and using that theorem, we fix the constants,

$$P_{N2}(\theta_1, \ldots, \theta_N) = \frac{1}{N!} \det \left[S_N(\theta_j - \theta_k) \right]_{j,k=1,\ldots,N}. \tag{10.1.9}$$

According to the same Theorem 5.2.1, the n-level correlation function is

$$R_n(\theta_1, ..., \theta_n) = \det \left[S_N(\theta_j - \theta_k) \right]_{j,k=1,\ldots,n} . \tag{10.1.10}$$

The level density is

$$R_1(\theta) = S_N(0) = \frac{N}{2\pi}, \tag{10.1.11}$$

as it should be. The two-level correlation function is

$$R_2(\theta, \varphi) = \left[S_N(0) \right]^2 - \left[S_N(\theta - \varphi) \right]^2, \tag{10.1.12}$$

and the two-level cluster function is

$$T_2(\theta - \varphi) = \left[S_N(\theta - \varphi) \right]^2 = \left(\frac{\sin(N(\theta - \varphi)/2)}{2\pi \, \sin((\theta - \varphi)/2)} \right)^2 . \tag{10.1.13}$$

On taking the limit $N \to \infty$, while keeping $N\theta = 2\pi\xi$ and $N\varphi = 2\pi\eta$ finite, we have

$$Y_2(\xi, \eta) = \lim \left(\frac{2\pi}{N} \right)^2 T_2(\theta, \varphi) = \left(\frac{\sin(\pi r)}{\pi r} \right)^2 , \tag{10.1.14}$$

$$r = |\xi - \eta| . \tag{10.1.15}$$

This result is identical to the two-point cluster function for the case of the unitary Gaussian ensemble, Eq. (5.2.20).

More generally, in the limit $N \to \infty$,

$$\lim \frac{2\pi}{N} S_N(\theta - \varphi) = \frac{\sin(\pi r)}{\pi r} \equiv s(r), \tag{10.1.16}$$

so that in this limit the n-level correlation and cluster functions for any finite n are identical to those for the Gaussian unitary ensemble; Section 5.2.

10.2. Unitary Ensemble. Level Spacings

For the spacing distribution one first finds the probability $A_2(\alpha)$ that a randomly chosen interval of length 2α will contain none of the angles θ_j:

$$A_2(\alpha) = \int_{\alpha}^{2\pi-\alpha} \cdots \int P_{N2}(\theta_1, ..., \theta_N) d\theta_1 \cdots d\theta_N \tag{10.2.1}$$

$$= \text{const} \times \int_{\alpha}^{2\pi-\alpha} \cdots \int \left|\det\left[e^{ip\theta_j}\right]\right|^2 d\theta_1 \cdots \theta_N.$$

$$= \text{const} \times \int_{\alpha}^{2\pi-\alpha} \cdots \int \det\left[\frac{1}{2\pi}\sum_{j=1}^{N} e^{i(p-q)\theta_j}\right] d\theta_1 \cdots d\theta_N \tag{10.2.2}$$

from (10.1.2), where p and q take values as in (10.1.3); or (*cf.* Appendix A.12)

$$A_2(\alpha) = \det\left[\frac{1}{2\pi}\int_{\alpha}^{2\pi-\alpha} d\theta e^{i(p-q)\theta}\right] \tag{10.2.3}$$

$$= \det\left[\delta_{pq} - \frac{1}{2\pi}\int_{-\alpha}^{\alpha} d\theta \cos(p-q)\theta\right]. \tag{10.2.4}$$

The constant was fixed by the requirement that $A_2(0) = 1$. Since

$$\sum_q \left[\delta_{pq} - \cos(p\theta)\cos(q\theta)\right]\left[\delta_{qr} - \sin(q\theta)\sin(r\theta)\right]$$
$$= \delta_{pr} - \cos(p-r)\theta, \tag{10.2.5}$$

we may factorize the expression (10.2.4),

$$A_2(\alpha) = A_1(\alpha) A_1'(\alpha), \tag{10.2.6}$$

with

$$
\begin{aligned}
A_1(\alpha) &= \det\left[\delta_{pq} - \frac{1}{2\pi}\int_{-\alpha}^{\alpha} d\theta \cos(p\theta)\cos(q\theta)\right] \\
&= \det\left[\delta_{pq} - \frac{1}{2\pi}\left(\frac{\sin(p-q)\alpha}{p-q} + \frac{\sin(p+q)\alpha}{p+q}\right)\right], \qquad (10.2.7)
\end{aligned}
$$

$$
\begin{aligned}
A_1'(\alpha) &= \det\left[\delta_{pq} - \frac{1}{2\pi}\int_{-\alpha}^{\alpha} d\theta \sin(p\theta)\sin(q\theta)\right] \\
&= \det\left[\delta_{pq} - \frac{1}{2\pi}\left(\frac{\sin(p-q)\alpha}{p-q} - \frac{\sin(p+q)\alpha}{p+q}\right)\right], \qquad (10.2.8)
\end{aligned}
$$

$$
p, q = -\left(m - \frac{1}{2}\right), -\left(m - \frac{3}{2}\right), \ldots, \left(m - \frac{1}{2}\right), \qquad (10.2.9)
$$

where we have taken $N = 2m$, even.

In the limit $N \to \infty$, $N\alpha = 2\pi t$, $Np\alpha = 2\pi\xi t$, $Nq\alpha = 2\pi\eta t$, all finite, $A_1(\alpha)$ becomes $B_1(t)$, the Fredholm determinant of the integral equation

$$
\lambda f(\xi) = \int_{-1}^{1} \bar{Q}(\xi, \eta) f(\eta) d\eta, \qquad (10.2.10)
$$

where

$$
\bar{Q}(\xi, \eta) = \frac{1}{2}\left[Q(\xi, \eta) + Q(\xi, -\eta)\right], \qquad (10.2.11)
$$

is the even part of

$$
Q(\xi, \eta) = \frac{\sin(\xi - \eta)\pi t}{(\xi - \eta)\pi}. \qquad (10.2.12)
$$

In other words,

$$
A_1(\alpha) \to B_1(t) = \prod_{i=0}^{\infty} (1 - \lambda_{2i}), \qquad (10.2.13)
$$

where λ_{2i} are the eigenvalues of the integral equation (5.3.14) corresponding to the even solutions.

Similarly,

$$
A_1'(\alpha) \to B_1'(t) = \prod_{i=0}^{\infty} (1 - \lambda_{2i+1}), \qquad (10.2.14)
$$

where λ_{2i+1} are the eigenvalues of the same integral equation (5.3.14) corresponding to the odd solutions.

Thus in the limit of large N,

$$A_2(\alpha) \to E_2(0;s) = E_1(0;s)E_1'(s) = \prod_{i=0}^{\infty}(1-\lambda_i), \tag{10.2.15}$$

which is identical to the spacing distribution for the Gaussian unitary ensemble.

This result was expected. Actually, the limiting n-point correlation function for any finite n being identical for the circular and Gaussian unitary ensembles, all their statistical properties involving a finite number of levels will coincide in the limit $N \to \infty$.

10.3. Orthogonal Ensemble. Correlation and Cluster Functions

The orthogonal ensemble is the most important from a practical point of view. It is also one of the most complicated from the mathematical point of view. We can use the identity

$$\left|e^{i\theta_j} - e^{i\theta_k}\right| = i^{-1}\exp\left(-\frac{i}{2}(\theta_j+\theta_k)\right)\left(e^{i\theta_k} - e^{i\theta_j}\right), \quad \text{if } \theta_j \le \theta_k, \tag{10.3.1}$$

to write

$$\prod_{1\le j<k\le N}\left|e^{i\theta_j} - e^{i\theta_k}\right| = i^{-N(N-1)/2}\exp\left(-\frac{i}{2}(N-1)\sum_1^N \theta_j\right) \times \prod_{1\le j<k\le N}\left(e^{i\theta_k} - e^{i\theta_j}\right), \tag{10.3.2}$$

where the θ_j are supposed to be ordered

$$-\pi \le \theta_1 \le \theta_2 \le \cdots \le \theta_N \le \pi. \tag{10.3.3}$$

Writing the product of differences in Eq. (10.3.2) as a Vandermonde determinant and multiplying the column containing the powers of $e^{i\theta_j}$ by $\exp[-i(N-1)\theta_j/2]$, we have

$$\prod_{1\le j<k\le N} \left|e^{i\theta_j}-e^{i\theta_k}\right| = i^{-N(N-1)/2}\det\left[e^{ip\theta_j}\right], \tag{10.3.4}$$

$$\begin{aligned} p &= \frac{1}{2}(1-N),\ \frac{1}{2}(3-N),\ \ldots,\ \frac{1}{2}(N-3),\ \frac{1}{2}(N-1); \\ j &= 1,2,\ldots,N; \end{aligned} \tag{10.3.5}$$

if the θ_j are ordered as in Eq. (10.3.3).

We want to write Eq. (10.3.4) as an $N\times N$ quaternion determinant. For this purpose let $S_N(\theta)$ be as in Eq. (10.1.6), and let

$$DS_N(\theta) = \frac{d}{d\theta}S_N(\theta) = \frac{1}{2\pi}\sum_p ipe^{ip\theta}, \tag{10.3.6}$$

$$IS_N(\theta) = \int_0^\theta S_N(\varphi)d\varphi, \tag{10.3.7}$$

so that

$$IS_N(\theta) = \frac{1}{2\pi i}\sum_p p^{-1}e^{ip\theta}, \qquad N \text{ even}, \tag{10.3.8}$$

$$= \frac{1}{2\pi i}{\sum_p}' p^{-1}e^{ip\theta} + \frac{\theta}{2\pi}, \qquad N \text{ odd} \tag{10.3.9}$$

the term $p=0$ being excluded in the last sum. In addition, let

$$JS_N(\theta) = -\frac{1}{2\pi i}\sum_q q^{-1}e^{iq\theta}, \tag{10.3.10}$$

where q takes the values

$$q = \pm\frac{1}{2}(N+1),\ \pm\frac{1}{2}(N+3),\ldots\,. \tag{10.3.11}$$

Then

$$IS_N(\theta) - JS_N(\theta) = \varepsilon_N(\theta) \tag{10.3.12}$$

is a step function whose character depends only on the parity of N. In fact, for any integer m with

$$2\pi m < \theta < 2\pi(m+1), \tag{10.3.13}$$

we have

$$\varepsilon_N(\theta) = \frac{1}{2}(-1)^m, \quad N \text{ even}; \qquad \varepsilon_N(\theta) = m + \frac{1}{2}, \quad N \text{ odd}. \tag{10.3.14}$$

At the points of discontinuity $\theta = 2\pi m$,

$$\varepsilon_N(\theta) = 0, \quad N \text{ even}; \qquad \varepsilon_N(\theta) = m, \quad N \text{ odd}. \tag{10.3.15}$$

The lack of uniform convergence of the series defining JS_N will not cause any difficulty. The function $S_N(\theta)$ is even in θ, while DS_N, IS_N, JS_N and ε_N are all odd in θ.

Define the quaternion

$$\sigma_{N1}(\theta) = \begin{bmatrix} S_N(\theta) & DS_N(\theta) \\ JS_N(\theta) & S_N(\theta) \end{bmatrix}, \tag{10.3.16}$$

and the $N \times N$ quaternion matrix

$$Q_{N1}(\theta_1, ..., \theta_N) = [\sigma_{N1}(\theta_j - \theta_k)]_{j,k=1,...,N}\,. \tag{10.3.17}$$

We will now verify that $P_{N1}(\theta_1, ..., \theta_N)$, Eq. (10.3.4), is proportional to det Q_{N1}, or that det $[C(Q_{N1})]$ is proportional to the square of $P_{N1}(\theta_1, ..., \theta_N)$ (*cf.* Section 6.2). Moreover, we will also verify that this Q_{N1} satisfies all the conditions of Theorem 6.2.1, allowing us to write down immediately the n-level correlation and cluster functions. We treat separately the cases N even and N odd. These considerations parallel those of Chapter 6.

10.3.1. THE CASE $N = 2m$, EVEN

The $2N \times 2N$ matrix

$$\begin{bmatrix} S_N(\theta_j - \theta_k) & DS_N(\theta_j - \theta_k) \\ IS_N(\theta_j - \theta_k) & S_N(\theta_j - \theta_k) \end{bmatrix} \tag{10.3.18}$$

is the product of two matrices

$$\begin{bmatrix} e^{ip\theta_j} \\ (ip)^{-1}e^{ip\theta_j} \end{bmatrix} \quad \text{and} \quad \frac{1}{2\pi}\left[e^{-ip\theta_k} \qquad ipe^{-ip\theta_k}\right], \tag{10.3.19}$$

of orders $2N \times N$ and $N \times 2N$, respectively. The rank of the matrix in Eq. (10.3.18) is thus N. The N rows

$$[\,IS_N(\theta_j - \theta_k) \qquad S_N(\theta_j - \theta_k)\,] \tag{10.3.20}$$

are linear combinations of the N rows

$$[\,S_N(\theta_j - \theta_k) \qquad DS_N(\theta_j - \theta_k)\,]\,. \tag{10.3.21}$$

Therefore, the determinant of $C(Q_{N1})$ is not changed when we subtract the rows (10.3.20) from the corresponding rows of $C(Q_{N1})$. This gives

$$\det\left[C(Q_{N1})\right] = \det\left[DS_N(\theta_j - \theta_k)\right]\det\left[\varepsilon_N(\theta_j - \theta_k)\right]. \tag{10.3.22}$$

Now by virtue of Eqs. (10.3.4) and (10.3.6),

$$\begin{aligned} \det\left[DS_N(\theta_j - \theta_k)\right] &= (2\pi)^{-N}\det\left[e^{ip\theta_j}\right]\det\left[ipe^{-ip\theta_j}\right] \\ &= \text{const} \times \left[P_{N1}(\theta_1, ..., \theta_N)\right]^2. \end{aligned} \tag{10.3.23}$$

The second determinant in (10.3.22), $\det\left[\varepsilon_N(\theta_j - \theta_k)\right]$ is (i) piecewise constant with possible discontinuities only at places where $\theta_j - \theta_k = 2\pi m$ with integer m, (ii) periodic with period 2π in each variable θ_j, and (iii) a symmetric function of $(\theta_1, ..., \theta_N)$. It follows from these three properties that this determinant is a constant independent of $\theta_1, ..., \theta_N$, except at the points of discontinuity where $P_{N1}(\theta_1, ..., \theta_N) = 0$.

10.3.2. THE CASE $N = 2m+1$, ODD

In this case zero appears as a value of p in Eq. (10.3.19). Replacing

$$(ip)^{-1}e^{ip\theta_j} \to \delta^{-1}, \qquad ipe^{ip\theta_k} \to \delta, \tag{10.3.24}$$

when $p = 0$, consider now instead of Eq. (10.3.18), the matrix

$$\begin{bmatrix} S_N(\theta_j - \theta_k) & DS_N(\theta_j - \theta_k) + \delta/2\pi \\ IS_N(\theta_j - \theta_k) + (2\pi\delta)^{-1} - \dfrac{\theta_j - \theta_k}{2\pi} & S_N(\theta_j - \theta_k) \end{bmatrix}, \tag{10.3.25}$$

of rank N. The determinant of

$$C_\delta(Q_{N1}) = \begin{bmatrix} S_N(\theta_j - \theta_k) & DS_N(\theta_j - \theta_k) + \delta/2\pi \\ JS_N(\theta_j - \theta_k) & S_N(\theta_j - \theta_k) \end{bmatrix} \tag{10.3.26}$$

is not changed by subtracting the N rows of Eq. (10.3.25) from the corresponding N rows of Eq. (10.3.26). Therefore,

$$\begin{aligned} \det\left[C_\delta(Q_{N1})\right] &= \det\left[DS_N(\theta_j - \theta_k) + \delta/2\pi\right] \\ &\quad \times \det\left[\varepsilon_N(\theta_j - \theta_k) + (2\pi\delta)^{-1} + (\theta_k - \theta_j)/(2\pi)\right]. \end{aligned} \tag{10.3.27}$$

The first factor is

$$\text{const} \times \delta \left(\det\left[e^{ip\theta_j}\right]\right)^2. \tag{10.3.28}$$

In the second factor we subtract the first column from each of the remaining columns, obtaining

$$(2\pi\delta)^{-1}\det\left[1_N + O(\delta),\ \varepsilon_N(\theta_j - \theta_k) - \varepsilon_N(\theta_j - \theta_1) + ((\theta_k - \theta_1)/2\pi)\right], \tag{10.3.29}$$

where 1_N means a single column of unit elements, and k labels the remaining columns from 2 to N. Passing to the limit $\delta \to 0$ in Eqs. (10.3.27), (10.3.28), and (10.3.29),

$$\det\left[C(Q_{N1})\right] = \text{const} \times \left(\det\left[e^{ip\theta_j}\right]\right)^2 d_N, \tag{10.3.30}$$

where now

$$d_N = \det\left[1_N, \quad \varepsilon_N(\theta_j - \theta_k) - \varepsilon_N(\theta_j - \theta_1)\right]. \tag{10.3.31}$$

The terms $(\theta_k - \theta_1)/2\pi$ in Eq. (10.3.29) contributing nothing to the determinant. By the same argument as was used for N even; d_N is independent of $\theta_1, ..., \theta_N$ except at places where $P_{N1}(\theta_1, ..., \theta_N) = 0$.

10.3.3. CONDITIONS OF THEOREM 6.2.1

From the parity of the functions S_N, DS_N, and JS_N and Eq. (2.4.17), it is evident that Q_{N1} is self-dual. Adopting the notation

$$(f * g)(\theta) = \int_0^{2\pi} f(\theta - \varphi) g(\varphi) d\varphi, \tag{10.3.32}$$

it is straightforward to verify that

$$S_N * S_N = S_N, \tag{10.3.33}$$

$$DS_N * S_N = S_N * DS_N = DS_N, \tag{10.3.34}$$

$$JS_N * S_N = S_N * JS_N = 0, \tag{10.3.35}$$

$$JS_N * DS_N = DS_N * JS_N = 0. \tag{10.3.36}$$

Hence

$$\sigma_{N1} * \sigma_{N1} = \sigma_{N1} + \lambda \sigma_{N1} - \sigma_{N1} \lambda, \tag{10.3.37}$$

with

$$\lambda = \frac{1}{2} \begin{bmatrix} 1 & 0 \\ 0 & -1 \end{bmatrix} = -\frac{i}{2} e_1. \tag{10.3.38}$$

Also

$$\int_0^{2\pi} \sigma_{N1}(0) d\theta = \int_0^{2\pi} S_N(0) d\theta = N. \tag{10.3.39}$$

Thus, all the conditions of Theorem 6.2.1 are satisfied. Integrating over all the variables, one can fix the constants:

$$P_{N1}(\theta_1, ..., \theta_N) = \frac{1}{N!} \det \left[\sigma_{N1}(\theta_j - \theta_k) \right]_{j,k=1,...,N}. \tag{10.3.40}$$

10.3.4. CORRELATION AND CLUSTER FUNCTIONS

As in Chapter 6, Theorem 6.2.1 now gives us with Eq. (10.3.40) the n-level correlation function

$$R_n(\theta_1, ..., \theta_n) = \det\left[\sigma_{N1}(\theta_j - \theta_k)\right]_{j,k=1,\ldots,n}, \tag{10.3.41}$$

and the n-level cluster function

$$T_n(\theta_1, ..., \theta_n) = \sum_P \sigma_{N1}(\theta_1 - \theta_2)\sigma_{N1}(\theta_2 - \theta_3)\cdots\sigma_{N1}(\theta_n - \theta_1), \tag{10.3.42}$$

the sum is taken over all $(n-1)!$ distinct cyclic permutations of the indices $(1, ..., n)$.

The level density is

$$R_1(\theta) = \sigma_{N1}(0) = S_N(0) = \frac{N}{2\pi}, \tag{10.3.43}$$

as was expected. In the limit $N \to \infty$, $N\theta = 2\pi\xi$, $N\varphi = 2\pi\eta$ finite, one gets

$$\lim \frac{2\pi}{N} S_N(\theta - \varphi) = s(r) = \frac{\sin(\pi r)}{\pi r}, \tag{10.3.44}$$

$$\lim \left(\frac{2\pi}{N}\right)^2 DS_N(\theta - \varphi) = Ds(r) = \frac{d}{dr}s(r), \tag{10.3.45}$$

$$\lim IS_N(\theta - \varphi) = Is(r) = \int_0^r s(t)dt, \tag{10.3.46}$$

$$\lim JS_N(\theta - \varphi) = Js(r) = Is(r) - \frac{1}{2}, \tag{10.3.47}$$

$$\lim \frac{2\pi}{N}\sigma_{N1}(\theta - \varphi) \approx \sigma_1(r) = \begin{bmatrix} s(r) & Ds(r) \\ Js(r) & s(r) \end{bmatrix}. \tag{10.3.48}$$

Thus in the limit of large N, the n-level correlation and cluster functions for the circular orthogonal ensemble are identical to those for the Gaussian orthogonal ensemble.

10.4. Orthogonal Ensemble. Level Spacings

To avoid unnecessary complications let us take N even, $N = 2m$. Let $u(\theta)$, $v(\theta)$ be functions defined over $(-\pi, \pi)$ and consider the average

$$\begin{aligned} H &= \left\langle \prod_{\text{alt}} u(\theta_j) \prod_{\text{alt}}{}' v(\theta_k) \right\rangle \\ &= \int_0^{2\pi} \cdots \int \prod_{\text{alt}} u(\theta_j) \prod_{\text{alt}}{}' v(\theta_k) P_{N1}(\theta_1, ..., \theta_N) d\theta_1 ... d\theta_N, \end{aligned} \tag{10.4.1}$$

taken with respect to the orthogonal ensemble $\beta = 1$, defined by Theorem 9.4.4. Here $\prod_{\text{alt}}$ denotes a product taken over a set of m alternate points θ_j as they lie on the unit circle and $\prod_{\text{alt}}'$ a product over the remaining m points. This average can again be calculated by integration over alternate variables, using Eq. (10.3.4). We define

$$f'_{pq} = \iint_{-\pi}^{\pi} u(\theta) v(\varphi) \varepsilon(\theta - \varphi) \left(e^{ip\varphi + iq\theta} - e^{ip\theta + iq\varphi} \right) d\theta d\varphi, \tag{10.4.2}$$

with

$$\varepsilon(\theta) = \frac{1}{2} \text{sgn}(\theta) = \begin{cases} |\theta| / (2\theta), & \theta \neq 0, \\ 0, & \text{otherwise.} \end{cases} \tag{10.4.3}$$

and do integrations step by step as in Section 6.5. The result is

$$H^2 = \text{const} \times \det \left[f'_{pq} \right], \quad p, q = -m + \frac{1}{2}, -m + \frac{3}{2}, ..., m - \frac{1}{2}. \tag{10.4.4}$$

By reversing the order of the columns we can write

$$H^2 = \text{const} \times \det \left[f_{pq} \right], \tag{10.4.5}$$

with

$$f_{pq} = \frac{ip}{4\pi} f'_{p,-q}$$

$$= \frac{ip}{4\pi} \int\int_{-\pi}^{\pi} u(\theta)v(\varphi)\varepsilon(\theta-\varphi)\left(e^{ip\varphi-iq\theta} - e^{ip\theta-iq\varphi}\right) d\theta d\varphi. \quad (10.4.6)$$

If the functions $u(\theta)$ and $v(\varphi)$ satisfy the relation

$$u(-\theta)v(-\varphi) = u(\theta)v(\varphi), \quad (10.4.7)$$

then

$$f_{-p,-q} = f_{p,q} \quad (10.4.8)$$

and there are further simplifications. As can be easily verified, we now have

$$\det\left[f_{p,q} + f_{-p,q}\right] = \det\left[f_{p,q} - f_{-p,q}\right], \quad p,q = \frac{1}{2}, \frac{3}{2}, \ldots, m - \frac{1}{2}, \quad (10.4.9)$$

and H may be written as a determinant

$$H = \text{const} \times \det\left[F_{pq}\right], \quad p,q = \frac{1}{2}, \frac{3}{2}, \ldots, m - \frac{1}{2}; \quad (10.4.10)$$

with

$$F_{pq} = f_{p,q} + f_{-p,q}$$

$$= \frac{p}{\pi} \int\int_{-\pi}^{\pi} u(\theta)v(\varphi)\varepsilon(\theta-\varphi)$$

$$\times \left[\cos(p\varphi)\sin(q\theta) - \cos(p\theta)\sin(q\varphi)\right] d\theta d\varphi. \quad (10.4.11)$$

The constant in Eq. (10.4.10) above can be fixed to be 1 by the condition that when $u(\theta) = v(\varphi) = 1$, then $H = 1$.

To get the probability that a randomly chosen interval of length 2α is empty of eigenvalues, put

$$u(\theta) = v(\theta) = \begin{cases} 1, & \text{if } \; -\pi + \alpha < \theta < \pi - \alpha, \\ 0, & \text{otherwise,} \end{cases} \tag{10.4.12}$$

in Eq. (10.4.1). We have chosen the center of the excluded interval to be at $\theta = \pi$ so that Eq. (10.4.7) is satisfied. From Eqs. (10.4.10) and (10.4.11) we then have

$$A_1(\alpha) = \det\left[F_{pq}\right], \quad p, q = \frac{1}{2}, \frac{3}{2}, \ldots, m - \frac{1}{2}, \tag{10.4.13}$$

where

$$\begin{aligned} F_{pq} &= \delta_{pq} - \frac{1}{\pi}\int_{-\alpha}^{\alpha} \cos(p\theta)\cos(q\theta)d\theta \\ &= \delta_{pq} - \frac{\sin(p-q)\alpha}{(p-q)\pi} - \frac{\sin(p+q)\alpha}{(p+q)\pi}. \end{aligned} \tag{10.4.14}$$

If we choose

$$\begin{aligned} &u(\theta) = v(\theta) = 1, && \text{if } \; -\pi + \alpha < \theta < \pi - \alpha, \\ &u(\theta) = 0, \;\; v(\theta) = 2, && \text{if } \; \pi - \alpha < \theta < \pi + \alpha, \end{aligned} \tag{10.4.15}$$

in Eq. (10.4.1), then we get the probability that a randomly chosen interval of length 2α will contain at most one eigenvalue. The choice $v(\theta) = 2$, rather than 1, in the interval $(\pi - \alpha, \pi + \alpha)$ arises from the fact that while ordering $-\pi \le \theta_1 \le \cdots \le \theta_{2m} \le \pi$, the interval $(\pi - \alpha, \pi + \alpha)$ becomes unattainable for half the levels belonging to the alternate series. Equations (10.4.10) and (10.4.11) then give

$$A_1'(\alpha) = \det\left[F_{pq}'\right], \quad p, q = \frac{1}{2}, \frac{3}{2}, \ldots, m - \frac{1}{2}; \tag{10.4.16}$$

$$\begin{aligned} F_{pq}' &= \delta_{pq} - \frac{1}{\pi}\int_{-\alpha}^{\alpha} \sin(p\theta)\sin(q\theta)d\theta \\ &= \delta_{pq} - \frac{\sin(p-q)\alpha}{(p-q)\pi} + \frac{\sin(p+q)\alpha}{(p+q)\pi}. \end{aligned} \tag{10.4.17}$$

When $m \to \infty$, while $m\alpha = \pi t$ remains finite, the limits of $A_1(\alpha)$ and $A_1'(\alpha)$ are $B_1(t)$ and $B_1'(t)$, respectively. These limits are obtained exactly as in Section 10.2.

Note that although $B_1'(t)$ is the probability that a randomly chosen interval of length $2t$ will not contain any of the eigenvalues belonging to the same alternate series, its second derivative is not a probability. The probability density for spacings between pairs of next nearest neighbours (i.e., pairs of levels having one level in between) is given instead (*cf.* Appendix A.8) by

$$p_1(1;s) = \frac{d^2}{ds^2}\left(B_1(t) + B_1'(t)\right), \quad s = 2t, \tag{10.4.18}$$

where $B_1(t) = E_1(0;s)$, the probability that the interval of length $2t$ will contain none of the eigenvalues, is given by Eq. (6.5.19) or Eq. (10.2.13) and $B_1'(t)$ is given by Eq. (10.2.14). Comparing with Eqs. (5.1.18), (6.5.19), and (6.7.21) we get the identity

$$B_1'(t) - B_1(t) = \prod_{\rho=0}^{\infty}(1-\lambda_{2\rho})\sum_{i=0}^{\infty}\frac{\lambda_{2i}}{1-\lambda_{2i}}f_{2i}(1)\int_{-1}^{1}f_{2i}(y)dy, \tag{10.4.19}$$

or with Eqs. (10.2.13) and (10.2.14),

$$\prod_{i=0}^{\infty}\left(\frac{1-\lambda_{2i+1}}{1-\lambda_{2i}}\right) = 1 + \sum_{i=0}^{\infty}\frac{\lambda_{2i}}{1-\lambda_{2i}}f_{2i}(1)\int_{-1}^{1}f_{2i}(y)dy, \tag{10.4.20}$$

where $f_i(x)$ are the normalized spheroidal functions defined by Eqs. (5.3.22) and (5.4.23). A direct proof of Eq. (10.4.20) by Gaudin is given in Appendix A.16, Eq. (A.16.6).

Proceeding as in Section 6.8, we can find lower and upper bounds to $B_1'(t)$,

$$\exp\left(-\left(\frac{\pi t}{2}\right)^2\right) \le B_1'(t) \le \left(1+\left(\frac{\pi t}{2}\right)^2\right)\exp\left(-\left(\frac{\pi t}{2}\right)^2\right). \tag{10.4.21}$$

10.5. Symplectic Ensemble. Correlation and Cluster Functions

The joint probability density function for the eigenvalue angles of a unitary self-dual random matrix taken from the symplectic ensemble was derived in Chapter 9 as

$$P_{N4}(\theta_1, ..., \theta_N) = \text{const} \times \prod_{j<k} \left| e^{i\theta_j} - e^{i\theta_k} \right|^4 . \tag{10.5.1}$$

To deal with integrals containing such an expression we write Eq. (10.5.1) as a confluent alternant. As in Eq. (10.3.2)

$$\prod_{j<k} \left| e^{i\theta_j} - e^{i\theta_k} \right|^4 = \exp\left(-2i(N-1) \sum_1^N \theta_j \right) \times \prod_{j<k} \left(e^{i\theta_j} - e^{i\theta_k} \right)^4 . \tag{10.5.2}$$

Note that because the power 4 is an even integer, ordering of the variables is no longer necessary. The fourth power of the product of differences expressed as a determinant [*cf.* Mehta, (1989), Section 7.1] is

$$\prod_{1 \le j < k \le N} \left(e^{i\theta_j} - e^{i\theta_k} \right)^4 = \det \left[e^{ik\theta_j} \quad k e^{i(k-1)\theta_j} \right]_{1 \le j \le N,\ 0 \le k \le 2N-1} . \tag{10.5.3}$$

If we multiply the $(2j-1)$th and the $(2j)$th columns by $\exp[-(2N-1)i\theta_j/2]$ and $\exp[-(2N-3)i\theta_j/2]$, respectively, we obtain

$$\exp\left(-2i(N-1) \sum_1^N \theta_j \right) \prod_{j<k} \left(e^{i\theta_j} - e^{i\theta_k} \right)^4$$
$$= \det \left[e^{ip\theta_j} \quad \left(p + N - \frac{1}{2} \right) e^{ip\theta_j} \right]$$
$$= \det \left[e^{ip\theta_j} \quad p e^{ip\theta_j} \right] , \tag{10.5.4}$$

where p varies over the half odd integers

$$-\left(N - \frac{1}{2} \right), -\left(N - \frac{3}{2} \right), ..., \left(N - \frac{1}{2} \right) . \tag{10.5.5}$$

We construct next an $N \times N$ quaternion self-dual matrix $Q_{N4}(\theta_1, ..., \theta_N)$ satisfying the conditions of Theorem 6.2.1 and such that its determinant be proportional to $P_{N4}(\theta_1, ..., \theta_N)$. With S_N, DS_N, and IS_N defined as in Eqs. (10.1.6), (10.3.6), and (10.3.7), let

$$\sigma_{N4}(\theta) = \frac{1}{2} \begin{bmatrix} S_{2N}(\theta) & DS_{2N}(\theta) \\ IS_{2N}(\theta) & S_{2N}(\theta) \end{bmatrix}, \tag{10.5.6}$$

and

$$Q_{N4}(\theta_1, ..., \theta_N) = [\sigma_{N4}(\theta_j - \theta_k)]_{j,k=1,...,N} \, . \tag{10.5.7}$$

With Eqs. (10.3.33), (10.3.34), and

$$DS_{2N} * IS_{2N} = IS_{2N} * DS_{2N} = S_{2N}, \tag{10.5.8}$$

it is straightforward to verify that

$$\sigma_{N4} * \sigma_{N4} = \sigma_{N4}. \tag{10.5.9}$$

Also

$$\int_0^{2\pi} \sigma_{N4}(0) d\theta = \frac{1}{2} \int S_{2N}(0) d\theta = \frac{N}{2\pi}. \tag{10.5.10}$$

Thus all the conditions of Theorem 6.2.1 are satisfied. It remains to verify that

$$\det [C(Q_{N4})] = \text{const} \times \prod_{1 \le j < k \le N} \left| e^{i\theta_j} - e^{i\theta_k} \right|^8. \tag{10.5.11}$$

For this, note that $C(Q_{N4})$ is the product of two matrices,

$$C(Q_{N4}) = \frac{1}{4\pi} \begin{bmatrix} e^{ip\theta_j} \\ (ip)^{-1} e^{ip\theta_j} \end{bmatrix} [e^{-ip\theta_k}, \quad ipe^{-ip\theta_k}], \tag{10.5.12}$$

and the two matrices on the right have, apart from a constant, the same determinant, the confluent alternant of Eq. (10.5.4).

We are now prepared to write down the correlation and cluster functions. Integrating over all the variables, we fix the constants,

$$P_{N4}(\theta_1, ..., \theta_N) = \frac{1}{N!} \det \left[\sigma_{N4}(\theta_j - \theta_k \right]_{j,k=1,...,N} . \tag{10.5.13}$$

The n-level correlation function is (*cf.* Theorem 6.2.1)

$$R_n(\theta_1, ..., \theta_n) = \det \left[\sigma_{N4}(\theta_j - \theta_k) \right]_{j,k=1,...,n} ; \tag{10.5.14}$$

and the n-level cluster function is

$$T_n(\theta_1, ..., \theta_n) = \sum_P \sigma_{N4}(\theta_1 - \theta_2)\sigma_{N4}(\theta_2 - \theta_3) \cdots \sigma_{N4}(\theta_n - \theta_1), \tag{10.5.15}$$

the sum being taken, as usual, over all $(n-1)!$ distinct cyclic permutations of the indices $(1, 2, ..., n)$.

Putting $n = 1$, we get the level density

$$R_1(\theta) = \sigma_{N4}(0) = \frac{1}{2} S_{2N}(0) = \frac{N}{2\pi}, \tag{10.5.16}$$

as expected. In the limit $N \to \infty$, $N\theta = 2\pi\xi$ finite,

$$\lim \frac{2\pi}{N} \sigma_{N4}(\theta) \approx \sigma_4(\xi) = \begin{bmatrix} s(2\xi) & Ds(2\xi) \\ Is(2\xi) & s(2\xi) \end{bmatrix}, \tag{10.5.17}$$

with s, Ds, and Is given by Eqs. (10.3.44)–(10.3.46) or (6.4.7)–(6.4.9), (6.4.11).

A comparison with Section 7.2 will show that in the limit of large N the n-level correlation and cluster functions for the circular symplectic ensemble coincide with those of the Gaussian symplectic ensemble.

10.6. Relation between Orthogonal and Symplectic Ensembles

With Eqs. (10.3.4) and (10.5.4) we will now prove the important result:

Theorem 10.6.1. *The statistical properties of N alternate angles θ_j, where $e^{i\theta_j}$ are the eigenvalues of a symmetric unitary random matrix of order $2N \times 2N$ taken from the orthogonal ensemble, are identical to those of the N angles φ_j, where $e^{i\varphi_j}$ are the eigenvalues of an $N \times N$ quaternion self-dual unitary random matrix taken from the symplectic ensemble.*

Proof. Suppose that $\theta_1 \leq \theta_2 \leq \cdots \leq \theta_{2N} \leq \theta_1 + 2\pi$. We write the joint probability density function

$$P_{2N,1}(\theta_1, ..., \theta_{2N}) = \text{const} \times \prod_{1 \leq j < k \leq 2N} \left| e^{i\theta_j} - e^{i\theta_k} \right|$$

$$= \text{const} \times \det \left[e^{ip\theta_j} \right], \tag{10.6.1}$$

$$p = -\left(N - \frac{1}{2}\right), -\left(N - \frac{3}{2}\right), ..., \left(N - \frac{1}{2}\right), \quad j = 1, 2, ..., 2N, \tag{10.6.2}$$

as in Eq. (10.3.4) and integrate over alternate variables $\theta_1, \theta_3, ... \theta_{2N-1}$, as in Section 6.5. The limits of integration for θ_{2j-1} are $(\theta_{2j-2}, \theta_{2j})$, except when $j = 1$. For θ_1 these limits are $(\theta_{2N} - 2\pi, \theta_2)$. Thus, the integration over the odd-indexed variables replaces the θ_1 column with

$$\int_{\theta_{2N}-2\pi}^{\theta_2} d\theta e^{ip\theta} = (ip)^{-1} \left(e^{ip\theta_2} + e^{ip\theta_{2N}} \right), \tag{10.6.3}$$

and the θ_{2j-1} column, for $j > 1$, with

$$\int_{\theta_{2j-2}}^{\theta_{2j}} d\theta e^{ip\theta} = (ip)^{-1} \left(e^{ip\theta_{2j}} - e^{ip\theta_{2j-2}} \right). \tag{10.6.4}$$

This later column can be changed, as in Section 6.5, to

$$(ip)^{-1} \left(e^{ip\theta_{2j}} + e^{ip\theta_{2N}} \right). \tag{10.6.5}$$

The $(2N-1)$th column is now simply $(ip)^{-1}2e^{ip\theta_{2N}}$, which allows us to drop the $e^{ip\theta_{2N}}$ term from every other column. The final result is

$$\int_{\theta_1 \le \theta_2 \le \cdots \le \theta_{2N} \le \theta_1 + 2\pi} \cdots \int d\theta_1 d\theta_3 \cdots d\theta_{2N-1} P_{2N,1}(\theta_1, ..., \theta_{2N})$$

$$= \text{const} \times \det \left[e^{ip\theta_{2j}}, \quad ipe^{ip\theta_{2j}} \right]$$
$$= \text{const} \times P_{N4}(\theta_2, \theta_4, ..., \theta_{2N}), \qquad (10.6.6)$$

which establishes the theorem.

10.7. Symplectic Ensemble. Level Spacings

We did not study in detail the level spacings for the Gaussian symplectic ensemble in Section 7.3. We have seen in Sections 10.1, 10.3, and 10.5 that in the limit of large N Gaussian ensembles are statistically equivalent to the corresponding circular ensembles. We have also seen in Section 10.6 that for any N, the eigenvalues of a random $N \times N$ matrix from the circular symplectic ensemble are statistically equivalent to the alternate eigenvalues of a random $2N \times 2N$ matrix from the circular orthogonal ensemble. We now use these equivalences to deduce the spacing probabilities for the symplectic ensemble, from those of the orthogonal one.

Let us denote by $B_4(t; y_1, ..., y_r)$ the probability that a randomly chosen interval of length $2t$ (measured in units of the local mean spacing) contains one level each at positions $y_1 < \cdots < y_r$ inside this interval and none others, the levels being the eigenvalues of a random matrix taken from the (Gaussian or circular) symplectic ensemble; let $B_1(t; y_1, ..., y_r)$ denote the same probability when the random matrix is taken from the (Gaussian or circular) orthogonal ensemble. The relation between symplectic and orthogonal ensembles, proved in Section 10.6 above, then tells us that $B_4(t; y_1, ..., y_r)$ is a sum of four terms, arising from the various possibilities of alternating $y_1, ..., y_r$ with the other eigenvalues

$z_1, \ldots z_s,\ s = r-1, r$ or $r+1$,

$$\begin{aligned}2B_4(t; y_1, \ldots, y_r) = &\int B_1(2t; y_1, z_1, y_2, z_2, \ldots, z_{r-1}, y_r)dz_1 \cdots dz_{r-1} \\ &+ \int B_1(2t; y_1, z_1, y_2, z_2, \ldots, y_r, z_r)dz_1 \cdots dz_r \\ &+ \int B_1(2t; z_0, y_1, z_1, y_2, \ldots, z_{r-1}, y_r)dz_0 dz_1 \cdots dz_{r-1} \\ &+ \int B_1(2t; z_0, y_1, z_1, y_2, \ldots, y_r, z_r)dz_0 dz_1 \cdots dz_r,\end{aligned} \tag{10.7.1}$$

where the integrations on the right hand side are carried with the restrictions

$$-2t < z_0 < y_1 < z_1 < y_2 < \cdots < y_r < z_r < 2t. \tag{10.7.2}$$

Note that on the right hand side we have $2t$ instead of t and an extra factor 2 on the left hand side, because the local mean spacings for orthogonal and symplectic ensembles differ by a factor 2.

It is straightforward, though somewhat cumbersome, to substitute the expressions of B_1 from Sections 6.6 and 6.7 and integrate over the alternate variables z_j. The result is much simpler if we integrate over all the variables in the region (10.7.2). Thus the probability $E_4(r; s)$, that a randomly chosen interval of length s contains exactly r eigenvalues of a self-dual quaternion random matrix taken from the symplectic ensemble is given by

$$E_4(0; s) = E_1(0; 2s) + \frac{1}{2}E_1(1; 2s), \tag{10.7.3}$$

$$\begin{aligned}E_4(r; s) = &E_1(2r; 2s) + \frac{1}{2}E_1(2r-1; 2s) \\ &+ \frac{1}{2}E_1(2r+1; 2s), \quad r > 0,\end{aligned} \tag{10.7.4}$$

where $E_1(r; s)$ is the same probability when the random matrix is chosen from the orthogonal ensemble. The $E_1(r; s)$ are given by Eqs. (6.5.19),

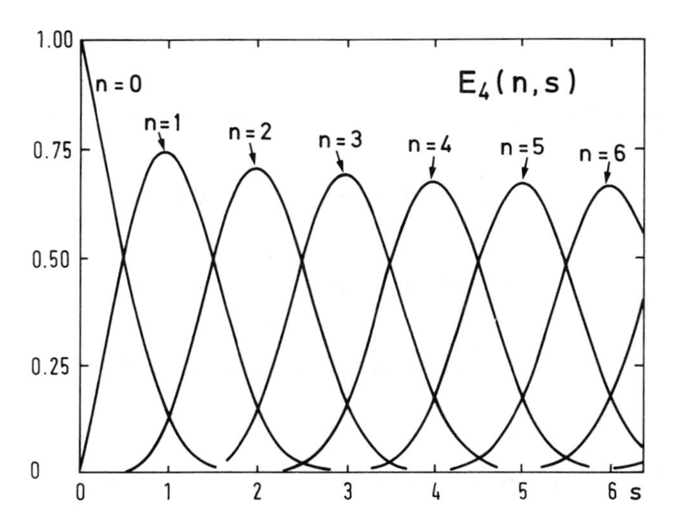

FIG. 10.1. The n-level spacings $E_4(n, s)$ for the Gaussian symplectic ensemble.

(6.6.12), and (6.7.21). Equation (10.7.3) can also be written as

$$\begin{aligned} E_4(0; s/2) &= \frac{1}{2}\left(D_+(t) + D_-(t)\right) \\ &= \frac{1}{2}\left(\prod_i (1 - \lambda_{2i}) + \prod_i (1 - \lambda_{2i+1})\right) \end{aligned} \tag{10.7.5}$$

$s = 2t$; where λ_i are the eigenvalues of the integral equation (5.3.14) and $D_\pm(t)$ are defined by Eqs. (5.5.9), (5.5.10).

Figure 10.1 shows the functions $E_4(n; s)$ for some values of n. Note that the various peaks are higher and narrower than the corresponding ones of $E_2(n; s)$ for the unitary ensemble.

10.8. Brownian Motion Model

Just as in Chapter 8, we can construct a Brownian motion model for the elements of our unitary matrices. Every matrix U taken from the ensemble $E_{\beta c}$ can be written as

$$U = VV^D, \tag{10.8.1}$$

where V is unitary and V^D is the transpose or the dual (*cf.* Chapter 2 and Appendix A.3) of V, according as β is 1 or 4. For the unitary ensemble, $\beta = 2$, V^D is unrelated to V, except that V^D is unitary and (10.8.1) holds. A permissible small change in U is then given by

$$\delta U = V(i\delta M)V^D, \tag{10.8.2}$$

where δM is an infinitesimal Hermitian matrix which is symmetric and hence real if $\beta = 1$, and self-dual if $\beta = 4$. Let us denote the independent real components of δM by δM_μ $(\equiv \delta M_{jk}^{(\lambda)})$; $\mu = 1, 2, ..., p$; $p = N + \frac{1}{2}N(N-1)\beta$ (or $\lambda = 0$ for $1 \le j = k \le N$, and $0 \le \lambda \le \beta - 1$ for $1 \le j < k \le N$). The isotropic and representation independent Brownian motion of U is defined by the statement that U at time t moves to $U + \delta U$ at time $t + \delta t$, where δU is given by Eq. (10.8.2) and the real parameters δM_μ are independent random variables with the moments

$$\langle \delta M_\mu \rangle = 0, \tag{10.8.3}$$

$$f \left\langle (\delta M_\mu)^2 \right\rangle = g_\mu kT\delta t, \tag{10.8.4}$$

where g_μ is given by Eq. (8.2.13).

The effect of the Brownian motion of U on its eigenvalues $\exp(i\theta_j)$ may again be found by choosing U to be diagonal at time t and calculating $\exp(i\theta_j + i\delta\theta_j)$ at time $t + \delta t$ by perturbation theory:

$$\delta\theta_j = \delta M_{jj}^{(0)} + \sum_{\substack{\ell \\ (\ell \ne j)}} \sum_{\lambda=0}^{\beta-1} \left(\delta M_{\ell j}^{(\lambda)}\right)^2 \left[\frac{1}{2}\cot\left(\frac{\theta_j - \theta_\ell}{2}\right)\right] + \cdots. \tag{10.8.5}$$

Equations (10.8.3) and (10.8.4) imply that the angles θ_j execute a Brownian motion with

$$f \langle \delta\theta_j \rangle = E(\theta_j)\delta t, \tag{10.8.6}$$

$$f \left\langle (\delta\theta_j)^2 \right\rangle = 2kT\delta t, \tag{10.8.7}$$

where

$$E(\theta_j) = \sum_{\substack{\ell \\ (\ell \ne j)}} \frac{1}{2}\cot\left(\frac{\theta_j - \theta_\ell}{2}\right). \tag{10.8.8}$$

This force $E(\theta_j)$ is exactly the component, tangential to the circle, of the electric field produced at $\exp(i\theta_j)$ by unit charges placed at all the other points $\exp(i\theta_\ell)$ at which U has eigenvalues. Thus

$$E(\theta_j) = -\frac{\partial W}{\partial \theta_j}, \tag{10.8.9}$$

$$W = -\sum_{\ell<j} \ln \left| e^{i\theta_\ell} - e^{i\theta_j} \right| . \tag{10.8.10}$$

One may write the corresponding Focker-Planck equation and find that the unique stationary probability density for the eigenvalue angles which corresponds to the completely diffused probability density of U is

$$P(\theta_1, ..., \theta_N) = c' \prod_{\ell<j} \left| e^{i\theta_\ell} - e^{i\theta_j} \right|^\beta . \tag{10.8.11}$$

This is again a new proof of Theorem 9.4.4.

10.9. Wigner's Method for the Orthogonal Circular Ensemble

A method of integration for the orthogonal circular ensemble was devised by Wigner. Though this method is far inferior to the one given in the previous sections of this chapter, it is of historical importance as it gave the normalization constant and the slope of the probability density of the spacings at zero spacing. The correct slope turned out to be $\pi^2/6$ rather than $\pi/2$ as given by the "Wigner surmise"

$$p_W(S) = \frac{\pi S}{2} \exp\left(-\frac{\pi}{4} S^2\right) . \tag{10.9.1}$$

This was how for the first time (10.9.1) was put in some "disrepute" by Wigner himself. The calculation was never published. Moreover, it was believed, incorrectly, to have been a calculation of the mean square spacing. This same slope was later calculated by Mehta and Dyson, using different methods (*cf.* Sections 6.5 and 10.4).

Wigner integrates $\exp\left[i\left(\nu_1\theta_1 + ... + \nu_N\theta_N\right)\right]$ over the region $2\varphi_1 \leq \theta_1 \leq \theta_2 \leq \cdots \leq \theta_N \leq 2\varphi_2$ and antisymmetrizes it in the variables $\nu_1, ..., \nu_N$ to get (*cf.* Appendix A.23)

Theorem 10.9.1. *The integral*

$$I(\varphi_1, \varphi_2) = \int\int d\theta_1 \cdots d\theta_{2m} \det \left[\exp\left(i\nu_j \theta_k\right)\right]_{j,k=1,2,\ldots,2m} \quad (10.9.2)$$

taken over the region $2\varphi_1 \leq \theta_1 \leq \theta_2 \leq \cdots \leq \theta_{2m} \leq 2\varphi_2$ *is given by*

$$I(\varphi_1, \varphi_2) = J(\varphi_1, \varphi_2) \det \left[\nu_j^{2k-2} \sin(\nu_j x) \quad \nu_j^{2k-1} \cos(\nu_j x)\right], \quad (10.9.3)$$

$$k = 1, \ldots, m; \quad j = 1, \ldots, 2m; \quad (10.9.3')$$

with

$$J(\varphi_1, \varphi_2) = (-i)^m 2^{2m} \exp\left[i(\varphi_1 + \varphi_2) \sum_1^{2m} \nu_k\right] (\nu_1 \cdots \nu_{2m})^{-1} \times \prod_{i \leq j < k \leq 2m} (\nu_j + \nu_k)^{-1} \quad (10.9.4)$$

and $x = \varphi_2 - \varphi_1$. *Note that the columns of the determinant in* (10.9.3) *are obtained, except for a possible change in sign, by successive differentiations of the first column, that is, of* $[\sin(\nu_j x)]$.

In Theorem 10.9.1 taking the limits $-\nu_{2j-1} = \nu_{2j} = j - 1/2$ for $j = 1, \ldots, m$, one gets (*cf.* Appendix A.23).

Theorem 10.9.2. *The integral*

$$\int\int d\theta_1 \cdots d\theta_{2m} \det \left[\exp(ip\theta_j)\right] \quad (10.9.5)$$

taken over the region $2\varphi_1 \leq \theta_1 \leq \theta_2 \leq \cdots \leq \theta_{2m} \leq 2\varphi_2$; *where* p *varies over the half-odd integers* $-m + 1/2$, $-m + 3/2$, ..., $m - 1/2$, *and* $j = 1$, 2, ..., $2m$, *is equal to*

$$i^m 2^{6m} (m!)^2 \left[\prod_{j=1}^{m} (2j)!\right]^{-2} \det\left[D(x)\right] \quad (10.9.6)$$

where $x = \varphi_2 - \varphi_1$ *and* $D(x)$ *is the matrix*

$$D(x) = \begin{bmatrix} \frac{\partial}{\partial q}\left(q^{2\ell-2}\sin qx\right) & q^{2\ell-2}\sin qx \\ \frac{\partial}{\partial q}\left(q^{2\ell-1}\cos qx\right) & q^{2\ell-1}\cos qx \end{bmatrix}, \tag{10.9.7}$$

$$\ell = 1,\ \ldots,\ m; \quad q = 1/2,\ 3/2,\ \ldots,\ m - 1/2. \tag{10.9.7$'$}$$

This is really a corollary of Theorem 10.9.1, but because of its importance we stated it as a theorem.

Writing $x = \pi - \alpha$ in Eqs. (10.9.5)–(10.9.7), we obtain the probability $E_1(0, 2\alpha)$ that there are no levels in an interval of length 2α,

$$\begin{aligned} E_1(0, 2\alpha) &= \int_{-\pi+\alpha}^{\pi-\alpha} \cdots \int_{-\pi+\alpha}^{\pi-\alpha} d\theta_1 \cdots \mathrm{d}\theta_{2m} \left(4^{2m}\pi^m m!\right)^{-1} \\ &\quad \times \prod_{1 \le j < k \le 2m} \left|e^{i\theta_j} - e^{i\theta_k}\right| \\ &= \frac{(2m)!}{2^{4m}\pi^m m!} i^{-m} \\ &\quad \times \int \cdots \int d\theta_1 \cdots d\theta_{2m} \det\left[e^{ip\theta_j}\right]_{\substack{j=1,\ \ldots,\ 2m \\ p=-m+1/2,\ \ldots,\ m-1/2}} \\ &= 2^{2m}\pi^{-m}\frac{m!}{(2m)!}\left[\prod_{j=1}^{m-1}(2j)!\right]^{-2} D(\alpha), \end{aligned} \tag{10.9.8}$$

where $D(\alpha)$ is the determinant

$$D(\alpha) = \begin{bmatrix} \frac{\partial}{\partial q}\left(q^{2\ell-2}\cos q\alpha\right) + \pi q^{2\ell-2}\sin q\alpha & q^{2\ell-2}\cos q\alpha \\ \frac{\partial}{\partial q}\left(q^{2\ell-1}\sin q\alpha\right) - \pi q^{2\ell-1}\cos q\alpha & q^{2\ell-1}\sin q\alpha \end{bmatrix}; \tag{10.9.9}$$

$$\begin{aligned} q &= 1/2,\ 3/2,\ \ldots,\ m-1/2, \\ \ell &= 1,\ 2,\ \ldots,\ m. \end{aligned} \tag{10.9.9'}$$

To expand $D(\alpha)$ in powers of α it will be convenient to simplify it further so that for small α it looks like a checkerboard. We replace the $(2\ell-1)$th row by the linear combination

$$r_{2\ell-1} + \frac{2\ell-2}{\pi} r_{2\ell-2} + \alpha r_{2\ell}, \tag{10.9.10}$$

where r_j denotes the jth row.

For small α, including powers up to, say, the second, Eq. (10.9.9) can therefore be written

$$D(\alpha) = \begin{bmatrix} \frac{1}{\pi}(2\ell-2)^2 q^{2\ell-3}\alpha + (2\ell-1)q^{2\ell-1}\alpha^2 & q^{2\ell-2} + \frac{1}{\pi}(2\ell-2)q^{2\ell-2}\alpha + \frac{1}{2}q^{2\ell}\alpha^2 \\ -\pi q^{2\ell-1} + 2\ell q^{2\ell-1}\alpha + \frac{1}{2}\pi q^{2\ell+1}\alpha^2 & q^{2\ell}\alpha \end{bmatrix}. \tag{10.9.11}$$

Following Wigner, one can compute the first few terms in the power-series expansion of $D(\alpha)$. For instance, if we write

$$D(\alpha) = D_0 + D_1\alpha + D_2\alpha^2 + \cdots, \tag{10.9.12}$$

then

$$\begin{aligned} D_0 &= D(0) = \det \begin{bmatrix} 0 & q^{2\ell-2} \\ -\pi q^{2\ell-1} & 0 \end{bmatrix}_{\substack{q=1/2,\ 3/2,\ \ldots,\ m-1/2 \\ \ell=1,\ 2,\ \ldots,\ m}} \\ &= \pi^m \det\left[q^{2\ell-2}\right] \det\left[q^{2\ell-1}\right] \\ &= \pi^m \prod_q q \prod_{p<q} \left(p^2 - q^2\right)^2 \\ &= \pi^m \frac{(2m)!}{2^{2m} m!} \prod_{j=1}^{m-1} [(2j)!]^2 \end{aligned} \tag{10.9.13}$$

and

$$D_1/D_0 = \sum_{\ell=1}^{m} \left(\frac{2\ell-2}{\pi}\right) + \sum_{\ell=1}^{m} \left(-\frac{2\ell}{\pi}\right) = -\frac{2m}{\pi}, \tag{10.9.14}$$

so that Eq. (10.9.12) gives

$$E(\alpha) = 1 - \frac{2m}{\pi}\alpha + \dots . \tag{10.9.15}$$

To calculate higher terms in the series Eq. (10.9.15) is more laborious. The calculation should show that the term in α^2 is absent,

$$D_2 = 0, \tag{10.9.16}$$

for the probability density of the spacings at zero spacing is zero. If one has enough patience, one can go a step further and get the slope of the probability density of the spacings at zero spacing.

However, to make any further progress seems to be difficult.

Summary of Chapter 10

Starting with the joint probability density $P(\theta_1, \dots, \theta_N)$ of the eigenvalues $\exp(i\theta_j)$, $j = 1, \dots, N$; for each of the three ensembles, orthogonal, unitary and symplectic, expressions are derived for the n-level correlation and cluster functions and for the spacings. The asymptotic results are identical to those for the corresponding Gaussian ensembles. (See the summaries of Chapters 5, 6, and 7.) The alternate eigenvalues of a matrix taken from the orthogonal (circular) ensemble behave as if they were the eigenvalues of a matrix from the symplectic (circular) ensemble; Section 10.6. From this one deduces the following expressions for the probability $E_4(n; s)$ that a randomly chosen interval of length s contains exactly n levels of a matrix from the symplectic ensemble:

$$E_4(0; s) = E_1(0; 2s) + \frac{1}{2}E_1(1; 2s) \tag{10.7.3}$$

and for $n > 0$,

$$E_4(n; s) = E_1(2n; 2s) + \frac{1}{2}E_1(2n-1; 2s) + \frac{1}{2}E_1(2n+1; 2s), \tag{10.7.4}$$

where $E_1(n;s)$ is the same probability for a matrix from the orthogonal ensemble.

A Brownian motion model starting with any initial values of the matrix elements and ultimately settling to their equilibrium distributions is considered in Section 10.8. The joint probability density for the eigenvalues also settles to

$$P(\theta_1, ..., \theta_N) \propto \prod_{1 \le j < k \le N} |\exp(i\theta_j) - \exp(i\theta_k)|^\beta . \tag{10.8.11}$$

where $\beta = 1$ for unitary symmetric matrices, $\beta = 2$ for unitary matrices and $\beta = 4$ for unitary symplectic matrices.

Finally, an old inefficient method of integration over the orthogonal ensemble due to Wigner is explained in Section 10.9.

11 / Circular Ensembles. Thermodynamics

In Chapter 10 we studied the correlation functions, cluster functions, and spacing distribution of the eigenvalues of a unitary matrix taken from Dyson's circular ensembles. In this chapter we calculate the partition function, the energy, the free energy, the entropy, and the specific heat of the energy levels defined in complete analogy with the classical mechanics. We then apply thermodynamic ideas to "derive" the leading terms in the asymptotic behavior of the spacing distribution for large spacings.

11.1. The Partition Function

As in Chapter 4, consider a thin circular conducting wire of radius unity; let N point charges be free to move on this wire. The universe is supposed to be two-dimensional. The charges repel one another according to the two-dimensional Coulomb law, so that the potential energy due to this electrostatic repulsion is

$$W = - \sum_{1 \le \ell < j \le N} \ln \left| e^{i\theta_\ell} - e^{i\theta_j} \right| . \tag{11.1.1}$$

As in Chapter 4, we discard the trivial velocity-dependent contributions. The positional partition function at a temperature kT is given by

$$\Psi_N(\beta) = (2\pi)^{-N} \int_0^{2\pi} \cdots \int_0^{2\pi} e^{-\beta W} d\theta_1 \cdots d\theta_N, \tag{11.1.2}$$

$$\beta = (kT)^{-1} \tag{11.1.3}$$

where W is given by Eq. (11.1.1). The aim of this section is to prove the following theorem.

Theorem 11.1.1. *For any positive interger N and a real or complex β the partition function $\Psi_N(\beta)$ is given by*

$$\Psi_N(\beta) = \Gamma(1 + \beta N/2)\left[\Gamma(1 + \beta/2)\right]^{-N}. \tag{11.1.4}$$

The proof may be divided into three parts :

1. The argument depends on the fact that W is bounded from below. It is intuitively clear (for a proof see Appendix A.24) that this minimum is attained when the N charges are situated at the corners of a regular polygon of N sides inscribed in the unit circle. The value of the minimum is easily calculated and gives

$$W \geq W_0 = -(N/2)\ln N. \tag{11.1.5}$$

Therefore we can write

$$\Psi_N(\beta) = \int_0^Y P(y) y^\beta dy, \tag{11.1.6}$$

where

$$Y = \exp(-W_0) = N^{N/2}, \tag{11.1.7}$$

and $P(y)$ is a positive weight function. In other words, $\Psi_N(\beta)$ is a moment function defined over a finite interval $(0, Y)$, and thus has special analytical properties (Shohat and Tamarkin, 1943). It is analytic in the upper half plane, Re $\beta > 0$, and in this region it satisfies the inequality

$$|\Psi_N(\beta)| < C\left|Y^\beta\right|. \tag{11.1.8}$$

Now the function

$$\psi_N(\beta) = \Gamma(1 + \beta N/2)\left[\Gamma(1 + \beta/2)\right]^{-N} \tag{11.1.9}$$

satisfies all these conditions. It is singular only on the negative real axis and for large $|\beta|$,

$$\psi_N(\beta) \approx N^{1/2}(\pi\beta)^{-(N-1)/2} Y^\beta. \tag{11.1.10}$$

The function

$$\Delta(\beta) = e^{2\beta W_0} \left[\Psi_N(2\beta) - \psi_N(2\beta) \right] \tag{11.1.11}$$

is therefore regular and bounded in the upper half plane, Re $\beta > 0$. At this point we can use a theorem of Carlson (Titchmarsh, 1939).

Carlson's Theorem. *If a function of β is regular and bounded in the upper half plane,* Re $\beta > 0$, *and is zero for* $\beta = 1, 2, 3, \ldots$, *then it is identically zero.*

Applying this theorem to the function $\Delta(\beta)$ we conclude that if the expressions (11.1.2) and (11.1.4) are equal for all even integers β, then they are identically equal.

2. The integral

$$(2\pi)^{-N} \int_0^{2\pi} \cdots \int_0^{2\pi} \prod_{\ell<j} \left| e^{i\theta_\ell} - e^{i\theta_j} \right|^{2k} d\theta_1 \cdots d\theta_N \tag{11.1.12}$$

is equal to the constant term in the series expansion, involving positive as well as negative powers of z_j; $j = 1, 2, ..., N$, of the product

$$\prod_{\substack{\ell, j \\ (\ell \neq j)}} \left(1 - \frac{z_\ell}{z_j} \right)^k . \tag{11.1.13}$$

To see this let $z_j = \exp(i\theta_j)$, so that

$$|z_j - z_\ell|^{2k} = (z_j - z_\ell)^k (z_j^{-1} - z_\ell^{-1})^k = \left(1 - \frac{z_\ell}{z_j} \right)^k \left(1 - \frac{z_j}{z_\ell} \right)^k \tag{11.1.14}$$

and note that any power other than zero of z_j; $j = 1, 2, ..., N$, vanishes on integration.

3. The constant term in the expansion of

$$\prod_{\substack{\ell, j \\ (\ell \neq j)}} \left(1 - \frac{z_j}{z_\ell} \right)^{a_j} \tag{11.1.15}$$

is given by

$$\frac{(a_1 + \cdots + a_N)!}{a_1! \cdots a_N!}. \tag{11.1.16}$$

A proof of this was first given by Wilson (1962); a shorter proof by Good (1970) is reproduced in Appendix A.25.

If we put $a_1 = a_2 = \cdots = a_N = k$ in Eqs. (11.1.15) and (11.1.16), we get the constant term in the expansion of Eq. (11.1.13) as

$$(Nk)!(k!)^{-N} = \Gamma(1 + kN) \left[\Gamma(1 + k)\right]^{-N}. \tag{11.1.17}$$

The proof of Theorem 11.1.1 is now complete.

4. The fact that the constant term in Eq. (11.1.14) is given by Eq. (11.1.17) can also be deduced from Selberg's integral, Chapter 17.7. But the statement in 3 above is stronger.

11.2. Thermodynamic Quantities

Theorem 11.1.1 specifies completely the thermodynamic properties of a finite Coulomb gas of N charges on the unit circle. For applications to the energy level series we are interested only in the special case of a very large $N : N \longrightarrow \infty$. In this section we study the statistical mechanics of an infinite Coulomb gas or, equivalently, that of an infinitely long series of eigenvalues.

The partition function (11.1.2) is normalized in a way that the energy of the gas is zero at infinite temperature ($\beta = 0$). The potential energy at zero temperature is then the ground-state energy,

$$W_0 = -(N/2) \ln N. \tag{11.2.1}$$

To obtain finite limits for the thermodynamic variables as $N \longrightarrow \infty$ we must first change the zero of the energy to the position W_0. By definition the gas then has zero energy at zero temperature and positive energy at any positive temperature. The partition function defined on the new energy scale is

$$\Phi_N(\beta) = (2\pi)^{-N} \int_0^{2\pi} \cdots \int_0^{2\pi} \exp\left[-\beta(W - W_0)\right] d\theta_1 \cdots d\theta_N. \tag{11.2.2}$$

The free energy per particle $F_N(\beta)$ is

$$F_N(\beta) = -(\beta N)^{-1} \ln\Phi_N(\beta) \tag{11.2.3}$$

$$= (1/2)\ln N - (\beta N)^{-1}\ln\Gamma(1+\beta N/2)$$

$$+ \beta^{-1}\ln\Gamma(1+\beta/2), \tag{11.2.4}$$

where we have used Theorem 11.1.1. Taking the limit $N \longrightarrow \infty$, we obtain the following theorem.

Theorem 11.2.1. *As $N \longrightarrow \infty$ the free energy per particle of the Coulomb gas at temperature $kT = \beta^{-1}$ tends to the limiting value*

$$F(\beta) = \beta^{-1}L(\beta/2) + [1 - \ln(\beta/2)]/2 \tag{11.2.5}$$

$$L(z) = \ln\Gamma(1+z). \tag{11.2.6}$$

The values of other thermodynamic quantities follow from Eq. (11.2.5):
Energy per particle:

$$U(\beta) = F + \beta\frac{\partial F}{\partial \beta} = (1/2)\left[L'(\beta/2) - \ln(\beta/2)\right]. \tag{11.2.7}$$

Entropy per particle:

$$S(\beta) = \beta^2\frac{\partial F}{\partial \beta} = (\beta/2)\left[L'(\beta/2) - 1\right] - L(\beta/2). \tag{11.2.8}$$

Specific heat per particle:

$$C(\beta) = -\beta^2\frac{\partial U}{\partial \beta} = -(\beta^2/4)L''(\beta/2) + \beta/2. \tag{11.2.9}$$

To calculate the values of these thermodynamic quantities for physically interesting values of β the following formulas [*cf.* Bateman (1953) or Abramowitz and Stegun (1965)] may be used:

$$L(z) = -\gamma z + \sum_{2}^{\infty} (-1)^n \frac{\zeta(n)}{n} z^n, \qquad |z| \leq 1, \tag{11.2.10}$$

$$L(z) = z\ln z - z + \frac{1}{2}\ln z + \frac{1}{2}\ln(2\pi) + \frac{1}{12z} + O\left(\frac{1}{z^2}\right), \qquad |z| \to \infty, \tag{11.2.11}$$

$$L(z+n) - L(z) = \sum_{r=1}^{n} \ln(z+r), \tag{11.2.12}$$

$$L'(z) = -\gamma + \sum_{2}^{\infty} (-1)^n \zeta(n) z^{n-1}, \qquad |z| < 1, \tag{11.2.13}$$

$$L'(z+n) - L'(z) = \sum_{r=1}^{n} (z+r)^{-1}, \tag{11.2.14}$$

$$L'(1/2) = 2 - \gamma - 2\ln 2, \tag{11.2.15}$$

$$L''(z) = \sum_{n=1}^{\infty} (z+n)^{-2}, \qquad z \neq -1,\ -2,\ \ldots \tag{11.2.16}$$

where γ is Euler's constant $\gamma = 0.5772\cdots$ and $\zeta(k)$ are the sums of the inverse powers of the integers

$$\zeta(k) = \sum_{n=1}^{\infty} n^{-k}. \tag{11.2.17}$$

In particular,

$$\zeta(2) = \sum_{1}^{\infty} n^{-2} = \frac{\pi^2}{6} \tag{11.2.18}$$

and

$$\sum_{1}^{\infty}(2n-1)^{-2} = \frac{\pi^2}{8}. \tag{11.2.19}$$

Table 11.1 summarizes this calculation.

11.3. Statistical Interpretation of U and C

If we denote the ensemble average by $\langle\ \rangle$,

$$\langle f \rangle = \frac{\int_0^{2\pi} \cdots \int_0^{2\pi} f(\theta_1, \cdots, \theta_N)\, e^{-\beta(W-W_0)} d\theta_1 \cdots d\theta_N}{\int_0^{2\pi} \cdots \int_0^{2\pi} e^{-\beta(W-W_0)} d\theta_1 \cdots d\theta_N}, \tag{11.3.1}$$

then from Eqs. (11.1.4), (11.1.5), (11.2.2), (11.2.7), and (11.2.9) we have

$$\langle W - W_0 \rangle = -\left[\Phi_N(\beta)\right]^{-1} \frac{\partial}{\partial \beta} \Phi_N(\beta) = NU \tag{11.3.2}$$

and

$$\begin{aligned} \left\langle (W - \langle W \rangle)^2 \right\rangle &= \left\langle (W - W_0)^2 \right\rangle - (\langle W - W_0 \rangle)^2 \\ &= \left[\Phi_N(\beta)\right]^{-1} \frac{\partial^2}{\partial \beta^2} \Phi_N(\beta) - (NU)^2 \\ &= \beta^{-2} NC, \end{aligned} \tag{11.3.3}$$

where W is the electrostatic energy given by Eq. (11.1.1) and $W_0 = -(1/2)N\ln N$ is the minimum value of W when the charges are uniformly spaced. Thus U is, apart from normalization, the ensemble average of the logarithm of the geometric mean of all distances between pairs of eigenvalues, and $\beta^{-2}C$ is the statistical mean square fluctuation of the same quantity.

Table 11.1

Thermodynamic quantities for the Coulomb gas model. The physically relevant values of the "temperature" are $(kT)^{-1} = \beta = 0,\ 1,\ 2,\ 4$ and ∞. For definitions of $L,\ F,\ U,\ S$, and C see the text.

	$\beta \to 0$	$\beta = 1$	$\beta = 2$	$\beta = 4$	$\beta \to \infty$
$L\left(\frac{\beta}{2}\right)$	$-\frac{\gamma\beta}{2} + \frac{\pi^2\beta^2}{48} + \cdots$	$\frac{1}{2}\ln\left(\frac{\pi}{4}\right) = -0.121$	0	$\ln 2 = 0.693$	$\frac{\beta}{2}\ln\left(\frac{\beta}{2}\right) - \frac{\beta}{2} + \frac{1}{2}\ln(\pi\beta) + \frac{1}{6\beta} + \cdots$
$L'\left(\frac{\beta}{2}\right)$	$-\gamma + \frac{\pi^2\beta}{12} + \cdots$	$2 - \gamma - \ln 4 = 0.036$	$1 - \gamma = 0.423$	$\frac{3}{2} - \gamma = 0.923$	$\ln\left(\frac{\beta}{2}\right) + \frac{1}{\beta} - \frac{1}{3\beta^2} + \cdots$
$L''\left(\frac{\beta}{2}\right)$	$\frac{\pi^2}{6} + \cdots$	$\frac{\pi^2}{2} - 4 = 0.935$	$\frac{\pi^2}{6} - 1 = 0.645$	$\frac{\pi^2}{6} - \frac{5}{4} = 0.395$	$\frac{2}{\beta} - \frac{2}{\beta^2} + \frac{4}{3\beta^3} + \cdots$
F	$\frac{1}{2}\left[1 - \gamma - \ln\left(\frac{\beta}{2}\right)\right] + \frac{\pi^2\beta}{48} + \cdots$	$\frac{1}{2}\left[1 + \ln\left(\frac{\pi}{2}\right)\right] = 0.726$	$\frac{1}{2} = 0.500$	$\frac{1}{2} - \frac{\ln 2}{4} = 0.327$	$\frac{1}{2\beta}\ln(\pi\beta) + \frac{1}{6\beta^2} + \cdots$
U	$\frac{1}{2}\left[-\gamma - \ln\left(\frac{\beta}{2}\right)\right] + \frac{\pi^2\beta}{24} + \cdots$	$1 - \frac{1}{2}(\gamma + \ln 2) = 0.365$	$\frac{1}{2}(1 - \gamma) = 0.211$	$\frac{3}{4} - \frac{1}{2}(\gamma + \ln 2) = 0.115$	$\frac{1}{2\beta} - \frac{1}{6\beta^2} + \cdots$
S	$-\frac{\beta}{2} + \frac{\pi^2\beta^2}{48} + \cdots$	$\frac{1}{2}(1 - \gamma - \ln\pi) = -0.361$	$-\gamma = -0.577$	$1 - 2\gamma - \ln 2 = -0.848$	$\frac{1}{2}[1 - \ln(\pi\beta)] - \frac{1}{3\beta} + \cdots$
C	$\frac{\beta}{2} - \frac{\pi^2\beta^2}{24} + \cdots$	$\frac{3}{2} - \frac{\pi^2}{8} = 0.266$	$2 - \frac{\pi^2}{6} = 0.355$	$7 - \frac{2}{3}\pi^2 = 0.420$	$\frac{1}{2} - \frac{1}{3\beta} + \cdots$

For analyzing the properties of observed eigenvalue series, W seems to be a good quantity. It has two great advantages over the other quantities such as F and S:

1. W can be computed from the eigenvalue pair-correlation function alone without analyzing higher order correlations.

2. The statistical uncertainty of W is known from the value of C.

Theorem 11.3.1. *Let $z_1, z_2, ..., z_N$ be the eigenvalues of a random unitary matrix taken from one of the ensembles E_1, E_2 or E_4. The quantity*

$$W - W_0 = \frac{1}{2} N \ln N - \sum_{i<j} \ln |z_i - z_j| \tag{11.3.4}$$

has the average value NU and the root mean square deviation $\beta^{-1}(NC)^{1/2}$ with the values of U and C listed in Table 11.1.

11.4. Continuum Model for the Spacing Distribution

In this section we exploit an argument of classical statistical mechanics to arrive at the asymptotic form of spacing distribution for large spacings.

As in Section 11.1, we write the joint probability density function for the eigenvalues $e^{i\theta_j}; j = 1, 2, ..., N$, of a random unitary $N \times N$ matrix as

$$P_{N\beta}(\theta_1, ..., \theta_N) = C'_{N\beta} e^{-\beta W}, \tag{11.4.1}$$

with

$$W = -\sum_{j<k} \ln \left| e^{i\theta_j} - e^{i\theta_k} \right|, \tag{11.4.2}$$

The probability that an arc of length 2α will contain none of the angles θ_j, that is, the $E_m(\alpha)$ of Eq. (10.2.1) or Eq. (10.4.13), is given by

$$E_\beta(0, \alpha) = \Psi_\beta(\alpha) / \Psi_\beta(0), \tag{11.4.3}$$

where

$$\Psi_\beta(\alpha) = \int_\alpha^{2\pi-\alpha} \cdots \int_\alpha^{2\pi-\alpha} e^{-\beta W} d\theta_1 \cdots d\theta_N \tag{11.4.4}$$

is the partition function of the analogous Coulomb gas of N charges compressed in a circular arc of length $2\pi - 2\alpha$, whereas $\Psi_\beta(0)$ is the partition function of the same gas on the whole unit circle. We write

$$E_\beta(0,\alpha) = \exp\left(-\beta\left[F_N(\beta,\alpha) - F_N(\beta,0)\right]\right) \tag{11.4.5}$$

where $F_N(\beta,\alpha)$ is the free energy of the Coulomb gas on the arc $2\pi - 2\alpha$.

The hypothesis that for large N the Coulomb gas forms a continuous electric fluid obeying the laws of thermodynamics may be put in the form of the following three assumptions:

1. There is a macroscopic charge density; that is there exists a smooth function $\sigma_\alpha(\theta)$ such that the average number of charges on the arc $(\theta, \theta + d\theta)$ is $\sigma_\alpha(\theta)d\theta$.

2. For a given density $\sigma_\alpha(\theta)$ the free energy is the sum of the two terms

$$F = V_1 + V_2, \tag{11.4.6}$$

where V_1 is the macroscopic potential energy

$$V_1 = -\frac{1}{2}\int\int_\alpha^{2\pi-\alpha} \sigma_\alpha(\theta)\sigma_\alpha(\phi)\ln\left|e^{i\theta} - e^{i\phi}\right| d\theta d\phi \tag{11.4.7}$$

and V_2 is the contribution from the individual arcs, depending only on the local density

$$V_2 = \int_\alpha^{2\pi-\alpha} \sigma_\alpha(\theta) f_\beta\left[\sigma_\alpha(\theta)\right] d\theta, \tag{11.4.8}$$

$f_\beta\left[\sigma\right]$ being the free energy per particle of a Coulomb gas having uniform density σ on the whole unit circle. The factor 1/2 in Eq. (11.4.7) is there because the interaction between two arc elements is counted twice.

3. In almost all cases the density σ_α adjusts itself in such a way that the free energy $F_N(\beta,\alpha)$ given by Eqs. (11.4.6), (11.4.7), and (11.4.8) is a minimum, subject to the condition

$$\int_\alpha^{2\pi-\alpha} \sigma_\alpha(\theta) d\theta = N. \tag{11.4.9}$$

This last equation expresses the fact that the total number of charges is fixed to N.

There is no rigorous mathematical justification for the above assumptions. But they are so much accepted by tradition that we make no apologies for adopting them.

The functional $f_\beta(\sigma)$ remains to be specified, and we write

$$f_\beta(\sigma) = U_\beta(\sigma) - \beta^{-1} S_\beta(\sigma), \tag{11.4.10}$$

where $U_\beta(\sigma)$ is the energy and $S_\beta(\sigma)$ is the entropy per particle for a uniform gas of

$$N' = 2\pi\sigma \tag{11.4.11}$$

charges on the whole unit circle.

As in Eq. (11.2.7), the energy per particle is

$$U_\beta(\sigma) = -\frac{1}{2}\ln N' + U(\beta). \tag{11.4.12}$$

The term $-\frac{1}{2}\ln N'$ is included, for we now have to take the total energy, including the ground-state energy $-\frac{1}{2}N'\ln N'$.

The entropy $S_\beta(\sigma)$, if calculated as in Section 11.2, is independent of N' for large N'. However, one thing should be noted. The calculation of the entropy in Section 11.2 was made for a gas of N distinguishable particles. The entropy so defined is not an extensive quantity. To make it extensive we must subtract $N!$ from the classical entropy, which amounts to treating the particles as undistinguishable. As we need the $S_\beta(\sigma)$ in Eq. (11.4.10) to be an extensive quantity, we write

$$S_\beta(\sigma) = \ln\left(\frac{N}{N'}\right) + S(\beta), \tag{11.4.13}$$

where $S(\beta)$ is given by Eq. (11.2.8). Putting

$$\sigma_\alpha(\theta) = \frac{N}{2\pi}\rho_\alpha(\theta) \tag{11.4.14}$$

and collecting Eq. (11.4.6) to (11.4.13), we have

$$\beta F_N(\beta, \alpha) = G_2 + G_1 + G_0, \tag{11.4.15}$$

where

$$G_2 = -\frac{\beta}{2}\left(\frac{N}{2\pi}\right)^2 \int\int_{\alpha}^{2\pi-\alpha} \rho_\alpha(\theta)\rho_\alpha(\phi)\ln\left|e^{i\theta}-e^{i\phi}\right| d\theta d\phi, \tag{11.4.16}$$

$$G_1 = (1-\beta/2)\left(\frac{N}{2\pi}\right)\int_{\alpha}^{2\pi-\alpha} \rho_\alpha(\theta)\ln\rho_\alpha(\theta)d\theta, \tag{11.4.17}$$

$$G_0 = \beta N\left[F(\beta)-\frac{1}{2}\ln N\right]; \tag{11.4.18}$$

and $F(\beta)$ given by Eq. (11.2.5). One has to minimize the quantity (11.4.15) under the restriction

$$\int_{\alpha}^{2\pi-\alpha} \rho_\alpha(\theta)d\theta = 2\pi. \tag{11.4.19}$$

When $\alpha = 0$, the equilibrium density $\rho_\alpha(\theta) = 1$, and $G_2 = G_1 = 0$, so that

$$\beta F_N(\beta,0) = G_0 = \beta N\left[F(\beta)-\frac{1}{2}\ln N\right] \tag{11.4.20}$$

and from Eq. (11.4.5)

$$E_\beta(0,\alpha) = \exp\left[-\min_\rho(G_2+G_1)\right]. \tag{11.4.21}$$

Using Lagrange's method to minimize (G_2+G_1) under the restriction (11.4.19), we get for $\beta \neq 2$

$$\begin{aligned} &-\beta\left(\frac{N}{2\pi}\right)^2\int_{\alpha}^{2\pi-\alpha} \rho_\alpha(\phi)\ln\left|e^{i\theta}-e^{i\phi}\right| d\phi \\ &\quad+\left(\frac{N}{2\pi}\right)(1-\beta/2)\ln\rho_\alpha(\theta) \\ &\quad+\left(\frac{N}{2\pi}\right)(1-\beta/2)-\lambda = 0, \\ &\qquad\qquad \alpha < \theta < 2\pi-\alpha, \end{aligned} \tag{11.4.22}$$

where λ is the undetermined constant. Letting

$$V_\alpha(\theta) = -\int_\alpha^{2\pi-\alpha} \rho_\alpha(\phi)\ln\left|e^{i\theta} - e^{i\phi}\right| d\phi, \tag{11.4.23}$$

we have

$$\beta\left(\frac{N}{2\pi}\right)(1-\beta/2)^{-1}V_\alpha(\theta) + \ln\rho_\alpha(\theta) = \text{constant}$$

or

$$\rho_\alpha(\theta) = Ae^{-\nu V_\alpha(\theta)}, \tag{11.4.24}$$

$$\nu = \beta\left(\frac{N}{2\pi}\right)(1-\beta/2)^{-1}. \tag{11.4.25}$$

When $\beta = 2$, G_1 is zero, and the minimization leads to

$$V_\alpha(\theta) = -\int_\alpha^{2\pi-\alpha} \rho_\alpha(\phi)\ln\left|e^{i\theta} - e^{i\phi}\right| d\phi = \text{constant}. \tag{11.4.26}$$

The $V_\alpha(\theta)$ is the electrostatic potential at the angle θ produced by all the other charges. If $\beta \neq 2$, these charges are in thermal equilibrium at an effective temperature

$$kT_\beta = \frac{1}{\beta}(1-\beta/2) = \frac{1}{\beta} - \frac{1}{2} \tag{11.4.27}$$

under the potential $V_\alpha(\theta)$ generated by themselves. If $\beta = 2$, the potential $V_\alpha(\theta)$ is constant and the charges are in electrostatic equilibrium on a conducting circular arc of length $2\pi - 2\alpha$.

The problem in the case of unitary ensemble $\beta = 2$ is the easiest to handle. The classical problem of charge distribution on a slotted conducting cylinder is well-known. Here we give only the result (Smythe, 1950). The solution of Eqs. (11.4.19) and (11.4.26) is

$$\rho_\alpha(\theta) = \sin\frac{\theta}{2}\left(\sin^2\frac{\theta}{2} - \sin^2\frac{\alpha}{2}\right)^{-1/2}, \tag{11.4.28}$$

$$V_\alpha = 2\pi\ \ln(\cos\frac{\alpha}{2}). \tag{11.4.29}$$

Equation (11.4.21) therefore gives

$$\ln E_2(0,\alpha) = -\min_\rho G_2 = N^2 \ln\left(\cos\frac{\alpha}{2}\right). \tag{11.4.30}$$

In the limit $N \longrightarrow \infty$, $s = 2t = 2\alpha N/(2\pi)$ finite, we obtain

$$\begin{aligned} \ln E_2(0;s) &= \lim[N^2 \ln\cos(\alpha/2)] \\ &= \lim\left(-\frac{1}{8}N^2\alpha^2\right) = -\frac{\pi^2}{2}t^2. \end{aligned} \tag{11.4.31}$$

When $\beta \neq 2$, one can apply perturbation theory to expand F in powers of αN. Since G_2 is of order $(\alpha N)^2$ and G_1 is of order αN, one can treat G_1 as a "small perturbation" over G_2. The first order contribution is

$$\beta F_1 = (1-\beta/2)\left(\frac{N}{2\pi}\right)\int_\alpha^{2\pi-\alpha} \bar{\rho}_\alpha(\theta)\,\ln\bar{\rho}_\alpha(\theta)\;\mathrm{d}\theta \tag{11.4.32}$$

where $\bar{\rho}_\alpha(\theta)$ is the unperturbed charge density (11.4.28).

To evaluate Eq. (11.4.32) make the following change of variables,

$$\sin^2\frac{\theta}{2} = \sin^2\frac{\alpha}{2} + \cos^2\frac{\alpha}{2}\sin^2\varphi. \tag{11.4.33}$$

Hence,

$$\begin{aligned} &\int_\alpha^{2\pi-\alpha} \bar{\rho}_\alpha(\theta)\ln\bar{\rho}_\alpha(\theta)d\theta \\ &= 2\int_\alpha^{\pi} \bar{\rho}_\alpha(\theta)\ln\bar{\rho}_\alpha(\theta)d\theta \\ &= 4\int_0^{\pi/2} d\varphi\left[\frac{1}{2}\ln\left(\sin^2\varphi + \tan^2\frac{\alpha}{2}\right) - \ln\sin\varphi\right] \\ &= 2\pi\,\ln\left(\tan\frac{\alpha}{2} + \sec\frac{\alpha}{2}\right). \end{aligned} \tag{11.4.34}$$

(For the last step one can look in a table of integrals, e.g., Gradshteyn and Rhizhik (1965).) In the limit $N \longrightarrow \infty$, $s = 2t = 2\alpha N/(2\pi)$ finite, this gives

$$\beta F_1 = (1 - \beta/2)\pi t, \tag{11.4.35}$$

and

$$\ln E_\beta(0; s) = -\frac{\pi^2}{4}\beta t^2 - (1 - \beta/2)\,\pi t. \tag{11.4.36}$$

In Chapter 12 we will derive an asymptotic series for $E_\beta(0; s)$ valid for large s. it will then be seen that the terms derived above from thermodynamic arguments are correct.

Summary of Chapter 11

Considering the eigenvalues of matrices from the circular ensembles as electric charges confined to the circumference of the unit circle, various thermodynamic quantities, such as the partition function, the free energy, the specific heat, ..., are calculated and their statistical interpretation is given (Sections 11.1–11.3). From such thermodynamic considerations the dominant terms in the asymptotic behavior of $E(0; s)$ is derived,

$$\ln E_\beta(0; s) = -\beta(\pi s)^2/16 - (1 - \beta/2)\pi s/2, \tag{11.4.36}$$

where $E_\beta(0; s)$ is the probability that a randomly chosen interval of length s contains no eigenvalues.

12 / Asymptotic Behavior of $E_\beta(0, s)$ for Large s

In Chapters 5, 6, and 10 we derived expressions for the probability $E_\beta(0, s)$ that an arbitrary interval of length s contains none of the eigenvalues of a random matrix chosen from the unitary $(\beta = 2)$, orthogonal $(\beta = 1)$, or symplectic $(\beta = 4)$ ensemble. They were expressed in terms of the infinite products $D_+(t) = \prod_{j=0}^{\infty}[1 - \lambda_{2j}(t)]$, and $D_-(t) = \prod_{j=0}^{\infty}[1 - \lambda_{2j+1}(t)]$, $s = 2t$, where $\lambda_j(t)$ are the eigenvalues of a certain integral equation. In other words, $D_\pm(t)$ are the Fredholm determinants of certain kernels over the interval $(-1, 1)$, or, with a change of variables, of other kernels over the interval $(0, t)$. The eigenfunctions of the integral equation in question also satisfy a certain second order differential equation depending on the parameter t. They are known as the prolate spheroidal functions in the mathematical literature. Their asymptotic behavior for large t can be ascertained from the differential equation, and from this knowledge one can derive the asymptotic behavior of $\lambda_j(t)$ and hence of $D_\pm(t)$ for large t.

Another way to get the asymptotic behavior of $D_\pm(t)$ is to find the same for the Toeplitz determinant [*cf.* Eq. (10.2.3)]

$$D_N(\alpha) = \det\left[C_{j-k}\right]_{j,k=1,\ldots,N}, \tag{12.0.1}$$

with

$$C_j \equiv C_j(\alpha) = \frac{1}{2\pi}\int_\alpha^{2\pi-\alpha} \exp(ij\theta)\, d\theta, \tag{12.0.2}$$

$$D_+(t)\, D_-(t) = \lim_{N\to\infty} D_N(2\pi t/N). \tag{12.0.3}$$

The behavior of $D_+(t)$ or $D_-(t)$ can be deduced from that of their product as we will see.

A third way to find the asymptotic behavior of $D_\pm(t)$ is to express them in terms of other Fredholm determinants $\Delta_\pm(t)$ defined over (t,∞). To construct these new convenient kernels over (t,∞), Dyson used the mathematical methods of the inverse scattering theory. The quantum theory of scattering deals with finding the phase shifts starting from a given potential, while the problem of inverse scattering consists in constructing the potential from the phase shifts. These problems can be solved by studying the wave functions either near the origin or near infinity and as Dyson noticed, involves certain Fredholm determinants. From the knowledge of $D_\pm(t)$ near $t \approx 0$, its dominant behavior for $t \to \infty$ and from the considerations of inverse scattering theory near the origin one finds a potential corresponding to the Fredholm determinants $D_\pm(t)$. For this potential one then constructs from the considerations near infinity other Fredholm determinants $\Delta_\pm(t)$ defined this time over (t,∞). Actually near the origin or at infinity one considers always a pair of potentials, one of them being the object of study while the other "comparison" potential is simple enough so that the corresponding solutions are either known or easily calculable. One can expand these new Fredholm determinants $\Delta_\pm(t)$ in inverse powers of t.

As a result of the above three methods one finally ends up with an asymptotic series for $\ln D_\pm(t)$, whose successive terms can in principle be computed with more and more effort.

12.1. Asymptotics of the $\lambda_n(t)$

From Eqs. (5.3.10), (6.5.11), and (10.7.5) we have

$$E_2(0,s) = D(s) \equiv D_+(t)\,D_-(t), \qquad E_1(0,s) = D_+(t), \tag{12.1.1}$$

and

$$E_4(0,s/2) = \frac{1}{2}[D_+(t) + D_-(t)], \tag{12.1.2}$$

with

$$D_+(t) = \prod_{n=0}^{\infty}[1-\lambda_{2n}(t)], \tag{12.1.3}$$

$$D_-(t) = \prod_{n=0}^{\infty}[1-\lambda_{2n+1}(t)], \tag{12.1.4}$$

$s = 2t$, and where λ_n are the eigenvalues of the integral equation

$$\lambda\, f(x) = \int_{-1}^{1} \frac{\sin (x-y)\,\pi t}{(x-y)\,\pi} f(y)\, dy, \tag{12.1.5}$$

arranged in decreasing order of magnitude

$$1 \geq \lambda_0 \geq \lambda_1 \geq \cdots \geq 0. \tag{12.1.6}$$

Though Eqs. (12.1.1)–(12.1.5) as they stand are complete, it is helpful to know that for certain discrete values of χ (depending on t, but independent of x), the solutions of the differential equation (5.3.22),

$$\frac{d}{dx}\left(1-x^2\right)\frac{df}{dx} + \left(\chi - \pi^2 t^2 x^2\right) f(x) = 0 \tag{12.1.7}$$

are regular at $x = \pm 1$ and then they are also the solutions of the integral equation (12.1.4). They are called prolate spheroidal functions, they have been studied extensively in the literature, they have been tabulated for t not too large, and their asymptotic behavior has been derived. The asymptotic formulas for $\lambda_n(t)$ with t large and any n are a little involved, since the order of n relative to that of t is arbitrary. They take a simpler form when n is finite and t is large. These formulas read as follows (*cf.* des Cloizeaux and Mehta, 1972, 1973).

First determine b_n by the implicit relation

$$\left(n+\frac{1}{2}\right)\frac{\pi}{2} = \pi t \int_0^{\min\left(1,\sqrt{\varepsilon}\right)} \left(\frac{\varepsilon - y^2}{1-y^2}\right)^{1/2} dy + \eta(b), \tag{12.1.8}$$

where

$$\varepsilon = 1 - \frac{2b}{\pi t}, \tag{12.1.9}$$

$$\eta(b) = \varphi(b) - \frac{b}{2}\left(\ln\left|\frac{b}{2}\right| - 1\right), \tag{12.1.10}$$

and $\varphi(b)$ is the phase of $\Gamma\left[(1+ib)/2\right]$,

$$\Gamma\left[(1+ib)/2\right] = \left(\frac{\pi}{\cosh(\pi b/2)}\right)^{1/2} e^{i\varphi(b)}. \tag{12.1.11}$$

Then

$$1-\lambda_n = (2\pi)^{1/2}(u/2e)^{u/2}\left[\Gamma\left(\frac{u+1}{2}\right)\right]^{-1} \times e^{-2\pi t\delta(\varepsilon)}\left[1+e^{\pi b}\right]^{-1}, \quad (12.1.12)$$

where

$$u = \pi t\varepsilon = \pi t - 2b \quad (12.1.13)$$

and the function $\delta(\varepsilon)$ is defined by

$$\delta(\varepsilon) = (1-\varepsilon)\left(\int_0^{\pi/2}\frac{\sin^2\alpha}{\left(\sin^2\alpha+\varepsilon\cos^2\alpha\right)^{1/2}}d\alpha - \frac{\pi}{4}\right), \quad \text{if } \varepsilon \le 1,$$
$$(12.1.14)$$
$$\delta(\varepsilon) = 0, \quad \text{if} \quad \varepsilon \ge 1.$$

When n is finite and $t \gg 1$, these formulas take the simpler form (*cf.* Slepian, 1965).

$$1-\lambda_n(t) = \pi^{n+1}2^{3n+2}t^{n+1/2}\exp(-2\pi t)/n!\ \left\{1+O\left(t^{-1}\right)\right\}, \quad (12.1.15)$$

so that

$$\frac{[1-\lambda_{2n}(t)]^{1/4}[1-\lambda_{2n+2}(t)]^{3/4}}{[1-\lambda_{2n+1}(t)]^{3/4}[1-\lambda_{2n+3}(t)]^{1/4}} \approx \left(1-\frac{1}{(2n+2)^2}\right)^{1/4}. \quad (12.1.16)$$

Also when $n \approx t \gg 1$, $b_n \approx \pi(n-2t)/[2\ln(2\sqrt{\pi t})]$, $1-\lambda_n \approx \left(1+e^{-\pi b_n}\right)^{-1} \approx 1$, and relation (12.1.15) is again valid. Thus Eq. (12.1.15) can be taken to be correct for $t \gg 1$ and any n. Hence

$$\frac{D_+(t)}{D_-(t)} = \frac{(1-\lambda_0)^{3/4}}{(1-\lambda_1)^{1/4}}\prod_{n=0}^{\infty}\frac{(1-\lambda_{2n})^{1/4}(1-\lambda_{2n+2})^{3/4}}{(1-\lambda_{2n+1})^{3/4}(1-\lambda_{2n+3})^{1/4}}$$
$$\approx \frac{(1-\lambda_0)^{3/4}}{(1-\lambda_1)^{1/4}}\prod_{n=0}^{\infty}\left(1-\frac{1}{(2n+2)^2}\right)^{1/4}. \quad (12.1.17)$$

From Eq. (12.1.15) and $\sin\theta = \theta\prod_{n=1}^{\infty}[1-\theta^2/(n^2\pi^2)]$ for $\theta = \pi/2$, this gives

$$\frac{D_+(t)}{D_-(t)} \approx \left(4\pi\sqrt{t}\, e^{-2\pi t}\right)^{3/4}\left(32\pi^2 t^{3/2} e^{-2\pi t}\right)^{-1/4}(2/\pi)^{1/4}$$

$$\approx \sqrt{2}\, e^{-\pi t}. \tag{12.1.18}$$

Thus we get a relation between the asymptotic behaviors of $D_+(t)$, $D_-(t)$ and their product $D(t) = D_+(t)\, D_-(t)$ for large t, namely

$$\ln D_\pm(t) \approx \frac{1}{2}\ln D(t) \mp \frac{\pi}{2}t. \tag{12.1.19}$$

Using Eqs. (12.1.8) to (12.1.14) and doing some analysis one can actually find the coefficients of t^2, t, and $\ln t$ in the asymptotic expansion of $\ln D(t)$. The details of this tedious analysis will not be reproduced here (*cf.* des Cloizeaux and Mehta, 1973), as we will be finding these terms in the next section from a theorem of Widom.

12.2. Asymptotics of Toeplitz Determinants

A theorem of Szegö (Szegö, 1959) is as follows.

Theorem 12.2.1. *If $f(\theta)$ is a positive function over $0 \le \theta \le 2\pi$, its derivative satisfies a Lipschitz condition and $D_N(f)$ is the $N \times N$ Toeplitz determinant*

$$D_N(f) = \det\left[\frac{1}{2\pi}\int_0^{2\pi} f(\theta)\exp[i(j-k)\theta]\mathrm{d}\theta\right]_{j,k=1,\cdots,N.} \tag{12.2.1}$$

Then as $N \longrightarrow \infty$

$$\ln D_N(f) \approx N f_0 + \frac{1}{4}\sum_{k=1}^{\infty} k\, f_k f_{-k} + O(1), \tag{12.2.2}$$

where f_k are the Fourier coefficients of $\ln f(\theta)$,

$$\ln f(\theta) = \sum_{k=-\infty}^{\infty} f_k e^{ik\theta}. \tag{12.2.3}$$

Widom (1971) extended this theorem for functions $f(\theta)$ which are positive only on an arc of the unit circle.

Theorem 12.2.2. *Let $f(\theta) = 1$, if $\alpha \le \theta \le 2\pi - \alpha$, and $f(\theta) = 0$ if either $\theta < \alpha$ or $\theta > 2\pi - \alpha$. Then as $N \longrightarrow \infty$*

$$\ln D_N(f) \approx N^2 \ln \cos\frac{\alpha}{2} - \frac{1}{4}\ln\left(N \sin\frac{\alpha}{2}\right) + \frac{1}{12}\ln 2 + 3\zeta'(-1) \tag{12.2.4}$$

where $\zeta'(z)$ is the derivative of the Riemann zeta function.

Taking the limit $\alpha N = 2\pi t \gg 1$, one has $(s = 2t)$,

$$\ln D(t) \approx -\frac{1}{2}(\pi t)^2 - \frac{1}{4}\ln(\pi t) + \frac{1}{12}\ln 2 + 3\zeta'(-1). \tag{12.2.5}$$

This together with Eq. (12.1.18) give

$$\ln D_\pm(t) \approx -\frac{(\pi t)^2}{4} \mp \frac{\pi t}{2} - \frac{1}{8}\ln(\pi t) + \left(\frac{1}{24} \pm \frac{1}{4}\right)\ln 2 + \frac{3}{2}\zeta'(-1). \tag{12.2.6}$$

12.3. Fredholm Determinants and the Inverse Scattering Theory

With the new variables $\xi = x\pi t$, $\eta = y\pi t$, Eq. (12.1.5) takes the form

$$\lambda f(\xi/\pi t) = \frac{1}{\pi}\int_{-\pi t}^{\pi t} \frac{\sin(\xi - \eta)}{\xi - \eta} f(\eta/\pi t)\, d\eta.$$

And as $f(-x) = \pm f(x)$, the λ_j are the eigenvalues of the integral equations

$$\lambda_{2j} g(x) = \int_0^{\pi t} K_+(x,y) g(y)\, dy\,, \tag{12.3.1}$$

$$\lambda_{2j+1} g(x) = \int_0^{\pi t} K_-(x,y) g(y)\, dy\,, \tag{12.3.2}$$

with

$$K_{\pm}(x,y) = \frac{1}{\pi}\left(\frac{\sin(x-y)}{x-y} \pm \frac{\sin(x+y)}{x+y}\right). \tag{12.3.3}$$

Hence $D_{\pm}(t)$ are the Fredholm determinants of the kernels (12.3.3) or of

$$K_{+}(x,y) = \frac{2}{\pi}\int_0^1 \cos(kx)\cos(ky)dk\ , \tag{12.3.4}$$

$$K_{-}(x,y) = \frac{2}{\pi}\int_0^1 \sin(kx)\sin(ky)dk\ , \tag{12.3.5}$$

over the interval $(0,\tau)$, $D_{\pm}(t) = \det\left[1 - K_{\pm}\right]_0^{\tau}$, $\tau = \pi t$.

Fredholm determinants occur also in the theory of inverse scattering. In the quantum theory of scattering the potential is given and phase shifts are calculated. The problem of inverse scattering deals with reconstructing the potential given the phase shifts. Two methods are available for this purpose. The Gel'fand-Levitan method constructs the potential by working from $x = 0$ upwards, while Marchenko method does it by working downwards from $x = \infty$. The first method is related to a Fredholm determinant over the interval $(0,\tau)$, while the second involves that over the interval (τ,∞). Both methods are equivalent and any one of them is sufficient to solve the inverse scattering problem.

Dyson took advantage of this equivalence and related the Fredholm determinants $D_{\pm}(t)$ defined over $(0,\pi t)$ to an inverse scattering problem applying the Gel'fand-Levitan method. He then applied Marchenko method to the same problem and expressed $D_{\pm}(t)$ in terms of other Fredholm determinants $\Delta_{\pm}(t)$ defined this time over $(\pi t,\infty)$. These later can be expanded in inverse powers of t.

As we are unable to do better than Dyson, we will reproduce his entire analysis. Sections 12.3–12.6 are taken almost verbatim from his paper "Fredholm determinants and inverse scattering problems" and reprinted with permission of Springer-Verlag Publishers, *Commun. Math. Phys.* **47**, 171–183 (1976).

In the problem of inverse scattering we are dealing with two potentials $V_1(x)$, $V_2(x)$ on the half-line $0 < x < \infty$, V_1 being supposed unknown and V_2 known. We have two corresponding families of wave-functions $u_1(k,x)$, $u_2(k,x)$ satisfying the wave equations

$$\left[(d^2/dx^2) + k^2 - V_j(x)\right]u_j(k,x) = 0\ , \quad j = 1,\ 2\ . \tag{12.3.6}$$

and generating spectral kernels $K_1,\ K_2$ defined by

$$\delta(x-y) - K_j(x,y) = \int u_j(k,x)u_j(k,y)dk \ . \tag{12.3.7}$$

In the Gel'fand–Levitan version (Gel'fand and Levitan, 1951) of inverse scattering theory, the wave functions $u_j(k,x)$ are subject to boundary conditions at $x = 0$, namely

$$u_1(k,0) = u_2(k,0) = gq(k) \ , \tag{12.3.8}$$

$$u_j'(k,0) = h_j q(k) \ , \tag{12.3.9}$$

where $q(k)$ is a given function and $g,\ h_1,\ h_2$ are given coefficients. The Fredholm determinants

$$D_j(t) = \det\,[1 - K_j]_0^\tau \ , \qquad \tau = \pi t \ , \tag{12.3.10}$$

are defined on the finite interval $[0,\tau]$, and satisfy the conditions

$$V_1(\tau) - V_2(\tau) = 2(d^2/d\tau^2)\,[\ln(D_1(t)/D_2(t))] \ , \tag{12.3.11}$$

$$h_1 - h_2 = g(d/d\tau)\,[\ln(D_1(t)/D_2(t))]_{\tau=0} \ . \tag{12.3.12}$$

(*cf.* Chadan and Sabatier, 1977).

In the Marchenko version (Marchenko, 1950) of the theory, the wave functions $u_j(k,x)$ are required to become asymptotically equal at infinity, thus

$$u_1(k,x) - u_2(k,x) \longrightarrow 0 \quad \text{as} \quad x \longrightarrow \infty \ , \tag{12.3.13}$$

uniformly in k. The potentials $V_1,\ V_2$ must approach each other closely enough so that the integral

$$\int |V_1(\tau) - V_2(\tau)|\,\tau d\tau \tag{12.3.14}$$

converges at infinity. The Fredholm determinants

$$\Delta_j\,(\tau) = \det\,[1 - K_j]_\tau^\infty \tag{12.3.15}$$

are defined on the infinite interval $[\tau, \infty]$, and

$$V_1(\tau) - V_2(\tau) = 2\left(d^2/d\tau^2\right)\{\ln[\Delta_1(\tau)/\Delta_2(\tau)]\}. \tag{12.3.16}$$

Equations (12.3.11) and (12.3.16) are important as they relate two sets of Fredholm determinants one defined on the finite interval $(0, \tau)$, and the other defined over the infinite interval (τ, ∞). What is crucial is the choice of the potentials. One of the potentials is imposed by the Fredholm determinant we want to study. The second comparison potential need not be the same near the origin and at large distances; it has to satisfy the boundary conditions (12.3.8) and (12.3.9) near the origin, and the convergence condition (12.3.14) at large distances. Also the choice should be simple enough so as to allow computation of the corresponding Fredholm determinants.

When the Gel'fand–Levitan and Marchenko formalisms are applied to the inverse scattering problem, it is customary to assume that the unknown wave functions $u_1(k, x)$ form a complete orthonormal set. Then Eq. (12.3.7) implies

$$K_1 = 0, \qquad D_1 = \Delta_1 = 1. \tag{12.3.17}$$

12.4. Application of the Gel'fand–Levitan Method

We apply the Gel'fand-Levitan formalism to the potentials

$$V_1(\tau) = W_\pm(\tau) = -2\left(d^2/d\tau^2\right)\ln D_\pm(t) - 1, \qquad \tau = \pi t, \tag{12.4.1}$$

with the comparison potential

$$V_2(\tau) = -1. \tag{12.4.2}$$

According to Eq. (12.2.6)

$$W_\pm(\tau) \sim -\frac{1}{4}\tau^{-2} \quad \text{as} \quad \tau \longrightarrow \infty. \tag{12.4.3}$$

We take $u_1(k, x)$ to be a complete orthonormal set of solutions of Eq. (12.3.6), so that Eq. (12.3.11) holds with

$$D_1(\tau) = 1, \qquad D_2(\tau) = D_\pm(t). \tag{12.4.4}$$

The wave-functions $u_2(k,x)$ must be cosines and sines of $\left(\left(k^2+1\right)^{1/2}x\right)$ in order to satisfy Eq. (12.3.6). We have to normalize these wave functions so that

$$K_2(x,y) = K_\pm(x,y), \tag{12.4.5}$$

with K_2 given by Eq. (12.3.7) and $K_\pm$ by Eq. (12.3.3). Thus we require

$$\int_0^\infty u_2(k,x)\,u_2(k,y)\;dk = (2/\pi)\int_1^\infty \begin{matrix}\cos\\ \\ \sin\end{matrix}\; kx \;\begin{matrix}\cos\\ \\ \sin\end{matrix}\; ky\; dk. \tag{12.4.6}$$

which is satisfied by choosing

$$u_2(k,x) = s(k)\begin{matrix}\cos\\ \\ \sin\end{matrix}\left(\left(k^2+1\right)^{1/2}x\right), \tag{12.4.7}$$

$$s(k) = (2/\pi)^{1/2}\left[k^2/\left(k^2+1\right)\right]^{1/4}. \tag{12.4.8}$$

We next have to determine the boundary conditions satisfied by $u_1(k,x)$ and $u_2(k,x)$ at $x=0$. In the even case, when $V_1 = W_+$, Eqs. (12.3.8) and (12.3.9) hold with

$$q(k) = s(k), \qquad g = 1, \qquad h_2 = 0. \tag{12.4.9}$$

In this case Eq. (12.3.12) gives

$$h_1 = -(d/d\tau)\left[\ln D_+(t)\right]_{\tau=0} = K_+(0,0) = (2/\pi), \quad \tau = \pi t. \tag{12.4.10}$$

Therefore the boundary conditions for u_1 are

$$u_1(k,0) = s(k), \qquad u_1'(k,0) = (2/\pi)\,s(k). \tag{12.4.11}$$

In the odd case, when $V_1 = W_-$, Eq. (12.3.12) gives

$$h_1 = h_2, \tag{12.4.12}$$

and the boundary conditions for u_1 are

$$u_1(k,0) = 0, \qquad u_1'(k,0) = \left(k^2+1\right)^{1/2} s(k). \tag{12.4.13}$$

We are now in a position to determine the behavior of $u_1(k,x)$ as $x \longrightarrow \infty$. Since the potential V_1 decreases according to Eq. (12.4.3) at infinity, and the $u_1(k,x)$ are an orthonormal system, we have

$$u_1(k,x) \sim (2/\pi)^{1/2} \begin{matrix} \cos \\ \\ \sin \end{matrix} [kx + \eta(k)]. \tag{12.4.14}$$

where the phase shift $\eta(k)$ must be calculated separately for the even and odd cases. Following Jost (1947), we define $J(k,x)$ to be the solution of Eq. (12.3.6) with potential V_1 and the asymptotic behavior

$$J(k,x) \sim \exp(ikx) \quad \text{as} \quad x \longrightarrow \infty. \tag{12.4.15}$$

The functions

$$a(k) = J(k,0), \qquad b(k) = J'(k,0) \tag{12.4.16}$$

are the boundary values of a function analytic in the half-plane (Im $k >$ 0), with the symmetry property

$$a(-k) = a^*(k), \qquad b(-k) = b^*(k), \qquad k \text{ real}, \tag{12.4.17}$$

and the asymptotic behavior

$$a(k) \sim 1, \qquad b(k) \sim ik, \qquad k \longrightarrow \infty. \tag{12.4.18}$$

The Wronskian

$$J(-k,x)\,J'(k,x) - J'(-k,x)\,J(k,x) \tag{12.4.19}$$

is independent of x. Equating its value at $x = 0$ with its value at $x = \infty$, we find

$$a^*(k)\,b(k) - a(k)\,b^*(k) = 2ik. \tag{12.4.20}$$

The comparison of Eq. (12.4.14) with (12.4.15) implies

$$u_1(k,x) = (2/\pi)^{1/2} \begin{matrix} \text{Re} \\ \\ \text{Im} \end{matrix} \{J(k,x)\exp[i\eta(k)]\}. \tag{12.4.21}$$

Consider first the even case. Then the boundary conditions (12.4.11) together with Eq. (12.4.21) imply

$$\mathrm{Re}\{\exp[i\eta(k)]a(k)\} = \left[k^2/\left(k^2+1\right)\right]^{1/4}. \tag{12.4.22}$$

$$\mathrm{Re}\{\exp[i\eta(k)]\left[b(k)-(2/\pi)\,a(k)\right]\} = 0. \tag{12.4.23}$$

The function

$$\varphi(k) = b(k) - (2/\pi)\,a(k) \tag{12.4.24}$$

is analytic in the upper half-plane and satisfies the same conditions (12.4.17), (12.4.18) as $b(k)$. According to Eq. (12.4.23)

$$\exp[i\eta(k)] = i\left[\varphi^*(k)/|\varphi(k)|\right]. \tag{14.4.25}$$

When we substitute Eq. (12.4.25) into (12.4.22) and make use of Eq. (12.4.20), we obtain

$$|\varphi(k)| = \left[k^2(k^2+1)\right]^{1/4}. \tag{12.4.26}$$

The only analytic function satisfying Eqs. (12.4.17), (12.4.18), and (12.4.26) is

$$\varphi(k) = \left[-k(k+i)\right]^{1/2}. \tag{12.4.27}$$

Putting Eq. (12.4.27) back into Eq. (12.4.25), we find

$$\exp[i\eta(k)] = \left[(k-i)/(k+i)\right]^{1/4}, \tag{12.4.28}$$

and so the phase shift is given by

$$k\tan 2\eta(k) = -1. \tag{12.4.29}$$

In the odd case, the boundary conditions (12.4.13) imply

$$\mathrm{Im}\{\exp[i\eta(k)]a(k)\} = 0, \tag{12.4.30}$$

$$\mathrm{Im}\{\exp[i\eta(k)]b(k)\} = \left[k^2\left(k^2+1\right)\right]^{1/4}. \tag{12.4.31}$$

Equation (12.4.30) implies

$$\exp[i\eta(k)] = \left[a^*(k)/|a(k)|\right], \tag{12.4.32}$$

and this with Eqs. (12.4.31) and (12.4.20) gives

$$|a(k)| = \left[k^2/\left(k^2+1\right)\right]^{1/4}. \tag{12.4.33}$$

The analytic function $a(k)$ is then

$$a(k) = [k/(k+i)]^{1/2}, \tag{12.4.34}$$

and Eq. (12.4.32) gives $\exp[i\eta(k)] = [(k+i)/(k-i)]^{1/4}$, or

$$k \tan 2\eta(k) = +1. \tag{12.4.35}$$

Both cases are included in the formula

$$\eta(k) = \mp\frac{1}{2}\arctan\left(k^{-1}\right). \tag{12.4.36}$$

The potentials $W_\pm(\tau)$ are uniquely determined (Levinson, 1949) by the property that they give scattering states (12.4.14) with the phase shifts (12.4.36), and no bound states.

It is a curious fact that the identity Eq. (5.5.14) can be written in the form

$$\left[\left(d^2/\mathrm{d}\tau^2\right) - 1 - W_-(\tau)\right] u_B(\tau) = 0, \tag{12.4.37}$$

with

$$u_B(\tau) = D_+(t)/D_-(t) \sim 2^{1/2}\mathrm{e}^{-\tau} \quad \text{as} \quad \tau \longrightarrow \infty. \tag{12.4.38}$$

Thus, $u_B(\tau)$ is an acceptable bound-state wave function with energy (-1) in the potential $W_-(\tau)$. However, u_B does not satisfy the boundary condition (12.4.13) for the odd case and is therefore irrelevant to the determination of W_-. Another alternative form of Eq. (5.5.14) is

$$\left[\left(d^2/\mathrm{d}\tau^2\right) - 1 - W_+(\tau)\right]\left[u_B(\tau)\right]^{-1} = 0. \tag{12.4.39}$$

The wave function $(u_B)^{-1}$ satisfies the correct boundary condition (12.4.11) for the even case at $\tau = 0$, but fails to converge at infinity, and is therefore also irrelevant to the determination of W_+.

12.5. Application of the Marchenko Method

We apply the Marchenko formalism to the potential $V_1(\tau)$ defined by Eq. (12.4.1). What should we choose for the comparison potential $V_2(\tau)$? It is possible to carry through the analysis with $V_2(\tau) = 0$. The calculations are then formally simple, but the integral (12.3.14) diverges in view of Eq. (12.4.3), and the results obtained are poorly convergent and of doubtful utility. The next most simple choice would be

$$V_2(\tau) = -\frac{1}{4}\tau^{-2}. \tag{12.5.1}$$

This makes the integral (12.3.14) converge. The wave functions $u_2(k,x)$ are now Bessel functions of the quantity (kx), and the phase shift produced by the potential $V_2(\tau)$ is $(\mp\pi/4)$ independent of k. The fact that this phase shift agrees with the phase shift (12.4.36) at $k=0$ reflects the fact that V_1 and V_2 have the same behavior at infinity. But the same argument carried one step further suggests a far better choice for V_2. The phase shift (12.4.36) behaves like

$$\eta(k) \sim \mp\left(\frac{1}{4}\pi - \frac{1}{2}k\right), \tag{12.5.2}$$

with an error of order k^3 for small k. A Bessel function of the quantity $k(x \mp 1/2)$ gives the phase shift (12.5.2) for all k. We therefore choose for V_2 the potential

$$V_2(\tau) = -\frac{1}{4}\left(\tau \pm \frac{1}{2}\right)^{-2}, \tag{12.5.3}$$

with the expectation that this will make the difference $(V_1 - V_2)$ decrease much more rapidly as $\tau \longrightarrow \infty$. The results of the calculation justify our expectation. The singularity of $V_2(\tau)$ at $\tau = 1/2$ (in the odd case) means that the analysis is valid only for $\tau > 1/2$. This is not a serious limitation, since our purpose is to study the behavior of the potentials for large τ.

The wave functions $u_2(k,x)$ are solutions of the equation

$$\left[\left(d^2/dx^2\right) + k^2 + \frac{1}{4}\left(x \pm \frac{1}{2}\right)^{-2}\right] u_2(k,x) = 0, \tag{12.5.4}$$

with asymptotic behavior determined by Eq. (12.3.13). It is convenient to use the notations

$$j(z) = z^{1/2} J_0(z), \quad y(z) = z^{1/2} Y_0(z), \tag{12.5.5}$$

$$h(z) = z^{1/2} H_0^1(z), \quad k(z) = z^{1/2} K_0(z), \tag{12.5.6}$$

for the Bessel functions. We take for the wave functions $u_1(k,x)$ the same complete orthonormal set that we studied in Section 12.4, with asymptotic behavior given by Eqs. (12.4.14) and (12.4.36). Then Eq. (12.3.13) implies

$$u_2(k,x) = \alpha(k)\, j(k(x \pm 1/2)) \pm \beta(k)\, y(k(x \pm 1/2)), \tag{12.5.7}$$

with

$$\alpha(k) = \cos[\theta(k)/2], \qquad \beta(k) = \sin[\theta(k)/2], \tag{12.5.8}$$

$$\theta(k) = (k - \arctan k). \tag{12.5.9}$$

The Marchenko formula (12.3.16) becomes

$$V_1(\tau) = W_\pm(\tau) = -\frac{1}{4}\left(\tau \pm \frac{1}{2}\right)^{-2} - 2\left(d^2/d\tau^2\right) \ln \Delta_\pm(t), \tag{12.5.10}$$

$$\Delta_\pm(\tau) = \det[1 - F_\pm]_\tau^\infty, \quad \tau = \pi t, \tag{12.5.11}$$

with the kernels $F_\pm$ defined by

$$F_\pm(x,y) = \delta(x-y) - \int_0^\infty u_2(k,x)\, u_2(k,y)\, dk. \tag{12.5.12}$$

Using the completeness relation

$$\delta(x-y) = \int_0^\infty j\left(k\left(x \pm \frac{1}{2}\right)\right) j\left(k\left(y \pm \frac{1}{2}\right)\right) dk. \tag{12.5.13}$$

we bring Eq. (12.5.12) to the form

$$F_\pm(x,y) = \frac{1}{2}\int_0^\infty \mathrm{Re}\big[\,(1-\exp[\mp i\theta(k)])$$

$$\times\, h(k(x\pm 1/2))h(k(y\pm 1/2))\big]\,dk$$

$$= \frac{1}{4}\left[\int_0^\infty - \int_{-\infty}^0\right](1-\exp[\mp i\theta(k)])$$
$$\times\, h(k(x\pm 1/2))h(k(y\pm 1/2))dk. \qquad (12.5.14)$$

The function

$$\exp[\mp i\theta(k)] = \exp(\mp ik)(1\pm ik)\left(k^2+1\right)^{-1/2} \qquad (12.5.15)$$

is analytic in the upper half-plane with a cut from $(+i)$ to $(+i\infty)$. In the even case, the exponential growth of Eq. (12.5.15) is compensated by the exponential decrease of the Hankel functions in Eq. (12.5.14) for all positive x and y. In the odd case Eq. (12.5.15) decreases exponentially in the upper half-plane, but we must require $x, y > 1/2$ so that the term in Eq. (12.5.14) not involving Eq. (12.5.15) has exponential decrease. With this proviso, we may move the path of integration in both parts of Eq. (12.5.14) to the positive imaginary axis by writing $k = iz$. The terms involving Eq. (12.5.15) cancel along the cut, and we are left with

$$F_\pm(x,y) = \left(2/\pi^2\right)\left[\int_0^\infty dz - \int_0^1 \exp[\pm\varphi(z)]dz\right]$$

$$\times\, k(z(x\pm 1/2))k(z(y\pm 1/2)), \qquad (12.5.16)$$

$$\varphi(z) = z - \operatorname{arctanh} z. \qquad (12.5.17)$$

The series expansion

$$\ln\Delta_\pm(\tau) = \mathrm{tr}\left[\ln(1-F_\pm)\right]_\tau^\infty = -\sum_1^\infty n^{-1}\mathrm{tr}\left[(F_\pm)^n\right]_\tau^\infty \qquad (12.5.18)$$

converges absolutely for all positve τ in the even case, and at least for

$$\tau > \frac{1}{2} + (4\pi)^{-1} \tag{12.5.19}$$

in the odd case. The formula (12.5.10) with (12.5.16) and (12.5.18) provides a practical method for computing the potentials $W_{\pm}(\tau)$, either by using the series expansion or by finding numerically the eigenvalues of the kernels $F_{\pm}$.

The relations (12.4.1) and (12.5.10), connecting $W_{\pm}(\tau)$ with $D_{\pm}(t)$ and $\Delta_{\pm}(\tau)$, have the consequence that the quantity

$$\ln D_{\pm}(t) + \frac{1}{4}\tau^2 + \frac{1}{8}\ln\left|\tau \pm \frac{1}{2}\right| - \ln \Delta_{\pm}(\tau), \quad \tau = \pi t, \tag{12.5.20}$$

is a linear function of t. But we know that as $t \longrightarrow \infty$ the behavior of $\ln D_{\pm}(t)$ is governed by Eq. (12.2.6), while $\ln \Delta_{\pm}(\tau)$ tends to zero. The asymptotic formula (12.2.6) can therefore be replaced by the identity

$$\ln D_{\pm}(t) = -\frac{1}{4}\tau^2 \mp \frac{1}{2}\tau - \frac{1}{8}\ln\left|\tau \pm \frac{1}{2}\right| \pm \frac{1}{4}\ln 2 + B + \ln \Delta_{\pm}(\tau). \tag{12.5.21}$$

with

$$B = \frac{1}{24}\ln 2 + \frac{3}{2}\zeta'(-1). \tag{12.5.22}$$

This identity establishes the desired connection between the Fredholm determinants defined on $[0, \tau]$ and those defined on $[\tau, \infty]$.

12.6. Asymptotic Expansions

We wish to obtain the asymptotic expansion of $\ln \Delta_{\pm}(\tau)$ in negative powers of $(\tau \pm 1/2)$. We can then use Eq. (12.5.21) to obtain the extension to negative powers of the asymptotic formula (12.2.6) for $\ln D_{\pm}(t)$. The expansion of $\ln \Delta_{\pm}(\tau)$ is derived from Eq. (12.5.18), the nth term of the sum giving contributions of order τ^{-3n} and higher. We expand the integrand of Eq. (12.5.16) in powers of z, so that

$$F_{\pm}(x, y) = \pm\frac{1}{3}F_3 \pm \frac{1}{5}F_5 - \frac{1}{18}F_6 \pm \frac{1}{7}F_7 + \cdots \tag{12.6.1}$$

with the term

$$F_m = (2/\pi^2) \int_0^\infty z^m k(z\,(x \pm 1/2)\,)k(z\,(y \pm 1/2)\,)dz \qquad (12.6.2)$$

homogeneous of degree $(-m-1)$ in $(x \pm 1/2)$ and $(y \pm 1/2)$. The error involved in extending the integral from 1 to ∞ is of order $e^{-\tau}$ and is negligible in an asymptotic expansion. When the trace of $(F_\pm)^n$ is calculated, we obtain terms proportional to $(\tau \pm 1/2)^{-N}$ by forming products

$$F_{m_1} F_{m_2} \cdots F_{m_n} \qquad (12.6.3)$$

with

$$m_1 + m_2 + \cdots + m_n = N. \qquad (12.6.4)$$

In Eq. (12.6.1) the odd terms carry the $\pm$ sign while the even terms do not. Equation (12.6.4) implies that terms with even N appear with the same coefficients in $\ln \Delta_+$ and $\ln \Delta_-$, while terms with odd N have equal and opposite coefficients. Therefore

$$\ln \Delta_\pm(\tau) = \sum (\pm 1)^m \, a_m \left(\tau \pm \frac{1}{2}\right)^{-m}. \qquad (12.6.5)$$

Moreover,

$$a_1 = a_2 = a_4 = 0, \qquad (12.6.6)$$

because no term of these orders can appear in any trace of $[F_\pm]^n$ according to Eq. (12.6.1). The formal relation

$$\ln \Delta_\pm(-\tau) = \ln \Delta_\mp(\tau) \qquad (12.6.7)$$

holds for the asymptotic expansions (12.6.5), but cannot hold as an identity. On the contrary, if $\Delta_\pm(-\tau)$ is defined by Eq. (12.5.21), using the convention

$$D_\pm(-t) = \det\left[1 + K_\pm\right]_0^\tau, \qquad (12.6.8)$$

which is the correct analytic continuation of Eq. (12.3.10), then Eqs. (12.6.7) is definitely false. There is nothing paradoxical in this failure of Eq. (12.6.7), since the expansion (12.6.5) is not convergent.

The coefficients a_m in Eq. (12.6.5) can all be computed as traces of products of kernels (12.6.2). But we can find the first two nonvanishing coefficients a_3 and a_5 without calculating any integrals, by using the identity (5.5.14). It is convenient to work with the derivatives

$$\begin{aligned} L_{\pm}[\tau] &= \frac{d}{d\tau} \ln D_{\pm}(t) \\ &= -\tau/2 \mp 1/2 - \frac{1}{8}(\tau \pm 1/2)^{-1} + \frac{d}{d\tau} \ln \Delta_{\pm}(\tau). \end{aligned} \tag{12.6.9}$$

and their sum and difference

$$U = L_+ + L_-, \qquad V = L_+ - L_-. \tag{12.6.10}$$

The identity (5.5.14) then becomes

$$U' + V^2 = 0. \tag{12.6.11}$$

Equations (12.6.9) and (12.6.5) give to order τ^{-6}

$$\begin{aligned} U &= -\tau - \frac{1}{4}\tau\left(\tau^2 - \frac{1}{4}\right)^{-1} + 12a_3\tau^{-5}, \\ V &= -1 + \frac{1}{8}\left(\tau^2 - \frac{1}{4}\right)^{-1} - 6a_3\tau^{-4} - (15a_3 + 10a_5)\tau^{-6}. \end{aligned} \tag{12.6.12}$$

When we substitute Eq. (12.6.12) into Eq. (12.6.11), we find that all terms vanish to order τ^{-6} provided that

$$a_3 = -(3/256), \tag{12.6.13}$$

$$a_5 = -(45/2048). \tag{12.6.14}$$

The vanishing of Eq. (12.6.11) provides a check of the consistency of our procedures. Unfortunately it is not possible to determine the coefficients beyond a_5 in this way, because Eq. (12.6.11) gives only one equation for each pair of unknowns (a_{2m}, a_{2m+1}).

Putting together Eqs. (12.6.5), (12.6.13), (12.6.14) we have the asymptotic expansion of $\ln \Delta_\pm(\tau)$ in powers of τ^{-1},

$$\ln \Delta_\pm(\tau) = \mp\frac{3}{256}\tau^{-3} + \frac{9}{512}\tau^{-4} \mp \frac{81}{2048}\tau^{-5} + \cdots \tag{12.6.15}$$

The potentials that give rise to the phase shifts (12.4.36) have the expansion

$$\begin{aligned} W_\pm(\tau) = & -\frac{1}{4}\tau^{-2} \pm \frac{1}{4}\tau^{-3} - \frac{3}{16}\tau^{-4} \\ & \pm \frac{13}{32}\tau^{-5} - \frac{25}{32}\tau^{-6} \pm \frac{1239}{512}\tau^{-7} - \cdots. \end{aligned} \tag{12.6.16}$$

Finally, the Fredholm determinants $D_\pm(t)$ have the asymptotic expansions

$$\begin{aligned} \ln D_\pm(t) = & -\frac{1}{4}\pi^2 t^2 \mp \frac{1}{2}\pi t - \frac{1}{8}\ln\left|\pi t \pm \frac{1}{2}\right| \pm \frac{1}{4}\ln 2 + B \\ & \mp \frac{3}{256}\left(\pi t \pm \frac{1}{2}\right)^{-3} \mp \frac{45}{2048}\left(\pi t \pm \frac{1}{2}\right)^{-5} + \cdots \end{aligned} \tag{12.6.17}$$

with

$$B = \frac{1}{24}\ln 2 + \frac{3}{2}\zeta'(-1) \approx -0.219\ 250\ 583. \tag{12.6.18}$$

The numerical value of B is discussed in McCoy and Wu (1973), Appendix B. Dyson gave a few more terms in the above expansion, namely

$$a_6 = \frac{63}{4096}, \quad a_7 = -\frac{3^2 \times 5^3 \times 7}{2^{16}}, \quad a_8 = \frac{3^2 \times 5^3 \times 23}{2^{17}}. \tag{12.6.19}$$

The numerical precision of Eq. (12.6.17) is very good both for $D_+(t)$ and $D_-(t)$ beyond $t \geq 1$.

Summary of Chapter 12

Three different methods of finding the asymptotic behavior of $E_\beta(0;s)$ for large s are described. These methods are complimentary and result in the asymptotic series

$$\ln D_{\pm}(t) = -\frac{1}{4}\pi^2 t^2 \mp \frac{1}{2}\pi t - \frac{1}{8}\ln\left|\pi t \pm \frac{1}{2}\right| \pm \frac{1}{4}\ln 2 + B$$
$$+ \sum_{m=3}^{\infty} (\pm 1)^m a_m \left(\pi t \pm \frac{1}{2}\right)^{-m}, \tag{12.6.17}$$

with

$$B = \frac{1}{24}\ln 2 + \frac{3}{2}\zeta'(-1) \approx -0.219\,250\,583, \tag{12.6.18}$$

and

$$a_3 = -\frac{3}{256}, \quad a_4 = 0, \quad a_5 = -\frac{45}{2048}, \quad a_6 = \frac{63}{4096}, \quad \dots .$$

The probability $E_\beta(0; s)$ that a randomly chosen interval of length $s = 2t$ contains no eigenvalue is given in terms of the $D_{\pm}(t)$ as

$$E_1(0; s) = D_+(t),$$

$$E_2(0; s) = D_+(t) D_-(t),$$

and

$$E_4(0; s/2) = \frac{1}{2}\,[D_+(t) + D_-(t)].$$

13 / Gaussian Ensemble of Antisymmetric Hermitian Matrices

This short chapter deals with the statistical properties of the eigenvalues of an anti-symmetric Hermitian matrix whose elements are Gaussian random variables. While of no immediate physical interest, the precise analytical results obtained merit some mention for their mathematical elegance.

13.1. Level Density. Correlation Functions

In Section 3.4 we saw that the eigenvalues of an antisymmetric Hermitian matrix always come in pairs $\pm x_i$, x_i real. Let $x_1, \ldots, x_N$ be the positive eigenvalues of H. The probability density $\exp\left(-\text{tr } H^2/2\right)$ for H implies for $x_1, \ldots, x_N$ the joint probability density (*cf.* Section 3.4),

$$P_{2N}\left(x_1, \ldots, x_N\right) = C \prod_{j=1}^{N} \exp\left(-x_j^2\right) \prod_{1 \le j < k \le N} \left(x_j^2 - x_k^2\right)^2, \qquad (13.1.1)$$

if H is of even order $2N$, and

$$P_{2N+1}\left(x_1, \ldots, x_N\right) = C \prod_{j=1}^{N} x_j^2 \exp\left(-x_j^2\right) \prod_{1 \le j < k \le N} \left(x_j^2 - x_k^2\right)^2, \qquad (13.1.2)$$

if H is of odd order, $2N+1$. The constant C in the above two equations is not the same. As in Chapters 5 and 6, it is convenient to introduce the oscillator functions

$$\varphi_j\left(x\right) = \left(2^j j! \sqrt{\pi}\right)^{-1/2} \exp\left(\frac{x^2}{2}\right) \left(-\frac{d}{dx}\right)^j e^{-x^2}, \qquad (13.1.3)$$

so that Eqs. (13.1.1) and (13.1.2) can be written as

$$
\begin{aligned}
P_{2N}(x_1, \ldots, x_N) &= \frac{2^N}{N!} \left(\det\left[\varphi_{2j-2}(x_k)\right]\right)^2 \\
&= \frac{1}{N!} \det\left[K_N^{+}(x_j, x_k)\right]_{j,k=1,\ldots,N}, \qquad (13.1.4)
\end{aligned}
$$

$$
\begin{aligned}
P_{2N+1}(x_1, \ldots, x_N) &= \frac{2^N}{N!} \left(\det\left[\varphi_{2j-1}(x_k)\right]\right)^2 \\
&= \frac{1}{N!} \det\left[K_N^{-}(x_j, x_k)\right]_{j,k=1,\ldots,N}, \qquad (13.1.5)
\end{aligned}
$$

where

$$K_N^{+}(x, y) = 2 \sum_{j=0}^{N-1} \varphi_{2j}(x) \varphi_{2j}(y), \qquad (13.1.6)$$

$$K_N^{-}(x, y) = 2 \sum_{j=0}^{N-1} \varphi_{2j+1}(x) \varphi_{2j+1}(y), \qquad (13.1.7)$$

and the constants have been chosen such that

$$
\begin{aligned}
&\int_0^\infty \cdots \int_0^\infty P_{2N}(x_1, \ldots, x_N)\, dx_1 \cdots dx_N \\
&= \int_0^\infty \cdots \int_0^\infty P_{2N+1}(x_1, \ldots, x_N)\, dx_1 \cdots dx_N \\
&= 1. \qquad (13.1.8)
\end{aligned}
$$

The $K_N^{\pm}(x, y)$ satisfy the conditions of Theorem 5.2.1, namely

$$\int_0^\infty K_N(x, x)\, dx = N, \qquad (13.1.9)$$

$$\int_0^\infty K_N(x, y) K_N(y, z)\, dy = K_N(x, z). \qquad (13.1.10)$$

Hence the n-level correlation function is

$$R_n(x_1, \ldots, x_N) = \det\left[K_N(x_j, x_k)\right]_{j,k=1,\ldots,n}. \tag{13.1.11}$$

In particular the level density is

$$\begin{aligned} \sigma_{2N}(x) &= 2\sum_{j=0}^{N-1} \varphi_{2j}^2(x) \\ &= 4\sqrt{N}\left(A_N(x) - \frac{1}{4x}\varphi_{2N}(x)\,\varphi_{2N-1}(x)\right), \end{aligned} \tag{13.1.12}$$

$$\begin{aligned} \sigma_{2N+1}(x) - \delta(x) &= 2\sum_{j=0}^{N-1} \varphi_{2j+1}^2(x) \\ &= 4\sqrt{N}\left(A_N(x) + \frac{1}{4x}\varphi_{2N}(x)\,\varphi_{2N-1}(x)\right), \end{aligned} \tag{13.1.13}$$

where

$$A_N(x) = \frac{1}{4\sqrt{N}}\sum_{j=0}^{2N-1} \varphi_j^2(x). \tag{13.1.14}$$

In the limit $N \longrightarrow \infty$, (*cf.* Appendices A.9 and A.10),

$$A_N(x) \approx \frac{1}{2\pi}\left(1 - \frac{x^2}{4N}\right)^{1/2}, \tag{13.1.15}$$

$$\frac{1}{4x}\varphi_{2N}(x)\,\varphi_{2N-1}(x) \approx -\frac{1}{2\pi}\frac{\sin(2\xi)}{2\xi}, \tag{13.1.16}$$

$$\begin{aligned} K_N^{\pm}(x,y) &= \sqrt{N}\left\{\frac{\varphi_{2N}(x)\,\varphi_{2N-1}(y) - \varphi_{2N}(y)\,\varphi_{2N-1}(x)}{x-y}\right. \\ &\quad \left.\mp \frac{\varphi_{2N}(x)\,\varphi_{2N-1}(y) + \varphi_{2N}(y)\,\varphi_{2N-1}(x)}{x+y}\right\} \\ &\approx \frac{2}{\pi}\sqrt{N}\left(\frac{\sin(\xi-\eta)}{\xi-\eta} \pm \frac{\sin(\xi+\eta)}{\xi+\eta}\right), \end{aligned} \tag{13.1.17}$$

where $\xi = 2x\sqrt{N}$, $\eta = 2y\sqrt{N}$.

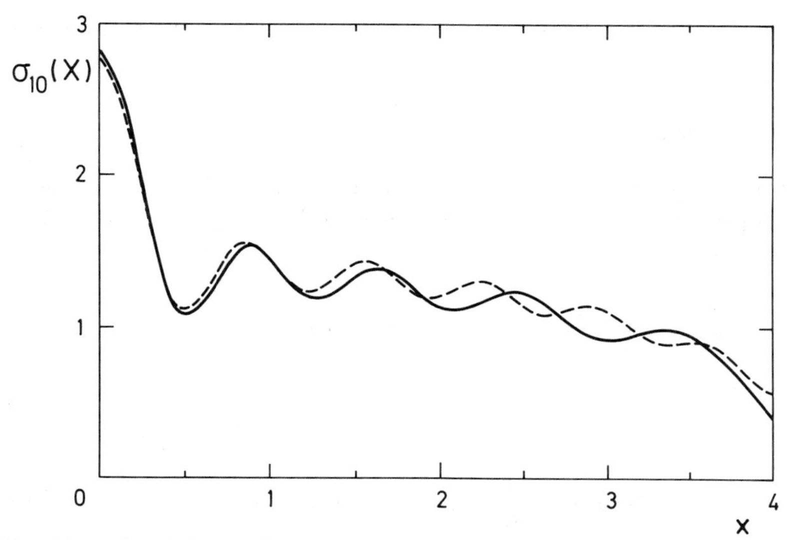

FIG. 13.1. Level density for 10×10 antisymmetric Hermitian matrices. The solid curve is the exact result, Eq. (13.1.12), while the dashed curve is its asymptotic expression, Eqs. (13.1.15) and (13.1.16). Reprinted with permission from Elsevier Science Publishers, Mehta, M. L., and Rosenzweig, N. Distribution laws for the roots of a random anti-symmetric Hermitian matrix, *Nucl. Phys.*, A109, 449–456 (1968).

It is curious to note that near the origin the level density contains a term $\sin(2\xi)/(2\xi)$, and therefore has large oscillations. Thus the average spacing has to be defined in an artificial way. Figure 13.1 is a plot of $\sigma_{2N}(x)$ for $N = 5$ and its asymptotic expression.

13.2. Level Spacings

In considering level spacings, one has to distinguish between two situations: (i) central spacings, i.e., the probability of having the smallest positive eigenvalue at $s/2$, and the largest negative one therefore at $-s/2$, and (ii) noncentral spacings, i.e., the probability of having two adjacent eigenvalues at α and β, $\beta \geq \alpha > 0$, and simultaneously therefore two other adjacent ones at $-\alpha$ and $-\beta$. This noncentral spacing distribution will depend not only on $(\beta - \alpha)$ but very strongly on $(\beta + \alpha)/2$ as well. This dependence is expected from the oscillatory behavior of the level density. As we move away from the origin, the dependence on $(\beta + \alpha)/2$

decreases and at about 20 to 30 mean spacings from the origin becomes negligible.

13.2.1. CENTRAL SPACINGS

For definiteness we shall take the matrices to be of even order $2N$. The probability that all positive eigenvalues are $\geq \alpha$, and therefore all negative ones $\leq -\alpha$, is

$$\int_{\alpha}^{\infty} \cdots \int_{\alpha}^{\infty} \frac{2^N}{N!} \left(\det \left[\varphi_{2j-2}(x_k)\right]\right)^2 dx_1 \cdots dx_N$$

$$= \det \left[2 \int_{\alpha}^{\infty} \varphi_{2j-2}(x) \varphi_{2k-2}(x) \, dx\right]_{j,k=1,2,\ldots,N} \tag{13.2.1}$$

$$= \det \left[\delta_{jk} - \int_{-\alpha}^{\alpha} \varphi_{2j-2}(x) \varphi_{2k-2}(x) \, dx\right]_{j,k=1,2,\ldots,N} \tag{13.2.2}$$

$$= D_N(\alpha) \underset{N \longrightarrow \infty}{\approx} D_+\left(\frac{2}{\pi} \alpha N^{1/2}\right), \tag{13.2.3}$$

where the function $D_+(t)$ is given by Eq. (5.5.9); i.e.,

$$D_+(t) = \prod_j (1 - \lambda_{2j}), \tag{13.2.4}$$

and the λ_{2j} are the eigenvalues of the integral equation (5.3.14) corresponding to the even solutions, i.e.,

$$2\pi\lambda f(x) = \int_{-t}^{t} \left[\frac{\sin(x+y)\pi}{x+y} + \frac{\sin(x-y)\pi}{x-y}\right] f(y) \, dy. \tag{13.2.5}$$

The probability that there is one eigenvalue at α $(\alpha \geq 0)$ and none in the interval $(0, \alpha)$, and therefore one at $-\alpha$ and none in $(-\alpha, 0)$, is obtained by differentiation. The result is

$$P(-\alpha, \alpha) = -\frac{1}{2} \frac{d}{d\alpha} D_N(\alpha)$$

$$\underset{N \longrightarrow \infty}{\approx} -\frac{1}{\pi} N^{1/2} D'_+\left(\frac{1}{\pi} S N^{1/2}\right), \tag{13.2.6}$$

where $S = 2\alpha$ is the spacing. The average value of the central spacings is

$$\begin{aligned}\langle S\rangle &= \int_0^\infty 2\alpha P(-\alpha,\alpha)d\alpha \\ &= -\int_0^\infty \alpha\frac{d}{d\alpha}D_N(\alpha)d\alpha \\ &\approx -\frac{\pi}{2\sqrt{N}}\int_0^\infty t\frac{d}{dt}D_+(t)dt = CN^{-1/2},\end{aligned} \tag{13.2.7}$$

where the constant C is

$$C = \frac{\pi}{2}\int_0^\infty D_+(t)\,\mathrm{d}t \approx 1.0097. \tag{13.2.8}$$

The cumulative distribution function of the central spacings is

$$F(S) = \int_0^S P(S_1)\,\mathrm{d}S_1 = 1 - D_+\left(\frac{C}{\pi}\frac{S}{\langle S\rangle}\right). \tag{13.2.9}$$

Note that the occurrence of levels at α and $-\alpha$ are not independent events. Thus we should not divide the result by the probability of having a level at $-\alpha$. Also a factor $1/2$ is introduced in Eq. (13.2.6) to take this into account.

13.2.2. NONCENTRAL SPACINGS

The probability that there are no eigenvalues in the interval (α,β), $\beta \ge \alpha > 0$, and hence none in the interval $(-\beta,-\alpha)$ is given by

$$\int_{\mathrm{out}(\alpha,\beta)}\cdots\int \frac{2^N}{N!}\left\{\det\left[\varphi_{2j-2}(x_k)\right]\right\}^2 dx_1\cdots dx_N \tag{13.2.10}$$

$$= \det\left[\delta_{jk} - 2\int_\alpha^\beta \varphi_{2j-2}(x)\,\varphi_{2k-2}(x)\,dx\right]_{j,k=1,2,\ldots,N} \tag{13.2.11}$$

where out (α,β) means that the integrations are carried from 0 to α and from β to ∞, the interval (α,β) being left out. The probability of having one level at α, another at β, and none in between can then be obtained (*cf.* Section 5.1) by two differentiations of the expression (13.2.11). In the limit of large N, expression (13.2.11) simplifies to

$$A\left(\frac{4}{\pi}\alpha N^{1/2}, \frac{4}{\pi}\beta N^{1/2}\right), \tag{13.2.12}$$

where

$$A(a,b) = \prod_j (1-\lambda_j), \tag{13.2.13}$$

and the λ_j are the eigenvalues of the integral equation

$$\pi\lambda f(x) = \int_a^b \left[\frac{\sin(x+y)\pi}{x+y} + \frac{\sin(x-y)\pi}{x-y}\right] f(y)\, dy. \tag{13.2.14}$$

However, if a and b are far removed from the origin, we may neglect $\sin[(x+y)\pi]/(x+y)$ in comparison with $\sin[(x-y)\pi]/(x-y)$. Thus for large values of $(a+b)/2$, Eq. (13.2.14) reduces to

$$\pi\lambda f(x) = \int_{-(b-a)/2}^{(b-a)/2} \frac{\sin(x-y)\pi}{x-y} f(y)\, dy, \tag{13.2.15}$$

whose solutions are known to be spheroidal functions (*cf.* Section 5.3). The final result is

$$A(a,b) \approx D_+(b-a)\, D_-(b-a) \tag{13.2.16}$$

where $D_+(b-a)$ and $D_-(b-a)$ are given by Eqs. (5.5.9) and (5.5.10).

Summary to Chapter 13

For a Gaussian ensemble of antisymmetric Hermitian matrices we give the level density, the n-point correlation function and the probability density for the nearest neighbor spacings. The results depend sensitively on whether the order of the matrices is even or odd and in what region of the level density lie the spacings considered.

14 / Another Gaussian Ensemble of Hermitian Matrices

In Chapters 5, 6, and 7 we studied three particular Gaussian ensembles.

(i) The ensemble of Hermitian matrices with equally probable real and imaginary parts (for quaternion matrices, equally probable self-dual and anti-self-dual parts), known as the Gaussian unitary ensemble (GUE).

(ii) The ensemble of real symmetric matrices, known as the Gaussian orthogonal ensemble (GOE).

(iii) The ensemble of self-dual quaternion matrices, known as the Gaussian symplectic ensemble (GSE).

These three ensembles are basic models for energy level fluctuations of complex systems. For time-reversal invariant systems, GOE or GSE is appropriate, depending on the properties of the Hamiltonian. On the other hand in the absence of time-reversal symmetry, the GUE is appropriate. For nuclear spectra GOE appears to be a good model, while for the zeros of the Riemann zeta function on the critical line GUE seems to apply.

Gaussian ensembles with an arbitrary ratio of the mean square values of their real and imaginary parts (of time-reversal invariant and non-invariant parts) are of interest due to a suggestion of Wigner, that the analysis of data in comparision with such ensembles may give an upper bound on the time-reversal breaking part of the nuclear forces. As we will see, the transition in the fluctuation properties for the ensemble going from GOE to GUE adequate for the above purpose, is very rapid and, in fact, discontinuous for infinite order matrices. The situation is similar for ensembles going from GSE to GUE.

In this chapter we will study two ensembles of Hermitian matrices, the elements of which are Gaussian random variables. In one case the real and imaginary parts of the matrix elements have mean square values $2v^2$ and $2v^2\alpha^2$, respectively. In the other case the self-dual and anti-self-dual parts of the (quaternion) matrix elements have mean square values $2v^2$ and $2v^2\alpha^2$, respectively. In particular, we will derive all correlation and cluster functions of the eigenvalues for finite or infinite order matrices. Though only the case $\alpha^2 \ll 1$ is relevant for immediate physical applications, these results are obtained with little extra effort and it is worthwhile to record them for their mathematical elegance.

In Sections 14.1 and 14.2 we describe the ensembles and give a summary of the results. Later sections deal with the proofs; in Section 14.3 we derive the joint probability density for the eigenvalues and in Section 14.4 the correlation and cluster functions for finite and infinite order matrices. Recurrence relations and integrals which are frequently used in this chapter will be found in Appendix A.26; the normalization integrals for the joint probability density of the eigenvalues is given in Appendices A.27 and A.28; Appendices A.29 and A.30 contain the verification of some equations of Section 14.4, while Appendix A.31 deals with some asymptotic forms when the order N of the matrices is large.

14.1. Summary of Results. Matrix Ensembles from GOE to GUE and Beyond

Consider an ensemble of $N \times N$ Hermitian matrices $[H_{jk}] = [R_{jk} + iS_{jk}]$, with R real symmetric and S real antisymmetric, i.e.,

$$R_{jk} = R^*_{jk} = R_{kj}, \qquad S_{jk} = S^*_{jk} = -S_{kj}. \tag{14.1.1}$$

The joint probability density for the matrix elements is taken to be

$$P(H) = c \exp\left[-\sum_{j,k}\left(\frac{R^2_{jk}}{4v^2} + \frac{S^2_{jk}}{4v^2\alpha^2}\right)\right], \tag{14.1.2}$$

$$dH = \prod_{j\le k} dR_{jk} \prod_{j<k} dS_{jk}, \tag{14.1.3}$$

where the normalization constant c is

$$c = 2^{-N/2}\alpha^{-N(N-1)/2}\left(2\pi v^2\right)^{-N^2/2}. \tag{14.1.4}$$

On the average

$$\frac{\left\langle \|\mathrm{Im}\ H\|^2 \right\rangle}{\left\langle \|\mathrm{Re}\ H\|^2 \right\rangle} = \frac{\left\langle \|S\|^2 \right\rangle}{\left\langle \|R\|^2 \right\rangle} = \frac{N-1}{N+1}\alpha^2 \approx \alpha^2 \quad \text{for large } N. \tag{14.1.5}$$

We shall choose the scale v^2 such that

$$2v^2\left(1+\alpha^2\right) = 1. \tag{14.1.6}$$

As special cases we have (i) $\alpha^2 = 0$, so that $S = 0$ with probability one and the matrices H form the GOE; (ii) $\alpha^2 = 1$, on the average R and S have the same magnitude for large N and the ensemble is GUE; (iii) $\alpha^2 \longrightarrow \infty$, S dominates R, and the ensemble of H may be referred to as the anti-symmetric Gaussian orthogonal ensemble (*cf.* Chapter 13).

The joint probability density for the eigenvalues $x_1, ..., x_N$ of H is

$$p\left(x_1, ..., x_N\right) = C_N \exp\left[-\frac{1}{2}\left(1+\alpha^2\right)\sum x_j^2\right] \times \Delta\left(x_1, ..., x_N\right) \mathrm{Pf}\left[F_{ij}\right], \tag{14.1.7}$$

where

$$\Delta(x) \equiv \Delta\left(x_1, ..., x_N\right) = \prod_{1\le i<j\le N}\left(x_i - x_j\right), \tag{14.1.8}$$

and $\mathrm{Pf}\left[F_{ij}\right]$ is the Pfaffian of a $2m \times 2m$ matrix defined below. If N is even, $2m = N$, we define

$$F_{ij} = f\left(x_i - x_j\right), \quad i, j = 1, 2, ..., N. \tag{14.1.9}$$

If N is odd, $2m = N+1$, we use the preceding definition and in addition

$$F_{i,N+1} = -F_{N+1,i} = 1, \quad i = 1, 2, ..., N, \quad F_{N+1,N+1} = 0 \tag{14.1.9$'$}$$

with

$$f(x) = \operatorname{erf}\left[\left(\frac{1-\alpha^4}{4\alpha^2}\right)^{1/2} x\right]$$
$$\equiv \left(\frac{1-\alpha^4}{\pi\alpha^2}\right)^{1/2} \int_0^x \exp\left(-\frac{(1-\alpha^4)y^2}{4\alpha^2}\right) dy. \qquad (14.1.10)$$

The $[F_{ij}]$ is an antisymmetric matrix of even order and its Pfaffian is, apart from a sign, the square root of its determinant. (Note: Let $m = [N/2]$ be the integral part of $N/2$. The Pfaffian (see, e.g., Mehta, 1989) of any $N \times N$ antisymmetric matrix $A = [a_{ij}]$ is the alternating multilinear form in the elements a_{ij} with $i < j$,

$$\text{Pf } A \equiv \text{Pf}\,[a_{ij}] = \sum \pm a_{i_1 i_2} a_{i_3 i_4} ... a_{i_{2m-1} i_{2m}},$$

where the sum is taken over all permutations $(i_1, i_2, ..., i_N)$ of $(1, 2, ..., N)$ with the restrictions $i_1 < i_2$, $i_3 < i_4$, ..., $i_{2m-1} < i_{2m}$, $i_1 < i_3 < \cdots < i_{2m-1}$, and the sign is $+$ or $-$ according to whether the permutation is even or odd. With this general definition $\text{Pf}\,[F_{ij}]$ is just the Pfaffian of the $N \times N$ antisymmetric matrix $[f(x_i - x_j)]$.) The normalization constant C_N is

$$C_N^{-1} = 2^{3N/2}\left(1-\alpha^2\right)^{N(N-1)/4}\left(1+\alpha^2\right)^{-N(N+1)/4} \prod_{j=1}^{N} \Gamma(1+j/2). \qquad (14.1.11)$$

Note that when $\alpha^2 > 1$, $f(x)$ and C_N are both pure imaginary but $p(x_1, ..., x_N)$ remains real positive. Moreover, for $\alpha^2 = 0, 1$ or ∞, $p(x_1, ..., x_N)$ and all the quantities derived below have well defined limits.

All the eigenvalue correlations can be expressed in terms of functions of two variables. The n-level correlation function (*cf.* Section 5.1) is

$$R_n(x_1, ..., x_n) = \left\{\det\left[\phi(x_i, x_j)\right]_{i,j=1,...,n}\right\}^{1/2} \qquad (14.1.12)$$

and the n-level cluster function (cumulant of the preceding) is

$$T_n(x_1, ..., x_n) = \frac{1}{2} \mathrm{tr} \Sigma \phi(x_1, x_2) \phi(x_2, x_3) \cdots \phi(x_n, x_1), \qquad (14.1.13)$$

where

$$\phi(x, y) = \begin{bmatrix} \xi_N(x, y) + S_N(x, y) & D_N(x, y) \\ J_N(x, y) & \xi_N(y, x) + S_N(y, x) \end{bmatrix} \qquad (14.1.14)$$

and the sum in Eq. (14.1.13) is taken over all $(n-1)!$ distinct cyclic permutations of the indices $(1, 2, ..., n)$. Note the interchange of x and y in the lower right-hand corner of Eq. (14.1.14). The two-point functions ξ_N, S_N, D_N, and J_N are given in terms of other two-point functions $I_N(x, y)$, $g(x, y)$, $\mu_N(x, y)$ and the one-point functions $\varphi_j(x)$, $\psi_j(x)$, $A_j(x)$, and $\varepsilon(x)$ defined below.

$$\xi_N(x, y) = \begin{cases} \dfrac{\varphi_{2m}(x) \exp\left(-\alpha^2 y^2/2\right)}{\displaystyle\int_{-\infty}^{\infty} \varphi_{2m}(t) \exp\left(-\alpha^2 t^2/2\right) dt}, & N = 2m+1 \text{ odd}, \\ 0, & N = 2m \text{ even}, \end{cases} \qquad (14.1.15)$$

$$S_N(x, y) = \sum_{j=0}^{N-1} \varphi_j(x) \varphi_j(y) + \left(1 - \alpha^2\right) \left(\frac{1+\alpha^2}{1-\alpha^2}\right)^N (N/2)^{1/2} \varphi_{N-1}(x) A_N(y) \qquad (14.1.16)$$

$$= \begin{cases} \displaystyle\sum_{j=0}^{m-1} \left[\varphi_{2j}(x) \varphi_{2j}(y) - \psi_{2j}(x) A_{2j}(y)\right], & N = 2m \text{ even}, \\ \displaystyle\sum_{j=0}^{m-1} \left[\varphi_{2j+1}(x) \varphi_{2j+1}(y) - \psi_{2j+1}(x) A_{2j+1}(y)\right], & N = 2m+1 \text{ odd} \end{cases} \qquad (14.1.17)$$

$$D_N(x,y) = \sum_{j=0}^{N-1} \varphi_j(x)\psi_j(y) + (1-\alpha^2)\left(\frac{1+\alpha^2}{1-\alpha^2}\right)^N (N/2)^{1/2}\, \varphi_{N-1}(x)\varphi_N(y) \quad (14.1.18)$$

$$= \begin{cases} \sum_{j=0}^{m-1} [\varphi_{2j}(x)\psi_{2j}(y) - \psi_{2j}(x)\varphi_{2j}(y)], & N = 2m \text{ even}, \\ \sum_{j=0}^{m-1} [\varphi_{2j+1}(x)\psi_{2j+1}(y) - \psi_{2j+1}(x)\varphi_{2j+1}(y)], & N = 2m+1 \text{ odd}, \end{cases} \quad (14.1.19)$$

$$J_N(x,y) = I_N(x,y) + g(x,y) + \mu_N(x,y) - \mu_N(y,x), \quad (14.1.20)$$

where

$$I_N(x,y) = \sum_{j=0}^{N-1} \varphi_j(x)A_j(y) + (1-\alpha^2)\left(\frac{1+\alpha^2}{1-\alpha^2}\right)^N (N/2)^{1/2}\, A_N(x)A_{N-1}(y) \quad (14.1.21)$$

$$= \begin{cases} \sum_{j=0}^{m-1} [\varphi_{2j}(x)A_{2j}(y) - A_{2j}(x)\varphi_{2j}(y)], & N = 2m \text{ even}, \\ \sum_{j=0}^{m-1} [\varphi_{2j+1}(x)A_{2j+1}(y) - A_{2j+1}(x)\varphi_{2j+1}(y)], & N = 2m+1 \text{ odd}, \end{cases} \quad (14.1.22)$$

$$g(x,y) = \frac{1}{2}\left(\frac{1+\alpha^2}{1-\alpha^2}\right)^{1/2} \exp\left[-\frac{1}{2}\alpha^2\left(x^2+y^2\right)\right] f(x-y) \tag{14.1.23}$$

$$\begin{aligned} &= \sum_{j=0}^{\infty} A_j(x)\varphi_j(y) = \frac{1}{2}\sum_{j=0}^{\infty}\left[A_j(x)\varphi_j(y) - \varphi_j(x)A_j(y)\right] \\ &= \sum_{j=0}^{\infty}\left[A_{2j}(x)\varphi_{2j}(y) - \varphi_{2j}(x)A_{2j}(y)\right] \\ &= \sum_{j=0}^{\infty}\left[A_{2j+1}(x)\varphi_{2j+1}(y) - \varphi_{2j+1}(x)A_{2j+1}(y)\right], \end{aligned} \tag{14.1.24}$$

$$\mu_N(x,y) = \begin{cases} \dfrac{\exp\left(-\alpha^2x^2/2\right) A_{2m}(y)}{\displaystyle\int_{-\infty}^{\infty} \varphi_{2m}(t)\exp\left(-\alpha^2t^2/2\right) dt}, & N = 2m+1 \text{ odd}, \\ 0, & N = 2m \text{ even}, \end{cases} \tag{14.1.25}$$

$$\varphi_j(x) = \left(2^j j!\sqrt{\pi}\right)^{-1/2} \exp\left(x^2/2\right)(-d/dx)^j \exp\left(-x^2\right), \tag{14.1.26}$$

$$\begin{aligned} \psi_j(x) = &\left(\frac{1+\alpha^2}{1-\alpha^2}\right)^j \exp\left(\alpha^2x^2/2\right) \\ &\times (d/dx)\left[\exp\left(-\alpha^2x^2/2\right)\varphi_j(x)\right], \end{aligned} \tag{14.1.27}$$

$$\begin{aligned} A_j(x) = &\left(\frac{1-\alpha^2}{1+\alpha^2}\right)^j \exp\left(-\alpha^2x^2/2\right) \\ &\times \int_{-\infty}^{\infty} \exp\left(\alpha^2t^2/2\right)\varepsilon(x-t)\varphi_j(t)dt, \end{aligned} \tag{14.1.28}$$

$$\varepsilon(x) = \frac{1}{2}\text{sgn } x = \begin{cases} 1/2, & x > 0, \\ 0, & x = 0, \\ -1/2, & x < 0. \end{cases} \tag{14.1.29}$$

To see the equivalence of the pairs of equations (14.1.16), (14.1.17); (14.1.18), (14.1.19); (14.1.21), (14.1.22); and (14.1.23), (14.1.24) see Appendix A.26.

Note that $\varphi_{2j}(x)$, $\psi_{2j+1}(x)$, and $A_{2j+1}(x)$ are even functions, while $\varphi_{2j+1}(x)$, $\psi_{2j}(x)$ and $A_{2j}(x)$ are odd functions of x.

For large N, the level density $R_1(x)$ becomes a semicircle for all α

$$R_1(x) \approx \pi^{-1}\left(2N - x^2\right)^{1/2}. \tag{14.1.30}$$

The n-level correlation and cluster functions for $n > 1$ are discontinuous in α^2 at $\alpha^2 = 0$. We have the GOE results for $\alpha^2 = 0$ and the GUE results for $\alpha^2 > 0$. On the other hand if (RMS being the abbreviation of root mean square)

$$\rho \equiv \rho(x) = \frac{\text{RMS value of } (\text{Im } H_{ij})}{\text{Local average spacing at } x} \approx \frac{\alpha}{\sqrt{2}} R_1(x) \tag{14.1.31}$$

remains finite when $\alpha^2 \longrightarrow 0$ and $N \longrightarrow \infty$, the n-level correlation and cluster functions have well defined limits for all ρ. With $x_i - x_j \longrightarrow 0$ and $(x_i - x_j)\, R_1(x_i) \longrightarrow r_{ij} = r_i - r_j$, the n-level correlation function becomes

$$\begin{aligned} R_n(r_1, \ldots, r_n; \rho) \equiv & \lim_{N\to\infty} \{R_1(x_1), \ldots, R_1(x_n)\}^{-1} \\ & \times R_n(x_1, \ldots, x_n) \\ = & \left\{\det\left[\sigma(r_{ij}; \rho)\right]_{i,j=1,\ldots,n}\right\}^{1/2} \end{aligned} \tag{14.1.32}$$

and the corresponding n-level cluster function is

$$Y_n(r_1, \ldots, r_n; \rho) \equiv \lim_{N\to\infty} \{R_1(x_1), \ldots, R_1(x_n)\}^{-1}$$

$$\times T_n(x_1, ..., x_n)$$

$$= \frac{1}{2} \mathrm{tr} \sum \sigma(r_{12}; \rho)\, \sigma(r_{23}; \rho), ..., \sigma(r_{n1}; \rho) \quad (14.1.33)$$

where the sum is taken, as in Eq. (14.1.13), over all $(n-1)!$ distinct cyclic permutations of the indices $(1, 2, ..., n)$ and

$$\sigma(r;\rho) = \begin{bmatrix} (\sin \pi r)/\pi r & D(r;\rho) \\ J(r;\rho) & (\sin \pi r)/\pi r \end{bmatrix}, \quad (14.1.34)$$

is considered as an ordinary 2×2 matrix (not a quaternion!), with

$$D(r;\rho) = -\frac{1}{\pi} \int_0^{\pi} k \sin kr \exp\left(2\rho^2 k^2\right) dk, \quad (14.1.35)$$

$$J(r;\rho) = -\frac{1}{\pi} \int_{\pi}^{\infty} \frac{\sin kr}{k} \exp\left(-2\rho^2 k^2\right) dk. \quad (14.1.36)$$

Note that as $\rho \longrightarrow \infty$, $D \longrightarrow \infty$, $J \longrightarrow 0$, while the product $JD \longrightarrow 0$, so that in the expressions for R_n and Y_n one may replace D and J by zeros. For the asymptotic forms of J and D see Appendix A.31.

If we put $\alpha = 0$, we get back the results of Chapter 6. Putting $\alpha = 1$ is more delicate, since it involves a limiting process. Similarly, $\alpha \longrightarrow \infty$ is delicate.

14.2. Matrix Ensembles from GSE to GUE and Beyond

As in Section 14.1, consider an ensemble of $N \times N$ Hermitian quaternion matrices

$$[\mathbf{H}_{jk}] = [\mathbf{R}_{jk} + i\mathbf{S}_{jk}], \quad (14.2.1)$$

where $\mathbf{R}$ and $\mathbf{S}$ are real self-dual and real anti-self-dual quaternion matrices respectively, i.e., with

$$\mathbf{R} = \mathbf{R}^0 + \sum_{\mu=1}^{3} \mathbf{R}^{\mu} e_{\mu}, \quad \mathbf{S} = \mathbf{S}^0 + \sum_{\mu=1}^{3} \mathbf{S}^{\mu} e_{\mu}; \quad (14.2.2)$$

$\mathbf{R}^0$ and $\mathbf{S}^{\mu}$ are real symmetric while $\mathbf{R}^{\mu}$ and $\mathbf{S}^0$ are real antisymmetric.

The probability density for the matrix elements will be taken as

$$\mathbf{P}(\mathbf{H}) = c \exp\left[-\sum_{j,k}\sum_{\mu=0}^{3}\left(\frac{\left(\mathbf{R}_{jk}^{\mu}\right)^2}{4v^2} + \frac{\left(\mathbf{S}_{jk}^{\mu}\right)^2}{4v^2\alpha^2}\right)\right], \quad (14.2.3)$$

$$d\mathbf{H} = \prod_{j\leq k}\left(d\mathbf{R}_{jk}^{0}\prod_{\mu=1}^{3} d\mathbf{S}_{jk}^{\mu}\right)\prod_{j<k}\left(d\mathbf{S}_{jk}^{0}\prod_{\mu=1}^{3} d\mathbf{R}_{jk}^{\mu}\right), \quad (14.2.4)$$

with the normalization constant

$$c = 2^{-2N}\alpha^{-N(2N+1)}\left(2\pi v^2\right)^{-2N^2}. \quad (14.2.5)$$

On the average the ratio of the anti-self-dual and self-dual parts of $\mathbf{H}$ is

$$\left\langle \|\mathbf{S}\|^2\right\rangle / \left\langle \|\mathbf{R}\|^2\right\rangle = \left[(2N+1)/(2N-1)\right]\alpha^2 \approx \alpha^2 \quad \text{for large } N. \quad (14.2.6)$$

We shall again choose the scale $2v^2\left(1+\alpha^2\right) = 1$ as in (14.1.6).

As previously, we have the particular cases: (i) $\alpha^2 = 0$, so that $\mathbf{S} = 0$ with probability one and the matrices $\mathbf{H}$ form the GSE; (ii) $\alpha^2 = 1$, on the average $\mathbf{R}$ and $\mathbf{S}$ have the same magnitude and the ensemble is GUE (of $2N \times 2N$ matrices); (iii) $\alpha^2 \longrightarrow \infty$, $\mathbf{S}$ dominates $\mathbf{R}$ and the ensemble of $\mathbf{H}$ may be referred to as the anti-self-dual Gaussian symplectic ensemble.

The joint probability density for the eigenvalues $x_1, ..., x_{2N}$ of $\mathbf{H}$ is

$$\mathbf{p}\left(x_1, ..., x_{2N}\right) = \mathbf{C}_N \exp\left(-\frac{1+\alpha^2}{2}\sum x_j^2\right) \times \Delta\left(x_1, ..., x_{2N}\right) \operatorname{Pf}\left[\mathbf{F}\left(x_i - x_j\right)\right]_{i,j=1,...,2N}, \quad (14.2.7)$$

where Δ is the product of differences of the $2N$ variables $x_1, ..., x_{2N}$, Eq. (14.1.8),

$$\mathbf{F}(x) = x \exp\left[-\left(1-\alpha^4\right)x^2/4\alpha^2\right] \quad (14.2.8)$$

and

$$\mathbf{C}_N^{-1} = 2^{-N(N-1)}\pi^N\alpha^{3N}\left(1-\alpha^2\right)^{N(N-1)}\left(1+\alpha^2\right)^{-N(N+1)}$$

$$\times (2N)!(N!)^{-1}\prod_1^N (2j)!. \tag{14.2.9}$$

The limits in the three cases $\alpha^2 = 0, 1$ and ∞ are again well-defined.

The n-level correlation and cluster functions are again given by Eqs. (14.1.12) and (14.1.13) with ϕ replaced by

$$\mathbf{\Phi}(x,y) = \begin{bmatrix} \mathbf{S}_N(x,y) & \mathbf{K}_N(x,y) \\ \mathbf{I}_N(x,y) & \mathbf{S}_N(y,x) \end{bmatrix} \tag{14.2.10}$$

with the new functions

$$\mathbf{S}_N(x,y) = \sum_{j=0}^{N-1} \left[\varphi_{2j+1}(x)\varphi_{2j+1}(y) - \mathbf{\Psi}_{2j+1}(x)\mathbf{A}_{2j+1}(y)\right] \tag{14.2.11}$$

$$= \sum_{j=0}^{2N} \varphi_j(x)\varphi_j(y) + \left(1+\alpha^2\right)\left(\frac{1-\alpha^2}{1+\alpha^2}\right)^{2N+1}$$

$$\times \left(N+1/2\right)^{1/2}\varphi_{2N}(x)\mathbf{A}_{2N+1}(y), \tag{14.2.12}$$

$$\mathbf{I}_N(x,y) = \sum_{j=0}^{N-1} \left[\varphi_{2j+1}(x)\mathbf{A}_{2j+1}(y) - \mathbf{A}_{2j+1}(x)\varphi_{2j+1}(y)\right] \tag{14.2.13}$$

$$= \sum_{j=0}^{2N} \varphi_j(x)\mathbf{A}_j(y) + \left(1+\alpha^2\right)\left(\frac{1-\alpha^2}{1+\alpha^2}\right)^{2N+1}$$

$$\times \left(N+1/2\right)^{1/2}\mathbf{A}_{2N+1}(x)\mathbf{A}_{2N}(y), \tag{14.2.14}$$

$$\mathbf{K}_N(x,y) = \mathbf{D}_N(x,y) + \mathbf{g}(x,y), \tag{14.2.15}$$

$$\mathbf{D}_N(x,y) = \sum_{j=0}^{N-1} \left[\varphi_{2j+1}(x)\mathbf{\Psi}_{2j+1}(y) - \mathbf{\Psi}_{2j+1}(x)\varphi_{2j+1}(y)\right] \tag{14.2.16}$$

$$= \sum_{j=0}^{2N} \varphi_j(x)\mathbf{\Psi}_j(y) + \left(1+\alpha^2\right)\left(\frac{1-\alpha^2}{1+\alpha^2}\right)^{2N+1} \times (N+1/2)^{1/2}\varphi_{2N}(x)\varphi_{2N+1}(y), \tag{14.2.17}$$

$$\mathbf{g}(x,y) = -\frac{\left(1+\alpha^2\right)\left(1-\alpha^4\right)}{4\alpha^3\sqrt{\pi}} \exp\left[-\frac{1}{2}\alpha^2\left(x^2+y^2\right)\right]\mathbf{F}(x-y) \tag{14.2.18}$$

$$= \sum_{j=0}^{\infty} \mathbf{\Psi}_j(x)\varphi_j(y) = \frac{1}{2}\sum_{j=0}^{\infty}\left[\mathbf{\Psi}_j(x)\varphi_j(y) - \varphi_j(x)\mathbf{\Psi}_j(y)\right]$$

$$= \sum_{j=0}^{\infty}\left[\mathbf{\Psi}_{2j}(x)\varphi_{2j}(y) - \varphi_{2j}(x)\mathbf{\Psi}_{2j}(y)\right]$$

$$= \sum_{j=0}^{\infty}\left[\mathbf{\Psi}_{2j+1}(x)\varphi_{2j+1}(y) - \varphi_{2j+1}(x)\mathbf{\Psi}_{2j+1}(y)\right], \tag{14.2.19}$$

where $\varphi_j(x)$ and $\varepsilon(x)$ are given by Eqs. (14.1.26), (14.1.29),

$$\mathbf{\Psi}_j(x) = \left(\frac{1-\alpha^2}{1+\alpha^2}\right)^j \exp\left(-\frac{1}{2}\alpha^2x^2\right)\frac{d}{dx}\left(\exp\left(\frac{1}{2}\alpha^2x^2\right)\varphi_j(x)\right), \tag{14.2.20}$$

$$\mathbf{A}_j(x) = \left(\frac{1+\alpha^2}{1-\alpha^2}\right)^j \exp\left(\frac{1}{2}\alpha^2x^2\right) \times \int_{-\infty}^{\infty} \exp\left(-\frac{1}{2}\alpha^2y^2\right)\varepsilon\left(x-y\right)\varphi_j(y)\mathrm{d}y. \tag{14.2.21}$$

For large N, we again have a semicircular level density

$$\mathbf{R}_1(x) \approx \pi^{-1}\left(4N - x^2\right)^{1/2} \tag{14.2.22}$$

whereas, the n-level functions for $n > 1$ are discontinuous in α^2 at $\alpha^2 = 0$. On the other hand when $\alpha \longrightarrow 0$ and

$$\rho \equiv \rho(x) \approx (\alpha/\sqrt{2})\mathbf{R}_1(x) \tag{14.2.23}$$

is finite, the functions have well defined limits. Equations (14.1.32) and (14.1.33) are unchanged but with σ replaced by

$$\sigma(r;\rho) = \begin{bmatrix} (\sin \pi r)/\pi r & \mathbf{K}(r;\rho) \\ \mathbf{I}(r;\rho) & (\sin \pi r)/\pi r \end{bmatrix} \tag{14.2.24}$$

where

$$\mathbf{I}(r;\rho) = -\pi^{-1}\int_0^{\pi} \frac{\sin kr}{k} \exp\left(2\rho^2 k^2\right) dk, \tag{14.2.25}$$

$$\mathbf{K}(r;\rho) = -\pi^{-1}\int_{\pi}^{\infty} k \sin kr \exp\left(-2\rho^2 k^2\right) dk. \tag{14.2.26}$$

Note that the limit of $\mathbf{K}(r;\rho)$ at $\rho = 0$ contains a term in $\delta(r)/r$. This is due to the fact that in the GSE the eigenvalues are doubly degenerate. For $\rho \longrightarrow \infty$, the product **IK** being zero, **I** and **K** can be replaced by zeros [see the remark following Eq. (14.1.36)].

14.3. Joint Probability Density for the Eigenvalues

In this section we derive the joint probability density for the eigenvalues, Eqs. (14.1.7) and (14.2.7), from that of the matrix elements, Eqs. (14.1.2) and (14.2.3). The matrix element probability densities depend on the eigenvalues and the angular variables characterizing the eigenvectors, and one has to integrate over these angular variables. For $\alpha^2 = 0$, 1, ∞ in both ensembles the matrix element probability densities depend only on the eigenvalues. Also the Jacobian separates into a product of two functions, one involving only the eigenvalues and the other involving only the eigenvectors (*cf.* Chapter 3); therefore the integral over the angular variables, giving only a constant, need not be calculated. For arbitrary α^2 this simplification is not there, but we know (*cf.* Appendix A.5) an integral over the group of unitary matrices U,

$$\int \exp\left[c \operatorname{tr}\left(A - U^\dagger B U\right)^2\right] dU$$

$$= \text{const} \times [\Delta(a)\Delta(b)]^{-1} \det\left[\exp\left\{c\left(a_i - b_j\right)^2\right\}\right] \quad (14.3.1)$$

valid for arbitrary Hermitian matrices A and B having eigenvalues (a_i) and (b_i), respectively, where Δ is the product of differences, Eq. (14.1.8). We also have two other results at our disposal.

(i) The convolution of two independent Gaussian distributions with variances σ_1^2 and σ_2^2 is again a Gaussian with the variance $\sigma_1^2 + \sigma_2^2$:

$$\int_{-\infty}^{\infty} \left(2\pi\sigma_1^2\right)^{-1/2} \exp\left(-\frac{x^2}{2\sigma_1^2}\right) \left(2\pi\sigma_2^2\right)^{-1/2} \exp\left(-\frac{(y-x)^2}{2\sigma_2^2}\right) dx$$

$$= \left[2\pi\left(\sigma_1^2 + \sigma_2^2\right)\right]^{-1/2} \exp\left[-y^2/\left(2\sigma_1^2 + 2\sigma_2^2\right)\right]. \quad (14.3.2)$$

(ii) For any sets of functions $\theta_i(x)$, $\tau_i(x)$, and $\chi_i(x)$ and for a suitable measure $d\mu(x)$, integrals of the form

$$I_1 = \int \cdots \int \prod_1^N d\mu\left(x_i\right) \det\left[\theta_i\left(x_j\right)\right] \operatorname{sgn} \Delta(x), \quad (14.3.3)$$

$$I_2 = \int \cdots \int \prod_1^m d\mu\left(x_i\right) \det\left[\theta_i\left(x_j\right), \tau_i\left(x_j\right)\right]_{\substack{i=1,\ldots,2m, \\ j=1,\ldots,m}}, \quad (14.3.4)$$

$$I_3 = \int \cdots \int \prod_1^m d\mu\left(x_i\right) \det\left[\theta_i\left(x_j\right), \tau_i\left(x_j\right), \chi_i\left(x_m\right)\right]_{\substack{i=1,\ldots,2m-1 \\ j=1,\ldots,m-1}} \quad (14.3.5)$$

can be evaluated by the method of integration over alternate variables and the theory of Pfaffians (*cf.* Appendix A.18). The result is

$$I_1 = N! \,\mathrm{Pf}\,[a_{ij}]_{i,j=1,\ldots,2m}, \tag{14.3.6}$$

$$I_2 = m! \,\mathrm{Pf}\,[b_{ij}]_{i,j=1,\ldots,2m}, \tag{14.3.7}$$

$$I_3 = (m-1)! \,\mathrm{Pf}\,[c_{ij}]_{i,j=1,\ldots,2m}, \tag{14.3.8}$$

where $2m = N$ if N is even and $2m = N+1$ if N is odd,

$$a_{ij} = \int\int_{x \le y} d\mu(x) d\mu(y) \, [\theta_i(x)\theta_j(y) - \theta_j(x)\theta_i(y)] \tag{14.3.9}$$

for $i, j = 1, 2, ..., N$. When N is odd, we have in addition

$$a_{i,N+1} = -a_{N+1,i} = \int d\mu(x)\theta_i(x) \tag{14.3.10}$$

for $i = 1, 2, ..., N$, and $a_{N+1,N+1} = 0$. Also

$$b_{ij} = \int d\mu(x) \, [\theta_i(x)\tau_j(x) - \tau_i(x)\theta_j(x)] \tag{14.3.11}$$

for $i, j = 1, 2, ..., 2m$,

$$c_{ij} = b_{ij}, \quad i, j = 1, 2, ..., 2m-1, \tag{14.3.12}$$

$$c_{i,2m} = -c_{2m,i} = \int d\mu(x)\chi_i(x), \tag{14.3.13}$$

$$c_{2m,2m} = 0. \tag{14.3.14}$$

We therefore proceed in three steps.

Firstly we write H as a sum, $H = A + B$. This convenient breaking will be different according to whether $\alpha^2 < 1$ or $\alpha^2 > 1$ and has nothing to do with the previous writing of H as a sum of its real and imaginary parts.

Secondly we use Eq. (14.3.1) to integrate over the unitary matrix, diagonalizing B. The result is some determinant containing the eigenvalues of H and of A, and the product of differences of these eigenvalues.

Finally we integrate over the eigenvalues of A using (14.3.3)–(14.3.4).

The treatment of the ensemble of Hermitian quaternion matrices will be parallel. Constants will be ignored in the intermediate steps, the final constants will be fixed by the normalization condition (see Appendices A.27 and A.28). The detailed working is given in the following subsections.

14.3.1. MATRICES FROM GOE TO GUE AND BEYOND

We write H as a sum of two Hermitian matrices

$$H = A + B \tag{14.3.15}$$

and choose B so that its real and imaginary parts have the same variance. If $\alpha^2 < 1$, the real part of H has a larger variance than its imaginary part and we choose A to be real symmetric. If $\alpha^2 > 1$ it is the imaginary part of H which has a larger variance and we choose A to be pure imaginary antisymmetric. In either case we adjust the variance of A and the common variance of the real and imaginary parts of B in a proper way. Thus our choices are:

$$P_1(A) \propto \exp\left\{-\text{tr } A^2/\left[4v^2\left(1-\alpha^2\right)\right]\right\},$$

$$P_2(B) \propto \exp\left(-\text{tr } B^2/4v^2\alpha^2\right),$$

if $\alpha^2 < 1$, and

$$P_1(A) \propto \exp\left\{-\text{tr } A^2/\left[4v^2\left(\alpha^2-1\right)\right]\right\},$$

$$P_2(B) \propto \exp\left(-\text{tr } B^2/4v^2\right),$$

if $\alpha^2 > 1$. We combine these equations as

$$A^T = A^* = A \text{ sgn}\left(1-\alpha^2\right),$$

$$P_1(A) \propto \exp\left[-\text{tr } A^2/\left(4v^2\left|1-\alpha^2\right|\right)\right], \tag{14.3.16}$$

$$P_2(B) \propto \exp\left(-\text{tr } B^2/4v^2\gamma^2\right), \tag{14.3.17}$$

where $\gamma^2 = \min\left(1,\alpha^2\right)$ and $\text{sgn}(x)$, the sign of x, is $+1$ or -1 according as x is positive or negative; for $x = 0$, $\text{sign}(x) = 0$. Though Eqs. (14.3.16) and (14.3.17) look alike, they have an important difference, tr B^2 contains two sums of squares, those of the real and imaginary parts of B, whereas tr A^2 contains only one of them. Equation (14.1.2) is now written in the equivalent form

$$P(H) = \int P_1(A)P_2(H-A)dA. \tag{14.3.18}$$

Let $x_1, ..., x_N$ be the eigenvalues of H

$$H = UXU^\dagger, \quad U^\dagger U = 1, \tag{14.3.19}$$

so that (see Chapter 3)

$$dH \propto \Delta^2(x)dUdx, \tag{14.3.20}$$

and the joint probability density for the $x_1, ..., x_N$ is

$$p(x) \equiv p\left(x_1, ..., x_N\right) \propto \int P_1(A)P_2(H-A)\Delta^2(x)dUdA. \tag{14.3.21}$$

We consider separately the cases of A symmetric and antisymmetric.

When A is real symmetric, its real eigenvalues $a_1, ..., a_N$ are in general distinct and from Eqs. (14.3.16), (14.3.17), (14.3.21), and (14.3.1) we have

$$p(x) \propto \int \exp\left(-\sum_1^N \frac{a_i^2}{4v^2\left(1-\alpha^2\right)}\right)$$

$$\times \exp\left(-\frac{\operatorname{tr}\left(UXU^\dagger - A\right)^2}{4v^2}\right)\Delta^2(x)dUdA \quad (14.3.22)$$

$$\propto \Delta(x)\int \exp\left(-\sum_1^N \frac{a_i^2}{4v^2\left(1-\alpha^2\right)}\right)$$

$$\times \det\left[\exp\left(-\frac{\left(x_i - a_j\right)^2}{4v^2}\right)\right]\frac{1}{\Delta(a)}dA. \quad (14.3.23)$$

As far as the dependence on the eigenvalues is concerned, one has (*cf.* Chapter 3)

$$dA \propto |\Delta(a)|\, da_1 \cdots da_N, \quad (14.3.24)$$

so that

$$p(x) \propto \Delta(x)\int \exp\left(-\sum_1^N \frac{a_i^2}{4v^2\left(1-\alpha^2\right)}\right)$$

$$\times \det\left[\exp\left(-\frac{\left(x_i - a_j\right)^2}{4v^2}\right)\right] \operatorname{sqn}\, \Delta(a)da_1\cdots da_N. \quad (14.3.25)$$

Using Eqs. (14.3.3), (14.3.6), and (14.3.9), we get

$$p(x) \propto \Delta(x)\exp\left(-\sum_1^N \frac{x_i^2}{4v^2}\right)\operatorname{Pf}\left[b_{ij}\right] \quad (14.3.26)$$

where

$$b_{ij} = \iint_{-\infty\le z_1\le z_2\le\infty} dz_1dz_2(A(z_1,x_i)A(z_2,x_j) - A(z_1,x_j)A(z_2,x_i)) \quad (14.3.27)$$

$$A(z,x) = \exp\left(\frac{\left[z-(1-\alpha^2)x\right]^2}{4v^2\alpha^2(1-\alpha^2)}\right) \quad (14.3.27')$$

for $i, j = 1, 2, ..., N$, when the order N of the matrices is even. For N odd, we have one more row and column

$$b_{N+1,N+1} = 0,$$

and for $i = 1, ..., N$,

$$b_{i,N+1} = -b_{N+1,i} = \int_{-\infty}^{\infty} dz A(z, x_i). \tag{14.3.28}$$

On introducing the new variables $\mathcal{S} = (z_2 + z_1)/\sqrt{2}$, $t = (z_2 - z_1)/\sqrt{2}$, the integration on $\mathcal{S}$ in (14.3.27) can be performed. A little algebra then gives

$$b_{ij} = \text{const} \times \text{erf}\left[\left(\frac{1-\alpha^2}{8v^2\alpha^2}\right)^{1/2} (x_i - x_j)\right], \quad i, j = 1, ..., N. \tag{14.3.29}$$

Equations (14.3.26), (14.3.28), and (14.3.29) give Eq. (14.1.7) when A is real symmetric.

When A is antisymmetric pure imaginary, its eigenvalues are real and come in pairs $\pm a_i$; zero is necessarily an eigenvalue if the order N of A is odd. As far as the dependence on the eigenvalues is concerned (*cf.* Chapter 3)

$$dA \propto \prod_{1 \le i < j \le m} \left(a_i^2 - a_j^2\right)^2 da_1 \cdots da_m, \tag{14.3.30}$$

$$\Delta(a) = 2^m \prod_1^m a_i \prod_{1 \le i < j \le m} \left(a_i^2 - a_j^2\right)^2, \tag{14.3.31}$$

for $N = 2m$ even; while

$$dA \propto \prod_1^m a_i^2 \prod_{1 \le i < j \le m} \left(a_i^2 - a_j^2\right)^2, \tag{14.3.32}$$

$$\Delta(a) = 2^m \prod_1^m a_i^3 \prod_{1 \le i < j \le m} \left(a_i^2 - a_j^2\right)^2, \tag{14.3.33}$$

for $N = 2m+1$ odd. (We take $a_1, ..., a_m$ as distinct positive numbers). Thus from Eqs. (14.3.16)–(14.3.18), (14.3.21), (14.3.1), (14.3.30), and (14.3.31) we get for $N = 2m$ even

$$p(x) \propto \Delta(x) \exp\left(-\sum_1^N x_i^2/(4v^2\alpha^2)\right) \int_0^\infty \cdots \int_0^\infty \frac{da_1 \cdots da_m}{a_1 \cdots a_m}$$
$$\times \det\left[\exp\left(-\frac{\alpha^2}{4v^2(\alpha^2-1)}\left(a_j - \frac{\alpha^2-1}{\alpha^2}x_i\right)^2\right)\right.$$
$$\left.\exp\left(-\frac{\alpha^2}{4v^2(\alpha^2-1)}\left(a_j + \frac{\alpha^2-1}{\alpha^2}x_i\right)^2\right)\right],$$
$$i = 1, ..., N; \qquad j = 1, ..., m. \tag{14.3.34}$$

For $N = 2m+1$ odd, instead of (14.3.30) and (14.3.31) we use (14.3.32) and (14.3.33). The result is again Eq. (14.3.34) in which the determinant contains one more column

$$\exp\left[-\frac{\alpha^2}{4v^2(\alpha^2-1)}\left(\frac{\alpha^2-1}{\alpha^2}x_i\right)^2\right] = \exp\left(-\frac{\alpha^2-1}{4v^2\alpha^2}x_i^2\right). \tag{14.3.35}$$

For N even, we use Eqs. (14.3.4), (14.3.7), and (14.3.11), to get from Eq. (14.3.34)

$$p(x) \propto \Delta(x) \exp\left(-\sum_1^N \frac{x_i^2}{4v^2\alpha^2}\right) \mathrm{Pf}\,[b_{ij}] \tag{14.3.36}$$

with

$$b_{ij} = \int_0^\infty \frac{dz}{z}\left(\exp\left\{-\frac{\alpha^2}{4v^2(\alpha^2-1)}\left[\left(z - \frac{\alpha^2-1}{\alpha^2}x_i\right)^2 + \left(z + \frac{\alpha^2-1}{\alpha^2}x_j\right)^2\right]\right\}\right.$$
$$\left. - \exp\left\{-\frac{\alpha^2}{4v^2(\alpha^2-1)}\left[\left(z + \frac{\alpha^2-1}{\alpha^2}x_i\right)^2 + \left(z - \frac{\alpha^2-1}{\alpha^2}x_j\right)^2\right]\right\}\right). \tag{14.3.37}$$

For N odd, we use Eqs. (14.3.5), (14.3.8), (14.3.12), and (14.3.13); the result is again Eq. (14.3.36) in which the Pfaffian contains one more row and a column

$$b_{i,N+1} = -b_{N+1,i} = \exp\left(-\frac{\alpha^2-1}{4v^2\alpha^2}x_i^2\right). \tag{14.3.38}$$

A little algebra gives for Eq. (14.3.37)

$$\begin{aligned} b_{ij} &\propto \exp\left(-\frac{\alpha^2-1}{4v^2\alpha^2}\left(x_i^2+x_j^2\right)\right)\int_0^\infty \frac{dt}{t}\mathrm{e}^{-t^2}\sinh\left[\left(\frac{\alpha^2-1}{2v^2\alpha^2}\right)^{1/2}(x_i-x_j)\,t\right] \\ &\propto \exp\left(-\frac{\alpha^2-1}{4v^2\alpha^2}\left(x_i^2+x_j^2\right)\right)\mathrm{erf}\left[\left(\frac{\alpha^2-1}{8v^2\alpha^2}\right)^{1/2}(x_i-x_j)\right]. \end{aligned} \tag{14.3.39}$$

Equations (14.3.36), (14.3.39), and (14.3.38) give Eq. (14.1.7) for antisymmetric A.

Equation (14.1.7) is therefore established in all cases except for the overall constant. This constant is evaluated in Appendix A.27.

14.3.2. MATRICES FROM GSE TO GUE AND BEYOND

We write the $N \times N$ quaternion matrix $\mathbf{H}$ as a sum

$$\mathbf{H} = \mathbf{A} + \mathbf{B}, \qquad \bar{\mathbf{A}} = \mathbf{A}\ \mathrm{sgn}\left(1-\alpha^2\right). \tag{14.3.40}$$

The quaternion matrices $\mathbf{A}$ and $\mathbf{B}$ are Hermitian; the self-dual and anti-self-dual parts of $\mathbf{B}$ have the same variance; $\mathbf{A}$ is self-dual quaternion real if $\alpha^2 < 1$, and anti-self-dual quaternion pure imaginary if $\alpha^2 > 1$. Due to Eq. (14.3.2), Eq. (14.2.3) is equivalent to Eq. (14.3.18) with

$$P_1(\mathbf{A}) \propto \exp\left[-\mathrm{tr}\ \mathbf{A}^2/\left(4v^2\left|1-\alpha^2\right|\right)\right], \tag{14.3.41}$$

$$P_2(\mathbf{B}) \propto \exp\left(-\mathrm{tr}\ \mathbf{B}^2/4v^2\gamma^2\right), \tag{14.3.42}$$

where, as before, $\gamma^2 = \min\left(1,\alpha^2\right)$. Let

$$\mathbf{H} = UXU^\dagger, \quad UU^\dagger = 1, \tag{14.3.43}$$

where $\mathbf{H}$, X, and U are $2N \times 2N$ ordinary matrices, U is unitary, X is diagonal and the diagonal elements $x_1, ..., x_{2N}$ of X are the eigenvalues of $\mathbf{H}$. As before (Chapter 3)

$$\mathrm{d}\mathbf{H} \propto \Delta^2 (x_1, ..., x_{2N})\, dx\, dU, \tag{14.3.44}$$

$$dx \equiv dx_1, ..., dx_{2N}. \tag{14.3.45}$$

We consider separately, as before, the cases of $\mathbf{A}$ self-dual and $\mathbf{A}$ anti-self-dual.

When $\mathbf{A}$ is self-dual, its eigenvalues are real numbers $a_1, ..., a_N$ each repeated twice. Since the right-hand side of Eq. (14.3.1) is now of the form 0/0, we cannot use it directly, but have to take limits when the eigenvalues of $\mathbf{A}$ become equal in pairs. The calculation gives

$$\int \exp\left[-\mathrm{tr}\left(\mathbf{A} - UXU^{\dagger}\right)^2 / 4v^2\alpha^2\right] dU \propto \left(\Delta(x)\Delta^4(a)\right)^{-1}$$
$$\times \det\left[\exp\left(-\frac{\left(a_i - x_j\right)^2}{4v^2\alpha^2}\right)\right.$$
$$\left.(a_i - x_j)\exp\left(-\frac{\left(a_i - x_j\right)^2}{4v^2\alpha^2}\right)\right] \tag{14.3.46}$$

where $\Delta(x) \equiv \Delta(x_1, ..., x_{2N})$ and $\Delta(a) \equiv \Delta(a_1, ..., a_N)$; Eq. (14.1.8).

Now as far as the dependence on the eigenvalues is concerned (*cf.* Chapter 3)

$$d\mathbf{A} \propto \Delta^4(a) da_1, ..., da_N \equiv \Delta^4(a) da, \tag{14.3.47}$$

so that

$$\mathbf{p}(x) \equiv \mathbf{p}(x_1, ..., x_{2N}) \propto \Delta(x) \int da\, \exp\left(-\sum_1^N \frac{a_i^2}{2v^2(1-\alpha^2)}\right)$$
$$\times \det\left[\exp\left(-\frac{\left(a_i - x_j\right)^2}{4v^2\alpha^2}\right)\right.$$
$$\left.(a_i - x_j)\exp\left(-\frac{\left(a_i - x_j\right)^2}{4v^2\alpha^2}\right)\right]. \tag{14.3.48}$$

Using Eqs. (14.3.4), (14.3.7), and (14.3.11)

$$\mathbf{p}(x) \propto \Delta(x)\exp\left(-\sum_{1}^{2N}\frac{x_i^2}{4v^2}\right)\mathrm{Pf}\left[h_{ij}\right], \tag{14.3.49}$$

where

$$\begin{aligned} h_{ij} = {} & (x_i - x_j)\exp\left(\frac{x_i^2 + x_j^2}{4v^2}\right)\int_{-\infty}^{\infty} dz\, \exp\left(-\frac{z^2}{2v^2\left(1-\alpha^2\right)}\right) \\ & \times \exp\left(-\frac{\left(z - x_i\right)^2}{4v^2\alpha^2} - \frac{\left(z - x_j\right)^2}{4v^2\alpha^2}\right). \end{aligned} \tag{14.3.50}$$

A little algebra now gives

$$h_{ij} \propto (x_i - x_j)\exp\left(-\frac{1-\alpha^2}{8v^2\alpha^2}\left(x_i - x_j\right)^2\right). \tag{14.3.51}$$

From Eqs. (14.3.49) and (14.3.51) we get Eq. (14.2.7) for the case $\alpha^2 < 1$.

When $\mathbf{A}$ is anti-self-dual, its eigenvalues are the real numbers $\pm a_1$, $\pm a_2, ..., \pm a_N$, and from Eq. (14.3.1)

$$\begin{aligned} & \int \exp\left(-\frac{\mathrm{tr}\left(\mathbf{A} - UXU^{\dagger}\right)^2}{4v^2}\right) dU \propto \left[\Delta(x)\Delta\left(\pm a_1, ..., \pm a_N\right)\right]^{-1} \\ & \quad \times \det\left[\exp\left(-\frac{\left(a_i - x_j\right)^2}{4v^2}\right) \qquad \exp\left(-\frac{\left(a_i + x_j\right)^2}{4v^2}\right)\right]. \end{aligned} \tag{14.3.52}$$

However,

$$\Delta\left(\pm a_1, ..., \pm a_N\right) = 2^N \prod_{1}^{N} a_i \left[\Delta\left(a_1^2, ..., a_N^2\right)\right]^2, \tag{14.3.53}$$

and as far as dependence on the eigenvalues is concerned (*cf.* Chapter 3.4)

$$d\mathbf{A} \propto \prod_{1}^{N} a_i^2 \left[\Delta\left(a_1^2, ..., a_N^2\right)\right]^{2,} \tag{14.3.54}$$

so that

$$\begin{aligned}\mathbf{p}(x) \propto \Delta(x) \int_0^\infty \cdots \int_0^\infty da_1 \cdots da_N \prod_1^N a_i \exp\left(-\sum_1^N \frac{a_i^2}{2v^2(\alpha^2-1)}\right) \\ \times \det\left[\exp\left(-\frac{(a_i-x_j)^2}{4v^2}\right) \quad \exp\left(-\frac{(a_i+x_j)^2}{4v^2}\right)\right] \\ \propto \Delta(x)\exp\left(-\sum_1^{2N} \frac{x_i^2}{4v^2}\right) \mathrm{Pf}\,[h_{ij}], \end{aligned} \tag{14.3.55}$$

where in the last step we have used Eqs. (14.3.4), (14.3.7), (14.3.11), and

$$\begin{aligned} h_{ij} = \exp\left(\frac{x_i^2+x_j^2}{4v^2}\right) \int_0^\infty dz\; z \exp\left(-\frac{z^2}{2v^2(\alpha^2-1)}\right) \\ \times \left[\exp\left(-\frac{(z-x_i)^2}{4v^2} - \frac{(z+x_j)^2}{4v^2}\right)\right. \\ \left. - \exp\left(-\frac{(z+x_i)^2}{4v^2} - \frac{(z-x_j)^2}{4v^2}\right)\right]. \end{aligned} \tag{14.3.56}$$

A little algebra now gives

$$h_{ij} \propto (x_i - x_j) \exp\left(-\frac{\alpha^2-1}{8v^2\alpha^2}(x_i-x_j)^2\right). \tag{14.3.57}$$

Equations (14.3.55) and (14.3.57) give (14.2.7) for $\alpha^2 > 1$.

Equation (14.2.7) is therefore established for all values of α^2, except for the overall constant, which is evaluated in Appendix A.28.

14.4. Correlation and Cluster Functions

To derive the correlation functions we shall use the theory of quaternion matrices, in particular the Eqs. (6.2.6) and (6.2.9).

Let us consider the first ensemble. Using Theorem 6.2.1 several times one can get the correlation function

$$R_n(x_1, ..., x_n) = \frac{N!}{(N-n)!} \int_{-\infty}^{\infty} \cdots \int_{-\infty}^{\infty} p(x_1, ..., x_N)\, dx_{n+1} \cdots dx_N \tag{14.4.1}$$

as a determinant of an $n \times n$ quaternion matrix; the derivation relies on the fact that the $p(x)$ of Eq. (14.1.7) can be written as an $N \times N$ quaternion determinant

$$p(x_1, ..., x_N) = (1/N)! \det [\phi(x_i, x_j)]_{i,j=1,...,N} \tag{14.4.2}$$

where the quaternion $\phi(x, y)$ [Eq. (14.1.14)] satisfies the equality

$$\int_{-\infty}^{\infty} \phi(x, y)\phi(y, z)dy = \phi(x, z) + \tau\phi(x, z) - \phi(x, z)\tau, \tag{14.4.3}$$

in which τ is the constant quaternion

$$\tau = \frac{1}{2}\begin{bmatrix} 1 & 0 \\ 0 & -1 \end{bmatrix}. \tag{14.4.4}$$

Equation (14.1.12) then follows from Theorem 6.2.1. The proof of Eq. (14.4.2) is given in Appendix A.29 and that of Eq. (14.4.3) in Appendix A.30.

Similarly one can get the correlation functions for the other ensemble, since

$$\mathbf{p}(x_1, ..., x_{2N}) = [(2N)!]^{-1} \det [\mathbf{\Phi}(x_i, x_j)]_{i,j=1,...,2N}\,. \tag{14.4.5}$$

where $\mathbf{p}$ is now given by Eq. (14.2.7) and $\mathbf{\Phi}$ by Eq. (14.2.10), and since this $\mathbf{\Phi}$ satisfies Eq. (14.4.3).

The proof of Eq. (14.4.5) is given in Appendix A.29. The $\mathbf{\Phi}$ satisfies Eq. (14.4.3) with a quaternion τ which is the negative of (14.4.4); see Appendix A.30.

Equation (14.1.13) is obtained from the observation that [*cf.* Eq. (5.1.3)) the cluster function T_n is the cumulant of the correlation function R_n and the expression on the right-hand side of Eq. (14.1.13) is the cumulant of that on the right hand side of Eq. (14.1.12) (consider it as a determinant of an $n \times n$ self-dual matrix with quaternion elements; Chapter 6.2).

To get the limits for large N of the correlation and cluster functions it suffices then to take the limits of $\phi(x,y)$ and $\mathbf{\Phi}(x,y)$. The limits of the functions $S_N(x,y)$, $D_N(x,y)$, $I_N(x,y)$ and $J_N(x,y)$ are derived as follows (Pandey and Mehta, 1983). For large N

$$S_N(x,y) \approx \sum_{j=0}^{N-1} \varphi_j(x)\varphi_j(y). \tag{14.4.6}$$

Thus from Appendix A.9 the asymptotic level density is

$$R_1(x) = \lim_{N\to\infty} S_N(x,x) \approx \frac{1}{\pi}(2N - x^2)^{1/2}. \tag{14.1.30}$$

Also, as $N \to \infty$, $x - y \to 0$, while

$$r = (x-y)R_1(x) \tag{14.4.7}$$

remains finite, we have from Eq. (14.4.6) and Appendix A.10

$$S_N(x,y) \to R_1(x)\sin(\pi r)/(\pi r). \tag{14.4.8}$$

The limits of $I_N(x,y)$ and $D_N(x,y)$ are derived in Appendix A.31. Those of $\mathbf{S}_N$, $\mathbf{D}_N$, and $\mathbf{I}_N$ are obtained from those of S_N, D_N, and I_N by changing ρ^2 to $-\rho^2$ and replacing N by $2N$. For $\mathbf{g}$ we have

$$\begin{aligned}[\mathbf{R}_1(x)]^{-2}\mathbf{g}(x,y) &\approx -\left(r/8\rho^3\sqrt{\pi}\right)\exp\left(-r^2/8\rho^2\right) \\ &\approx -\pi^{-1}\int_0^\infty k\sin kr\exp\left(-2\rho^2k^2\right)dk, \end{aligned} \tag{14.4.9}$$

from which we get the limit of $\mathbf{K}_N$.

This completes the proof of the statements listed in Section 14.1. Some technical details are collected in Appendices A.26 to A.31.

Summary of Chapter 14

In this chapter we derive all correlation and cluster functions for the eigenvalues of the Hermitian matrices $H = H_1 + H_2$ when the probability density of the matrix elements is proportional to the product of $\exp\left(-\text{tr } H_1^2/(2v^2)\right)$ and $\exp\left(-\text{tr } H_2^2/(2v^2\alpha^2)\right)$, where H_1 and H_2 are either the symmetric and antisymmetric parts of H or they are the self-dual and anti-self-dual parts of H.

15 / Matrices with Gaussian Element Densities but with No Unitary or Hermitian Conditions Imposed

An ensemble of matrices whose elements are complex, quaternion, or real numbers, but with no other restrictions as to their Hermitian or unitary character, is of no immediate physical interest, for their eigenvalues may lie anywhere on the complex plane. However, an effort (Ginibre, 1965) has been made to investigate them and the results are interesting in their own right.

To define a matrix ensemble one has to specify two things: the space T on which the matrices vary and a probability density function over T. For the case to be tractable, these choices have to be reasonable. Following Ginibre (1965) we will take T to be the set of all $N \times N$ matrices and a Gaussian probability density for the matrix elements.

15.1. Complex Matrices

Let T be the set of all $N \times N$ complex matrices. The probability that a matrix from the set T will lie in $(S,\ S + \mathrm{d}S)$ is $P\,(S)\,\mu\,(dS)$, where $\mu\,(dS)$ is the linear measure

$$\mu\,(dS) = \prod_{j,k} dS_{jk}^{(0)}\,dS_{jk}^{(1)} \tag{15.1.1}$$

and $S_{jk}^{(0)}$, $S_{jk}^{(1)}$ are the real and imaginary parts of the matrix element

$$S_{jk} = S_{jk}^{(0)} + iS_{jk}^{(1)}. \tag{15.1.2}$$

For the function $P(S)$ we may choose, for example (Ginibre, 1965).

$$P(S) = \exp\left[-\mathrm{tr}\left(S^{\dagger}S\right)\right]. \tag{15.1.3}$$

We denote the ensemble so defined as T_c. It is visibly invariant under all unitary transformations.

To get any information about the eigenvalues, we must first find their joint probability density. This can be done, as in Chapter 3, by changing the variables from S_{jk} to the (complex) eigenvalues z_j of S and the auxiliary variables p_j. Since tr $\left(S^{\dagger}S\right)$ is not only a function of z_j but contains other variables p_j as well, these variables have to be chosen carefully to facilitate later integrations. Let the eigenvalues of S be distinct; the case when S has multiple eigenvalues need not be considered for the same reasons as in Chapter 3. Also, let X be the $N \times N$ matrix whose columns are the eigenvectors of S so that X is nonsingular and $X^{-1}SX = E$ is diagonal. From $S = XEX^{-1}$ we obtain by differentiation

$$dS = X\left(dE + dA\, E - E\, dA\right)X^{-1}, \tag{15.1.4}$$

with

$$dA = X^{-1}dX. \tag{15.1.5}$$

Equation (15.1.4) reads in terms of its components:

$$\left(X^{-1}dSX\right)_{jj}^{(0)} = dz_j^{(0)} = dx_j, \quad \left(X^{-1}dSX\right)_{jj}^{(1)} = dz_j^{(1)} = dy_j, \tag{15.1.6}$$

$$\left(X^{-1}dSX\right)_{jk}^{(0)} = \left(x_k - x_j\right)dA_{jk}^{(0)} - \left(y_k - y_j\right)dA_{jk}^{(1)}$$

$$\left(X^{-1}dSX\right)_{jk}^{(1)} = \left(y_k - y_j\right)dA_{jk}^{(0)} + \left(x_k - x_j\right)dA_{jk}^{(1)}, \quad j \neq k, \tag{15.1.7}$$

where x_j, y_j are the real and imaginary parts of z_j, the diagonal elements of E, whereas $dA_{jk}^{(0)}$ and $dA_{jk}^{(1)}$ are the real and imaginary parts of dA_{jk}. Whenever any set of differentials is expressed linearly in terms of those of the others, the ratio of the volume elements is equal to the Jacobian. The volume element in (15.1.1) is therefore given by

$$\begin{aligned}\mu(dS) &= \mu\left(X^{-1}dSX\right) \\ &= \prod_{j\neq k}\left|z_k - z_j\right|^2 dA_{jk}^{(0)}dA_{jk}^{(1)}\prod_i dx_i dy_i.\end{aligned} \tag{15.1.8}$$

One still has to evaluate the integral

$$J = \int \exp\left[-\operatorname{tr}\left(S^{\dagger}S\right)\right] \prod_{j \neq k} dA_{jk}^{(0)} dA_{jk}^{(1)}, \tag{15.1.9}$$

which can be done by a careful choice of the new variables of integration and by using properties of determinant expansions (see Appendix A.32 or A.35). The result is as follows.

The joint probability density for the eigenvalues of S belonging to the ensemble T_c of all complex matrices is given by

$$P_c\left(z_1, z_2, ..., z_N\right) = K_c \exp\left(-\sum_{1}^{N} |z_i|^2\right) \prod_{1 \leq i < j \leq N} |z_i - z_j|^2, \tag{15.1.10}$$

where K_c is the normalization constant given later by Eq. (15.1.17).

With this joint probability density function one can determine various quantities of interest as easily as in Chapter 5 or 10. For example, the probability that all the eigenvalues z_i will lie outside a circle of radius α centered at $z = 0$ is

$$E_{N_c}(\alpha) = \int \cdots \int_{|z_i| \geq \alpha} P_c\left(z_1, ..., z_N\right) \prod_i dx_i dy_i. \tag{15.1.11}$$

By writing

$$\begin{aligned}
\prod_{i<j} |z_i - z_j|^2 &= \prod_{i<j} \left(z_i - z_j\right)\left(z_i^* - z_j^*\right) \\
&= \det \begin{bmatrix} 1 & \dots & 1 \\ z_1 & \dots & z_N \\ \dots & \dots & \dots \\ z_1^{N-1} & \dots & z_N^{N-1} \end{bmatrix} \\
&\quad \times \det \begin{bmatrix} 1 & \dots & 1 \\ z_1^* & \dots & z_N^* \\ \dots & \dots & \dots \\ z_1^{*N-1} & \dots & z_N^{*N-1} \end{bmatrix}
\end{aligned}$$

and multiplying the two determinants row by row we get

$$E_{N_c}(\alpha) = K_c \int \cdots \int_{|z_i| \geq \alpha} \left(\prod_i dx_i dy_i \right) \exp\left(-\sum_1^N |z_i|^2 \right)$$
$$\times \det \begin{bmatrix} N & \sum_i z_i & \cdots & \sum_i z_i^{N-1} \\ \sum_i z_i^* & \sum_i z_i^* z_i & \cdots & \sum_i z_i^* z_i^{N-1} \\ \cdots & \cdots & \cdots & \cdots \\ \sum_i z_i^{*N-1} & \sum_i z_i^{*N-1} z_i & \cdots & \sum_i z_i^{*N-1} z_i^{N-1} \end{bmatrix}. \tag{15.1.12}$$

Since the integrand is symmetric in all the z_i, we can replace the first row with 1, z_1, z_1^2, ..., z_1^{N-1} and multiply the result by N; z_1 can now be eliminated from the other rows by subtracting a suitable multiple of the first row. The resulting determinant is symmetric in the $N-1$ variables z_2, z_3, ..., z_N; therefore we replace the second row with z_2^*, $z_2^* z_2$, ..., $z_2^* z_2^{N-1}$ and multiply the result by $N-1$. The process can be repeated and we get

$$E_{N_c}(\alpha) = K_c N! \int \cdots \int_{|z_i| \geq \alpha} \left\{ \prod_i \mathrm{d}x_i \mathrm{d}y_i \right\} \exp\left(-\sum_1^N |z_i|^2 \right)$$
$$\times \det \begin{bmatrix} 1 & z_1 & \cdots & z_1^{N-1} \\ z_2^* & z_2^* z_2 & \cdots & z_2^* z_2^{N-1} \\ \cdots & \cdots & \cdots & \cdots \\ z_N^{*N-1} & z_N^{*N-1} z_N & \cdots & z_N^{*N-1} z_N^{N-1} \end{bmatrix}. \tag{15.1.13}$$

Since the various rows now depend on distinct variables, we can integrate them separately with the exponential factor. By changing to polar coordinates and performing the angular integrations first we see that

$$\int_{|z| \geq \alpha} \mathrm{e}^{-|z|^2} z^{*j} \, z^k \, dxdy = \pi \, \delta_{jk} \Gamma\left(j+1, \alpha^2\right) \tag{15.1.14}$$

so that

$$E_{N_c}(\alpha) = K_c N! \pi^N \prod_{j=1}^{N} \Gamma\left(j, \alpha^2\right) \tag{15.1.15}$$

where $\Gamma\left(j, \alpha^2\right)$ is the incomplete gamma function

$$\Gamma\left(j, \alpha^2\right) = \int_{\alpha^2}^{\infty} e^{-x} x^{j-1} dx = \Gamma(j) e^{-\alpha^2} \sum_{\ell=0}^{j-1} \frac{\alpha^{2\ell}}{\ell!}. \tag{15.1.16}$$

Since $E_{N_c}(0) = 1$, the constant K_c can be determined from Eq. (15.1.15) as

$$K_c^{-1} = \pi^N \prod_{j=1}^{N} j! \tag{15.1.17}$$

and therefore

$$E_{N_c}(\alpha) = \prod_{j=1}^{N} \left(e^{-\alpha^2} \sum_{\ell=0}^{j-1} \frac{\alpha^{2\ell}}{\ell!} \right). \tag{15.1.18}$$

It is easy to convince oneself that $E_{N_c}(\alpha)$ tends to a well-defined limit as $N \longrightarrow \infty$. For small values of α one may expand $E_{N_c}(\alpha)$ in a power series:

$$E_{N_c}(\alpha) = 1 - \alpha^2 + \frac{1}{2}\alpha^6 - \frac{5}{12}\alpha^8 + \frac{7}{24}\alpha^{10} - \cdots . \tag{15.1.19}$$

To get the coefficient of α^{2i} in the above power series one may replace $\mathrm{e}^{-\alpha^2} \sum_{\ell=0}^{j-1} \alpha^{2\ell}/\ell!$ by unity for all $j > i$. In fact, one can even get for $E_c(\alpha) = \lim_{N\longrightarrow\infty} E_{N_c}(\alpha)$ a series of upper bounds and a series of lower bounds converging toward each other. We have the obvious inequality

$$0 < \prod_{j=r}^{N-1} \left[e^{-\alpha^2} a_j\left(\alpha^2\right) \right] \leq 1, \quad r \geq 0, \tag{15.1.20}$$

where

$$a_j(x) = \sum_{\ell=0}^{j} \frac{x^\ell}{\ell!} \tag{15.1.21}$$

is the truncated exponential series. On the other hand, the identity

$$e^{-\alpha^2} a_j\left(\alpha^2\right) = \exp\left(-\int_0^{\alpha^2} \frac{a_j(x) - a_{j-1}(x)}{a_j(x)}\, dx\right) \tag{15.1.22}$$

and the inequality $a_j(x) \geq a_\ell(x)$ for $j \geq \ell$ give us

$$\begin{aligned} \prod_{j=r}^{N-1} \left[e^{-\alpha^2} a_j\left(\alpha^2\right)\right] &= \exp\left[-\sum_{j=r}^{N-1} \int_0^{\alpha^2} \frac{a_j(x) - a_{j-1}(x)}{a_j(x)}\, dx\right] \\ &\geq \exp\left[-\sum_{j=r}^{N-1} \int_0^{\alpha^2} \frac{a_j(x) - a_{j-1}(x)}{a_r(x)}\, dx\right] \\ &= \exp\left[-\int_0^{\alpha^2} \frac{a_{N-1}(x) - a_{r-1}(x)}{a_r(x)}\, dx\right]. \end{aligned} \tag{15.1.23}$$

Taking the limit $N \longrightarrow \infty$, the inequalities (15.1.20) and (15.1.23) give us finally

$$\begin{aligned} 0 < F_s\left(\alpha^2\right) f_s\left(\alpha^2\right) &\leq F_r\left(\alpha^2\right) f_r\left(\alpha^2\right) \\ &\leq E_c(\alpha) \leq F_{r'}\left(\alpha^2\right) \leq F_{s'}\left(\alpha^2\right) \leq 1, \\ &\text{if } r > s > 0, \quad r' > s' > 0, \end{aligned} \tag{15.1.24}$$

where

$$F_r\left(\alpha^2\right) = \prod_{j=0}^{r-1} \left[e^{-\alpha^2} a_j\left(\alpha^2\right)\right] \equiv E_{rc}(\alpha) \tag{15.1.25}$$

and

$$f_r\left(\alpha^2\right) = \exp\left[-\int_0^{\alpha^2} \frac{e^x - a_{r-1}(x)}{a_r(x)}\, dx\right]. \tag{15.1.26}$$

Equation (15.1.24) gives us in particular

$$\exp\left(-\alpha^2 - \int_0^{\alpha^2} \frac{\mathrm{e}^x - 1}{1 + x}\, dx\right) \leq E_c(\alpha) \leq \left(1 + \alpha^2\right) e^{-2\alpha^2}. \tag{15.1.27}$$

To get the n-point correlation function

$$R_n(z_1, ..., z_n) = \frac{N!}{(N-n)!} \int \cdots \int P_c(z_1, ..., z_N) \prod_{i=n+1}^{N} dx_i dy_i \tag{15.1.28}$$

we proceed exactly as in Section 5.2 or 10.1. Equation (15.1.14) corresponds to the orthogonality property of φ_k in Section 5.2. The result is

$$R_n(z_1, ..., z_n) = \pi^{-n} \exp\left(-\sum_{j=1}^{n} |z_j|^2\right) \det\left[K_N(z_i, z_j)\right]_{i,j=1,...,n}, \tag{15.1.29}$$

where

$$K_N(z_i, z_j) = \sum_{\ell=0}^{N-1} \frac{(z_i z_j^*)^\ell}{\ell!} \tag{15.1.30}$$

As $N \longrightarrow \infty$ the correlation functions tend to well-defined limits:

$$R_n(z_1, ..., z_N) \simeq \pi^{-n} \exp\left(-\sum_1^n |z_i|^2\right) \det\left[e^{z_i z_j^*}\right]_{i,j=1,2,...,n}. \tag{15.1.31}$$

From Eqs. (15.1.29) and (15.1.30) the density of the eigenvalues is

$$R_1(z) = \pi^{-1} e^{-|z|^2} \sum_{\ell=0}^{N-1} \frac{|z|^{2\ell}}{\ell!}. \tag{15.1.32}$$

This density is isotropic and depends only on $|z| = r$, which was to be expected. It is constant $R_1(z) \approx 1/\pi$ for $r^2 \ll N$ and $R_1(z) \approx 0$ for $r^2 \gg N$. The sum in Eq. (15.1.32) can be estimated in an elementary way. From the inequalities

$$\begin{aligned} e^{r^2} - \sum_0^{N-1} \frac{r^{2\ell}}{\ell!} &= \sum_N^{\infty} \frac{r^{2\ell}}{\ell!} \leq \frac{r^{2N}}{N!} \sum_0^{\infty} \frac{r^{2\ell}}{(N+1)^\ell} \\ &= \frac{r^{2N}}{N!} \frac{N+1}{N+1-r^2}, \quad \text{for } r^2 < N \end{aligned} \tag{15.1.33}$$

and

$$\sum_{0}^{N-1} \frac{r^{2\ell}}{\ell!} \leq \frac{r^{2(N-1)}}{(N-1)!} \sum_{0}^{N-1} \left(\frac{N-1}{r^2}\right)^{\ell}$$
$$= \frac{r^{2(N-1)}}{(N-1)!} \frac{r^2}{r^2 - N + 1}, \quad \text{for } r^2 > N \tag{15.1.34}$$

we get

$$1 - \pi R_1(z) \leq e^{-r^2} \frac{r^{2N}}{N!} \frac{N+1}{N+1-r^2} \quad \text{for } r^2 < N \tag{15.1.35}$$

and

$$\pi R_1(z) \leq e^{-r^2} \frac{r^{2N}}{N!} \frac{N+1}{r^2+1-N} \quad \text{for } r^2 > N. \tag{15.1.36}$$

One can also estimate how fast the eigenvalue density falls from $1/\pi$ to 0 around $r^2 = N$. Putting $r = N^{1/2} \pm u$, $1 \lesssim u \ll N$, the leading term in Eqs. (15.1.35) and (15.1.36) is $\mathrm{e}^{-u^2}/2u\sqrt{\pi}$.

The two-point correlation function in the limit $N \longrightarrow \infty$ is

$$R_2(z_1, z_2) = \pi^{-2}\left[1 - \exp\left(-|z_1 - z_2|^2\right)\right] \tag{15.1.37}$$

and depends only on the distance between the eigenvalues.

15.2. Quaternion Matrices

In this section we consider matrices the elements of which are real quaternions (*cf.* Chapter 2). All four quaternion components of each matrix element are random variables. To proceed any further one has to know about the diagonalization of these matrices. The eigenvalue equation may be written as

$$SY = Y\lambda, \tag{15.2.1}$$

where Y is a vector with N quaternion components (the eigenvector) and λ is a quaternion number (the eigenvalue). There is no reason *a priori* for (real quaternion) solutions to Eq. (15.2.1) to exist. Fortunately, they do and in sufficient number (Appendix A.33). Writing Eq. (15.2.1) as

$$SY\mu = Y\mu\left(\mu^{-1}\lambda\mu\right), \tag{15.2.2}$$

we see that if λ is an eigenvalue then so is $\mu^{-1}\lambda\mu$ for arbitrary μ. Thus the eigenvalues of a given matrix are not just discrete points but describe closed curves, and one has to talk about the distribution of these eigencurves in the four-dimensional space. Even if one chooses to describe these curves by some fixed point or points on them, only one-sided linear independence of the corresponding eigenvector rays can be established by the usual methods. Although, for a given quaternion real matrix S another such X can be found (in the favorable circumstance of distinct eigencurves) which diagonalizes it,

$$S = XEX^{-1} \tag{15.2.3}$$

[E diagonal and real (Appendix A.33)], it seems difficult to establish it by purely quaternion means.

In view of these difficulties, from now on we shall employ the matrix representation of quaternions (*cf.* Chapter 2), thus doubling the size of the matrix S, and use well-known results on matrices with complex elements. Thus, in reality, this section does not deal with the quaternion matrices as such but with even-order complex matrices having a special structure; the elements of S satisfy the relations

$$S_{2i,2j} = S^*_{2i-1,2j-1}, \qquad S_{2i-1,2j} = -S^*_{2i,2j-1} \tag{15.2.4}$$

or, in the matrix notation,

$$SZ = ZS^*, \tag{15.2.5}$$

where Z is the antisymmetric, real, unitary matrix (2.4.1).

If X is the $2N \times 2N$ matrix whose columns are the eigenvectors of S, the eigenvalues being all distinct, then

$$S = XEX^{-1}, \tag{15.2.6}$$

where E is diagonal. The diagonal elements of E occur in complex conjugate pairs z_j, z_j^*; $j = 1, 2, ..., N$. The linear measure is

$$\mu(dS) = \prod_{\substack{1 \le i \le j \le N \\ \lambda=0,1}} dS_{2i,2j}^{(\lambda)} dS_{2i-1,2j}^{(\lambda)} \tag{15.2.7}$$

where $dS_{ij}^{(0)}$ and $dS_{ij}^{(1)}$ are the real and imaginary parts of dS_{ij}. For $P(S)$ we take

$$P(S) = \exp\left[-\frac{1}{2}\mathrm{tr}\left(S^\dagger S\right)\right]; \tag{15.2.8}$$

the factor 1/2 is there to compensate for the artificial doubling of the size of S. We denote the ensemble so defined by T_Q. Equations (15.1.4) and (15.1.5) are valid. If we write Eq. (15.1.4) in terms of the various components, the volume element of dS is

$$\mu(\mathrm{d}S) = \prod_i |z_i - z_i^*|^2 \prod_{i \ne j} \left(|z_i - z_j|^2 \, |z_i - z_j^*|^2\right) \times \prod_{\substack{i<j \\ \lambda=0,1}} \mathrm{d}A_{2i,2j}^{(\lambda)} \mathrm{d}A_{2i-1,2j}^{(\lambda)}. \tag{15.2.9}$$

The integration corresponding to Eq. (15.1.9) for this case is carried out in appendix A.34 or A.35. The result is as follows.

The joint probability density function for the eigenvalues of S belonging to the ensemble T_Q of all complex matrices satisfying Eq. (15.2.4) is given by

$$P_Q\left(z_1, ..., z_N\right) = K_Q \exp\left(-\sum_1^N |z_i|^2\right) \prod_i |z_i - z_i^*|^2 \times \prod_{i<j} \left(|z_i - z_j|^2 \, |z_i - z_j^*|^2\right), \tag{15.2.10}$$

where K_Q is the normalization constant given by Eq. (15.2.15).

With this joint probability density function one can determine the various quantities of interest with almost the same ease as in the unitary case. The method to be followed in all such calculations is to express $P_Q\left(z_1,...,z_N\right)$ as a quaternion determinant and use the integration method developed in Chapter 6 and Appendix A.18.

Let us write a $2N \times 2N$ Vandermonde determinant of the variables z_j, z_j^*, $j = 1, 2, ..., N$; that is, the determinant whose $(2j-1)$th column consists of the successive powers of z_j, $\left(1, z_j, z_j^2, ..., z_j^{2N-1}\right)$, and whose $2j$th column consists of the successive powers of z_j^*, $\left(1, z_j^*, z_j^{*2}, ..., z_j^{*2N-1}\right)$, for $j = 1, 2, ..., N$. We can clearly see that this determinant is nothing but

$$\prod_i \left(z_i^* - z_i\right) \prod_{i<j} \left(\left|z_i - z_j\right|^2 \left|z_i - z_j^*\right|^2\right). \tag{15.2.11}$$

Thus we are led to define with $z = x + iy$,

$$f_{ij}(u) = \int\int e^{-|z|^2} \left(z - z^*\right) u(z) \left(z^i z^{*j} - z^j z^{*i}\right) dxdy, \tag{15.2.12}$$

and the average value of $\prod_i u\left(z_i\right)$ (see Appendix A.18) is

$$\begin{aligned}\left\langle \prod_i u\left(z_i\right) \right\rangle &= \int \cdots \int P_Q\left(z_1, ..., z_N\right) \prod_i u\left(z_i\right) dx_i dy_i \\ &= K_Q N! \left(\det\left[f_{ij}\right]_{i,j=0,1,...,2N-1}\right)^{1/2}. \end{aligned} \tag{15.2.13}$$

Putting $u(z) = 1$ and equating the average Eq. (15.2.13) to unity, we get the value of K_Q :

$$f_{ij}(1) = 2\pi \left[j!\delta_{i+1,j} - i!\delta_{j+1,i}\right], \tag{15.2.14}$$

$$K_Q^{-1} = N!(2\pi)^N \prod_1^N \Gamma(2j). \tag{15.2.15}$$

Next we put

$$\begin{aligned} u(z) &= 0, \quad \text{if} \quad |z| < \alpha \\ &= 1, \quad \text{if} \quad |z| \geq \alpha, \end{aligned} \tag{15.2.16}$$

and obtain an expression for $E_{NQ}(\alpha)$, the probability that no eigenvalue will lie inside a circle of radius α centered at the origin

$$E_{NQ}(\alpha) = \prod_{j=1}^{N} \frac{\Gamma\left(2j, \alpha^2\right)}{\Gamma\left(2j, 0\right)}, \tag{15.2.17}$$

where the incomplete gamma functions $\Gamma\left(j, \alpha^2\right)$ are defined by Eq. (15.1.16). Corresponding to Eqs. (15.1.19), (15.1.20), and (15.1.26), we now have

$$E_Q(\alpha) = \lim_{N \to \infty} E_{NQ}(\alpha) = 1 - \frac{1}{2}\alpha^4 + \frac{1}{3}\alpha^6 - \frac{1}{6}\alpha^8 + \frac{1}{15}\alpha^{10} - \cdots, \tag{15.2.18}$$

$$f_s(\alpha) \leq f_r\left(\alpha^2\right) \leq E_Q(\alpha) \leq F_{r'}\left(\alpha^2\right) \leq F_{s'}\left(\alpha^2\right), \tag{15.2.19}$$

$$\text{for} \quad r > s > 0, \; r' > s' > 0,$$

where now

$$F_r\left(\alpha^2\right) = \prod_{j=0}^{r-1} \left(\mathrm{e}^{-\alpha^2} a_{2j+1}\left(\alpha^2\right)\right) \tag{15.2.20}$$

and

$$f_r\left(\alpha^2\right) = F_r\left(\alpha^2\right) \exp\left[-\int_0^{\alpha^2} \frac{\left(e^x - e^{-x}\right)/2 - a_{2r-1}(x)}{a_{2r+1}}\, dx\right], \tag{15.2.21}$$

with $a_j(x)$ given by Eq. (15.1.21).

To get the correlation functions $R_n\left(z_1, ..., z_n\right)$, Eq. (5.1.2), we proceed as in Sections 6.2–6.3 (see also Appendix A.17). Consider the $N \times N$ quaternion matrix

$$Q\left(z_1, ..., z_N\right) = \left[\sigma_N\left(z_i, z_j\right)\right]_{i,j=1,...,N}, \tag{15.2.22}$$

with

$$\sigma_N(z,\zeta) = \begin{bmatrix} \phi_N(\zeta, z^*) & \phi_N(\zeta^*, z^*) \\ \phi_N(z,\zeta) & \phi_N(z,\zeta^*) \end{bmatrix}, \tag{15.2.23}$$

and

$$\begin{aligned} \phi_N(u,v) &= -\phi_N(v,u) \\ &= \frac{1}{2\pi} \sum_{0\le i\le j\le N-1} \frac{2^j j!}{2^i i!(2j+1)!} \left(u^{2i}v^{2j+1} - v^{2i}u^{2j+1}\right). \end{aligned} \tag{15.2.24}$$

Lemma 15.2.1. *The matrix Q has the following properties:*

(i) Q is self-dual, its (i,j) element $\sigma_N(z_i,z_j)$ depends only on z_i and z_j;

(ii)

$$\begin{aligned} &\int \sigma_N(z,z)(z-z^*)\exp\left(-|z|^2\right) d^2z \\ &\quad = \int \sigma_N(z,z)(z-z^*)\exp\left(-|z|^2\right) dxdy = N; \end{aligned} \tag{15.2.25}$$

(iii)

$$\int \sigma_N(z,\zeta)\sigma_N(\zeta,\xi)(\zeta-\zeta^*)\exp\left(-|\zeta|^2\right) d^2\zeta = \sigma_N(z,\xi), \tag{15.2.26}$$

(iv)

$$\det Q = \text{const} \times \prod_{i=1}^{N} (z_i^* - z_i) \prod_{1\le i<j\le N} \left(|z_i - z_j|^2 \left|z_i - z_j^*\right|^2\right), \tag{15.2.27}$$

or

$$\det C(Q) = \text{const} \times \prod_{i=1}^{N} (z_i^* - z_i)^2 \prod_{1\le i<j\le N} \left(|z_i - z_j|^4 \left|z_i - z_j^*\right|^4\right), \tag{15.2.28}$$

where $C(Q)$ is the $2N \times 2N$ ordinary matrix corresponding to the $N \times N$ quaternion matrix Q (cf. Section 6.2).

Proof. Property (i) is almost obvious. Verification of (ii) is straightforward using the orthogonality property Eq. (15.1.14)

$$\int e^{-|z|^2} z^i z^{*j} d^2 z = \pi j! \, \delta_{ij} \tag{15.2.29}$$

For (iii) it is convenient first to note that

$$\begin{aligned}
&\int \phi_N(z, \zeta^*) \, \phi_N(\zeta, \xi) \, (\zeta - \zeta^*) \, e^{-|\zeta|^2} d^2\zeta \\
&\quad = -\int \phi_N(z, \zeta) \phi_N(\zeta^*, \xi) \, (\zeta - \zeta^*) \, e^{-|\zeta|^2} d^2\zeta \\
&\quad = \int \phi_N(z, \zeta) \phi_N(\xi, \zeta^*) e^{-|\zeta|^2} d^2\zeta = \frac{1}{2} \phi_N(z, \xi).
\end{aligned} \tag{15.2.30}$$

To verify (iv) requires a little more work. Consider the $2N \times 2N$ Vandermonde determinant $\det \left[u_i^{j-1} \right]$. In this determinant if we replace the rows of even powers as

$$u^{2j} \longrightarrow \xi_{2j}(u) = \sum_{i=0}^{j} \frac{2^j j!}{2^i i!} u^{2i}, \tag{15.2.31}$$

and the rows of odd powers as

$$u^{2j+1} \longrightarrow \xi_{2j+1}(u) = u^{2j+1} / (2j+1)! \tag{15.2.32}$$

the determinant is multiplied by a known constant. Therefore,

$$\prod_{1 \le i < j \le 2N} (u_i - u_j)^2 = \text{const} \times \det \left[\xi_{2j}(u_i) \quad \xi_{2j+1}(u_i) \right] \begin{bmatrix} \xi_{2j+1}(u_k) \\ -\xi_{2j}(u_k) \end{bmatrix},$$

$$i, k = 1, ..., 2N, \quad j = 0, 1, ..., N-1;$$

or

$$\prod_{i<j}(u_i - u_j)^2 = \text{const} \times \det\left[\sum_{j=0}^{N-1}[\xi_{2j}(u_i)\,\xi_{2j+1}(u_k) - \xi_{2j+1}(u_i)\,\xi_{2j}(u_k)]\right]$$

$$= \text{const} \times \det[\phi_N(u_i, u_k)], \tag{15.2.33}$$

This is essentially Eq. (15.2.28). End of proof.

Hence the n-point correlation function, according to Theorem 6.2.1, is

$$R_n(z_1, ..., z_n) = \prod_{i=1}^{n}\left(e^{-|z_i|^2}(z_i - z_i^*)\right)\det[\sigma_N(z_i, z_j)]_{i,j=1,...,n} \tag{15.2.34}$$

with the quaternion σ_N given by Eq. (15.2.23). The constant is fixed by the requirement that

$$\int\cdots\int P_Q(z_1, ..., z_N)\, d^2z_1, ..., d^2z_N = 1.$$

In the limit $N \longrightarrow \infty$,

$$\phi_N(u, v) \longrightarrow \phi(u, v)$$

$$= \frac{1}{(2\pi)^{1/2}} \sum_{0\le k\le i<\infty} [k!\,\Gamma(i + 3/2)]^{-1}$$

$$\times \left\{\left(\frac{u^2}{2}\right)^k \left(\frac{v^2}{2}\right)^{i+1/2} - \left(\frac{u^2}{2}\right)^{i+1/2} \left(\frac{v^2}{2}\right)^k\right\}. \tag{15.2.35}$$

Taking k and $i - k = k_1$ as independent summation indices,

$$\phi(u, v) = (2\pi)^{-1/2} \sum_{k_1=0}^{\infty} I_{k_1+1/2}(uv) \left\{\left(\frac{v}{u}\right)^{k_1+1/2} - \left(\frac{u}{v}\right)^{k_1+1/2}\right\} \tag{15.2.36}$$

where $I_{k_1+1/2}$ is the Bessel function

$$I_{k_1+1/2}(x) = \sum_{k=0}^{\infty} [k!.\Gamma(k + k_1 + 3/2)]^{-1} (x/2)^{2k+k_1+1/2}. \qquad (15.2.37)$$

By using the recurrence relation for Bessel functions, we obtain the following simpler equation (*cf.* Appendix A.36)

$$\phi(u, v) = \frac{1}{2\pi}(v - u)e^{uv} \int_0^1 \exp\left(\frac{1}{2}(u - v)^2 x\right) \frac{dx}{\sqrt{1 - x}}. \qquad (15.2.38)$$

Equation (15.2.34) now becomes

$$R_n(z_1, ..., z_n) = \prod_{i=1}^{n} \left(e^{-|z_i|^2} (z_i - z_i^*)\right)$$
$$\times \left\{ \det \begin{bmatrix} \phi(z_i, z_j) & \phi(z_i, z_j^*) \\ \phi(z_i^*, z_j) & \phi(z_i^*, z_j^*) \end{bmatrix} \right\}^{1/2}. \qquad (15.2.39)$$

15.3. Real Matrices

A matrix with real elements does not necessarily possess a sufficient number of real solutions to the eigenvalue equation (15.2.1). This is perhaps the reason for the great difficulties experienced in the investigation of random matrices with real elements. If all the eigenvalues are real, then, taking

$$P(S)\mu(dS) = e^{-\mathrm{tr}(S^\dagger S)} \prod_{i,j} dS_{i,j}, \qquad (15.3.1)$$

Ginibre has shown that the probability density function for the eigenvalues is identical to that of Gaussian orthogonal ensembles (3.1.17) (*cf.* Appendix A.37 or A.35). If some of the eigenvalues are not real, one has only a complicated integral expression for the joint probability density function of the eigenvalues, which has not yet been simplified.

Summary to Chapter 15

Three ensembles of random matrices S are considered: (i) the elements of S are complex numbers, (ii) they are real quaternions, and (iii) they are real numbers. The joint probability density of the matrix elements is taken to be proportional to $\exp[-\mathrm{tr}\,(S^\dagger S)]$ in each case. An explicit expression for the n-point correlation function is derived when the elements of S are either complex numbers or real quaternions, Eqs. (15.1.31) and (15.2.39), respectively. When the elements of S are real numbers, this has not yet been possible.

16 / Statistical Analysis of a Level Sequence

Experimentally one observes a finite stretch of energy levels of a system, giving rise to a list of numbers. This may be the set of resonance levels of a nucleus, the possible energies of a free particle on an odd shaped billiard table, or the zeros of the Riemann zeta function on the critical line. Suppose these numbers are $(E_1 \leq E_2 \leq \cdots \leq E_n)$ occupying an interval of length $2L$. The question is how well these numbers agree with the predictions of the statistical theory.

As we said in Section 1.4, one may draw for example, a histogram of the nearest neighbor spacings. For this one computes $S_i = E_{i+1} - E_i$, the average $D = \langle S_i \rangle$ of the S_i, the normalized spacings $s_i = S_i / \langle S_i \rangle = (E_{i+1} - E_i)/D$, then sorts these spacings according to their size, and draws a graph whose height between i/m and $(i+1)/m$ is proportional to the number of s_j such that $i/m \leq s_j < (i+1)/m$ for some convenient integer m, say, 10 or 20 depending on n, and for $i = 0, 1, 2, \ldots$. Or one may draw a histogram of the two level correlation function. And compare such histograms with the curves given by the theory. If n is fairly large, one may choose a reasonably big integer m and the histogram gets finer. But as French said once (1971, private conversation), "the statistical theory gives us the limiting curve of, say, the nearest neighbor spacings, but not its thickness. And one cannot give a quantitative estimate of how good a histogram fits a given curve."

The statistical theory makes the hypothesis that there exists a mean level spacing D and a very large number N, such that the statistical behaviour of $x_j = \pi E_j/(D\sqrt{2N})$, $j = 1, \ldots, n$, is the same as that of n consecutive real numbers $(x_1, \ldots, x_n)$ all much smaller than $(2N)^{1/2}$ of a much longer seris $(x_1, \ldots, x_N)$. The $(x_1, \ldots, x_N)$ are distributed on the real line with the joint probability density

$$P(x_1, ..., x_N) = \text{const} \times \exp\left(-\beta \sum_j x_j^2\right) \prod_{i<j} |x_i - x_j|^\beta, \qquad (16.0.1)$$

$\beta = 1$, 2, or 4. Or another theory makes the hypothesis that there exists a mean level spacing D and a very large N, such that the statistical properties of the n angles $\theta_j = 2\pi E_j/(ND)$, $j = 1, ..., n$, are the same as that of n consecutive angles of a much longer series $(\theta_1, ..., \theta_N)$. The $(\theta_1, ..., \theta_N)$ are distributed on the whole circle $0 \le \theta_j \le 2\pi$ with the probability density

$$P(\theta_1, ..., \theta_N) = \text{const} \times \prod_{j<k} \left|e^{i\theta_j} - e^{i\theta_k}\right|^\beta, \qquad (16.0.2)$$

$\beta = 1, 2,$ or 4. We did prove in Chapter 10 that the statistical behavior of the n numbers $(x_1, ..., x_n)$ is identical to that of the n angles $(\theta_1, ..., \theta_n)$ in the limit $N \to \infty$. For the analysis of the experiments, one needs to know the statistical properties of such a partial series.

For small n we already have at our disposal the cluster functions and level spacings studied earlier, though the "thickness" of the curves is lacking. In this chapter we will assume that $1 \ll n \ll N$. That is to say, the number n of observed levels is large enough to be described in statistical terms, but is still very small compared with the number of unobserved levels. The procedure will be to search for convenient statistics of the observed level series. The word "statistic" is here used in the technical sense customary among statisticians. A "statistic" is a number W which can be computed from an observed sequence of levels alone without other information, and for which the average value $\langle W \rangle$ and the variance

$$V_W = \left\langle (W - \langle W \rangle)^2 \right\rangle \qquad (16.0.3)$$

are known from the theoretical model. A convenient statistic should satisfy two conditions : (i) the computation of W from the observed data should be simple, and (ii) the theoretical figure of merit

$$\Phi_W = \left[V_W / \langle W \rangle^2\right] \qquad (16.0.4)$$

should be as small as possible. This Φ_W is the mean square fractional deviation which is to be expected between the observed W and the theoretical $\langle W \rangle$, if the theoretical model is a valid one.

A statistic may serve either of two purposes. If $\langle W \rangle$ involves some unknown parameter contained in the theoretical model, then W provides a measurement of this parameter with fractional error proportional to $\Phi_W^{1/2}$. If $\langle W \rangle$ is independent of parameters, then W provides a test of the theory. In the second case, the theory has clearly failed if $(W - \langle W \rangle)^2$ is found to be much larger than V_W.

In the theoretical model described above, there is only one parameter, namely the "ideal level spacing" D. The integer N is not an effective parameter, since all properties of the model become independent of N as $N \longrightarrow \infty$. In the analysis of observations, one needs only one statistic to measure D, and any additional statistic will provide a test of the model.

In practice one often has level series which are mixtures of two or more superimposed series with different values of angular momentum and parity. For the analysis of such series one must consider a generalized theoretical model. The generalized model consists of m uncorrelated level series superimposed on the same interval of length $2L$, their theoretical level spacings being $[D_1, ..., D_m]$. This generalized model contains m free parameters D_μ. It is more convenient to take for the parameters the overall mean level spacing

$$D = \left[\sum D_\mu^{-1} \right]^{-1} \tag{16.0.5}$$

and the fractions

$$a_\mu = (D/D_\mu) \ , \quad \sum a_\mu = 1, \tag{16.0.6}$$

a_μ being the proportion of levels belonging to the μth series. Equation (16.0.5) expresses the fact that the overall density of the levels is the sum of the partial level densities.

When one is dealing with a double series ($m = 2$), it is possible with two statistics to measure D and $a_1 = 1 - a_2$, and with a third statistic to obtain a meaningful test of the model. When the series is multiple ($m > 2$), the information that can be obtained in this way is necessarily more fragmentary.

We describe now the various types of statistics found useful in applications. They have been extensively applied in the analysis of nuclear resonance levels, energy levels of quantum chaotic systems and the zeros of the Riemann zeta function on the critical line. More recently, they are finding applications in analyzing the elastodynamic properties of structural materials, architectural acoustics, and so on. (*cf.* Weaver, 1989.) Any article on the statistical properties of such sequences will contain a few graphs of the number variance, of the least square statistic Δ and values of the covariance of the nearby spacings in addition to the histograms of the two level correlation function and of the probability densities $p(k;s)$ of the kth neighbor spacings for small k, (*cf.* Chapter 5.1). The F statistic, Section 16.5, has sometimes been used to detect slight imperfections in the experimental nuclear data (see Liou et al., 1972a, b). The energy statistic Q, Section 16.3, the Λ and other statistics, Sections 16.6, 16.7 did not receive much popularity.

16.1. Linear Statistics or the Number Variance

The simplest statistic for the measurement of the average spacing D is

$$W = n, \quad \langle W \rangle = \frac{2L}{D} = s. \tag{16.1.1}$$

The variance of this W, $V_n \equiv \mu_2 = \langle (n - \langle n \rangle)^2 \rangle = \langle n^2 \rangle - \langle n \rangle^2 = (\delta n)^2$, from Appendix A.38, is

$$\begin{aligned} V_n = {} & \frac{2}{\pi^2}[\ln(2\pi s) + \gamma - \frac{\pi^2}{8} - \mathrm{Ci}(2\pi s)] \\ & + \frac{4s}{\pi}\int_{\pi s}^{\infty}\left(\frac{\sin \xi}{\xi}\right)^2 d\xi + \frac{1}{\pi^2}\left(\int_{\pi s}^{\infty}\frac{\sin \xi}{\xi}d\xi\right)^2 \\ = {} & \frac{2}{\pi^2}\left(\ln(2\pi s) + 1 + \gamma - \frac{\pi^2}{8}\right) + O(s^{-1}), \quad \beta = 1, \end{aligned} \tag{16.1.2}$$

$$\begin{aligned} V_n & = \frac{1}{\pi^2}[\ln(2\pi s) + \gamma - \mathrm{Ci}(2\pi s)] + \frac{2s}{\pi}\int_{\pi s}^{\infty}\left(\frac{\sin \xi}{\xi}\right)^2 d\xi \\ & = \frac{1}{\pi^2}(\ln(2\pi s) + 1 + \gamma) + O(s^{-1}), \quad \beta = 2, \end{aligned} \tag{16.1.3}$$

and

$$V_n = \frac{1}{2\pi^2}\left[\ln(4\pi s) + \gamma - \mathrm{Ci}(4\pi s)\right]$$

$$+ \frac{2s}{\pi}\int_{2\pi s}^{\infty}\left(\frac{\sin\xi}{\xi}\right)^2 d\xi + \frac{1}{4\pi^2}\left(\int_0^{2\pi s}\frac{\sin\xi}{\xi}d\xi\right)^2$$

$$= \frac{1}{2\pi^2}\left(\ln(4\pi s) + 1 + \gamma + \frac{\pi^2}{8}\right) + O(s^{-1}), \quad \beta = 4, \quad (16.1.4)$$

where $\mathrm{Ci}(x)$ is the cosine integral

$$\mathrm{Ci}(x) = \ln x + \gamma - \int_0^x \frac{1-\cos\xi}{\xi}d\xi = -\int_x^{\infty}\frac{\cos\xi}{\xi}\,d\xi \qquad (16.1.5)$$

The figure of merit for this statistic n is $\Phi_n = V_n/s^2$, and is quite small in practice.

This n is one of a general class of linear statistics

$$W = \sum_{i=1}^{n} f(E_i), \quad \langle W\rangle = \frac{1}{D}\int_{-L}^{L} f(E)dE, \qquad (16.1.6)$$

where $f(x)$ is any function defined on the interval $(-L, L)$, the zero of energy is chosen for convenience at the center of the observed interval. The choice $f(x) = 1$ gives the statistic (16.1.1). As $f(x) = 1$ has discontinuities at the end points, this is not the linear statistic giving a minimum Φ_W. In fact any smooth function $f(x)$ will give a Φ_W lower than that for an $f(x)$ having discontinuities. And

$$f(x) = \left(1 - (x/L)^2\right)^{1/2} \qquad (16.1.7)$$

gives the minimum having a constant variance independent of s (*cf.* Appendix A.39). But in the practical situations with about a hundred observed levels, the precision gained is not worth the extra trouble.

Equations (16.1.1) and (16.1.6) hold both for simple and multiple series. In the case of multiple series, the variance of a linear statistic is the sum of the variances for each individual series. Thus, for example, Eq. (16.1.2) becomes

$$V_n = \frac{2m}{\pi^2}\left(\ln(2\pi s) + 1 + \gamma - \frac{\pi^2}{8}\right) + \frac{2}{\pi^2}\sum_{\mu} a_\mu \ln a_\mu + O(s^{-1}), \quad \beta = 1, \tag{16.1.8}$$

where m is the number of independent series and a_μ is the fraction of the levels belonging to the μth series. Thus a rare series contributes as much as a dense series to the error in the measurement of D. This effect of a rare series may be of some practical importance.

Figures 16.1 to 16.6 show the number variance for the nuclear levels, for energies of a chaotic system, for ultrasonic resonance frequencies of an aluminium block and for the zeros of the Riemann zeta function $\zeta(z)$ on the critical line Re $z = 1/2$. Note that for energy intervals less than or equal to about 10 times the mean spacing, the number variance agrees quite well with Eq. (16.1.2) or (16.1.3). For larger energy intervals, the number variance does not rise as required by those equations, but rather oscillates indefinitely around a roughly constant value, thus announcing the failure of the random matrix theory. This seems to happen for energy intervals when the classical periodic orbits gain importance on the classical chaotic motions. We do not yet have a clear understanding of the transition from an integrable system to a chaotic one and the consequent changes in the energy level statistics. But surprisingly enough, in many cases one can apparently devise an adhoc formula which reproduces the oscillations in the number variance. For example, for the Riemann zeta zeros, Berry (1988) proposed from heuristic arguments the formula

$$V(L) = V_2(L) + \frac{1}{\pi^2}[\mathrm{Ci}(4\pi L\tau) - \ln(4\pi L\tau) - \gamma] + \frac{2}{\pi^2}\sum_{p,r} \frac{\sin^2[2\pi Lr \ \ln p/\ln\ (E/2\pi)]}{r^2 p^r} \tag{16.1.9}$$

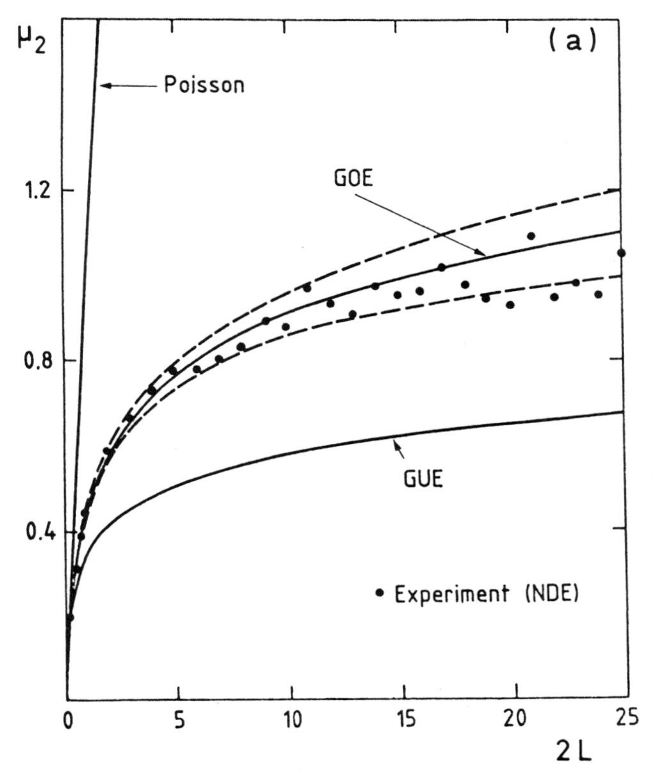

FIG. 16.1. Number variance $\mu_2 = \langle (n - \langle n \rangle)^2 \rangle$ for the nuclear data ensemble. For a detailed explanation of the data analyzed here see the caption of Figure 1.4, Chapter 1. The solid curves correspond to a Poisson process, the Gaussian orthogonal ensemble (GOE) and the Gaussian unitary ensemble (GUE). The dashed curves correspond to one standard deviation from the GOE curve. Points represent the experimental data. Reprinted with permission of The American Physical Society, Haq, R. U., Pandey, A., and Bohigas, O. Fluctuation properties of nuclear energy levels: do theory and experiment agree? *Phys. Rev. Lett.* 48 (1982) 1086–1089.

where $V_2(L)$ is given by Eq. (16.1.3), τ is a constant of the order of $1/4$, Ci is the cosine integral, Eq. (16.1.5), and the sum is taken over all primes p and all positive integers r such that $p^r < (E/2\pi)^\tau$. Here the zeros of $\zeta(z)$ taken into consideration are $z = 1/2 + i\gamma$ with $E - L \leq \gamma \leq E + L$.

Formula (16.1.9) reproduces the empirical variance of the zeros of $\zeta(z)$ very well for all values of L (see Figures 16.4–16.6) and is quite insensitive to small changes in the value of τ. This only shows that the

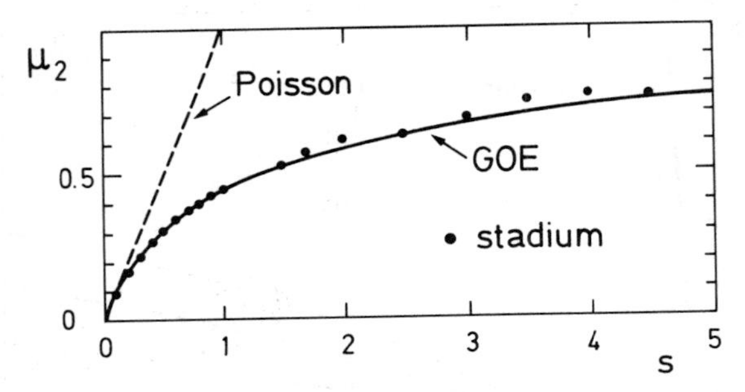

FIG. 16.2. Number variance for the possible energies of a particle free to move on the stadium consisting of a rectangle of size 1×2 with semicircular caps of radius 1. The stadium can be defined by the inequalities $|y| \leq 1$, and either $|x| \leq 1/2$ or $(x \pm 1/2)^2 + y^2 \leq 1$; *cf.* Figure 6.7. The solid curve corresponds to the Gaussian orthogonal ensemble (GOE). Supplied by O. Bohigas, from Bohigas et al. (1984c).

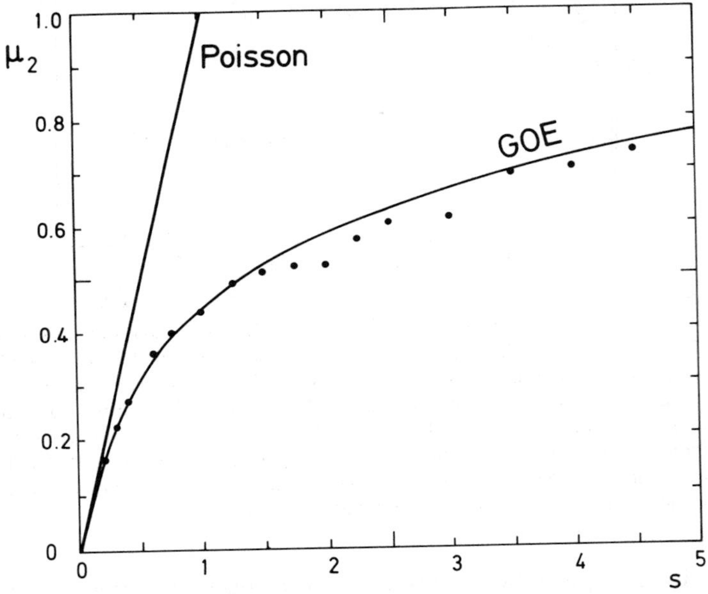

FIG. 16.3. Number variance for the ultrasonic resonance frequencies of an aluminium block. The linearly rising curve corresponds to the Poisson process, while the other one corresponds to the Gaussian orthogonal ensemble (GOE). From Weaver (1989). Reprinted with permission from American Institute of Physics, Weaver, R. L. Spectral statistics in elastodynamics, *J. Acoust. Soc. Amer.*, 85, 1005–1013 (1989).

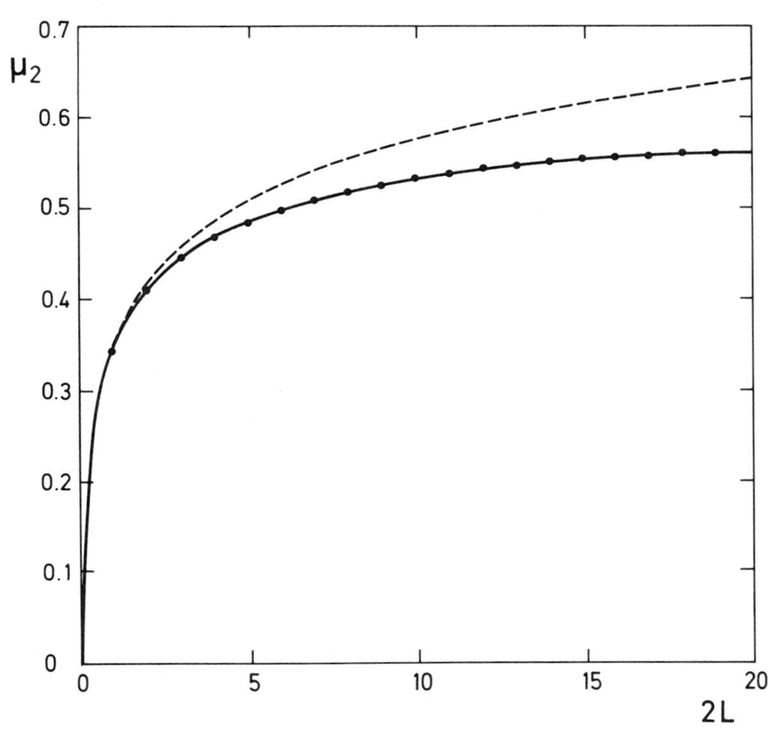

FIG. 16.4. Number variance for the zeta zeros. The dashed curve corresponds to the Gaussian unitary ensemble (GUE). The solid curve passes smoothly through the numerical data. From Odlyzko (1989). Copyright ©, 1989, American Telephone and Telegraph Company reprinted with permission.

zeros of $\zeta(z)$ have a deep relation with the primes and powers of primes, which we do not yet fully understand.

16.2. Least Square Statistic

It is customary to represent the experimental level distributions graphically by a plot of $N(E)$ against E, where $N(E)$ is the number of levels having energy between zero and E. The resulting graphs are in appearance like staircases, and the staircase gives a good visual impression of the overall regularity of the level series (*cf.* Figure 16.7). It is customary to measure the average level spacing by drawing a line having the

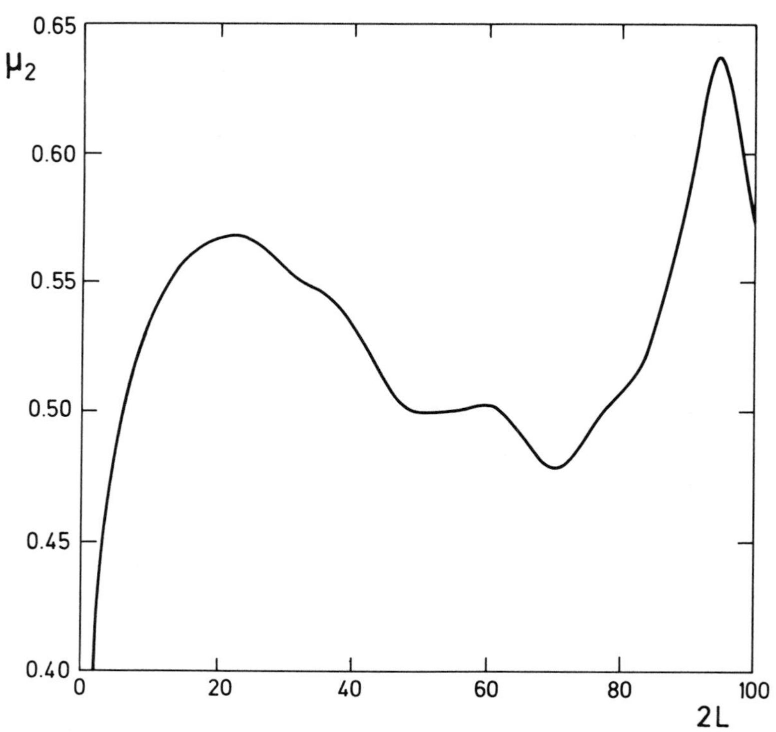

FIG. 16.5. Number variance for the zeta zeros (continued). The difference between the curve (16.1.9) and the numerical data is imperceptible on this figure and Figure 16.6. From Odlyzko (1989). Copyright ©, 1989, American Telephone and Telegraph Company reprinted with permission.

same average slope as the staircase. Fluctuations in the level series then appear as deviations of the staircase from the straight line.

The aim of this section is to make a quantitative study of this "staircase" method of analysis of data, and to find out precisely what can be learned from it. It should be possible to construct, from the deviations between the staircase and the straight line, a statistic which will serve to test whether the overall irregularity of the observed level series is in agreement with the theoretical model or not.

It is convenient to take the observed energy interval as $[-L, +L]$, with zero in the center. For negative E, $N(E)$ is defined as *minus* the number of levels between E and zero. The problem is to analyze the deviation

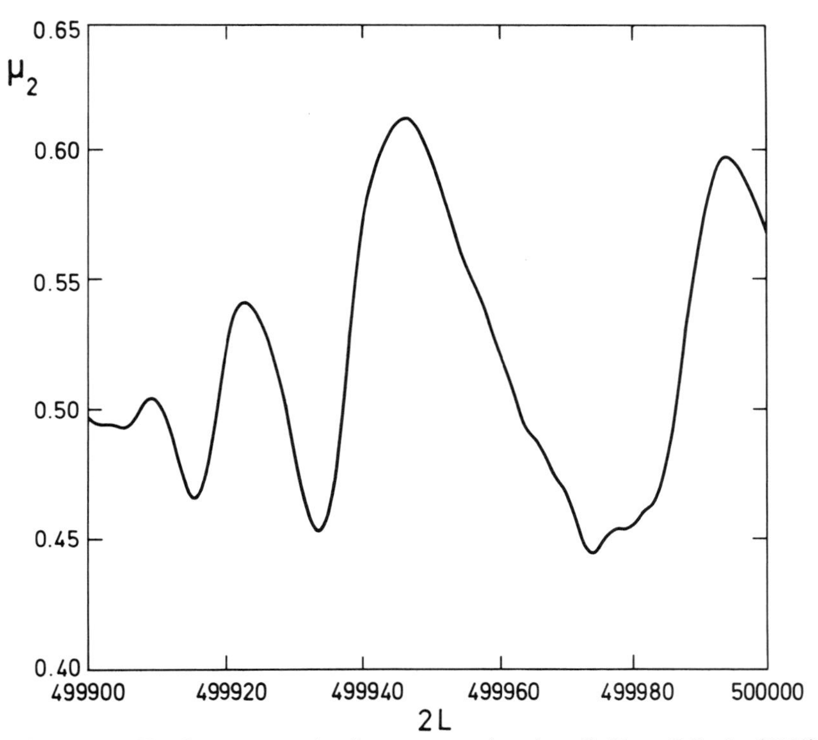

FIG. 16.6. Number variance for the zeta zeros (continued). From Odlyzko (1989). Copyright ©, 1989, American Telephone and Telegraph Company reprinted with permission.

of the staircase graph

$$y = N(E) \tag{16.2.1}$$

from a suitably chosen straight line

$$y = AE + B. \tag{16.2.2}$$

A suitable method of analysis is to fit Eqs. (16.2.2) to (16.2.1) by a least-square criterion. The mean-square deviation of Eq. (16.2.1) from the best fit Eq. (16.2.2) is

$$\Delta = \underset{A,B}{\text{Min}} \left(\frac{1}{2L} \int_{-L}^{L} [N(E) - AE - B]^2 \, dE \right). \tag{16.2.3}$$

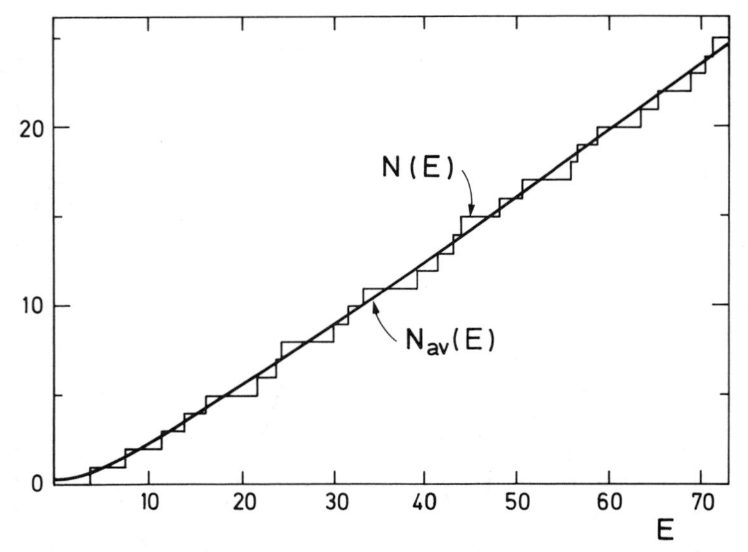

FIG. 16.7. $N(E)$ for the stadium; *cf.* Figure 16.2. Supplied by O. Bohigas, from Bohigas et al. (1984d).

We choose this Δ as the statistic which measures the irregularity of the level series.

Originally three possible ways of choosing the straight line were considered (Dyson and Mehta, 1962a, 1963), but the experience shows that it is better to let A and B vary independently and make a two-parameter fit. For applications, one should know the mean and the variance of Δ.

The mean value of Δ is (*cf.* Appendix A.40)

$$\langle\Delta\rangle = \frac{1}{\pi^2}\left(\ln(2\pi s) + \gamma - \frac{5}{4} - \frac{\pi^2}{8}\right) + O(s^{-1}), \qquad \beta = 1, \qquad (16.2.4)$$

$$= \frac{1}{2\pi^2}\left(\ln(2\pi s) + \gamma - \frac{5}{4}\right) + O(s^{-1}), \qquad \beta = 2, \qquad (16.2.5)$$

$$= \frac{1}{4\pi^2}\left(\ln(4\pi s) + \gamma - \frac{5}{4} + \frac{\pi^2}{8}\right) + O(s^{-1}), \qquad \beta = 4. \qquad (16.2.6)$$

It is instructive to compare these values with the average Δ for a random sequence without correlations. For such a sequence (Poisson

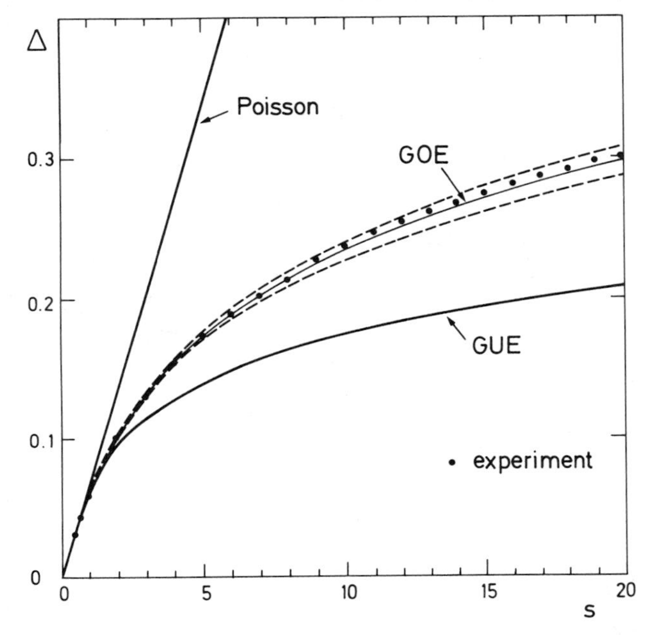

FIG. 16.8. Δ for the nuclear data ensemble; *cf.* Figure 16.1 or 1.4. Supplied by O. Bohigas; reprinted with permission of The American Physical Society, Bohigas, O., Haq, R. U., and Pandey, A. Higher order correlations in spectra of complex systems, *Phys. Rev. Lett.* 54 (1985) 1645–1648.

process),

$$\langle \Delta \rangle = s/15. \tag{16.2.7}$$

To calculate the variance of Δ is more lengthy, as it involves three- and four-level correlation functions as well. We will not do it, but content ourselves to say that the variance is a small constant independent of s.

Figures 16.8 to 16.11 give graphs of Δ for the nuclear energy levels, for energies of chaotic systems, and for the ultrasonic resonance frequencies of aluminium blocks.

For values of the Δ corresponding to various atomic spectra in the third long period of the periodic table (where total spin and parity are good quantum numbers), see Camarda and Georgopulos (1983). They also give the histogram for the nearest neighbor spacings (as in Figure 1.8c) and the covariances of two consecutive spacings (see Section 16.4

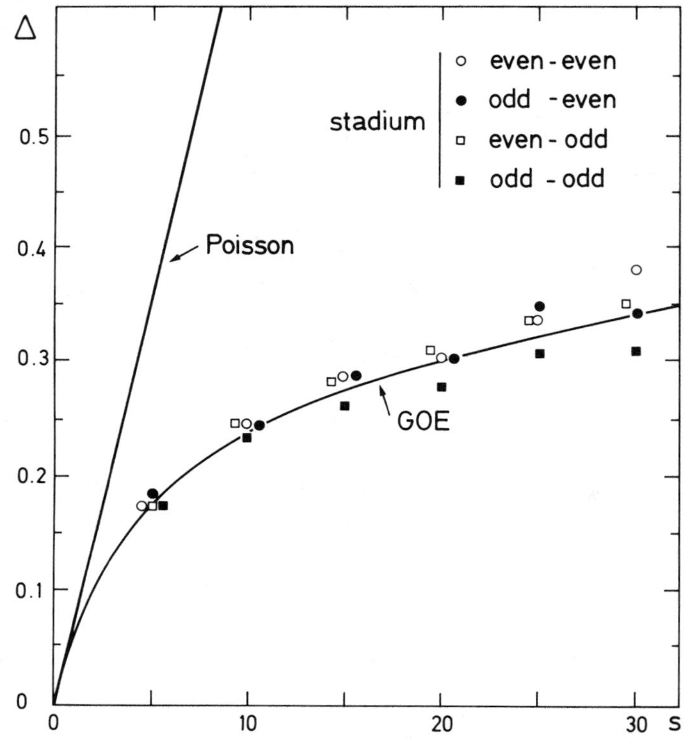

FIG. 16.9. Δ for the stadium, *cf.* Figure 16.2. The computed possible energies of the particle were divided into four categories according to the symmetry of its wave function and each category was analyzed separately. Supplied by O. Bohigas, from Bohigas et al. (1984c).

for the definition). The agreement with the GOE predictions is quite good.

16.3. Energy Statistic

We saw in Section 11.3 that the total energy

$$W = \sum_{1 \le j < k \le N} \ln \left| e^{i\theta_j} - e^{i\theta_k} \right|$$

is a good statistic, since its mean and variance are known from the theory. By means of W one could decide how accurately a given sequence

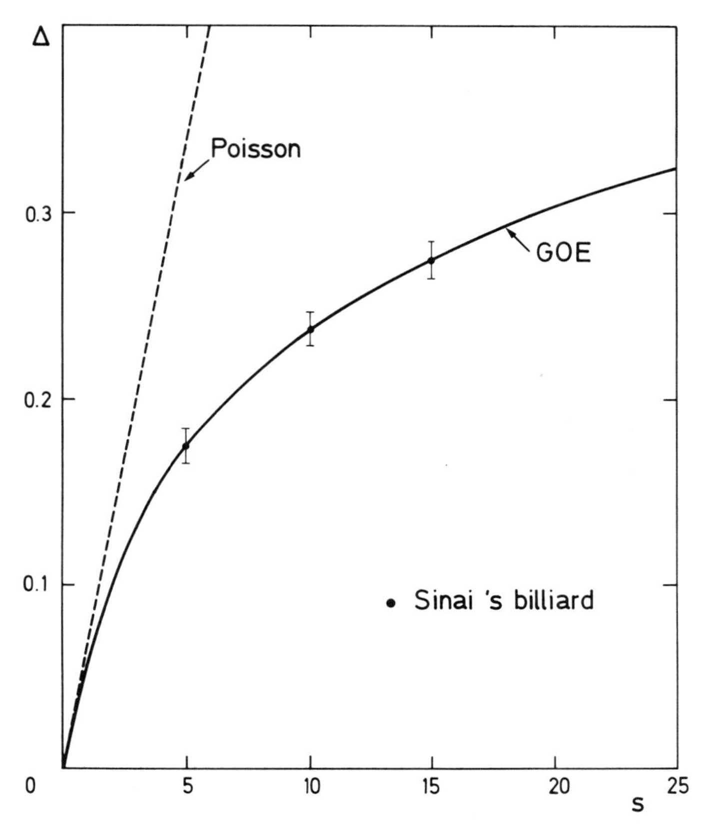

FIG. 16.10. Δ for the Sinai billiard; *cf.* Figure 6.8. Supplied by O. Bohigas, from Bohigas et al. (1984b).

of angles $0 \leq \theta_1, \theta_2, ..., \theta_N < 2\pi$ is or is not in agreement with the theoretical probability density (9.4.5). However, in practice we are given only a partial sequence of levels $(E_1, ..., E_n)$ lying in the interval $(-L, L)$. One could construct a statistic Q for this partial sequence similar to W and calculate its mean and variance. Take for example,

$$Q = -\sum_{i<j} f(E_i, E_j) \ln |(E_i - E_j)/R| + \frac{2L}{D} \sum_i U(E_i),$$

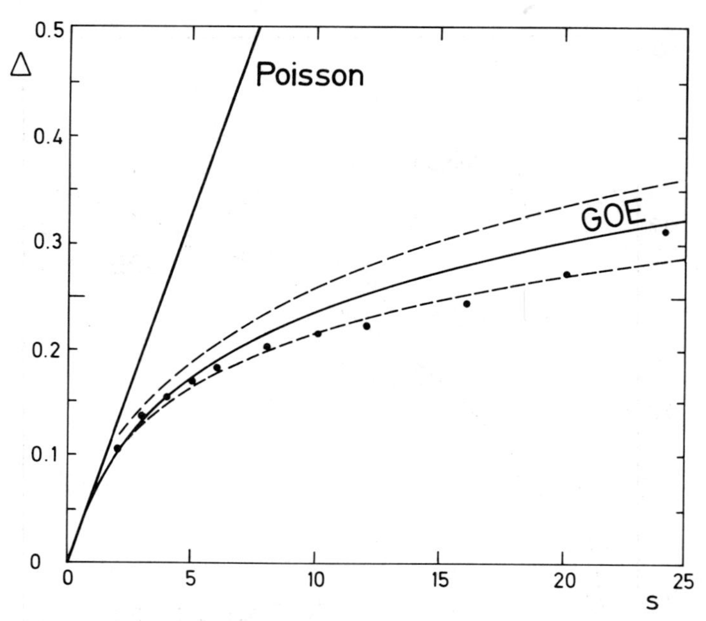

FIG. 16.11. Δ for the ultrasonic resonance frequencies of an aluminium block; *cf.* Figure 16.3. Dashed curves are one standard deviation for the GOE. Reprinted with permission from American Institute of Physics, Weaver, R. L. Spectral statistics in elastodynamics, *J. Acoust. Soc. Amer.*, *85*, 1005–1013 (1989).

where

$$\begin{aligned} f(E,E') &= 1, \quad \text{if} \quad |E-E'|<R, \quad |E|<L, \quad |E'|<L, \\ &= 0, \quad \text{otherwise,} \end{aligned}$$

$$\begin{aligned} U(E) &= -R/L, \qquad \text{if} \quad |E|<L-R, \\ &= -\frac{R}{2L}+\frac{L-|E|}{R}\left[1-\ln\left(\frac{L-|E|}{R}\right)\right], \\ &\qquad\qquad\qquad\qquad \text{if } L-R<|E|<L, \\ &= 0, \qquad \text{otherwise} \end{aligned}$$

and R is chosen to be 2 or 3 times the mean spacing D. In this example, apart from the first term similar to the free energy, an extra term has been added so that its variation with respect to each E_i is quite small. The mean value of Q can be evaluated using the two-level correlation function. To find its variance is more difficult, since it involves three- and four-level correlations as well. We will not do these calculations here, but refer the interested reader to the original paper (Dyson and Mehta, 1963).

16.4. Covariance of Two Consecutive Spacings

To have some idea of how consecutive spacings are correlated, one sometimes computes their covariance. Let s_1, s_2, s_3, ... be the distances between successive eigenvalues, (measured in units of the local mean spacing), $E(n;s)$ the probability that a randomly chosen interval of length s contains exactly n eigenvalues, and $p(n;s)ds$ the probability that the sum $s_1 + \cdots + s_{n+1}$ has a value between s and $s + ds$, i.e., $p(n;s)ds$ is the probability that two eigenvalues separated by n other eigenvalues be at a distance between s and $s + ds$. From

$$(s_1 + \cdots + s_{n+1})^2 = \sum_{j=1}^{n+1} s_j^2 + 2 \sum_{1 \le j < k \le n+1} s_j s_k \tag{16.4.1}$$

taking averages and using Eqs. (5.1.16)–(5.1.18) we get on two partial integrations

$$\langle s_1^2 \rangle = 2 \int_0^\infty E(0;s)ds, \tag{16.4.2}$$

and

$$\langle s_1 s_{n+1} \rangle = \int_0^\infty E(n;s)ds, \quad n > 0. \tag{16.4.3}$$

provided that the first two derivatives of the $E(n;s)$ decrease fast for large s, i.e., provided that $sdE(n;s)/ds$ and $s^2 d^2 E(n;s)/ds^2$ tend to 0 as $s \to \infty$, for all n. Also from $\left\langle (s_1 - s_{n+1})^2 \right\rangle \ge 0$, one has the inequality

$$\int_0^\infty E(n;s)ds \le 2 \int_0^\infty E(0;s)ds. \tag{16.4.4}$$

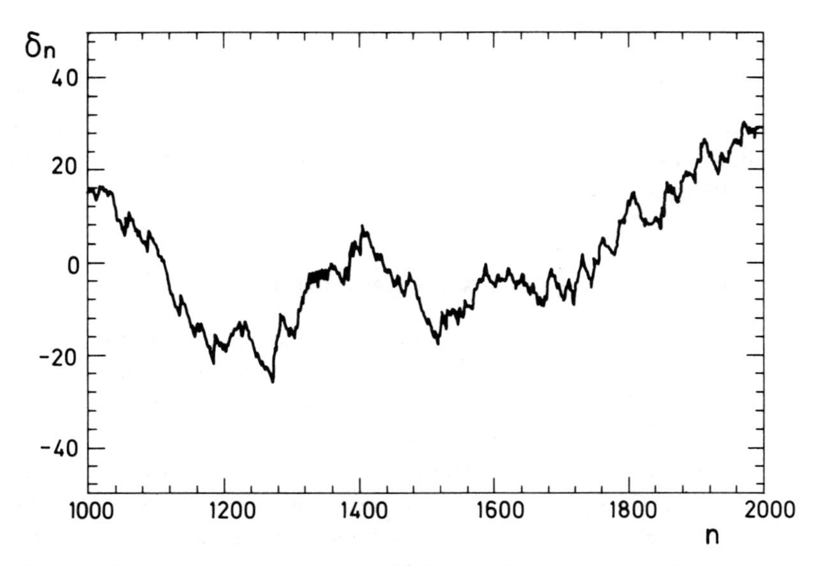

FIG. 16.12. $\delta_n = (s_1 + \cdots + s_n - n)^2$ for the Poisson process. Supplied by O. Bohigas, from Bohigas, Tomsovic and Ullmo.

The covariance of s_1, s_{n+1} is defined by

$$\begin{aligned} \operatorname{cov}(s_1, s_{n+1}) &= \frac{\langle (s_1 - \langle s_1 \rangle)(s_{n+1} - \langle s_{n+1} \rangle) \rangle}{\left(\left\langle (s_1 - \langle s_1 \rangle)^2 \right\rangle \left\langle (s_{n+1} - \langle s_{n+1} \rangle)^2 \right\rangle \right)^{1/2}} \\ &= \frac{\langle s_1 s_{n+1} \rangle - 1}{\langle s_1^2 \rangle - 1} = \frac{\int_0^\infty E(n; s) ds - 1}{2 \int_0^\infty E(0; s) ds - 1}. \end{aligned} \tag{16.4.5}$$

From the tables of the functions $E(n; s)$, Appendix A.14, one can estimate the integrals of $E(n; s)$ numerically, getting

$$2 \int_0^\infty E_\beta(0; s) ds = \begin{cases} 1.286, & \beta = 1, \\ 1.180, & \beta = 2, \\ 1.105, & \beta = 4; \end{cases} \tag{16.4.6}$$

$$\int_0^\infty E_\beta(1; s) ds = \begin{cases} 0.922, & \beta = 1, \\ 0.944, & \beta = 2, \\ 0.964, & \beta = 4. \end{cases} \tag{16.4.7}$$

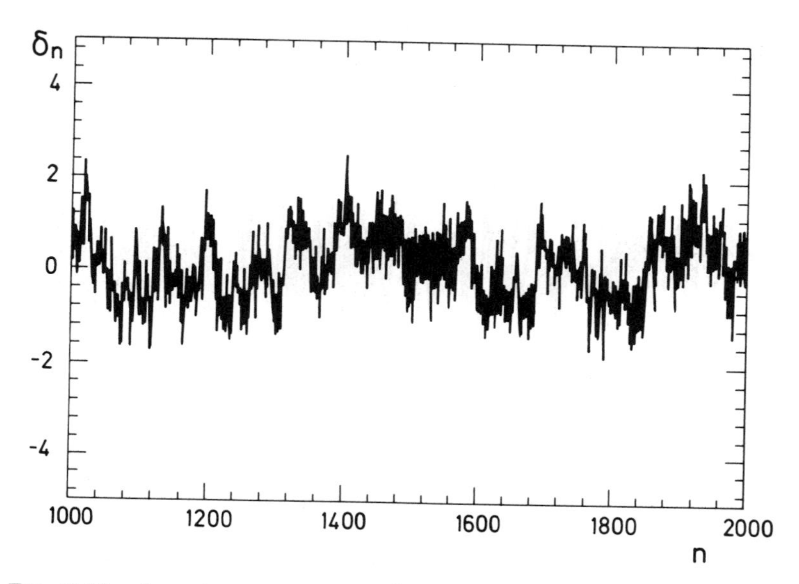

FIG. 16.13. $\delta_n = (s_1 + \cdots + s_n - n)^2$ for GOE (i.e., the Gaussian orthogonal ensemble.) Supplied by O. Bohigas, from Bohigas, Tomsovic and Ullmo.

Hence,

$$\begin{aligned} \mathrm{cov}(s_1, s_2) &= -0.27, \quad \beta = 1 \text{ (orthogonal ensemble)}, \\ &= -0.31, \quad \beta = 2 \text{ (unitary ensemble)}, \\ &= -0.34, \quad \beta = 4 \text{ (symplectic ensemble)}. \end{aligned} \tag{16.4.8}$$

At this point one may think of a simpler statistic $\delta_n = (s_1 + \cdots + s_n - n)^2$, plotted on Figures 16.12 to 16.15 for the Poisson process (level sequence with no correlations), Gaussian orthogonal ensemble (GOE), energies of a chaotic sytem and for those of an integrable system. The graphs for the GOE and chaotic systems look similar, while the contrast with the others is evident. For the Poisson process $E(n; s) = s^n \exp(-s)/n!$ and the mean value

$$\langle \delta_n^2 \rangle \equiv \langle (s_1 + \cdots + s_n - n)^2 \rangle = 2 \sum_{j=0}^{n-1} (n-j) \int_0^\infty E(j; s) ds - n^2$$

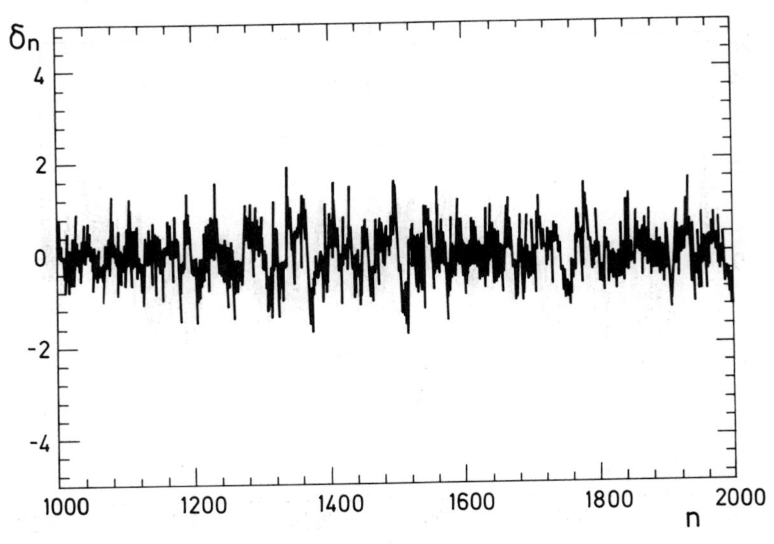

FIG. 16.14. $\delta_n = (s_1 + \cdots + s_n - n)^2$ for a chaotic system, stadium or Sinai's billiard; *cf.* Figures 6.9 and 6.10. Supplied by O. Bohigas, from Bohigas, Tomsovic and Ullmo.

is n. For the matrix ensembles $\langle \delta_n^2 \rangle$ is not known, specially for large n. It is expected that

$$I(j) = \int_0^\infty E(j; s) ds$$

approaches one as j increases, but how fast? According to a conjecture of Dyson (unpublished, quoted by Odlyzko, 1987)

$$I(j) \approx 1 - (\beta \pi^2 j^2)^{-1}, \qquad \beta = 1,\ 2 \quad \text{or} \quad 4.$$

If this conjecture is correct, then $\langle \delta_n^2 \rangle$ should rise as n, which does not seem to be the case.

16.5. The F Statistic

In order to detect a few missing or spurious levels in an otherwise pure sequence, Dyson introduced the so called F statistic. It is defined by

$$F_i = \sum_{j(\neq i)} f\big((x_i - x_j) / mD \big), \tag{16.5.1}$$

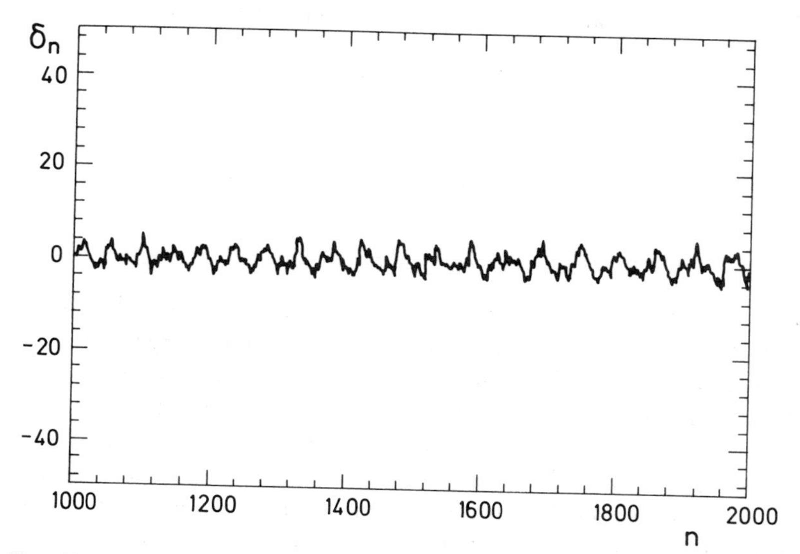

FIG. 16.15. $\delta_n = (s_1 + \cdots + s_n - n)^2$ for an integrable system, corresponding to the possible energies of a particle free to move on a billiard table of semicircular shape. Supplied by O. Bohigas, from Bohigas, Tomsovic and Ullmo.

where x_1, x_2, ... are the positions of the successive levels, D their average spacing and m is a free parameter. The "optimum" function $f(x)$ having the smallest figure of merit is found to be

$$f(x) = \begin{cases} \frac{1}{2} \ln\left(\frac{1+\sqrt{1-x^2}}{1-\sqrt{1-x^2}}\right), & \text{if} \quad |x| \leq 1, \\ 0, & \text{if} \quad |x| \geq 1. \end{cases} \tag{16.5.2}$$

Thus the sum in Eq. (16.5.1) extends over all x_j within $x_i \pm mD$. The average value of F depends only on the two-point cluster function,

$$\begin{aligned} \langle F \rangle &= \int_0^\infty [1 - Y_2(r)]\, f(r/m) dr \\ &= \int_0^m [1 - Y_2(r)] \ln\left(\frac{m+\sqrt{m^2-r^2}}{m-\sqrt{m^2-r^2}}\right) dr \\ &= m\pi - 2\int_0^m dx \int_0^x dr \frac{Y_2(r)}{\sqrt{x^2-r^2}}. \end{aligned} \tag{16.5.3}$$

Thus for a level from the Gaussian orthogonal ensemble

$$\langle F_i \rangle = m\pi - \ln(8m\pi) + 2 - \gamma,$$

$$\gamma = 0.577..., \quad \text{Euler's constant},$$

and

$$\langle F_i \rangle = m\pi,$$

for a spurious level. The variance of F involves three- and four-point correlations as well, and has been estimated to be $\ln(m\pi)$ for the Gaussian orthogonal ensemble. Choosing a reasonable value of m, say, about 10, the sequence of the numbers F_j will roughly be constant if the level series is pure. For a spurious level x_i, F_i will rise to a peak; if a level is missing, then F_i will dip at the the two adjacent levels. This statistic seems to be good for an almost pure level series; as the number of missing or spurious levels increases, it gets certainly unreliable.

16.6. The Λ Statistic

Another statistic sometimes used to detect missing or spurious levels was introduced by Monahan and Rosenzweig (1972). Comparing the theoretical level spacing distribution $\Psi(s)$, Section 5.1.3, with the experimental staircase graph $\Psi^*(s)$ deduced from a set of $(n+1)$ levels or n consecutive spacings, they define

$$\Lambda(n) = n \int_0^\infty [\Psi(x) - \Psi^*(x)]^2 dx. \tag{16.6.1}$$

Actually, they use the Wigner surmise

$$\Psi(x) = 1 - \exp(-\pi x^2/4),$$

which is simpler and close enough to the $\Psi(x)$ of the Gaussian orthogonal ensemble (GOE). Monahan and Rosenzweig computed the average and the variance of $\Lambda(n)$ by Monte-Carlo calculations of randomly generated matrices of a certain order from the GOE. If the successive spacings were statistically independent, each of them following the Wigner curve, then the average of $\Lambda(n)$ will be a constant, 0.293, independent of n. The long range order of the levels makes the average of $\Lambda(n)$ a decreasing function of n.

16.7. Statistics Involving Three- and Four-Level Correlations

Most of the statistics described yet depend exclusively on the two point correlations. To test whether a given sequence of levels have higher correlations as required by the theory, one has to consider statistics involving these correlations. Two such statistics are the skewness and the excess of the number of levels in a given energy interval.

Let $\mu_j = \langle (n - \langle n \rangle)^j \rangle$, which involves ν-level correlations for all $\nu \leq j$. We have already considered the number variance μ_2 above; Section 16.1. The skewness γ_1 and the excess γ_2 are defined as

$$\gamma_1 = \mu_3 \mu_2^{-3/2}, \qquad \gamma_2 = \mu_4 \mu_2^{-2} - 3. \tag{16.7.1}$$

One can express μ_2, μ_3, and μ_4 in terms of the averages

$$\begin{aligned}
\langle n^2 \rangle &= \int\int R_2(x,y)dxdy + \int R_1(x)dx, \\
\langle n^3 \rangle &= \int\int\int R_3(x,y,z)dxdydz \\
&\quad + 3\int\int R_2(x,y)dxdy + \int R_1(x)dx, \\
\langle n^4 \rangle &= \int\int\int\int R_4(x,y,z,t)dxdydzdt \\
&\quad + 6\int\int\int R_3(x,y,z)dxdydz \\
&\quad + 7\int\int R_2(x,y)dxdy + \int R_1(x)dx,
\end{aligned} \tag{16.7.2}$$

all integrals being taken from $-L$ to L. [The various coefficients 1, 3, 6, 7, etc. are the Stirling numbers of the first kind; they appear in the expansions $n^j = n_j + c_1 n_{j-1} + \cdots + c_{j-1} n_1 + c_j$, where $n_k = n(n-1)\cdots(n-k+1)$.]

For a sequence of levels without correlations, Poisson process, $\mu_2 = \mu_3 = s$ and $\mu_4 = 3s^2 + s$, so that $\gamma_1 = s^{-1/2}$ and $\gamma_2 = s^{-1}$. For matrix ensembles it is easier to compute them using numerical tables of $E(n; s)$,

$$\mu_j = \sum_{n=0}^{\infty} (n-s)^j E(n; s). \tag{16.7.3}$$

The γ_1 and γ_2 for the nuclear data ensemble are shown on the Figures 16.16a and 16.16b respectively, for chaotic systems on Figures 16.17a and 16.17b respectively, and for the zeros of the zeta function of Riemann on Figures 16.18a and 16.18b respectively. For comparison we have also given the curves of γ_1 and γ_2 for the Poisson process, for the Gaussian orthogonal ensemble (GOE) and for the Gaussian unitary ensemble (GUE). One sees that the nuclear data ensemble and energies of chaotic systems are consistent with GOE while the zeta zeros are consistent with GUE. Though not shown on the figures, the results for $5 \leq s \leq 25$ were also calculated and the above consistencies persist. In other words, the three and four level correlations of the considered level sequences are in good agreement with those required by the theory.

16.8. Other Statistics

From various other statistics considered in the literature, we will mention two more. One is the variance of the rth neighbor spacing probability density, $p(r-1; s)$; Section 5.1.2. Actually as r increases, $p(r-1; s)$ tends to a Gaussian centered at r and only its width $\sigma(r)$ has any significance. This $\sigma(r)$ is closely related to the number variance $(\delta n)^2 = \langle n^2 \rangle - \langle n \rangle^2$; Section 16.1 (see Brody et al., 1981). The other one is the repulsion parameter. Generalizing the Poisson distribution and the Wigner surmise, Brody (1973) writes the spacing probability density $p(0; s)$ as

$$p(0; s) = A s^{\omega} \exp\left(-B s^{\omega+1}\right), \tag{16.8.1}$$

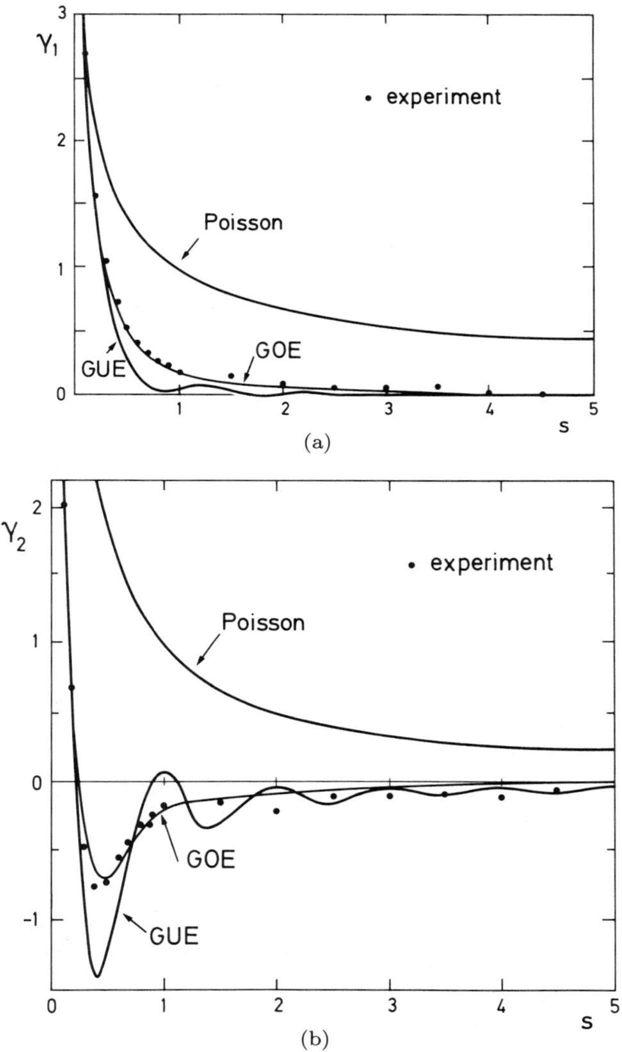

FIG. 16.16. The skewness γ_1 (a) or the excess γ_2 (b) for nuclear data ensemble; *cf.* Figure 16.1 or 1.4. From Bohigas et al. (1985). 16.16(a). Reprinted with permission of the American Physical Society, Bohigas, O., Haq, R. U., and Pandey, A. Higher order correlations in spectra of complex systems, *Phys. Rev. Lett.* 54 (1985) 1645–1648. 16.16(b). Reprinted with permission of the American Physical Society, Bohigas, O., Haq, R. U., and Pandey, A. Higher order correlations in spectra of complex systems, *Phys. Rev. Lett.* 54 (1985) 1645–1648.

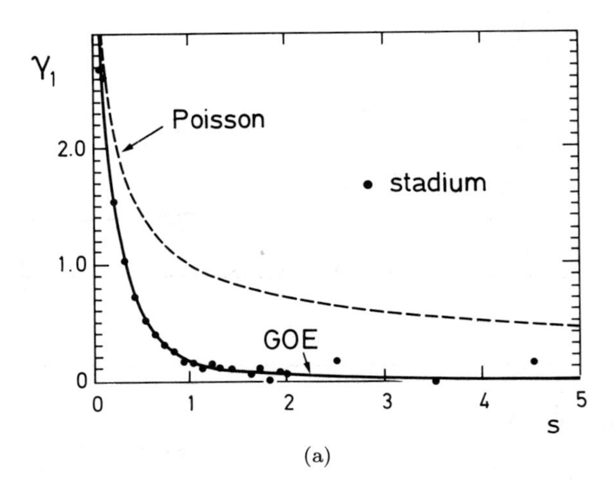

(a)

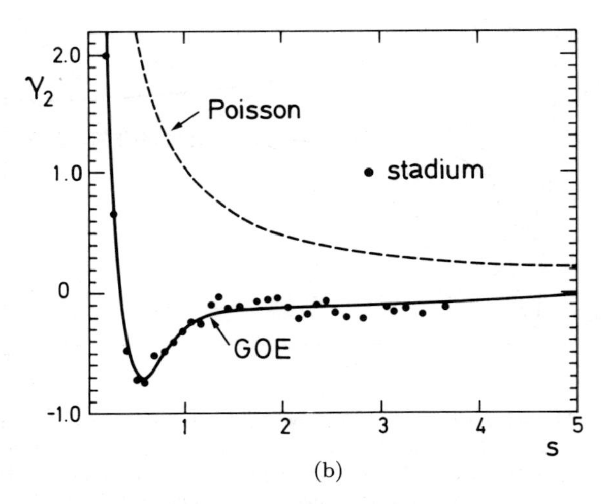

(b)

FIG. 16.17. The skewness γ_1 (a) or the excess γ_2 (b) for chaotic systems, stadium or Sinai's billiard. 16.17(a) and (b) supplied by O. Bohigas, from Bohigas et al. (1984c).

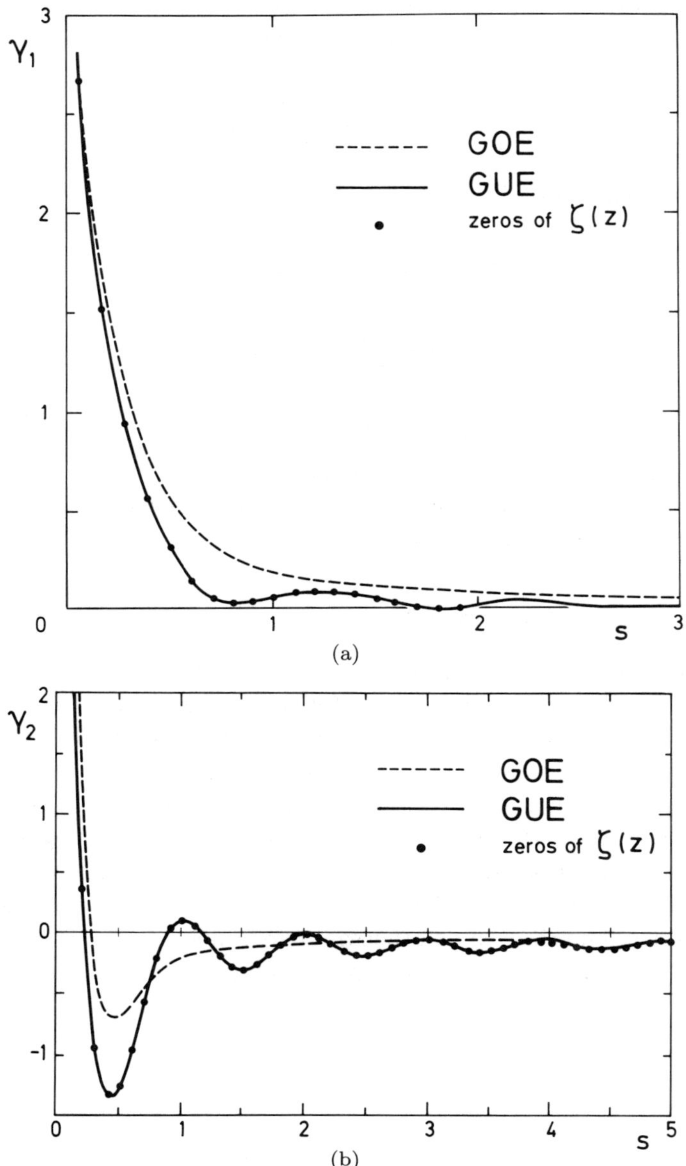

FIG. 16.18. The skewness γ_1 (a) and excess γ_2 (b) for the zeros of the Riemann zeta function on the critical line Re $z = 1/2$. Supplied by A. Odylzko and O. Bohigas. Figure 16.18 is based on the work of A. Odlyzko.

the two constants A and B are fixed in terms of ω by the requirement that $p(0;s)$ is normalized and the mean value of s is unity:

$$A = B(1+\omega), \qquad B = \left[\Gamma\left(\frac{\omega+2}{\omega+1}\right)\right]^{1+\omega}. \tag{16.8.2}$$

The kth moment of s is

$$\begin{aligned} \mu_k &= \int_0^\infty s^k A s^\omega \exp\left(-Bs^{\omega+1}\right) ds \\ &= \Gamma\left(\frac{\omega+k+1}{\omega+1}\right)\left(\Gamma\left(\frac{\omega+2}{\omega+1}\right)\right)^{-k}. \end{aligned}$$

The parameter ω measures the degree of repulsion between adjacent levels. It is determined either by comparing some moment of s, say μ_2, to that given by the data, or better by fitting the curve (16.8.1) to the experimental histogram by least squares.

Summary to Chapter 16

To estimate how well a given sequence of levels agrees with the eigenvalues of a random matrix, several statistics are considered. A statistic is a quantity that can be computed for the given sequence of levels and whose average and mean square scatter are known from the statistical theory of random matrices. Most popular of these statistics are the number variance and the best straight-line-fit Δ statistic along with the two-point correlation function $R_2(r)$ and the nearest neighbor spacing probability density $p(0;s)$. The F statistic has sometimes been used to detect small impurities in an otherwise pure sequence. The other statistics like the Q, Λ, or the repulsion parameter ω have not been so widely used.

17 / Selberg's Integral and Its Consequences

Selberg's integral is a generalization of Euler's beta integral. Aomoto found a slight generalization of Selberg's integral, thus providing an alternative proof of it. We give here both derivations essentially following the original papers. When Selberg's integral was not so widely known, some of its consequences were thought of as conjectures by various authors. Macdonald established a connection between this integral, simple Lie algebras and reflection groups. Some of the considerations of Macdonald, circulated for a long time as conjectures, have only recently been settled. Here we describe these developments in some detail.

17.1. Selberg's Integral

Theorem *For any positive integer n, let $dx \equiv dx_1 \cdots dx_n$,*

$$\Delta(x) \equiv \Delta(x_1, ..., x_n) = \prod_{1\le j<\ell\le n} (x_j - x_\ell), \quad \text{if} \quad n > 1, \qquad (17.1.1)$$

and

$$\Delta(x) = 1, \qquad \text{if} \quad n = 1,$$

$$\Phi(x) \equiv \Phi(x_1, ..., x_n) = |\Delta(x)|^{2\gamma} \prod_{j=1}^{n} x_j^{\alpha-1} (1 - x_j)^{\beta-1}. \qquad (17.1.2)$$

Then

$$\begin{aligned} I(\alpha, \beta, \gamma, n) &\equiv \int_0^1 \cdots \int_0^1 \Phi(x) dx \\ &= \prod_{j=0}^{n-1} \frac{\Gamma(1 + \gamma + j\gamma)\Gamma(\alpha + j\gamma)\Gamma(\beta + j\gamma)}{\Gamma(1 + \gamma)\Gamma(\alpha + \beta + (n + j - 1)\gamma)}, \end{aligned} \qquad (17.1.3)$$

and for $1 \le m \le n$,

$$\int_0^1 \cdots \int_0^1 x_1 x_2 \cdots x_m \Phi(x) dx$$

$$= \prod_{j=1}^{m} \frac{\alpha + (n-j)\gamma}{\alpha + \beta + (2n-j-1)\gamma} \int_0^1 \cdots \int_0^1 \Phi(x) dx, \qquad (17.1.4)$$

valid for integer n *and complex* α, β, γ *with*

$$\text{Re } \alpha > 0, \quad \text{Re } \beta > 0, \quad \text{Re } \gamma > -\min\left(\frac{1}{n}, \frac{\text{Re } \alpha}{(n-1)}, \frac{\text{Re } \beta}{(n-1)}\right). \qquad (17.1.5)$$

Equation (17.1.3) is Selberg's integral and Eq. (17.1.4) is Aomoto's extension of it.

17.2. Selberg's Proof of Equation (17.1.3)

(i) Let γ be a positive integer.

We will use the facts that $|\Delta(x)|^{2\gamma}$ is a homogeneous polynomial in $x_1, ..., x_n$ of degree $\gamma n(n-1)$ and that $|\Delta(x_1, ..., x_n)|^{2\gamma}$ is divisible by $|\Delta(x_1, ..., x_m)|^{2\gamma}$ for $1 \le m \le n$.

For $n = 1$, $\Delta(x) = 1$, and Eq. (17.1.3) is just Euler's integral

$$\int_0^1 x^{\alpha-1}(1-x)^{\beta-1} dx = \frac{\Gamma(\alpha)\Gamma(\beta)}{\Gamma(\alpha+\beta)}. \qquad (17.2.1)$$

For $n > 1$, let

$$|\Delta(x)|^{2\gamma} = \sum_{(j)} c_{j_1 \cdots j_n} x_1^{j_1} x_2^{j_2} \cdots x_n^{j_n}, \qquad (17.2.2)$$

where $c_{(j)} \equiv c_{j_1 \cdots j_n}$ are integers and the integers $j_1, ..., j_n$ satisfy $j_1 + \cdots + j_n = \gamma n(n-1)$. Without loss of generality we may suppose that $j_1 \le j_2 \le \cdots \le j_n$, so that $j_n \ge \gamma(n-1)$. Also for $1 \le m \le n$,

$$|\Delta(x_1, ..., x_n)|^{2\gamma} = |\Delta(x_1, ..., x_m)|^{2\gamma} P(x_1, ..., x_n)$$

with P a polynomial. Therefore,

$$\sum_{(j)} c_{j_1 \cdots j_n} x_1^{j_1} x_2^{j_2} \cdots x_n^{j_n} = P(x_1, ..., x_n) \sum_{(q)} c'_{q_1 \ldots q_m} x_1^{q_1} x_2^{q_2} \cdots x_m^{q_m}.$$

As $P(x)$ is a polynomial, one has $j_1 \geq q_1, j_2 \geq q_2, ..., j_m \geq q_m$. Also counting powers, $q_1 + ... + q_m = \gamma m(m-1)$; and at least one of the q, say q_i, is $\geq \gamma(m-1)$. Hence for $1 \leq m \leq n$, one has

$$j_m \geq j_i \geq q_i \geq \gamma(m-1). \tag{17.2.3}$$

On the other hand

$$|\Delta(x)|^{2\gamma} = (x_1 \cdots x_n)^{2\gamma(n-1)} \left| \Delta\left(\frac{1}{x}\right) \right|^{2\gamma},$$

so that

$$\sum_{(j)} c_{j_1 \ldots j_n} x_1^{j_1} x_2^{j_2} \cdots x_n^{j_n}$$

$$= (x_1 \cdots x_n)^{2\gamma(n-1)} \sum_{(p)} c_{p_1 \ldots p_n} x_1^{-p_1} x_2^{-p_2} \cdots x_n^{-p_n}.$$

Examine the set $p_1 = 2\gamma(n-1) - j_n$, $p_2 = 2\gamma(n-1) - j_{n-1}$, ..., $p_n = 2\gamma(n-1) - j_1$. For this set $p_1 \leq p_2 \leq \cdots \leq p_n$, if $j_1 \leq j_2 \leq \cdots \leq j_n$. Therefore, as above, for $1 \leq m \leq n$, $p_m \geq \gamma(m-1)$, or $2\gamma(n-1) - j_m \geq \gamma(n-m)$. Combining with Eq. (17.2.3),

$$\gamma(m-1) \leq j_m \leq 2\gamma(n-1) - \gamma(n-m) = \gamma(n+m-2). \tag{17.2.4}$$

Now substitute Eq. (17.2.2) on the left-hand side of Eq. (17.1.3), integrate term by term and use Eq. (17.2.1), to obtain

$$\begin{aligned} I(\alpha, \beta, \gamma, n) &= \sum_{(j)} c_{j_1 \ldots j_n} \prod_{m=1}^{n} \frac{\Gamma(\alpha + j_m)\Gamma(\beta)}{\Gamma(\alpha + \beta + j_m)} \\ &= \prod_{m=1}^{n} \frac{\Gamma(\alpha + \gamma(m-1))\Gamma(\beta)}{\Gamma(\alpha + \beta + \gamma(n+m-2))} \sum_{(j)} c_{j_1 \ldots j_n} Q_{(j)}(\alpha, \beta), \end{aligned}$$

where

$$Q_{(j)}(\alpha,\beta) = \prod_{m=1}^{n} \frac{\Gamma(\alpha+j_m)}{\Gamma(\alpha+\gamma(m-1))} \frac{\Gamma(\alpha+\beta+\gamma(n+m-2))}{\Gamma(\alpha+\beta+j_m)}$$

is, from Eq. (17.2.4), a polynomial in α and β. The degree of $Q_{(j)}$ in β is

$$\sum_{m=1}^{n} [\gamma(n+m-2) - j_m]$$

$$= \gamma(n-2)n + \gamma n(n+1)/2 - \gamma n(n-1) = \gamma n(n-1)/2.$$

Thus

$$I(\alpha,\beta,\gamma,n) = \prod_{m=1}^{n} \frac{\Gamma(\alpha+\gamma(m-1))\Gamma(\beta+\gamma(m-1))}{\Gamma(\alpha+\beta+\gamma(n+m-2))} \frac{R(\alpha,\beta)}{S(\beta)},$$

with

$$R(\alpha,\beta) = \sum_{(j)} c_{j_1\cdots j_n} Q_{(j)}(\alpha,\beta),$$

and

$$S(\beta) = \prod_{m=1}^{n} \frac{\Gamma(\beta+\gamma(m-1))}{\Gamma(\beta)}.$$

The $R(\alpha,\beta)$ is a polynomial in α and β; its degree in β being at most $\gamma n(n-1)/2$, while $S(\beta)$ is a polynomial in β of degree $\sum_{m=1}^{n} \gamma(m-1) = \gamma n(n-1)/2$.

As $\Delta(1-x) = \pm\Delta(x)$, the integral I is symmetric in α and β, so that$R(\alpha,\beta)/S(\beta) = R(\beta,\alpha)/S(\alpha)$ is a polynomial in β. Therefore $R(\alpha,\beta)$ is divisible by $S(\beta)$. Since the degree of $R(\alpha,\beta)$ in β is at most equal to that of $S(\beta)$, $R(\alpha,\beta)/S(\beta)$ is independent of β. By symmetry it is also independent of α. It may still depend on γ and n. Therefore

$$I(\alpha,\beta,\gamma,n) = c(\gamma,n) \prod_{m=1}^{n} \frac{\Gamma(\alpha+\gamma(m-1))\Gamma(\beta+\gamma(m-1))}{\Gamma(\alpha+\beta+\gamma(n+m-2))}. \quad (17.2.5)$$

To determine $c(\gamma, n)$, put $\alpha = \beta = 1$:

$$
\begin{aligned}
I(1,1,\gamma,n) &= \int_0^1 \cdots \int_0^1 |\Delta(x)|^{2\gamma}\, dx_1 \cdots dx_n \\
&= c(\gamma,n) \prod_{m=1}^{n} \frac{[\Gamma(1+\gamma(m-1))]^2}{\Gamma(2+\gamma(n+m-2))}. \qquad (17.2.6)
\end{aligned}
$$

Let y be the largest of the $x_1, ..., x_n$ and replace the other x_j by $x_j = yt_j$, $0 \leq t_j \leq 1$. Then

$$
|\Delta(x_1, ..., x_n)|^{2\gamma} = y^{\gamma n(n-1)} |\Delta(t_1, ..., t_{n-1})|^{2\gamma} \prod_{j=1}^{n-1} (1-t_j)^{2\gamma}
$$

and

$$
\begin{aligned}
I(1,1,\gamma,n) &= n \int_0^1 dy\; y^{n-1} y^{\gamma n(n-1)} I(1, 2\gamma+1, \gamma, n-1) \\
&= \frac{c(\gamma, n-1)}{\gamma(n-1)+1} \\
&\quad \times \prod_{m=1}^{n-1} \frac{\Gamma(1+\gamma(m-1))\Gamma(2\gamma+1+\gamma(m-1))}{\Gamma(1+2\gamma+1+\gamma(n-1+m-2))}.
\end{aligned}
\qquad (17.2.7)
$$

Comparing Eqs. (17.2.6) and (17.2.7) one gets on simplification,

$$
\frac{c(\gamma,n)}{c(\gamma,n-1)} = \frac{\Gamma(1+\gamma n)}{\Gamma(1+\gamma)},
$$

or

$$
c(\gamma,n) = c(\gamma,1) \prod_{m=2}^{n} \frac{\Gamma(1+\gamma m)}{\Gamma(1+\gamma)},
$$

and from Eqs. (17.2.1) and (17.2.5), $c(\gamma, 1) = 1$.

This completes the proof when γ is a non-negative integer.

(ii) When γ is a complex number, consider

$$f(\gamma) = \int_0^1 \cdots \int_0^1 |\Delta(x)|^{2\gamma} \prod_{j=1}^{n} x_j^{\alpha-1} (1 - x_j)^{\beta-1}\, dx_j$$

and

$$\begin{aligned} g(\gamma) &= \prod_{j=0}^{n-1} \frac{\Gamma(1+\gamma+j\gamma)\Gamma(\alpha+j\gamma)\Gamma(\beta+j\gamma)}{\Gamma(1+\gamma)\Gamma(\alpha+\beta+(n+j-1)\gamma)} \\ &= \prod_{j=1}^{n} \frac{\Gamma(1+j\gamma)\Gamma(\alpha+(j-1)\gamma)\Gamma(\beta+(n-j)\gamma)}{\Gamma(1+\gamma)\Gamma(\alpha+\beta+(n+j-2)\gamma)}, \end{aligned}$$

as functions of the complex variable γ, for fixed α and β, Re $\alpha > 0$, Re $\beta > 0$. The functions $f(\gamma)$ and $g(\gamma)$ are regular and analytic in the upper half plane. For $0 \le x_j \le 1$, $|\Delta(x)| \le 1$, so that $f(\gamma)$ is bounded for Re $\gamma \ge 0$. As for $g(\gamma)$ we may use Stirling's formula

$$\Gamma(z) \to \sqrt{2\pi}\, z^{z-1/2} e^{-z}, \qquad |z| \to \infty, \qquad \text{Re } z \ge 0,$$

to estimate it when $|\gamma| \to \infty$. After simplification we have

$$\begin{aligned} &\frac{\Gamma(1+j\gamma)\Gamma(\alpha+(j-1)\gamma)\Gamma(\beta+(n-j)\gamma)}{\Gamma(1+\gamma)\Gamma(\alpha+\beta+(n+j-2)\gamma)} \\ &\quad \approx \left(\frac{2\pi j}{\gamma}\right)^{1/2} \frac{(j-1)^{\alpha-1/2}(n-j)^{\beta-1/2}}{(n+j-2)^{\alpha+\beta-1/2}} \\ &\quad \times \left(\frac{j^j (j-1)^{j-1} (n-j)^{n-j}}{(n+j-2)^{n+j-2}}\right)^{\gamma}. \end{aligned}$$

when $|\gamma| \to \infty$, Re $\gamma \ge 0$. As $j^j (j-1)^{j-1} (n-j)^{n-j} (n+j-2)^{-(n+j-2)} \le 1$ for $2 \le j \le n-1$, $|g(\gamma)|$ is bounded for Re $\gamma \ge 0$ and goes to 0 as $|\gamma| \to \infty$. Thus in the upper half γ-plane, Re $\gamma \ge 0$, the function $h(\gamma) = f(\gamma) - g(\gamma)$ is regular, bounded, and takes the value 0 at non-negative integer values of γ. At this point we can apply a theorem of Carlson (Titchmarsh, 1939).

Carlson's Theorem. *If a function of* γ *is regular and bounded in the half plane* Re $\gamma \geq 0$, *and is zero for* $\gamma = 0, 1, 2, 3, \ldots$, *then it is identically zero.*

Applying this theorem to the function $h(\gamma)$, we deduce that Eq. (17.1.3) holds for all complex values of γ, Re $\gamma \geq 0$.

(iii) To extend the validity of Eq. (17.1.3) to slightly negative values of Re γ, as indicated in Eq. (17.1.5), one needs to examine in detail the convergence and analytic nature of the integral $f(\gamma)$ and the function $g(\gamma)$ in that region. We will not do it here and refer to the original paper (Selberg, 1944).

17.3. Aomoto's Proof of Equation (17.1.4)

For brevity let us write

$$\langle f(x_1, \ldots, x_n) \rangle = \frac{\displaystyle\int_0^1 \cdots \int_0^1 f(x_1, \ldots, x_n) \Phi(x) dx}{\displaystyle\int_0^1 \cdots \int_0^1 \Phi(x) dx}. \tag{17.3.1}$$

Integrating between 0 and 1 the identity

$$\begin{aligned} \frac{d}{dx_1} (x_1^a x_2 \ldots x_m \Phi) &= (a + \alpha - 1) x_1^{a-1} x_2 \ldots x_m \Phi \\ &\quad - (\beta - 1) \frac{x_1^a x_2 \cdots x_m}{1 - x_1} \Phi + 2\gamma \sum_{j=2}^{n} \frac{x_1^a x_2 \cdots x_m}{x_1 - x_j} \Phi \end{aligned} \tag{17.3.2}$$

we get, for $a = 1$ and $a = 2$,

$$0 = \alpha \langle x_2 \cdots x_m \rangle - (\beta - 1) \left\langle \frac{x_1 x_2 \cdots x_m}{1 - x_1} \right\rangle + 2\gamma \sum_{j=2}^{n} \left\langle \frac{x_1 x_2 \cdots x_m}{x_1 - x_j} \right\rangle, \tag{17.3.3}$$

and

$$0 = (\alpha+1)\langle x_1 x_2 \cdots x_m\rangle - (\beta-1)\left\langle \frac{x_1^2 x_2 \cdots x_m}{1-x_1}\right\rangle$$

$$+ 2\gamma \sum_{j=2}^{n} \left\langle \frac{x_1^2 x_2 \cdots x_m}{x_1 - x_j}\right\rangle. \tag{17.3.4}$$

Now

$$\left\langle \frac{x_1 x_2 \cdots x_m}{1-x_1}\right\rangle - \left\langle \frac{x_1^2 x_2 \cdots x_m}{1-x_1}\right\rangle = \langle x_1 x_2 \cdots x_m\rangle ; \tag{17.3.5}$$

interchanging x_1 and x_j and observing the symmetry

$$\left\langle \frac{x_1 x_2 \cdots x_m}{x_1 - x_j}\right\rangle = -\left\langle \frac{x_j x_2 \cdots x_m}{x_1 - x_j}\right\rangle$$

$$= \begin{cases} 0, & \text{if} \quad 2 \le j \le m, \\ \frac{1}{2}\langle x_2 \cdots x_m\rangle, & \text{if} \quad m < j \le n, \end{cases} \tag{17.3.6}$$

and

$$\left\langle \frac{x_1^2 x_2 \cdots x_m}{x_1 - x_j}\right\rangle = -\left\langle \frac{x_j^2 x_2 \cdots x_m}{x_1 - x_j}\right\rangle$$

$$= \begin{cases} \frac{1}{2}\langle x_1 x_2 \cdots x_m\rangle, & \text{if} \quad 2 \le j \le m, \\ \langle x_1 x_2 \cdots x_m\rangle, & \text{if} \quad m < j \le n. \end{cases} \tag{17.3.7}$$

Subtracting Eq. (17.3.4) from Eq. (17.3.3) and using the above, we get

$$[\alpha + 1 + \gamma(m-1) + 2\gamma(n-m) + \beta - 1]\langle x_1 x_2 \cdots x_m\rangle$$

$$= [\alpha + \gamma(n-m)]\langle x_2 \cdots x_m\rangle, \tag{17.3.8}$$

or repeating the process

$$\langle x_1 x_2 \cdots x_m \rangle = \frac{\alpha + \gamma(n-m)}{\alpha + \beta + \gamma(2n-m-1)} \langle x_1 \cdots x_{m-1} \rangle$$

$$= \prod_{j=1}^{m} \frac{\alpha + \gamma(n-j)}{\alpha + \beta + \gamma(2n-j-1)}, \tag{17.3.9}$$

which is Eq. (17.1.4).

For $m = n$, Eq. (17.1.4) or (17.3.9) can be written as

$$\frac{I(\alpha+1, \beta, \gamma, n)}{I(\alpha, \beta, \gamma, n)} = \langle x_1 \cdots x_n \rangle = \prod_{j=1}^{n} \frac{\alpha + \gamma(n-j)}{\alpha + \beta + \gamma(2n-j-1)} \tag{17.3.10}$$

or for a positive integer α,

$$I(\alpha, \beta, \gamma, n) = \prod_{j=1}^{n} \frac{\Gamma(\alpha + \gamma(n-j))\Gamma(1 + \beta + \gamma(2n-j-1))}{\Gamma(\alpha + \beta + \gamma(2n-j-1))\Gamma(1 + \gamma(n-j))}$$

$$\times I(1, \beta, \gamma, n). \tag{17.3.11}$$

As

$$I(\alpha, \beta, \gamma, n) = I(\beta, \alpha, \gamma, n), \tag{17.3.12}$$

we get

$$I(\alpha, \beta, \gamma, n) = \prod_{j=1}^{n} \frac{\Gamma(\alpha + \gamma(n-j))\Gamma(\beta + \gamma(n-j))}{\Gamma(\alpha + \beta + \gamma(2n-j-1))} c(\gamma, n), \tag{17.3.13}$$

where $c(\gamma, n)$ is independent of α and β. This is Eq. (17.2.5). The rest of the proof follows Selberg's reasonings.

Note that Eq. (17.3.13) is derived here for integer α, β and complex γ, while Selberg gets Eq. (17.2.5) for complex α, β and integer γ. One then completes the argument by analytical continuation using Carlson's theorem.

A slight change of reasoning due to Askey gives Eq. (17.3.13) directly for complex α, β and γ as follows. Equations (17.3.10) and (17.3.12) give the ratio of $I(\alpha,\beta,\gamma,n)$ and $I(\alpha,\beta+m,\gamma,n)$ for any integer m,

$$I(\alpha,\beta,\gamma,n) = \prod_{j=1}^{n} \frac{[\alpha+\beta+\gamma(2n-j-1)]_m}{[\beta+\gamma(n-j)]_m} I(\alpha,\beta+m,\gamma,n), \tag{17.3.14}$$

where we have used the notation

$$(a)_m = \Gamma(a+m)/\Gamma(a), \quad m \geq 0; \tag{17.3.15}$$

i.e.,

$$(a)_0 = 1,$$

and

$$(a)_m = a(a+1)\cdots(a+m-1)\ , \quad m \geq 1.$$

Now

$$\begin{aligned} I(\alpha,\beta+m,\gamma,n) &= \int_0^1 \cdots \int_0^1 |\Delta(x)|^{2\gamma} \\ &\quad \times \prod_{j=1}^{n} x_j^{\alpha-1} (1-x_j)^{\beta+m-1}\, dx_j \\ &= m^{-\alpha n-\gamma n(n-1)} \int_0^m \cdots \int_0^m |\Delta(x)|^{2\gamma} \\ &\quad \times \prod_{j=1}^{n} x_j^{\alpha-1} \left(1-\frac{x_j}{m}\right)^{\beta+m-1} dx_j \end{aligned} \tag{17.3.16}$$

and

$$\frac{(a)_m}{(b)_m}\, m^{b-a} = \frac{\Gamma(b)}{\Gamma(a)} \frac{\Gamma(a+m)}{\Gamma(b+m)}\, m^{b-a}, \tag{17.3.17}$$

so in the limit $m \to \infty$,

$$I(\alpha,\beta,\gamma,n) = \mathcal{I}(\alpha,\gamma,n) \prod_{j=1}^{n} \frac{\Gamma(\beta+\gamma(n-j))}{\Gamma(\alpha+\beta+\gamma(2n-j-1))} \tag{17.3.18}$$

where

$$\mathcal{I}(\alpha, \gamma, n) = \int_0^\infty \cdots \int_0^\infty |\Delta(x)|^{2\gamma} \prod_{j=1}^{n} x_j^{\alpha-1} \exp(-x_j)\, dx_j \,. \quad (17.3.19)$$

Equations (17.3.12) and (17.3.18) give (17.3.13).

17.4. Other Averages

Taking polynomials other than $x_1^a x_2 \cdots x_m$ in Eq. (17.3.2), and/or differentiating with respect to x_2 or x_3, one can evaluate other integrals. For example, for $n \geq m > 1$,

$$[\alpha + \beta + \gamma(2n - m - 2)] \left\langle x_1^2 x_2 \cdots x_m \right\rangle$$

$$= [\alpha + \gamma(n - m)] \left\langle x_1^2 x_2 \cdots x_{m-1} \right\rangle + \gamma \left\langle x_1 x_2 \cdots x_m \right\rangle ,$$

and

$$[\alpha+\beta+1+2\gamma(n-1)] \left\langle x_1^2 \right\rangle = [\alpha+1+2\gamma(n-1)] \left\langle x_1 \right\rangle - \gamma(n-1) \left\langle x_1 x_2 \right\rangle , \quad n \geq 1,$$

these together with Eq. (17.3.9) give the value of $\left\langle x_1^2 x_2 \cdots x_m \right\rangle$.

17.5. Other Forms of Selberg's Integral

Selberg's integral can also be written in various other forms. We note two more of them here:

$$I(\alpha, \beta, \gamma, n) = \int_0^\infty \cdots \int_0^\infty |\Delta(x)|^{2\gamma}$$

$$\times \prod_{j=1}^{n} x_j^{\alpha-1} (1 + x_j)^{-\alpha-\beta-2\gamma(n-1)} dx_j, \quad (17.5.1)$$

obtained from Eq. (17.1.3) by the change of variables $x_j = y_j/(1+y_j)$; and

$$J(a,b,\alpha,\beta,\gamma,n) = \int_{-\infty}^{\infty} \cdots \int_{-\infty}^{\infty} |\Delta(x)|^{2\gamma} \times \prod_{j=1}^{n} (a+ix_j)^{-\alpha}(b-ix_j)^{-\beta} dx_j$$

$$= \frac{(2\pi)^n}{(a+b)^{(\alpha+\beta)n-\gamma n(n-1)-n}} \cdot \times \prod_{j=0}^{n-1} \frac{\Gamma(1+\gamma+j\gamma)\Gamma(\alpha+\beta-(n+j-1)\gamma-1)}{\Gamma(1+\gamma)\Gamma(\alpha-j\gamma)\Gamma(\beta-j\gamma)}. \tag{17.5.2}$$

Equations corresponding to Eq. (17.1.4) are

$$\int_{0}^{\infty} \cdots \int_{0}^{\infty} [\Delta(x)]^{2\gamma} \prod_{j=1}^{m} \left(x_j(1+x_j)^{-1}\right) \times \prod_{j=1}^{n} x_j^{\alpha-1}(1+x_j)^{-\alpha-\beta-2\gamma(n-1)} dx_j$$

$$= \prod_{j=n-m}^{n-1} \frac{\alpha+j\gamma}{\alpha+\beta+(n+j-1)\gamma} I(\alpha,\beta,\gamma,n) \tag{17.5.3}$$

and

$$\int_{-\infty}^{\infty} \cdots \int_{-\infty}^{\infty} \prod_{j=1}^{m} (a+ix_j) |\Delta(x)|^{2\gamma} \times \prod_{j=1}^{n} (a+ix_j)^{-\alpha}(b-ix_j)^{-\beta} dx_j$$

$$= (a+b)^m \prod_{j=n-m}^{n-1} \left(\frac{\alpha-j\gamma-1}{\alpha+\beta-(n+j-1)\gamma-2}\right) \times J(a,b,\alpha,\beta,\gamma,n), \tag{17.5.4}$$

where m and n are integers with $1 \le m \le n$ while $a,\ b,\ \alpha,\ \beta$ and γ are complex numbers. Equations (17.5.1) and (17.5.3) are valid for $\alpha,\ \beta,\ \gamma$ satisfying (17.1.5), while Eqs. (17.5.2) and (17.5.4) are valid for Re a, Re b, Re α, Re $\beta > 0$, Re $(\alpha+\beta) > 1$, and

$$-\frac{1}{n} < \text{Re } \gamma < \min\left(\frac{\text{Re } \alpha}{n-1}, \frac{\text{Re } \beta}{n-1}, \frac{\text{Re } (\alpha+\beta+1)}{2(n-1)}\right). \tag{17.5.5}$$

(Here $i = \sqrt{-1}$, and not an index!)

In Eqs. (17.5.2) and (17.5.4) one can make sure of the dependence on a and b by the change of variables $2x_j = i(a-b) + (a+b)y_j$. A proof of Eq. (17.5.2) can be supplied following Selberg's reasoning, Section 17.2, by starting with the Cauchy integral

$$\int_{-\infty}^{\infty} (a+ix)^{-\alpha}(b-ix)^{-\beta} dx = \frac{2\pi}{(a+b)^{\alpha+\beta-1}} \frac{\Gamma(\alpha+\beta-1)}{\Gamma(\alpha)\Gamma(\beta)}, \tag{17.5.6}$$

instead of the Euler integral (17.2.1). Or one can first establish Eq. (17.5.4) following Aomoto's reasoning, Section 17.3. Then taking $m = n$ in Eq. (17.5.4) and iterating, one gets from the symmetry in $(a,\ \alpha) \Leftrightarrow (b,\ \beta)$ the equation corresponding to Eq. (17.2.5) or Eq. (17.3.13) as

$$(2\pi)^{-n} J(a,b,\alpha,\beta,\gamma,n) = C(\gamma,n)(a+b)^{-(\alpha+\beta)n+\gamma n(n-1)+n} \times \prod_{j=0}^{n-1} \frac{\Gamma(\alpha+\beta-(n+j-1)\gamma-1)}{\Gamma(\alpha-j\gamma)\Gamma(\beta-j\gamma)}, \tag{17.5.7}$$

where $C(\gamma,n)$ does not depend on $a,\ b,\ \alpha$ or β. To determine $C(\gamma,n)$, put $a = b = 1$, $\alpha = \beta = \gamma(n-1)+1$, and make the change of variables

$x_j = \tan(\theta_j/2)$. This gives

$$C(\gamma, n) = (2\pi)^{-n} 2^{\gamma n(n-1)+n}$$

$$\times \prod_{j=0}^{n-1} \Gamma(1+j\gamma) J(1, 1, \gamma(n-1)+1, \gamma(n-1)+1, \gamma, n)$$

$$= (2\pi)^{-n} 2^{\gamma n(n-1)} \prod_{j=0}^{n-1} \Gamma(1+j\gamma)$$

$$\times \int_{-\pi}^{\pi} \cdots \int_{-\pi}^{\pi} \prod_{1 \le j < k \le n} |\sin(\theta_j - \theta_k)/2|^{2\gamma} \prod_{j=1}^{n} d\theta_j.$$

The integral in the last line above is (see, for example, Section 11.1 or 17.7)

$$\int \cdots \int \prod |\sin(\theta_j - \theta_k)/2|^{2\gamma} \prod d\theta_j$$

$$= \int \cdots \int \prod |e^{i\theta_j} - e^{i\theta_k}|^{2\gamma} \prod d\theta_j$$

$$= (2\pi)^n \Gamma(1+n\gamma)/[\Gamma(1+\gamma)]^n. \tag{17.5.8}$$

This gives

$$C(\gamma, n) = 2^{\gamma n(n-1)} \prod_{j=1}^{n} \frac{\Gamma(1+j\gamma)}{\Gamma(1+\gamma)}. \tag{17.5.9}$$

For more details see Morris (1982).

17.6. Some Consequences of Selberg's Integral

By changing variables of integration, choosing various appropriate values for α and β or taking limits we may derive a few Selberg type integrals related to all classical orthogonal polynomials.

Jacobi,

$$x_j = (1 - y_j)/2, \quad 1 - x_j = (1 + y_j)/2,$$

$$|\Delta(x)| = 2^{-n(n-1)/2} |\Delta(y)|, \quad dx_j = dy_j/2,$$

$$\int_{-1}^{1} \cdots \int_{-1}^{1} |\Delta(x)|^{2\gamma} \prod_{j=1}^{n} (1 - x_j)^{\alpha - 1} (1 + x_j)^{\beta - 1} dx_j$$

$$= 2^{\gamma n(n-1) + n(\alpha+\beta-1)} \prod_{j=0}^{n-1} \frac{\Gamma(1 + \gamma + j\gamma)\Gamma(\alpha + \gamma j)\Gamma(\beta + \gamma j)}{\Gamma(1 + \gamma)\Gamma(\alpha + \beta + \gamma(n + j - 1))}. \tag{17.6.1}$$

Gegenbauer, $\alpha = \beta = \lambda + 1/2$ in (17.6.1):

$$\int_{-1}^{1} \cdots \int_{-1}^{1} |\Delta(x)|^{2\gamma} \prod_{j=1}^{n} \left(1 - x_j^2\right)^{\lambda - 1/2} dx_j$$

$$= 2^{\gamma n(n-1) + 2\lambda n} \prod_{j=0}^{n-1} \frac{\Gamma(1 + \gamma + j\gamma)[\Gamma(\lambda + \gamma j + 1/2)]^2}{\Gamma(1 + \gamma)\Gamma(2\lambda + \gamma(n + j - 1) + 1)}. \tag{17.6.2}$$

Legendre, $\alpha = \beta = 1$ in Eq. (17.6.1):

$$\int_{-1}^{1} \cdots \int_{-1}^{1} |\Delta(x)|^{2\gamma} \prod_{j=1}^{n} dx_j$$

$$= 2^{\gamma n(n-1) + n}$$

$$\times \prod_{j=0}^{n-1} \frac{\Gamma(1 + \gamma + j\gamma)[\Gamma(\gamma j + 1)]^2}{\Gamma(1 + \gamma)\Gamma(\gamma(n + j - 1) + 2)}. \tag{17.6.3}$$

Chebyshev, $\alpha = \beta = 3/2$ or $\alpha = \beta = 1/2$ in Eq. (17.6.1):

$$\int_{-1}^{1} \cdots \int_{-1}^{1} |\Delta(x)|^{2\gamma} \prod_{j=1}^{n} \left(1 - x_j^2\right)^{\pm 1/2} dx_j$$

$$= 2^{\gamma n(n-1) + n \pm n} \cdot \prod_{j=0}^{n-1} \frac{\Gamma(1 + \gamma + j\gamma)(\Gamma(\gamma j + 1 \pm 1/2)^2}{\Gamma(1 + \gamma)\Gamma(\gamma(n + j - 1) + 2 \pm 1)}. \tag{17.6.4}$$

Laguerre, put $x_j = y_j/L$, $\beta = L+1$, and take the limit $L \to \infty$ in Eq. (17.1.3):

$$\int_0^\infty \cdots \int_0^\infty |\Delta(x)|^{2\gamma} \prod_{j=1}^{n} x_j^{\alpha-1} \exp(-x_j) dx_j$$

$$= \prod_{j=0}^{n-1} \frac{\Gamma(1+\gamma+j\gamma)\Gamma(\alpha+\gamma j)}{\Gamma(1+\gamma)}. \tag{17.6.5}$$

Set $x_j = y_j^2/2$ in Eq. (17.6.5) to get

$$\int_{-\infty}^\infty \cdots \int_{-\infty}^\infty \prod_{1 \le j < \ell \le n} \left|x_j^2 - x_\ell^2\right|^{2\gamma} \prod_{j=1}^{n} |x_j|^{2\alpha-1} \exp(-x_j^2/2) dx_j$$

$$= 2^{\alpha n + \gamma n(n-1)} \prod_{j=1}^{n} \frac{\Gamma(1+j\gamma)\Gamma(\alpha+\gamma(j-1))}{\Gamma(1+\gamma)}. \tag{17.6.6}$$

Hermite, put $x_j = y_j/L$, $\alpha = \beta = aL^2 + 1$, and take the limit $L \to \infty$ in Eq. (17.6.1):

$$\int_{-\infty}^\infty \cdots \int_{-\infty}^\infty |\Delta(x)|^{2\gamma} \prod_{j=1}^{n} \exp(-ax_j^2) dx_j$$

$$= (2\pi)^{n/2} (2a)^{-n(\gamma(n-1)+1)/2} \prod_{j=1}^{n} \frac{\Gamma(1+j\gamma)}{\Gamma(1+\gamma)}. \tag{17.6.7}$$

For integer γ this last equation can also be written as a finite algebraic identity. From

$$\int_{-\infty}^\infty \exp\left(-a^2x^2 - 2iax\lambda\right) x^n dx$$

$$= \left(\frac{i}{2a}\frac{d}{d\lambda}\right)^n \int_{-\infty}^\infty \exp\left(-a^2x^2 - 2iax\lambda\right) dx$$

$$= \left(\frac{i}{2a}\frac{d}{d\lambda}\right)^n \exp(-\lambda^2) \int_{-\infty}^\infty \exp\left(-(ax+i\lambda)^2\right) dx$$

$$= \frac{\sqrt{\pi}}{a} \left(\frac{i}{2a}\frac{d}{d\lambda}\right)^n \exp(-\lambda^2),$$

we can replace the integration by a differentiation at the point $x_j = 0$,

$$\int_{-\infty}^{\infty} \exp\left(-ax_j^2\right) x_j^n dx_j = \sqrt{\frac{\pi}{a}} \left(\frac{i}{2\sqrt{a}}\frac{\partial}{\partial x_j}\right)^n \exp(-x_j^2)\Big|_{x_j=0}.$$

Equation (17.6.7) therefore takes the form

$$\left(\frac{i}{2\sqrt{a}}\right)^{\gamma n(n-1)} \prod_{1\le p<q\le n} \left(\frac{\partial}{\partial x_p} - \frac{\partial}{\partial x_q}\right)^{2\gamma} \exp\left(-\sum_{j=1}^{n} x_j^2\right)\Bigg|_{x_j=0} = (2a)^{-\gamma n(n-1)/2} \prod_{j=1}^{n} \frac{(j\gamma)!}{\gamma!} \tag{17.6.8}$$

Replacing the exponential by its power series expansion, one notes that the term $\left(-\Sigma x_j^2\right)^\ell$ gives zero on differentiation if $\ell < \gamma n(n-1)/2$, and leaves a homogeneous polynomial of order $\ell - \gamma n(n-1)/2$ in the variables $x_1, ..., x_n$, if $\ell > \gamma n(n-1)/2$. On setting $x_j = 0$, $j = 1, ..., n$, one sees therefore that there is only one term, corresponding to $\ell = \gamma n(n-1)/2$, which gives a nonzero contribution. Equation (17.6.8) is now

$$\prod_{1\le p<q\le n} \left(\frac{\partial}{\partial x_p} - \frac{\partial}{\partial x_q}\right)^{2\gamma} \left(\sum_{j=1}^{n} x_j^2\right)^{\ell} = 2^\ell\, \ell! \prod_{j=1}^{n} \frac{(j\gamma)!}{\gamma!}, \qquad \ell = \gamma n(n-1)/2.$$

If $P(x) \equiv P(x_1, ..., x_n)$ and $Q(x) \equiv Q(x_1, ..., x_n)$ are homogeneous polynomials in $x \equiv (x_1, ..., x_n)$ of the same degree, then a little reflection shows that $P(\partial/\partial x)Q(x)$ is a constant which is also equal to $Q(\partial/\partial x)P(x)$. Thus one can interchange the roles of x_j and $\partial/\partial x_j$ to get

$$\left(\sum_{j=1}^{n} (\partial^2/\partial x_j^2)\right)^{\ell} \prod_{1\le i<j\le n} (x_i - x_j)^{2\gamma} = 2^\ell\, \ell! \prod_{j=1}^{n} \frac{(j\gamma)!}{\gamma!}, \qquad \ell = \gamma n(n-1)/2. \tag{17.6.9}$$

17.7. Normalization Constant for the Circular Ensembles

Put $\alpha = -\gamma(n-1)+\varepsilon$ in Eq. (17.1.3) and take the limit $\varepsilon \to 0$. When one substitutes the expansion (17.2.2) in the left hand side integrand of Eq. (17.1.3), the dominant (i.e., the most divergent) contribution comes from the term $(x_1 x_2 \cdots x_n)^{\gamma(n-1)}$. Using $\Gamma(-j+\varepsilon) = \Gamma(\varepsilon)\prod_{p=1}^{j}(-p+\varepsilon)^{-1}$, one gets the following result.

If one expands $(\Delta(x))^{2\gamma}$ in a power seris of the variables $x_1, \ldots, x_n$, the coefficient of $(x_1 \cdots x_n)^{\gamma(n-1)}$ is $(-1)^{\gamma n(n-1)/2}(n\gamma)!/(\gamma!)^n$.

Or,

If one expands $[\Delta(x)\Delta(x^{-1})]^\gamma$ in positive and negative powers of $x_1, \ldots, x_n$, the constant term is $(n\gamma)!/(\gamma!)^n$.

One can express this result also as

$$\int_0^{2\pi} \cdots \int_0^{2\pi} \prod_{1\le j<k\le n} \left|e^{i\theta_j} - e^{i\theta_k}\right|^{2\gamma} \prod_{j=1}^{n} d\theta_j/(2\pi)$$

$$= \int_0^{2\pi} \cdots \int_0^{2\pi} \prod_{1\le j<k\le n} \left\{e^{i\theta_j} - e^{i\theta_k}\right\}^\gamma \left\{e^{-i\theta_j} - e^{-i\theta_k}\right\}^\gamma$$

$$\times \prod_{j=1}^{n} d\theta_j/(2\pi)$$

$$= (n\gamma)!/(\gamma!)^n; \tag{17.7.1}$$

since if one expands the integrand in the above equation in positive and negative powers of $\exp(i\theta_j)$, only the constant term survives on integration.

By computing the subdominant terms one can get a few more coefficients in such expansions.

17.8. Averages with Laguerre or Hermite Weights

Let us write

$$\langle f(x)\rangle = \int_0^\infty \cdots \int_0^\infty f(x)\Phi(x)dx \Big/ \int_0^\infty \cdots \int_0^\infty \Phi(x)dx, \tag{17.8.1}$$

where

$$\Phi(x) \equiv \Phi(x_1, ..., x_n) = |\Delta(x)|^{2\gamma} \prod_{j=1}^{n} x_j^{\alpha-1} \exp(-x_j) dx_j.$$

Integrating

$$\frac{\partial}{\partial x_1} f(x)\Phi(x)$$

$$= \frac{\partial f}{\partial x_1}\Phi(x) + f(x)\left(\frac{\alpha-1}{x_1} - 1 + 2\gamma\sum_{j=2}^{n}\frac{1}{x_1 - x_j}\right)\Phi(x),$$

choosing various polynomials for $f(x)$, and applying Aomoto's argument one gets, for example,

$$\langle x_1 \cdots x_m \rangle = [\alpha+\gamma(n-m)] \langle x_1 \cdots x_{m-1} \rangle = \prod_{i=1}^{m} [\alpha+\gamma(n-i)], \qquad (17.8.2)$$

$$\left\langle x_1^2 \cdots x_j^2 x_{j+1} \cdots x_m \right\rangle$$

$$= [\alpha + 1 + \gamma(2n - m - j)] \left\langle x_1^2 \cdots x_{j-1}^2 x_j \cdots x_m \right\rangle$$

$$= \prod_{i=1}^{j} [\alpha + 1 + \gamma(2n - m - i)] \prod_{k=1}^{m} [\alpha + \gamma(n - k)], \quad (17.8.3)$$

$$\left\langle x_1^3 \cdots x_i^3 x_{i+1}^2 \cdots x_j^2 x_{j+1} \cdots x_m \right\rangle$$

$$= [\alpha + 2 + \gamma(2n - j - i)] \left\langle x_1^3 \cdots x_{i-1}^3 x_i^2 \cdots x_j^2 x_{j+1} \cdots x_m \right\rangle$$

$$+ \gamma(n - m) \left\langle x_1^3 \cdots x_{i-1}^3 x_i^2 \cdots x_{j-1}^2 x_j \cdots x_m x_{m+1} \right\rangle, \quad (17.8.4)$$

etc.

Taking $m = n$ in Eq. (17.8.2) one can express the (Laguerre) integral (17.6.5) with parameter α as a known multiple of that with parameter $\alpha \pm j$, j a positive integer. Instead of thus decreasing α, Askey and Richards (1989) increased it and took the limit $\alpha \to \infty$. In this way they related the integral (17.6.5) to the (Hermite) integral (17.6.7) and with a fine analysis determined them both, without recourse to Eq. (17.1.3). We will not reproduce the details here.

If in the above we replace $\Phi(x)$ by

$$\Phi(x) = |\Delta(x)|^{2\gamma} \prod_{j=1}^{n} \exp(-a x_j^2),$$

then similar arguments give

$$\langle x_1 \cdots x_{2m} \rangle = \left(-\frac{\gamma}{2a}\right) (2m-1) \langle x_1 \cdots x_{2m-2} \rangle = \left(-\frac{\gamma}{2a}\right)^m \frac{(2m)!}{2^m m!}, \tag{17.8.5}$$

[a result known to Ullah (1986) for $\gamma = 1/2$],

$$\begin{aligned} 2a \left\langle x_1^2 \cdots x_j^2 x_{j+1} \cdots x_m \right\rangle &= [1 + \gamma(n-m)] \left\langle x_1^2 \cdots x_{j-1}^2 x_j \cdots{}_{m-1} \right\rangle \\ &\quad - \gamma(j-1) \left\langle x_1^2 \cdots x_{j-2}^2 x_{j-1} \cdots x_m \right\rangle, \end{aligned} \tag{17.8.6}$$

$$\begin{aligned} 2a \left\langle x_1^3 \cdots x_j^3 x_{j+1} \cdots x_m \right\rangle &= [2 + \gamma(2n-m-j)] \left\langle x_1^3 \cdots x_{j-1}^3 x_j \cdots x_m \right\rangle \\ &\quad - \gamma(j-1) \left\langle x_1^3 \cdots x_{j-2}^3 x_{j-1}^2 x_j^2 x_{j+1} \cdots x_m \right\rangle, \end{aligned} \tag{17.8.7}$$

etc., so that , e.g.,

$$\begin{aligned} \left\langle x_1^2 x_2 \cdots x_{2m+1} \right\rangle &= \frac{1}{2a}[1 + \gamma(n-2m-1)] \langle x_1 \cdots x_{2m} \rangle \\ &= \frac{1}{2a}[1 + \gamma(n-2m-1)] \left(-\frac{\gamma}{2a}\right)^m \frac{(2m)!}{2^m m!}, \end{aligned} \tag{17.8.8}$$

$$\langle x_1^2 x_2^2 x_3 \cdots x_{2m} \rangle = \left(\frac{1}{2a}\right)^2 [1+\gamma(n-2m)][1+\gamma(n-2m+1)]$$
$$\times \left(-\frac{\gamma}{2a}\right)^{m-1} \frac{(2m-2)!}{2^{m-1}(m-1)!} + \left(-\frac{\gamma}{2a}\right)^{m+1} \frac{(2m)!}{2^m m!}, \tag{17.8.9}$$

$$\langle x_1^3 x_2 \cdots x_{2m} \rangle = \frac{1}{2a}[2+\gamma(2n-2m-1)] \left(-\frac{\gamma}{2a}\right)^m \frac{(2m)!}{2^m m!}, \tag{17.8.10}$$

etc.

17.9. Connection with Finite Reflection Groups

Macdonald (1982) observed that $x_i - x_j = 0$, $1 \le i < j \le n$, are the equations of the reflecting (hyper)planes of the finite group A_n. There are other finite groups generated by reflections. In the n-dimensional Euclidean space $\mathcal{R}^n$, consider a certain number of planes all passing through the origin. If the angles between these planes are properly chosen, then the group generated by reflections in these planes is finite. These groups were enumerated and classified by Coxeter as $A_n (n \ge 1)$, $B_n (n \ge 2)$, $D_n (n \ge 4)$, E_6, E_7, E_8, F_4, G_2, H_3, H_4, and $I_2(p)$, $(p \ge 5)$. $(G_2 \equiv I_2(6))$. The first three sequences are known as the classical groups, while the rest are the exceptional groups. Let G be one of these groups and let $P(x)$ be the product of distances of the point $x \equiv (x_1, ..., x_n)$ from all the reflecting planes belonging to G. In other words, if the equations of the planes are $\sum_{i=1}^n a_i^{(\alpha)} x_i = 0$ with $\sum_{i=1}^n \left(a_i^{(\alpha)}\right)^2 = 1$, then $P(x) = \prod_\alpha \left(\sum_i a_i^{(\alpha)} x_i\right)$. Let N be the degree of $P(x)$, i.e., the number of reflecting planes in G.

Not knowing Selberg's integral, Macdonald conjectured (1982) that

$$\int_{-\infty}^{\infty} \cdots \int_{-\infty}^{\infty} \exp\left(-\sum x_i^2/2\right) |P(x)|^{2\gamma} \, dx_1 ... dx_n$$
$$= 2^{-N\gamma} (2\pi)^{n/2} \prod_{j=1}^n \frac{\Gamma(1+\gamma d_j)}{\Gamma(1+\gamma)}, \tag{17.9.1}$$

where d_i are the degrees of a basic set of polynomials invariant under G. The n integers $m_i = d_i - 1$, $i = 1, \ldots, n$, are known also as the (Coxeter) indices of G and have a few other remarkable properies (*cf.* Mehta 1989, Section 7.11).

For the group A_n Eq. (17.9.1) is the same as Eq. (17.6.7). Note the difference in normalization; the reflecting planes are $(x_i - x_j)/\sqrt{2} = 0$ in Eq. (17.9.1) and not $x_i - x_j = 0$ as in Eq. (17.6.7). For the groups B_n, and D_n, Eq. (17.9.1) is the same as Eq. (17.6.6) with $\alpha = \gamma + 1/2$ and $\alpha = 1/2$, respectively. For the two-dimensional discrete rotation group $I_2(p)$,

$$P(x) = \prod_{j=0}^{p-1} (x_1 \sin(j\pi/p) - x_2 \cos(j\pi/p)), \tag{17.9.2}$$

and Eq. (17.9.1) can be directly verified by introducing polar coordinates. Opdam recently (1989) gave a uniform proof of Eq. (17.9.1) for the groups A_n, B_n, D_n, E_6, E_7, E_8, F_4, and $I_2(6)$; the proof depends on the existence of a crystallographic lattice corresponding to each of these groups. Garvan (1989) gave a computer proof of Eq. (17.9.1) for the groups H_3 and F_4. Opdam now announces a proof for the groups H_3 and H_4. Thus, Eq.(17.9.1) is no longer a conjecture but a theorem.

We give here a convenient list of reflecting (hyper)planes for the exceptional (nonfactorizable finite reflection) groups mentioned above, together with the degree of $P(x)$ and the list of the integers d_i. The enteries are arranged as follows.

(i) Coxeter symbol of the group, the subscript indicating the dimension of the space in which it operates, i.e., the number n of the variables in the integral (17.9.1).

(ii) Number N of the reflecting (hyper)planes or the degree of the polynomial $P(x)$ in Eq. (17.9.1).

(iii) The integers d_i, the degrees of the polynomials in a basic invariant set (invariant under reflections of the group).

(iv) A list of reflecting (hyper)planes or the linear factors of the polynomial $P(x)$, not normalized in general.

Group $\mathbf{E_6}$: $N = 36$; $d_i = (2, 5, 6, 8, 9, 12)$; factors of $P(x)$: x_1, x_2, x_3, x_4, $x_1 \pm x_2 \pm x_3 \pm x_4$, $x_1 \pm x_2 \pm \sqrt{2}x_5$, $x_3 \pm x_4 \pm \sqrt{2}x_5$, $x_1 \pm x_3 \pm$

$\frac{1}{\sqrt{2}}(x_5 - \sqrt{3}x_6)$, $x_2 \pm x_4 \pm \frac{1}{\sqrt{2}}(x_5 - \sqrt{3}x_6)$, $x_1 \pm x_4 \pm \frac{1}{\sqrt{2}}(x_5 + \sqrt{3}x_6)$, $x_2 \pm x_3 \pm \frac{1}{\sqrt{2}}(x_5 - \sqrt{3}x_6)$.

Group $\mathbf{E_7}$: $N = 63$; $d_i = (2, 6, 8, 10, 12, 14, 18)$; factors of $P(x)$: x_i, $i = 1, 2, ..., 7$ and $x_i \pm x_j \pm x_k \pm x_\ell$ with $(i, j, k, \ell) = (1, 2, 3, 4)$, $(1, 2, 5, 6)$, $(1, 3, 5, 7)$, $(1, 4, 6, 7)$, $(2, 3, 6, 7)$, $(2, 4, 5, 7)$, and $(3, 4, 5, 6)$.

Group $\mathbf{E_8}$: $N = 120$; $d_i = (2, 8, 12, 14, 18, 20, 24, 30)$; factors of $P(x)$: x_i, $i = 1, 2, ..., 8$ and $x_i \pm x_j \pm x_k \pm x_\ell$ with $(i, j, k, \ell) = (1, 2, 3, 4)$, $(1, 2, 5, 6)$, $(1, 3, 5, 7)$, $(1, 4, 6, 7)$, $(2, 3, 6, 7)$, $(2, 4, 5, 7)$, $(3, 4, 5, 6)$, $(5, 6, 7, 8)$, $(3, 4, 7, 8)$, $(2, 4, 6, 8)$, $(2, 3, 5, 8)$, $(1, 4, 5, 8)$, $(1, 3, 6, 8)$, and $(1, 2, 7, 8)$.

Group $\mathbf{F_4}$: $N = 24$; $d_i = (2, 6, 8, 12)$; factors of $P(x)$: x_1, x_2, x_3, x_4, $x_1 \pm x_2 \pm x_3 \pm x_4$, $x_i \pm x_j$, $1 \leq i < j \leq 4$.

Group $\mathbf{H_3}$: $N = 15$; $d_i = (2, 6, 10)$; factors of $P(x)$: x_1, x_2, x_3, $ax_1 \pm bx_2 \pm cx_3$, and those obtained by the cyclic permutations of (a, b, c), i.e., $ax_2 \pm bx_3 \pm cx_1$ and $ax_3 \pm bx_1 \pm cx_2$, where $a = \cos(\pi/5) = (\sqrt{5}+1)/4$, $b = \cos(2\pi/5) = (\sqrt{5} - 1)/4$ and $c = \cos(\pi/3) = 1/2$.

Group $\mathbf{H_4}$: $N = 60$; $d_i = (2, 12, 20, 30)$; factors of $P(x)$: x_1, x_2, x_3, x_4, $x_1 \pm x_2 \pm x_3 \pm x_4$, $ax_1 \pm bx_2 \pm cx_3$, $ax_1 \pm bx_3 \pm cx_4$, $ax_1 \pm bx_4 \pm cx_2$, $ax_2 \pm bx_4 \pm cx_3$, and those obtained by the cyclic permutations of (a, b, c) in the last four sets. The numbers a, b, c are, respectively, the cosines of the angles $\pi/5$, $2\pi/5$, and $\pi/3$ as for the group H_3.

17.10. A Second Generalization of the Beta Integral

A second generalization of the beta-integral by Selberg is as follows:

$$\int \cdots \int y_1 \cdots y_m \, |\Delta(y)|^{2\gamma} \left(1 - \sum_{i=1}^{n} y_i\right)^{\beta-1} \prod_{j=1}^{n} y_j^{\alpha-1} dy_j$$

$$= \frac{\Gamma(\beta)}{\Gamma(\beta + m + \alpha n + \gamma n(n-1))} \prod_{i=1}^{m} [\alpha + \gamma(n-i)]$$

$$\times \prod_{j=1}^{n} \frac{\Gamma(\alpha + \gamma(n-j))\Gamma(1+\gamma j)}{\Gamma(1+\gamma)}, \tag{17.10.1}$$

where the integral is taken over $0 \leq y_i,\ y_1 + \cdots + y_n \leq 1$. This is to be compared with the long-known generalization

$$\int \cdots \int \left(1 - \sum_{i=1}^{n} y_i\right)^{\beta-1} \prod_{j=1}^{n} y_j^{\alpha_j - 1} dy_j$$

$$= \Gamma(\beta) \prod_{j=1}^{n} \Gamma(\alpha_j) / \Gamma(\alpha_1 + \cdots + \alpha_n + \beta). \qquad (17.10.2)$$

the integral being again taken over $0 \leq y_i,\ y_1 + \cdots + y_n \leq 1$.

Proof. From Eqs. (17.6.5) and (17.8.2) one has by a change of variables

$$\int_0^\infty \cdots \int_0^\infty x_1 \cdots x_m \, |\Delta(x)|^{2\gamma} \prod_{j=1}^{n} x_j^{\alpha-1} \exp(-\lambda x_j) dx_j$$

$$= \lambda^{-m-\alpha n - \gamma n(n-1)} \prod_{i=1}^{m} (\alpha + \gamma(n-i))$$

$$\times \prod_{j=1}^{n} \frac{\Gamma(\alpha + \gamma(n-j)) \Gamma(1+\gamma j)}{\Gamma(1+\gamma)}.$$

Multiplying both sides by $\lambda^{\beta-1} e^{-\lambda}$ and integrating over λ from 0 to ∞, one gets

$$\int_0^\infty \cdots \int_0^\infty x_1 \cdots x_m \, |\Delta(x)|^{2\gamma} \left(1 + \sum_{i=1}^{n} x_i\right) \prod_{j=1}^{n} x_j^{\alpha-1} dx_j$$

$$= \frac{\Gamma(\beta - m - \alpha n - \gamma n(n-1))}{\Gamma(\beta)} \prod_{i=1}^{m} [\alpha + \gamma(n-i)]$$

$$\times \prod_{j=1}^{n} \frac{\Gamma(\alpha + \gamma(n-j)) \Gamma(1+\gamma j)}{\Gamma(1+\gamma)},$$

provided that the variables appearing in the gamma functions all have positive real parts. Now make a change of variables,

$$x_j = y_j \left(1 - \sum_{i=1}^{n} y_i\right)^{-1}, \quad y_j = x_j \left(1 + \sum_{i=1}^{n} x_i\right)^{-1}.$$

Setting $u = \left(1 - \sum_{i=1}^{n} y_i\right)^{-1}$, the Jacobian is

$$\det\left[\frac{\partial x_i}{\partial y_j}\right] = \det\left[u\delta_{ij} + y_i u^2\right].$$

This determinant can be calculated by bordering it with an extra row and a column,

$$\begin{aligned}\det\left[u\delta_{ij} + u^2 y_i\right] &= \det\begin{bmatrix} 1 & 0 \\ y_i & u\delta_{ij} + u^2 y_i \end{bmatrix} = \det\begin{bmatrix} 1 & -u^2 \\ y_i & u\delta_{ij} \end{bmatrix} \\ &= u^n + u^2 u^{n-1} \sum_{i=1}^{n} y_i = u^{n+1}.\end{aligned}$$

The limits of integration for y_i are $0 \le y_i, \quad \sum_{i=1}^{n} y_i \le 1$. Hence,

$$\begin{aligned}&\int_0^\infty \cdots \int_0^\infty x_1 \cdots x_m |\Delta(x)|^{2\gamma} \left(1 + \sum_{i=1}^{n} x_i\right)^{-\beta} \prod_{j=1}^{n} x_j^{\alpha-1} dx_j \\ &= \int \cdots \int y_1 \cdots y_m |\Delta(y)|^{2\gamma} \left(1 - \sum_{i=1}^{n} y_i\right)^{\beta-m-\alpha n-\gamma n(n-1)-1} \\ &\quad \times \prod_{j=1}^{n} y_j^{\alpha-1} dy_j\end{aligned}$$

where the integral is taken over $0 \le y_i$, $y_1 + \cdots + y_n \le 1$. Or redefining

β, we get

$$\int \cdots \int y_1 \cdots y_m \, |\Delta(y)|^{2\gamma} \left(1 - \sum_{i=1}^{n} y_i\right)^{\beta-1} \prod_{j=1}^{n} y_j^{\alpha-1} dy_j$$

$$= \frac{\Gamma(\beta)}{\Gamma(\beta + m + \alpha n + \gamma n(n-1))} \prod_{i=1}^{m} [\alpha + \gamma(n-i)]$$

$$\times \prod_{j=1}^{n} \frac{\Gamma(\alpha + \gamma(n-j))\Gamma(1+\gamma j)}{\Gamma(1+\gamma)}. \tag{17.10.1}$$

End of proof.

17.11. Some Related Difficult Integrals

Following Macdonald one may replace $\Delta(x)$ in Eq. (17.1.2) by the polynomial $P(x)$ of Section 17.9; no evaluation or guess is known for

$$\int_0^1 \cdots \int_0^1 |P(x)|^{2\gamma} \prod_{i=1}^{n} x_i^{\alpha-1}(1-x_i)^{\beta-1} dx_i. \tag{17.11.1}$$

or for

$$\int_{-1}^1 \cdots \int_{-1}^1 |P(x)|^{2\gamma} \prod_{i=1}^{n} (1+x_i)^{\alpha-1}(1-x_i)^{\beta-1} dx_i. \tag{17.11.1$'$}$$

As other possible extensions consider the integrals

$$A(n,s,\gamma) = \pi^{-ns/2} \int \cdots \int |\Delta(\vec{r})|^{2\gamma} \exp\left(\sum_{i=1}^{n} [-\vec{r}_i^{\,2}]\right) \prod_{i=1}^{n} d\vec{r}_i, \tag{17.11.2}$$

$$B(n,s,\gamma) = \pi^{-ns/2} \prod_{j=1}^{n} \Gamma(j+s/2) \int_{|\vec{r}_i|\le 1} \cdots \int |\Delta(\vec{r})|^{2\gamma} \, d\vec{r}_1 \cdots d\vec{r}_n, \tag{17.11.3}$$

with

$$\Delta(\vec{r}) = \prod_{1\le i<j\le n} |\vec{r}_i - \vec{r}_j|, \tag{17.11.4}$$

where $\vec{r}_1, ..., \vec{r}_n$ are vectors in the s-dimensional Euclidean space $\mathcal{R}^s$; they vary over all the space $\mathcal{R}^s$ in Eq. (17.11.2) and inside the unit sphere $|\vec{r}_i| \le 1$, $i = 1, ..., n$, in Eq. (17.11.3). Or consider the integral

$$C(n, s, \gamma) = 2^{n(n+1)/2}\pi^{n(s+1)/2}\left[\Gamma\left(\frac{s+1}{2}\right)\right]^{-n}$$

$$\times \int\cdots\int_{|\vec{r}_i|=1} |\Delta(\vec{r})|^{2\gamma}\, d\Omega_1 \cdots d\Omega_n, \tag{17.11.5}$$

where $\vec{r}_1, ..., \vec{r}_n$ are vectors in $\mathcal{R}^{s+1}$ and they vary over the s-dimensional surface of the sphere $|\vec{r}_i| = 1$, $i = 1, ..., n$ in $\mathcal{R}^{s+1}$. Note that here s is the dimension of the surface of the unit sphere in $s+1$ dimensions.

The case $s = 1$ of these integrals was considered up to now [*cf.* Eqs. (17.6.7), (17.6.1), and (11.1.4)]. For $s = 2$ and $\gamma = 1$, the integrals (17.11.2) and (17.11.3) can be dealt with as in Chapter 15 with the result

$$A(n, 2, 1) = B(n, 2, 1) = \prod_{j=1}^{n} j! \tag{17.11.6}$$

The integral (17.11.5) for $s = 2, \gamma = 1$ was evaluated by Caillol (1981) using a similar method, as explained below.

In terms of the Cayley–Klein parameters

$$\alpha_j = \cos(\theta_j/2)\exp(i\varphi_j/2), \quad \beta_j = \sin(\theta_j/2)\exp(-i\varphi_j/2),$$

where θ_j, φ_j are the polar angles of $\vec{r}_j$, one has

$$|\vec{r}_j - \vec{r}_k|^2 = 4\,|\alpha_j\beta_k - \alpha_k\beta_j|^2 = 4\,|\beta_j\beta_k|^2 \left|\frac{\alpha_j}{\beta_j} - \frac{\alpha_k}{\beta_k}\right|^2,$$

so that

$$|\Delta(\vec{r})|^2 = 2^{n(n-1)} \prod_{j=1}^{n} |\beta_j|^{2(n-1)} \prod_{1\le j<k\le n} \left|\frac{\alpha_j}{\beta_j} - \frac{\alpha_k}{\beta_k}\right|^2.$$

Also

$$\int d\Omega \left(\frac{\alpha}{\beta}\right)^j \left(\frac{\alpha}{\beta}\right)^{*k} |\beta|^{2(n-1)}$$

$$= \int_0^{2\pi} d\varphi \int_0^{\pi} d\theta \, \sin\theta [\cos(\theta/2)]^{j+k}$$

$$\times [\sin(\theta/2)]^{2(n-1)-j-k} \exp[i(j-k)\varphi]$$

$$= 2\pi\delta_{jk} \int_0^{\pi} d\theta \, \sin\theta [\cos(\theta/2)]^{2j} [\sin(\theta/2)]^{2(n-1-j)}$$

$$= 4\pi \frac{j!(n-j-1)!}{n!} \delta_{jk}. \tag{17.11.7}$$

Therefore, if one writes

$$K_n(\vec{r}_j, \vec{r}_k) = \sum_{\ell=0}^{n-1} \frac{n!}{4\pi.\ell!(n-\ell-1)!} \left(\frac{\alpha_j}{\beta_j}\right)^\ell \left(\frac{\alpha_k}{\beta_k}\right)^{*\ell} (\beta_j \beta_k^*)^{n-1}, \tag{17.11.8}$$

then

$$|\Delta(\vec{r})|^2 = 2^{n(n-1)}(4\pi)^n$$

$$\times \prod_{\ell=0}^{n-1} \frac{\ell!(n-\ell-1)!}{n!} \det [K_n(\vec{r}_j, \vec{r}_k)]_{j,k=1,\ldots,n}$$

$$= 2^{n(n-1)}(4\pi)^n \prod_{\ell=0}^{n-1} \frac{(\ell!)^2}{n!} \det [K_n(\vec{r}_j, \vec{r}_k)]_{j,k=1,\ldots,n},$$

$$\int K_n(\vec{r}_j, \vec{r}_j) d\Omega_j = n,$$

$$\int K_n(\vec{r}_j, \vec{r}_k) K_n(\vec{r}_k, \vec{r}_p) d\Omega_k = K_n(\vec{r}_j, \vec{r}_p).$$

So that from Theorem 5.2.1,

$$
\begin{aligned}
C(n,2,1) &= \int \cdots \int_{|\vec{r}_j|=1} |\Delta(\vec{r})|^2 \, d\Omega_1 \cdots d\Omega_n \\
&= 2^{n(n-1)} n! \, (4\pi)^n \prod_{\ell=0}^{n-1} \frac{(\ell!)^2}{n!}.
\end{aligned}
\tag{17.11.9}
$$

If one writes

$$
P(\vec{r}_1, ..., \vec{r}_n) = \frac{1}{n!} \det \left[K_n(\vec{r}_j, \vec{r}_k) \right]_{j,k=1,\ldots,n},
$$

then this P is normalized and can be interpreted as a probability density. The m-point correlation function, from Theorem 5.2.1, is

$$
\begin{aligned}
&\frac{n!}{(n-m)!} \int \cdots \int P(\vec{r}_1, ..., \vec{r}_n) d\Omega_{m+1} \cdots d\Omega_n \\
&= \det \left[K_n(\vec{r}_j, \vec{r}_k) \right]_{j,k=1,\ldots,m}.
\end{aligned}
$$

Integrals (17.11.2), (17.11.3), and (17.11.5) for $s > 2$ or $s = 2, \ \gamma > 1$ are not known. Of course, one can compute them with some effort for $\gamma = 1$, any s and small values of n, say, $n = 2, 3$, or 4. For example,

$$
A(1,s,1) = B(1,s,1) = C(1,s,1) = C(2,s,1) = 1,
$$

$$
\begin{aligned}
A(2,s,1) = B(2,s,1) &= s, \\
C(3,s,1) &= s(s+2)(s+1)^{-2}, \\
A(3,s,1) &= s(s+2)(2s-1)/2, \\
A(4,s,1) &= s(s+2)^2(2s^3+4s^2-9s+4)/2; \\
B(3,s,1) &= s(s^2+2s-2), \\
C(4,s,1) &= s(s^2+3s-1)(s+1)^{-3},
\end{aligned}
$$

etc.

For applications in polymer theory one would like to know the value of

$$\int \cdots \int |\Delta(\vec{r})|^{2\gamma} \exp\left[-\sum_{j=1}^{n} (\vec{r}_{j-1} - \vec{r}_j)^2\right] \prod_{j=1}^{n} |\vec{r}_j|^{2\gamma}\, d\vec{r}_j, \qquad (17.11.10)$$

with $\vec{r}_0 = 0$, specially in $s = 3$ dimensions. But this has a different structure.

It is curious to note that with a constant vector $\vec{a}$ say of unit length, the integral

$$I(\alpha_0, \alpha_1, ..., \alpha_n) = \int \cdots \int |\vec{a} - \vec{r}_1 - \cdots - \vec{r}_n|^{-2\alpha_0} \prod_{j=1}^{n} |\vec{r}_j|^{-2\alpha_j}\, d\vec{r}_j \qquad (17.11.11)$$

can be evaluated without much difficulty, viz.,

$$I(\alpha_0, ..., \alpha_n) = \pi^{ns/2} \frac{\Gamma(\alpha_0 + \cdots + \alpha_n - ns/2)}{\Gamma((n+1)s/2 - \alpha_0 - \cdots - \alpha_n)} \times \prod_{j=0}^{n} \frac{\Gamma(s/2 - \alpha_j)}{\Gamma(\alpha_j)}, \qquad (17.11.12)$$

valid for values of α_j such that the arguments of all the gamma functions have a positive real part.

Proof. (Selberg). Taking $\vec{r}_1 + \vec{r}_2$ instead of $\vec{r}_2$ as the new variable,

$$I(\alpha_0, ..., \alpha_n) = \int \cdots \int |\vec{a} - \vec{r}_2 - \cdots - \vec{r}_n|^{-2\alpha_0} \times |\vec{r}_1|^{-2\alpha_1} |\vec{r}_2 - \vec{r}_1|^{-2\alpha_2} \cdots |\vec{r}_n|^{-2\alpha_n}\, d\vec{r}_1 \cdots d\vec{r}_n. \qquad (17.8.9)$$

Setting here $\vec{r}_1 = |\vec{r}_2|\, \vec{r}\,'_1$, we get

$$I(\alpha_0, ..., \alpha_n) = I(\alpha_2, \alpha_1) I(\alpha_0, \alpha_1 + \alpha_2 - s/2, \alpha_3, ..., \alpha_n).$$

Repeating the process we can express $I(\alpha_0, ..., \alpha_n)$ in terms of $I(\beta_1, \beta_2)$.

For $I(\alpha_0, \alpha_1)$, writing

$$|\vec{r}|^{-2\alpha} = [\Gamma(\alpha)]^{-1} \int_0^\infty t^{\alpha-1} \exp\left(-t\,|\vec{r}|^2\right) dt,$$

one has

$$\begin{aligned} I(\alpha_0, \alpha_1) &= \int |\vec{r}|^{-2\alpha_1} |\vec{a} - \vec{r}|^{-2\alpha_0} \, d\vec{r} \\ &= \frac{1}{\Gamma(\alpha_0)\Gamma(\alpha_1)} \int_0^\infty dt_0 \int_0^\infty dt_1 t_0^{\alpha_0 - 1} t_1^{\alpha_1 - 1} \\ &\quad \times \int d\vec{r} \exp\left[-t_0 |\vec{a} - \vec{r}|^2 - t_1 |\vec{r}|^2\right]. \end{aligned}$$

The Gaussian integral over $\vec{r}$ gives

$$\pi^{s/2} (t_0 + t_1)^{-s/2} \exp[-t_0 t_1 / (t_0 + t_1)].$$

Now make a change of variables

$$t_0 = uv, \quad t_1 = (1-u)v, \quad \frac{\partial(t_0, t_1)}{\partial(u, v)} = v,$$

so that

$$\begin{aligned} \Gamma(\alpha_0)\Gamma(\alpha_1) I(\alpha_0, \alpha_1) &= \pi^{s/2} \int_0^1 du \int_0^\infty dv \; u^{\alpha_0 - 1} (1-u)^{\alpha_1 - 1} \\ &\quad \times v^{\alpha_0 + \alpha_1 - s/2 - 1} \exp[-vu(1-u)]. \end{aligned}$$

Integration over v leaves us with a beta integral.

Collecting the results we get Eq. (17.11.12).

Summary of Chapter 17

The main object of this chapter is to present the evaluation of certain definite multiple integrals. These integrals were encountered in different contexts by various investigators and the discovery of their similarity was a pleasant surprise. The most significant of them is Eq. (17.1.3) or its slight generalization Eq. (17.1.4). The other important related integrals or consequences being Eqs. (17.6.1) to (17.6.6), (17.7.1), (17.9.1), and (17.10.1).

18 / Gaussian Ensembles. Level Density in the Tail of the Semicircle

The density of nuclear levels increases steeply almost like an exponential in the experimentally observed energy range. On the other hand, the eigenvalue density for the Gaussian ensembles is a semicircle in the first approximation:

$$\sigma(x) \approx \pi^{-1}(A - x^2)^{1/2}, \qquad A = 2N. \qquad (18.0.1)$$

Therefore one might think that near the lower end, $x = -A^{1/2}$, this density looks like the actual rise in nuclear level density. Although the deviations must be small compared with the dominant behavior (18.0.1), the tail might still contain an infinite number of eigenvalues. For example, $\sigma(x)$ may be proportional to $N^{1/3}$ in a region extending to $N^{1/6}$, so that the number of eigenvalues not accounted for by Eq. (18.0.1) will be proportional to $N^{1/2}$, increasing rapidly with the dimension number N of the matrices of the set. If this were the case, we should expect that the correlation functions and the spacing distribution in the tail part would be nearer the actual situation. However, the following calculation (Bronk, 1964a, b; 1965) shows that this is not true. Near $x = \pm A^{1/2}$ the eigenvalue density $\sigma_N(x)$ is $\sim N^{1/6}$ in a region of extent $\sim N^{-1/6}$, so that the total number of eigenvalues in the tail part remains finite and amounts to only a few, even when $N \longrightarrow \infty$.

18.1. Level Density near the Inflection Point

For the Gaussian unitary ensemble the eigenvalue density $\sigma_N(x)$ is the sum [*cf.* Eq. (5.2.16) and Appendix A.9]

$$\sigma_N(x) = \sum_{j=0}^{N-1} \varphi_j^2(x) \equiv \sum_{j=0}^{N-1} \left\{2^j j! \pi^{1/2}\right\}^{-1} e^{-x^2} \{H_j(x)\}^2 \quad (18.1.1)$$

$$= (N/2)^{1/2} \left[\varphi_N'(x)\varphi_{N-1}(x) - \varphi_N(x)\varphi_{N-1}'(x)\right], \quad (18.1.2)$$

where φ' is the derivative of φ. To have any similarity with the exponential the function $\sigma_N(x)$ must be convex from below, and we will be interested only in that region. Let us therefore determine the inflection point x_0 such that $\sigma_N(x)$ is convex from below for all $x > x_0$. Differentiating Eq. (18.1.2) and substituting from the differential equation

$$\varphi_j''(x) + (2j + 1 - x^2)\varphi_j(x) = 0, \quad (18.1.3)$$

satisfied by the harmonic oscillator function $\varphi_j(x)$, we get

$$\sigma_N'(x) = -(2N)^{1/2}\varphi_N(x)\varphi_{N-1}(x).$$

Differentiating once more, we obtain

$$\sigma_N''(x) = -(2N)^{1/2} \left[\varphi_N'(x)\varphi_{N-1}(x) + \varphi_N(x)\varphi_{N-1}'(x)\right]. \quad (18.1.4)$$

We are interested in the location of the largest zero of $\sigma_N''(x)$. For $x \geq (2N)^{1/2}$, $\varphi_N(x)$ and $\varphi_{N-1}(x)$ are both positive and decreasing so that $\varphi_N'(x)$ and $\varphi_{N-1}'(x)$ are both negative. Because the outermost maxima of $\varphi_j(x)$ move out with the increase of j, $\varphi_{N-1}'(x)$ is negative and $\varphi_N(x)$ is positive when $\varphi_N'(x)$ first becomes zero. As we decrease x across the value $(2N)^{1/2}$, $\varphi_N(x)$ will attain its maximum value and then decrease to zero, whereas $\varphi_{N-1}(x)$ will always remain positive. Thus $\sigma_N''(x)$ changes sign as x varies from the largest zero of $\varphi_N'(x)$ to the largest zero of $\varphi_N(x)$ and therefore must vanish somewhere in between. These largest zeros lie very near each other and their location is known (Szegö, 1959):

$$x_0 \approx (2N)^{1/2} - 1.856(2N)^{-1/6}. \quad (18.1.5)$$

We are interested in estimating the number of eigenvalues larger than x_0 :

$$\int_{x_0}^{\infty} \sigma_N(x)dx. \tag{18.1.6}$$

To estimate $\varphi_j(x)$ near the transition point put

$$x = (2j+1)^{1/2} - 2^{-1/2}3^{-1/3}j^{-1/6}t \tag{18.1.7}$$

so that the differential equation (18.1.3) for $\varphi_j(x)$ is transformed to

$$\frac{d^2}{dt^2}\tilde{\phi}_j(t) + \frac{1}{3}t\tilde{\phi}_j(t) = 0, \qquad tj^{-2/3} \ll 1 \ll j; \tag{18.1.8}$$

$$\varphi_j(x) = \tilde{\phi}_j(t). \tag{18.1.9}$$

This is Airy's equation. The solution that goes to zero for $t \longrightarrow \infty$ is given by

$$\tilde{\phi}_j(t) = iaj^{-1/12}(-t)^{1/2}e^{i(\pi/6)}H_{1/3}^{(1)}\left(2i\left(\frac{-t}{3}\right)^{3/2}\right), \tag{18.1.10a}$$

if $t \leq 0$, and

$$\tilde{\phi}_j(t) = 2(3)^{-1/2}aj^{-1/12}t^{1/2}$$

$$\times \left\{J_{-1/3}\left[2\left(\frac{t}{3}\right)^{3/2}\right] + J_{1/3}\left[2\left(\frac{t}{3}\right)^{3/2}\right]\right\}, \tag{18.1.10b}$$

if $t \geq 0$, where $H_j^{(1)}(x)$ is the Hankel function of the first kind and $J_j(x)$ is the Bessel function (*cf.* Abramowitz and Stegun, 1965, Chapter 10.4). The ratio of the normalizations in Eqs. (18.1.10a) and (18.1.10b) is such that the two forms of $\tilde{\phi}_j(t)$ join smoothly at $t = 0$. The constant a may be determined from the condition that the average value of $\left|\tilde{\phi}_j(t)\right|^2$ over an interval for $t > 0$, $t \gg 1$ coincides with the classical approximation of the quantum mechanical probability density for a harmonic oscillator in that region. The asymptotic form of Eq. (18.1.10b) to be used for this purpose is

$$\tilde{\phi}_j(t) \propto t^{-1/4}\cos(\beta t^{3/2} - \frac{1}{4}\pi), \qquad t > 0, \tag{18.1.11}$$

$$\beta = 2(3)^{-3/2}. \tag{18.1.12}$$

For our purposes it is sufficient to note that a is a small constant of the order of 0.3 and it does not depend on j. Using the power series for $J_{-1/3}$ and $J_{1/3}$ in Eq. (18.1.10b), we obtain an approximate expression for $\tilde{\phi}_j(t)$ for $|t| < 1$:

$$\tilde{\phi}_j(t) \approx \bar{\phi}_j(t) = c_1 j^{-1/12} e^{c_2 t}, \tag{18.1.13}$$

$$c_1 \approx 1.477a, \qquad c_2 \approx 0.506. \tag{18.1.14}$$

As $t \longrightarrow 0$, $\tilde{\phi}_j(t) \longrightarrow \bar{\phi}_j(t)$. We can actually prove that, for $t < 1.856 \times 6^{1/3} j^{-1/6}$, $\tilde{\phi}_j(t) \leq \bar{\phi}_j(t)$. However, we shall evade this issue and be satisfied with the approximation of replacing $\tilde{\phi}_j\,(t)$ by $\bar{\phi}_j\,(t)$ in the entire region $x > x_0$. Realizing that the only terms contributing appreciably to the summation (18.1.1) are those with a large j, we set

$$j = N - 1 - \mu, \tag{18.1.15}$$

$$(2N)^{1/2} + y = (2j+1)^{1/2} - 2^{-1/2} 3^{-1/3} j^{-1/6} t, \tag{18.1.16}$$

expand in powers of μ/N, and keep only the dominant terms to get

$$\begin{aligned} &\left\{\bar{\phi}_{N-\mu-1}\,(t)\right\}^2 \\ &\approx c_1^2 N^{-1/6} \exp\left[-2^{3/2} 3^{1/3} c_2 N^{1/6}\left(y + \frac{\mu}{(2N)^{1/2}}\right)\right] \\ &\approx 2.18\, a^2 N^{-1/6} \exp\left\{-2.06\ N^{1/6}\left[y + \mu\,(2N)^{-1/2}\right]\right\}. \end{aligned} \tag{18.1.17}$$

Putting this in Eq. (18.1.1),

$$\sigma_N(x) \approx 2.18\ a^2 N^{-1/6} \exp\left\{-2N^{1/6} y\right\}$$

$$\times \int_0^N \exp\left(-2^{1/2} N^{-1/3} \mu\right) d\mu$$

$$\approx 1.54\ a^2 N^{1/6} \exp\left(-2N^{1/6} y\right). \tag{18.1.18}$$

Thus the eigenvalue density in the tail varies as $N^{1/6}$. The total number of eigenvalues in the tail is

$$\int_{x_0}^{\infty} \sigma_N(x)\, dx \approx 1.54\ a^2 N^{1/6}$$

$$\times \int_{-1.856(2N)^{-1/6}}^{\infty} \exp\left(-2N^{1/6} y\right) dy$$

$$\approx 0.77\ a^2 \exp\left(2^{5\times 1.856/6}\right) \approx 14.3\ a^2 \le 2. \tag{18.1.19}$$

The eigenvalue density (6.4.3) for the Gaussian orthogonal ensemble has one more term. To get it in the tail of the semicircle we need an estimation of

$$I = m^{1/2} \int_{x_0}^{\infty} \mathrm{d}x \left[\varphi_{2m-1}(x) \int_0^x \varphi_{2m}(y)\, \mathrm{d}y\right]. \tag{18.1.20}$$

where $N = 2m$ and x_0 has the same meaning as before. We transform the second integral on the right as follows:

$$\int_0^x \varphi_{2m}(y)\, dy = \int_0^{\infty} \varphi_{2m}(y)\, dy - \int_x^{\infty} \varphi_{2m}(y)\, dy$$

$$= \frac{1}{2} \int_{-\infty}^{\infty} \varphi_{2m}(y)\, dy - \int_x^{\infty} \varphi_{2m}(y)\, dy.$$

But [*cf.* Appendix A.26.5, Eq. (A.26.12)]

$$\int_{-\infty}^{\infty} \varphi_{2m}(y)dy = \left(2\sqrt{\pi}\ (2m)!\right)^{1/2} \left(2^{m+1}m!\right)^{-1} \approx (4m)^{-1/4}, \qquad m \gg 1,$$

so that

$$I = I_1 + I_2, \tag{18.1.21}$$

with

$$I_1 \approx (m/4)^{1/4} \int_{x_0}^{\infty} \varphi_{2m-1}(x)dx, \tag{18.1.22}$$

$$I_2 = -m^{1/2} \int_{x_0}^{\infty} dx \left[\varphi_{2m-1}(x) \int_{x}^{\infty} \varphi_{2m}(y)\, dy\right]. \tag{18.1.23}$$

Now one can use the approximation (18.1.13) to see that the right-hand side of Eq. (18.1.21) is a small constant independent of m.

For the Gaussian symplectic ensemble, the level density is, [Eq. (7.2.2), case $n = 1$],

$$\sigma_{N4}(x) = \frac{1}{\sqrt{2}} \sum_{0}^{2N} \varphi_j^2(x\sqrt{2}).$$

Its estimation in the tail of the semicircle follows from what we have already said before.

Summary of Chapter 18

For the Gaussian ensembles the dominant part of the eigenvalue density is a "semicircle. "How good is this estimate in the tail part? It is found that the decrease is quite sharp; in the tail part the eigenvalue density $\sigma_N(x)$ is $N^{1/6}$ extending to $N^{-1/6}$ beyond the "semicircle" limit point, so that the total number of eigenvalues lying outside the "semicircle" remains finite when $N \to \infty$.

Probably the same situation prevails for a much larger class of matrix ensembles.

19 / Restricted Trace Ensembles. Ensembles Related to the Classical Orthogonal Polynomials

As mentioned toward the end of Chapter 2, Gaussian ensembles are unsatisfactory because the various matrix elements $H_{ij}^{(\lambda)}$ are not equally weighted. Apart from Dyson's method, efforts have been made to equalize this weighting in a straightforward manner. For example, by diagonalizing on a computer a large number of random matrices (Porter and Rosenzweig, 1960a), the elements of which can be made to conform to a given probability law, we can learn a lot about their eigenvalue distributions. Such knowledge, although useful, is purely empirical, and we restrict ourselves in this chapter to only those cases in which these empirical findings can be put on a firmer footing.

19.1. Fixed Trace Ensemble

When working with large but finite dimensional Hermitian matrices, we cannot allow the elements to grow indefinitely because then one would be unable to normalize. Gaussian ensembles overcome this difficulty by giving exponentially vanishing weights to large values of matrix elements. Another method will be to apply a cut-off. Proceeding from the analogy of a fixed energy in classical statistical mechanics, Rosenzweig (1963) defines his "fixed trace" ensemble by the requirement that the trace of H^2 be fixed to a number r^2 with no other constraint. The number r is called the strength of the ensemble. The joint probability density function for the matrix elements of H is therefore given by

$$P_r(H) = K_r^{-1}\delta\left[\frac{1}{r^2} \text{ tr } H^2 - 1\right] \tag{19.1.1}$$

with

$$K_r = \int_{-\infty}^{\infty} \cdots \int_{-\infty}^{\infty} \delta\left(\frac{1}{r^2} \text{ tr } H^2 - 1\right) \prod_{\lambda} \prod_{i \le j} dH_{ij}^{(\lambda)}. \tag{19.1.2}$$

This probability density function is invariant under a change of basis

$$H' = W^R H W, \quad W^R W = 1, \tag{19.1.3}$$

where W is an orthogonal, unitary, or symplectic matrix according to the three possibilities noted in Chapter 2, and W^R is the transpose, Hermitian conjugate or the dual of W. This is evident from the fact that under such a transformation the volume element $dH = \prod_\lambda \prod_{i \le j} dH_{ij}^{(\lambda)}$ and the quantity tr H^2 are invariant.

The important thing to be noted about these ensembles is their moment equivalence with Gaussian ensembles of large dimensions. More precisely, if we choose the constant a in Eq. (2.6.18) to give

$$\langle \text{tr } H^2 \rangle_G \approx K_G^{-1} \int_{-\infty}^{\infty} \text{tr } H^2 e^{-a \text{ tr } H^2} dH = r^2 \tag{19.1.4}$$

then for any fixed value of the sum

$$s = \sum_{\lambda} \sum_{i \le j} \eta_{ij}^{(\lambda)}, \quad \eta_{ij}^{(\lambda)} \ge 0, \tag{19.1.5}$$

the ratio of the moments

$$M_r(N, \eta) = \left\langle \prod_{\lambda} \prod_{i \le j} \left(H_{ij}^{(\lambda)}\right)^{\eta_{ij}^{(\lambda)}} \right\rangle_r, \tag{19.1.6}$$

and

$$M_G(N, \eta) = \left\langle \prod_{\lambda} \prod_{i \le j} \left(H_{ij}^{(\lambda)}\right)^{\eta_{ij}^{(\lambda)}} \right\rangle_G, \tag{19.1.7}$$

tends to unity as the number of dimensions N tends to infinity. The subscripts r and G denote that the average is taken in the fixed trace and Gaussian ensembles, respectively.

Notice the analogy with the assumption

$$\langle E \rangle_{\text{grand canonical}} = E_{\text{canonical}} , \tag{19.1.8}$$

made in classical statistical mechanics to prove the equivalence there.

From (3.3.8)–(3.3.10) we get, with a little manipulation,

$$K_G = a^{-N/2-\beta N(N-1)/4} C_{N\beta} , \tag{19.1.9}$$

so that a partial differentiation with respect to a gives

$$\left\langle \text{tr } H^2 \right\rangle_G = [N/2 + \beta N(N-1)4] a^{-1} . \tag{19.1.10}$$

Therefore we make the choice

$$a = (2r^2)^{-1} N \left[1 + \beta(N-1)/2\right] . \tag{19.1.11}$$

Next, to calculate $M_r(N, \eta)$, put

$$H_{ij}^{(\lambda)} = a^{1/2} r h_{ij}^{(\lambda)} \xi^{-1/2} , \tag{19.1.12}$$

where ξ is a parameter. This gives

$$\begin{aligned} &M_r(N,\eta) \left(\frac{\xi}{ar^2}\right)^{N/2+\beta N(N-1)/4+s/2} \\ &= K_r^{-1} \int_{-\infty}^{\infty} \cdots \int_{-\infty}^{\infty} \delta\left(\frac{a}{\xi} \text{ tr } h^2 - 1\right) \\ &\times \prod_{\lambda} \prod_{i \le j} \left[\left(h_{ij}^{(\lambda)}\right)^{\eta_{ij}^{(\lambda)}} dh_{ij}^{(\lambda)} \right] . \end{aligned} \tag{19.1.13}$$

Multiplying both sides by $e^{-\xi}$ and integrating (first!) on ξ from 0 to ∞, we get

$$M_r(N,\eta)\Gamma\left(L+s/2\right)L^{-L-s/2}$$

$$= K_r^{-1}\int_{-\infty}^{\infty}\cdots\int_{-\infty}^{\infty} e^{-a\ \mathrm{tr}\ h^2}\prod_{\lambda}\prod_{i\le j}\left[\left(h_{ij}^{(\lambda)}\right)^{\eta_{ij}^{(\lambda)}} dh_{ij}^{(\lambda)}\right], \tag{19.1.14}$$

where we have put

$$L = ar^2 = N/2 + \beta N(N-1)/4, \tag{19.1.15}$$

or

$$M_r(N,\eta) = \frac{L^{L+s/2}}{\Gamma(L+s/2)}\frac{K_G}{K_r}M_G(N,\eta). \tag{19.1.16}$$

Setting $\eta_{ij}^{(\lambda)} = 0$ in the above and using the normalization condition $M_r(N,0) = M_G(N,0) = 1$, we get the ratio of the constants K_G and K_r. Substituting this ratio we then obtain

$$M_r(N,\eta) = \frac{L^{s/2}\Gamma(L)}{\Gamma(L+s/2)}M_G(N,\eta). \tag{19.1.17}$$

As $N \longrightarrow \infty$, $L \longrightarrow \infty$, and we can use Stirling's formula for the gamma functions

$$\Gamma(x) = x^{x-1/2}e^{-x}(2\pi)^{1/2}\left[1+O(1/x)\right]$$

to prove the asymptotic equality of all the finite moments $s \ll N$.

It is not very clear whether this moment equivalence implies that all local statistical properties of the eigenvalues in the two sets of ensembles are identical. This is so because these local properties of eigenvalues may not be expressible only in terms of finite moments of the matrix elements.

Except for the level density (see the next section), which is a semicircle, one does not know how to get, for example, the two-level correlation function for the fixed trace ensemble.

19.2. Bounded Trace Ensemble

Instead of keeping the trace constant, we might require it to be bounded (Bronk, 1964a). We would then obtain a bounded trace ensemble defined by the joint probability density function

$$P_B(H) = \begin{cases} \text{constant}, & \text{if tr } H^2 \leq r^2, \\ 0\ , & \text{if tr } H^2 > r^2. \end{cases} \tag{19.2.1}$$

This probability density is again invariant under a change of basis, Eq. (19.1.3).

The density of eigenvalues for this ensemble can easily be found in the existing literature. A theorem of Stieltjes (1914) states that if there are N unit masses located at the variable points x_1, x_2, ..., x_N in the interval $[-\infty, \infty]$ such that their moment of inertia is bounded by

$$\sum_{i=1}^{N} x_i^2 \leq N(N-1)/2\ , \tag{19.2.2}$$

the unique maximum of the function

$$V(x_1, ..., x_N) = \prod_{1 \leq i < j \leq N} |x_i - x_j|^\beta \tag{19.2.3}$$

will be obtained when the x_i are the zeros of the Hermite polynomial

$$H_N(x) = e^{x^2} \left(-\frac{d}{dx} \right)^N e^{-x^2}. \tag{19.2.4}$$

Thus, making the usual assumption of classical statistical mechanics that the actual eigenvalue density makes the logarithm of $V(x_1, ..., x_N)$ a maximum, we get the following result.

The eigenvalue density for the bounded trace ensemble is identical to the density of zeros of the Hermite-like polynomial

$$e^{N(N-1)x^2/2r^2} \left(-\frac{d}{dx} \right)^N e^{-N(N-1)x^2/2r^2} \tag{19.2.5}$$

and for large N is given by

$$\sigma(x) \cong \frac{N^2}{2\pi r^2}\left(\frac{4r^2}{N} - x^2\right)^{1/2}, \qquad \text{if} \quad |x| < 2rN^{-1/2}$$

$$\cong 0, \qquad \text{if} \quad |x| > 2rN^{-1/2}. \tag{19.2.6}$$

To work out the eigenvalue spacing distribution is much more difficult.

19.3. Matrix Ensembles and Classical Orthogonal Polynomials

One cannot but notice in Chapters 5, 6, and 7, for example, that the Gaussian ensembles are closely related to the Hermite polynomials. Orthogonal polynomials other than the Hermite polynomials have been extensively investigated (Bateman, 1953), and some authors (Fox and Kahn, 1964; Leff, 1963) have tried to take advantage of this fact. We can define a matrix ensemble by giving the joint probability density function for its eigenvalues arbitrarily:

$$P(x_1, ..., x_N) = \prod_{i=1} f(x_i) \prod_{i<j} |x_i - x_j|^\beta, \tag{19.3.1}$$

where the function $f(x)$ can be chosen to suit the needs.

A series of orthogonal polynomials is uniquely defined, apart from a phase factor, by the range (a, b) of the variable and the weight function $f(x) \geq 0$. The construction of these polynomials amounts to an application of Schmidt's orthonormalization procedure to the series of powers 1, x, x^2, ... with the scalar product

$$(\varphi_1, \varphi_2) = \int_a^b f(x)\varphi_1(x)\varphi_2(x)dx, \tag{19.3.2}$$

and gives the following set of polynomials (Bateman, 1953):

$$P_r(x) = (A_r A_{r-1})^{-1/2} \det \begin{bmatrix} c_0 & c_1 & \dots & c_r \\ c_1 & c_2 & \dots & c_{r+1} \\ \dots & \dots & \dots & \dots \\ c_{r-1} & c_r & \dots & c_{2r-1} \\ 1 & x & \dots & x^r \end{bmatrix}, \tag{19.3.3}$$

with

$$A_r = \det\left[c_{i+j}\right]_{i,j=0,1,\ldots,r}, \tag{19.3.4}$$

and

$$c_i = \int_a^b f(x)x^i dx. \tag{19.3.5}$$

The choices $f(x) = \exp\left(-\beta x^2/2\right)$, $-\infty < x < \infty$, and $f(x) = 1$, $x = e^{i\theta}$, $0 \leq \theta \leq 2\pi$ correspond to the Gaussian and the circular ensembles, respectively. The choice

$$f(x) = (1-x)^\mu (1+x)^\nu \ ; \quad \mu, \quad \nu > -1 \ ; \quad -1 \leq x \leq 1 \ , \tag{19.3.6}$$

gives an ensemble corresponding to the Jacobi polynomials. Suitable values of μ, ν will give ensembles corresponding to various classical orthogonal polynomials; for example, $\mu = \nu = 0$ gives the (Legendre) ensemble studied by Leff (1963, 1964a, b) and Vo-Dai and Derome (1975). Another choice studied by Bronk (1964a) and by Fox and Kahn (1964) is

$$f(x) = x^{\alpha-1}e^{-x} \ , \quad \alpha > -1 \ , \quad 0 \leq x < \infty \ . \tag{19.3.7}$$

This arises as follows. As in Section 15.1 consider the ensemble of all $N \times N$ complex matrices S with the joint probability density $P(S) \propto \exp(-\text{tr}\ S^\dagger S)$, i.e., the real and imaginary parts of the matrix elements are independent Gaussian random variables. Any such S can be written as (*cf.* Mehta, 1989, Section 4.8)

$$S = U^\dagger \Lambda V, \tag{19.3.8}$$

where U and V are unitary matrices and Λ is a diagonal matrix with real non-negative elements. The nonzero diagonal elements of Λ are known as the singular values of S; their squares are the (nonzero) eigenvalues of $SS^\dagger$ or of $S^\dagger S$ (they are the same). In some physical applications one may need the distribution of the eigenvalues not of S, but of $SS^\dagger$, the joint probability density of which can be derived from Eq. (19.3.8) to be

$$\prod_i x_i.\exp(-x_i) \prod_{i<j} (x_i - x_j)^2 \ . \tag{19.3.9}$$

This is a special case of Eq. (19.3.7).

Though from the point of view of applications the power $\beta = 2$ is the least interesting, it is mathematically the easiest to handle. The eigenvalue density $\sigma_N(x)$ can be expressed in terms of the related orthonormal polynomials. Thus,

$$\sigma_N(x) = \sum_{j=0}^{N-1} p_j^2(x) , \tag{19.3.10}$$

where $p_j(x)$ is the normalized polynomial corresponding to the weight function $f(x)$ and the interval $[a, b]$. The n-point correlation function

$$R_n(x_1, ..., x_N) = \frac{N!}{(N-n)!} \int \cdots \int P(x_1, ..., x_N)\, dx_{n+1} \cdots dx_N \tag{19.3.11}$$

is given by

$$R_n(x_1, ..., x_N) = \det\left[K_N(x_j, x_k)\right]_{j,k=1,2,...,n} , \tag{19.3.12}$$

where

$$K_N(x, y) = \sum_{j=0}^{N-1} p_j(x) p_j(y) . \tag{19.3.13}$$

The sum $K_N(x, y)$ can be expressed in closed form by using the Christoffel-Darboux formula (Bateman, 1953), but it is not very useful in asymptotic evaluations.

As we said above the Legendre ensembles defined by Eq. (19.3.1) and $f(x) = 1$, $-1 \le x \le 1$, were studied by Vo-Dai and Derome (1975) and later by Mehta (1976). The conclusion is that all the local properties of the eigenvalues are the same as those of the corresponding Gaussian ensembles. The unitary ensembles, $\beta = 2$, Eq. (19.3.1), with $f(x)$ the weight function for other classical orthogonal polynomials, including Laguerre, were studied by Fox and Kahn (1964).

Summary of Chapter 19

Three types of Hermitian (and/or symmetric or self-dual) matrix ensembles are considered.

(1) Fixed trace, in which the trace of H^2 is kept constant with no other restricton on the matrix elements of H. These ensembles are shown to

be equivalent to the Gaussian ensembles as far as finite moments of the matrix elements are concerned.

(2) Bounded trace, in which the trace of H^2 is less than or equal to a given constant. For these ensembles the eigenvalue density is a "semicircle," just as for the Gaussian ensembles.

(3) Ensembles related to the classical orthogonal polynomials. Gaussian ensembles are related to the Hermite polynomials. Instead of the Hermite ones if we choose other classical orthogonal polynomials, we get other ensembles. The eigenvalue density can be calculated in each case and is different for different polynomials. However, the spacing probability density seems not to depend on this choice.

20 / Bordered Matrices

In some physical situations it is instructive to consider an ensemble of matrices that is slightly more difficult than the diagonal ones. Matrices that have their elements on the principal diagonal and a few neighboring super- or subdiagonals distributed at random, while all other matrix elements are zero, are of importance, for example, in the theory where glass is represented as a collection of random nets. It is required to determine the distribution of the characteristic frequencies and of the absolute value square of the characteristic amplitudes of such random nets. The one-dimensional problem of such a linear chain with nearest neighbor interactions was first solved by Dyson (1953). The problem is simple enough to be treated analytically. It corresponds to an ensemble of Hermitian matrices whose only nonzero elements are those that lie in the layers immediately above and immediately below the principal diagonal. Later Wigner (1955, 1957a) treated the case of real symmetric matrices whose diagonal elements were equispaced. Elements in a few layers on each side of this principal diagonal had the same nonzero magnitude with a random sign, whereas all other elements were zero. In this chapter we present briefly some important features of these investigations.

In the physical situation of a disordered linear chain in which each atom interacts with many of its neighbors one has to deal with matrices with many layers of nonzero random elements on both sides of the principal diagonal (Engleman, 1958). Attempts have also been made to treat the two- and the three-dimensional lattices of random oscillators. A nice review containing references to the earlier work is due to Maradudin et al. (1958).

20.1. Random Linear Chain

Consider a chain of N masses, each connected to its immediate neighbors by springs that obey Hooke's law. The masses and the spring constants are random variables with known average characteristics. The problem is to determine the probability density function of the normal frequencies of this chain. This theory applies equally well to an electric transmission line composed of alternating capacitances and inductances with random characteristics. We will be interested in the limit $N \longrightarrow \infty$.

By simple algebraic manipulations it can be shown (Dyson, 1953) that the normal frequencies of such a chain are the eigenvalues of the $(2N-1) \times (2N-1)$ Hermitian matrix with elements

$$H_{j+1,j} = -H_{j,j+1} = i\lambda_j^{1/2}, \quad j = 1, ..., 2N-2, \tag{20.1.1}$$

where the λ_j are given in terms of the masses m_j and the spring constants K_j (connecting m_j and m_{j+1}) by

$$\lambda_{2j-1} = \frac{K_j}{m_j}, \qquad \lambda_{2j} = \frac{K_j}{m_{j+1}}. \tag{20.1.2}$$

All other elements of H are zero.

Because H is antisymmetric having an odd order its determinant is zero and hence one of its eigenvalues vanishes; this corresponds to the degenerate motion in which all the masses have exactly the same displacement. All other eigenvalues occur in pairs ω_j, $-\omega_j$. Let $M(\mu)$ be the distribution function defined as the proportion of the eigenvalues ω_j for which $\omega_j^2 \leq \mu$, so that a probability density function can be defined as

$$D(\mu) = \frac{dM(\mu)}{d\mu}. \tag{20.1.3}$$

It is required to find either $M(\mu)$ or $D(\mu)$ in the limit $N \longrightarrow \infty$ when the distribution of the λ_j is given.

Dyson instead considers

$$\begin{aligned} \Omega(z) &= \lim_{N \to \infty} (2N-1)^{-1} \sum_j \ln\left(1 + z\omega_j^2\right) \\ &= \int_0^\infty \ln(1 + z\mu)\, D(\mu) d\mu \end{aligned} \tag{20.1.4}$$

as a function of the complex variable z. That branch of the logarithm is taken which is real for real positive z. Then the integral (20.1.4) is convergent and defines an analytical function of z for the whole z-plane except for the negative real axis. As z tends from above to a point $-x$ on the negative real axis, the imaginary part of $\ln(1+z\mu)$ tends to zero if $x\mu < 1$ and to π if $x\mu > 1$. Hence Eq. (20.1.4) gives

$$\begin{aligned} \mathrm{Re}\left[\frac{1}{i\pi}\lim_{\epsilon\to 0}\ \Omega(-x+i\epsilon)\right] &= \int_{1/x}^{\infty} D(\mu)d\mu \\ &= 1 - M\left(\frac{1}{x}\right), \end{aligned} \tag{20.1.5}$$

or on differentiating

$$D\left(\frac{1}{x}\right) = -x^2 \mathrm{Re}\left[\frac{1}{i\pi}\lim_{\epsilon\to 0}\ \Omega'(-x+i\epsilon)\right]. \tag{20.1.6}$$

Therefore, once $\Omega'(z)$ is known, $D(\mu)$ is determined by its limiting values on the negative real axis. Most of the time, however, it is not possible to express $\Omega'(z)$ as a closed analytical expression and direct analytical continuation becomes impossible. For use in such cases Dyson derived the formula

$$\begin{aligned} D(\mu) = {} & \left(2\pi^2\mu\right)^{-1}\int_{-\infty}^{\infty} d\alpha(\cosh \pi\alpha) \\ & \times\left\{\int_0^{\infty} dx(x\mu)^{-1}\cos\left[\alpha\ \ln(x\mu)\right]\Omega'(x)\right\} \end{aligned} \tag{20.1.7}$$

to express $D(\mu)$ in terms of the values of $\Omega'(z)$ for the real positive values of z. The details of the derivation will not be given. Thus, even if $\Omega'(z)$ is known only numerically or approximately on the positive real axis, $D(\mu)$ can be evaluated by numerical integration.

Dyson derives an explicit formula for $\Omega(z)$ in terms of the λ_j. The derivation is based on expanding $\ln(1+z\omega_j^2)$ in powers of z, replacing the sums of powers of ω_j^2 by the traces of the even powers of H and counting the terms that give a nonzero contribution to such traces. We will content ourselves by giving only the results and refer the interested reader to the original paper (Dyson, 1953).

For an arbitrary chain with given coefficients λ_j, $\Omega(z)$ is given by

$$\Omega(z) = \lim_{N \longrightarrow \infty} \frac{1}{N} \sum_{a=1}^{2N-1} \ln[1 + \xi(a)] , \tag{20.1.8}$$

where $\xi(a)$ is the continued fraction

$$\xi(a) = \frac{z\lambda_a}{1+} \; \frac{z\lambda_{a+1}}{1+} \; \frac{z\lambda_{a+2}}{1+} \; \cdots . \tag{20.1.9}$$

Various assumptions can now be made as to what extent the λ_j are random and the consequences for the function $\Omega(z)$ can be derived. For example, if all the λ_j were the same, then all the ξ would be the same and

$$\xi = \frac{z\lambda}{1+\xi} ; \qquad \xi \longrightarrow 0, \quad \text{as} \quad z \longrightarrow 0.$$

Therefore,

$$\xi = \frac{1}{2}\left[(1 + 4z\lambda)^{1/2} - 1\right] \tag{20.1.10}$$

and by Eq. (20.1.8)

$$\Omega(z) = 2 \ln \left[\frac{1}{2}(1 + 4z\lambda)^{1/2} + \frac{1}{2}\right] , \tag{20.1.11}$$

On differentiation we get

$$\Omega'(z) = z^{-1}\left[1 - (1 + 4z\lambda)^{-1/2}\right] . \tag{20.1.12}$$

Continuing through the upper half plane to real negative values of $z < -(4\lambda)^{-1}$ we find

$$\Omega'(z) = z^{-1}\left[1 + i(1 + 4z\lambda)^{-1/2}\right] . \tag{20.1.13}$$

Hence Eq. (20.1.6) gives

$$\begin{aligned} D(\mu) &= \frac{1}{\pi}\left(4\lambda\mu - \mu^2\right)^{-1/2} , \qquad \mu < 4\lambda, \\ &= 0 , \qquad \mu > 4\lambda, \end{aligned} \tag{20.1.14}$$

a result that in this simple case can be checked directly.

Two more special cases have been analytically treated by Dyson:

1. All the λ_j are independent random variables with a given probability density function $G(\lambda)$. In this case the $\xi(a)$ also have a probability density function $F(\xi)$, the same for all a. An integral equation for $F(\xi)$ can be derived by equating the probabilities on both sides of the equality

$$\xi(a) = \frac{z\lambda_a}{1+\xi(a+1)}.$$

The kernel of the integral equation contains $G(\lambda)$. Once this integral equation is solved by iteration or otherwise and $F(\xi)$ so obtained is normalized,

$$\int_0^\infty F(\xi)d\xi = 1, \tag{20.1.15}$$

$\Omega(z)$ is given by Eq. (20.1.8):

$$\Omega(z) = 2\int_0^\infty F(\xi)\ln(1+\xi)d\xi\ . \tag{20.1.16}$$

$D(\mu)$ can then be calculated.

2. The λ_j are correlated in the following manner. The spring constants K_j are fixed and equal and the masses m_j are independent random variables with a given probability density function $G(m)$. In this case the variables $\eta_j = [\xi(2j)]^{-1}$ are uncorrelated and an integral equation for their probability density function can be derived from the recurrence formula

$$\eta_j = \frac{m_{j+1}}{zK} + \frac{\eta_{j+1}}{1+\eta_{j+1}} \qquad (K_j = K)\ . \tag{20.1.17}$$

Equation (20.1.17) follows from Eqs. (20.1.9) and (20.1.2). Also from (20.1.9) we have

$$[1+\xi(2j)]\,[1+\xi(2j-1)] = 1+\xi(2j)+\frac{zK}{m_j} \tag{20.1.18}$$

relating the probability density functions for $\xi(2j)$ and $\xi(2j+1)$, which will in general be different. Once these probability density functions are known as solutions of the above equations, the function $\Omega(z)$, and hence $D(\mu)$, can be calculated.

20.2. Bordered Matrices

Let us consider only the real symmetric matrices. The diagonal elements are integers ..., -2, -1, 0, 1, 2, ... The elements H_{jk} for which $|j-k| > m$ are zero, whereas the elements H_{jk} with $|j-k| \le m$ all have the same magnitude h :

$$\begin{aligned} H_{jk} &= \pm h\ , \qquad \text{if} \quad |j-k| \le m\ , \\ &= 0\ , \qquad \text{if} \quad |j-k| > m\ . \end{aligned} \tag{20.2.1}$$

Subject to the symmetry condition $H_{jk} = H_{kj}$, the signs of the H_{jk} are random. Let us denote the eigenvalues and the normalized eigenvectors of H by λ and $\psi^{(\lambda)}$, respectively:

$$H\psi^{(\lambda)} = \lambda\psi^{(\lambda)} \qquad \text{or} \qquad \sum_k H_{jk}\psi_k^{(\lambda)} = \lambda\psi_j^{(\lambda)}\ , \tag{20.2.2}$$

and for a λ lying between x and $x + \delta x$ find the expectation value of

$$\left(\psi_0^{(\lambda)}\right)^2\ , \tag{20.2.3}$$

where $\psi_0^{(\lambda)}$ is a particular component of $\psi^{(\lambda)}$. This expectation value will be written as $\sigma(x)\delta x$, where $\sigma(x)$ is named "the strength function." As the absorption of an energy-level depends, under certain conditions, only on the square of a definite component of the corresponding eigenstate, the function $\sigma(x)$ represents the strength of absorption around the energy value x. The problem is to find $\sigma(x)$ and the distribution of the eigenvalues λ.

When there is a single border, $m = 1$ (i.e., when H_{jk} are zero for $|j-k| > 1$), the problem is simple enough for an explicit evaluation of the eigenvalues and the eigenvectors. Such an H can be transformed by a diagonal matrix S with diagonal elements ± 1 to a matrix H_1 and the signs of the diagonal elements of S can be so chosen that all the off-diagonal elements of H_1 will have the negative sign, whereas the diagonal elements are the same as in H. In doing so neither the eigenvalues nor the squares of the components of the eigenvectors undergo any change.

The resulting matrix H_1 can be transformed to $-H_1$ by interchanging jth and $-j$th row and column and by transforming with an S whose diagonal elements are alternately $+1$ and -1. Furthermore, H_1 can be changed to H_1+1 by renumbering the rows and columns (the dimension of H should be infinite for the argument to apply here). Thus, along with λ_k, $-\lambda_k$ and λ_k+1 are also eigenvalues. By the continuity in h and the condition that for $h=0$ the eigenvalues of H are all integers we see that the eigenvalues of H always consist of all the integers:

$$\lambda_k = k \ , \quad k = 0, \ \pm 1, \ \pm 2, \ \dots \ . \tag{20.2.4}$$

Denoting the corresponding eigenvector by $\psi^{(k)}$, one sees from the above remarks about changing H_1 to H_1+1 that $\psi_\ell^{(k)} = \psi_{\ell+1}^{(k+1)}$, that is, that the ℓth component of $\psi^{(k)}$ depends only on the difference $(\ell - k)$:

$$\psi_\ell^{(k)} = \psi_{\ell-k}^{(0)}. \tag{20.2.5}$$

Again from the remarks about changing H_1 to $-H_1$, it follows that

$$\psi_{-\ell}^{(0)} = c(-1)^\ell \psi_\ell^{(0)} \ ,$$

where $c=\pm 1$. The continuity from $h=0$ again gives $c=1$. The equation $H_1\psi = \lambda\psi$ can now be written as

$$-h\psi_{\ell-1}^{(0)} + \ell\psi_\ell^{(0)} - h\psi_{\ell+1}^{(0)} = 0 \ , \tag{20.2.6}$$

and compared with the recursion formula for the Bessel functions

$$-J_{\ell-1}(z) + \frac{2\ell}{z} J_\ell(z) - J_{\ell+1}(z) = 0 \ . \tag{20.2.7}$$

We infer that

$$\psi_\ell^{(0)} = J_\ell(2h) = \psi_{\ell+k}^{(k)} \ . \tag{20.2.8}$$

The irregular Bessel functions also satisfy Eq. (20.2.7), but if the eigenvector components are taken as a linear combination of the regular and irregular Bessel functions then one cannot normalize.

The case of thick borders, $m \gg 1$, and large off-diagonal elements, $h \gg 1$, has been treated by Wigner under the condition that $h^2/m = q$

remains constant. Some of the remarks that apply for a singly bordered matrix also apply here, and from such considerations it can be deduced that $\sigma(x)$ is an even function. The average number of eigenvalues per unit interval at x is a periodic function of x with the period 1. Wigner then calculates the moments of $\sigma(x)$ and after long calculations derives an integral equation for it, which is difficult to solve except in certain limits.

Summary of Chapter 20

We consider matrices in which all matrix elements farther than a certain distance from the main diagonal are zero. The nonzero elements in and around the diagonal are, apart from symmetry or self-duality or Hermiticity, random numbers. If only one layer near the diagonal is nonzero, a linear chain, then this case is amenable to an analytical treatment. The more interesting case of a larger nonzero ribbon around the diagonal is very difficult.

21 / Invariance Hypothesis and Matrix Element Correlations

The entire theory of Gaussian ensembles is based on the two assumptions put forward in Chapter 2:

1. The ensemble is statistically invariant under a change of basis.
2. The matrix elements are statistically independent, hence uncorrelated

As mentioned in that chapter, Assumption 1 is quite natural, whereas Assumption 2 is somewhat artificially introduced to simplify the calculations.

There have been efforts (Ullah, 1964; Ullah and Porter, 1963) to determine what kind of correlations among the various matrix elements are implied by Assumption 1 alone. We shall see here that it leads to the vanishing of the ensemble averages of the following quantities:

1. Any odd power of an off-diagonal element.
2. The product of an odd power of an off-diagonal element and any power of a diagonal or another off-diagonal element.

What other relations, if any, on the eigenvalues or otherwise, are implied, is not clear.

We illustrate the method in relation to the orthogonal ensemble, though it is equally applicable to the unitary or the symplectic ensemble. Thus we consider a set of real symmetric matrices that is statistically invariant under real orthogonal transformations

$$H = H^* = H^T ,$$

$$P(H)dH = P(H')dH' ,$$

if

$$H' = RHR^{-1} , \quad RR^T = R^T R = 1 .$$

Choosing some complete set of basic functions, the eigenvalue equation can be written as

$$\sum_k H_{jk} a_{ik} = \theta_i a_{ij} \ , \tag{21.0.1}$$

where θ_i is the eigenvalue and a_{ij} is the jth component of the corresponding eigenvector. The eigenvectors are orthogonal and form a complete set:

$$\sum_j a_{ij} a_{kj} = \sum_j a_{ji} a_{jk} = \delta_{ik} \ . \tag{3.1.6}$$

Using (3.1.6) and (21.0.1), we can express the matrix elements as

$$H_{jk} = \sum_i \theta_i a_{ij} a_{ik} \ . \tag{3.1.5}$$

Now because the joint probability density function $P(H)$ is invariant under orthogonal transformations of the basis, it must depend essentially only on the eigenvalues θ_i and that, too, in a symmetric manner. The discussion leading to Eq. (3.1.16) is valid, and the joint probability density function can be written as a product of functions depending on mutually exclusive sets of variables:

$$P(\theta_i, p) = P(\theta_i) f(p), \tag{21.0.2}$$

the parameters p depending on the eigenvectors. We are interested in the averages of products of the matrix elements given by Eq. (3.1.5) above. Because of the separable nature of Eq. (21.0.2), this averaging can be done separately over the θ_i and the parameters p to get the result

$$\langle H_{jk} H_{\ell m} \cdots \rangle = \sum_{i_1, i_2, \ldots = 1}^{N} \langle \theta_{i_1} \theta_{i_2} \cdots \rangle \langle a_{i_1 j} a_{i_1 k} a_{i_2 \ell} a_{i_2 m} \cdots \rangle \ . \tag{21.0.3}$$

21.1. Random Orthonormal Vectors

Thus we have to find the averages of the products of components of a set of orthogonal unit vectors randomly oriented in the N-dimensional space. These averages may be written as the ratio of two integrals:

$$\langle Q\left(a_{i_1}, a_{i_2} \ldots\right) \rangle = \frac{\mathcal{N}}{\mathcal{D}} \ , \tag{21.1.1}$$

where

$$\mathcal{N} = \int_{-\infty}^{\infty} \int_{-\infty}^{\infty} Q\,\delta\left(\sum_{j=1}^{N} a_{i_1 j}^2 - 1\right)\delta\left(\sum_{j=1}^{N} a_{i_1 j} a_{i_2 j}\right)$$
$$\times\,\delta\left(\sum_{j=1}^{N} a_{i_2 j}^2 - 1\right)\cdots\prod_{j=1}^{N} da_{i_1 j}\, da_{i_2 j}\cdots \tag{21.1.2}$$

and $\mathcal{D}$ is the same integral without the Q inside.

The integral $\mathcal{N}$ can be evaluated when it involves only one or two vectors, whereas its evaluation in general is not at all easy. However, from the symmetry arguments we can conclude that

$$\left\langle a_{i_1 j} a_{i_1 k} \cdots a_{i_{2m+1} j} a_{i_{2m+1} k}\right\rangle = 0\,, \quad j \neq k, \tag{21.1.3}$$

implying that $\left\langle H_{jk}^{2m+1}\right\rangle = 0$. By similar reasonings we can convince ourselves that

$$\left\langle H_{jk}^{2m+1} H_{\ell m}^{r}\right\rangle = 0 \tag{21.1.4}$$

if $j \neq k$ and the pair (jk) is distinct from the pair (ℓm) or $(m\ell)$.

We can actually evaluate the ratio (21.1.1)–(21.1.3) for a simple expression Q, involving up to two matrix elements (Ullah, 1964). Let us first take only one N-dimensional unit vector with random components $u_1, u_2, ..., u_N$. Equation (21.1.1) then gives

$$\langle Q(u)\rangle = \frac{\int_{-\infty}^{\infty}\cdots\int_{-\infty}^{\infty} Q(u)\delta\left(\sum_1^N u_i^2 - 1\right)\prod_1^N du_i}{\int_{-\infty}^{\infty}\cdots\int_{-\infty}^{\infty} \delta\left(\sum_1^N u_i^2 - 1\right)\prod_1^N du_i}. \tag{21.1.5}$$

For Q we substitute $\prod_{i=1}^N u_i^{2m_i}$ and evaluate the integral

$$I = \int_{-\infty}^{\infty}\cdots\int_{-\infty}^{\infty} \delta\left(\sum_1^N u_i^2 - 1\right)\prod_1^N u_i^{2m_i}\, du_i\,. \tag{21.1.6}$$

Replacement of u_i with $u_i/\sqrt{r}$ gives

$$I r^{N/2+\Sigma_i m_i - 1} = \int_{-\infty}^{\infty}\cdots\int_{-\infty}^{\infty} \delta\left(\sum_1^N u_i^2 - r\right)\prod_1^N u_i^{2m_i}\, du_i\,. \tag{21.1.7}$$

Multiplying on both sides with e^{-r} and integrating (first!) over r from 0 to ∞, and then over the u_i, we get

$$\begin{aligned} I\Gamma\left(N/2+\sum_i m_i\right) &= \int_{-\infty}^{\infty}\cdots\int_{-\infty}^{\infty}\exp\left(-\sum_i u_i^2\right)\prod_i u_i^{2m_i}du_i \\ &= \prod_{i=1}^{N}\left(\int_{-\infty}^{\infty}e^{-u^2}u^{2m_i}du\right) \\ &= \prod_{i=1}^{N}\Gamma\left(m_i+1/2\right)\ . \end{aligned} \tag{21.1.8}$$

To get the denominator we put $m_i = 0$, $i = 1, 2, ..., N$. Thus the average of $\prod_{i=1}^{N} u_i^{2m_i}$ is

$$\left\langle\prod_{i=1}^{N}u^{2m_i}\right\rangle = \frac{\prod_{i=1}^{N}\Gamma(m_i+1/2)}{\Gamma\left(N/2+\sum_{i=1}^{N}m_i\right)}\frac{\Gamma(N/2)}{[\Gamma(1/2)]^N}\ . \tag{21.1.9}$$

Extending this method to two vectors $u = (u_1, ..., u_N)$ and $v = (v_1, ..., v_N)$, we can write

$$\begin{aligned} Q(u,v) &= \frac{\int_{-\infty}^{\infty}\cdots\int_{-\infty}^{\infty}Q\delta\left(\sum_i u_i^2-1\right)\delta\left(\sum_i u_iv_i\right)\delta\left(\sum_i v_i^2-1\right)\prod_i du_idv_i}{\int_{-\infty}^{\infty}\cdots\int_{-\infty}^{\infty}\delta\left(\sum_i u_i^2-1\right)\delta\left(\sum_i u_iv_i\right)\delta\left(\sum_i v_i^2-1\right)\prod_i du_idv_i} \\ &= \frac{\mathcal{N}}{\mathcal{D}}\ . \end{aligned} \tag{21.1.10}$$

Let us calculate the denominator. The evaluation of the numerator is similar, though somewhat lengthy. Replacing u_i with $u_i/\sqrt{r_1}$ and v_i with $v_i/\sqrt{r_2}$, we obtain

$$\begin{aligned} \mathcal{D}(r_1r_2)^{(N-3)/2} = \int_{-\infty}^{\infty}\cdots\int_{-\infty}^{\infty}\delta\left(\sum_i u_i^2-r_1\right)\delta\left(\sum_i u_iv_i\right) \\ \times\,\delta\left(\sum_i v_i^2-r_2\right)\prod_i du_idv_i. \end{aligned} \tag{21.1.11}$$

Multiplying by $e^{-r_1}e^{-r_2}$ and integrating on r_1, r_2 from 0 to ∞ as before, we now have

$$\mathcal{D}\left[\Gamma\left((N-1)/2\right)\right]^2 = \int_{-\infty}^{\infty} \cdots \int_{-\infty}^{\infty} \exp\left(-\sum_i \left(u_i^2 + v_i^2\right)\right) \times \delta\left(\sum_i u_i v_i\right) \prod_i du_i dv_i \ . \tag{21.1.12}$$

We substitute

$$u_i = \frac{1}{\sqrt{2}}\left(p_i + q_i\right) \ , \quad v_i = \frac{1}{\sqrt{2}}\left(p_i - q_i\right)$$

to get

$$\mathcal{D}\left[\Gamma\left((N-1)/2\right)\right]^2 = 2\int_{-\infty}^{\infty} \ldots \int_{-\infty}^{\infty} \exp\left[-\sum_i \left(p_i^2 + q_i^2\right)\right] \times \delta\left[\sum_i \left(p_i^2 - q_i^2\right)\right] \prod_i dp_i dq_i \ . \tag{21.1.13}$$

Introducing the spherical polar coordinates and integrating over the angles the right-hand side gives

$$2(NV_N)^2 \int_0^{\infty} \int_0^{\infty} e^{-(p^2+q^2)} \delta\left(p^2 - q^2\right) (pq)^{N-1} dp dq \ , \tag{21.1.14}$$

where

$$V_N = \pi^{N/2}\left[\Gamma\left(1 + N/2\right)\right]^{-1} = (2/N)\pi^{N/2}\left[\Gamma(N/2)\right]^{-1} \tag{21.1.15}$$

is the volume of the N-dimensional unit sphere. The remaining integrals are elementary. We finally get

$$\begin{aligned} \mathcal{D} &= (NV_N)^2\Gamma(N-1)(1/2)^N\left[\Gamma((N-1)/2)\right]^{-2} \\ &= 2^{N-2}\pi^{N-1}\left[\Gamma(N-1)\right]^{-1} \ . \end{aligned} \tag{21.1.16}$$

After lengthy algebra, Ullah obtained a complicated expression for the average of $\prod_i u_i^{2m_i} v_i^{2n_i}$. To get further results by this method for three or more vectors seems to be extremely difficult.

Such calculations may be used to compute some simple correlations; for example the correlation coefficient of two diagonal elements (Ullah and Porter, 1963)

$$C_{jk} = \frac{\langle \delta H_j \delta H_k \rangle}{\left[\langle (\delta H_j)^2 \rangle \langle (\delta H_k)^2 \rangle\right]^{1/2}} , \quad \delta H_j = H_{jj} - \langle H_{jj} \rangle , \qquad (21.1.17)$$

can be obtained from the averages

$$\langle H_{jj} \rangle = \langle \theta_1 \rangle , \quad \langle H_{jj}^2 \rangle = \frac{1}{N+2} \left[3 \langle \theta_1^2 \rangle + (N-1) \langle \theta_1 \theta_2 \rangle\right] ,$$

$$\langle H_{jj} H_{kk} \rangle = \frac{1}{N+2} \left[\langle \theta_1 \rangle^2 + (N+1) \langle \theta_1 \theta_2 \rangle\right] , \quad j \neq k , \qquad (21.1.18)$$

so that

$$C_{jk} = \frac{1 + (N+1)C}{3 + (N-1)C} , \qquad (21.1.19)$$

where

$$C = \frac{\langle \theta_1 \theta_2 \rangle}{\left[\langle \theta_1^2 \rangle \langle \theta_2^2 \rangle\right]^{1/2}} . \qquad (21.1.20)$$

To get information on the level density or the eigenvalue correlations is very difficult, if not impossible.

Summary of Chapter 21

If an ensemble of matrices is invariant under a change of basis and no other restriction, then the ensemble average of an odd power of any off-diagonal element is zero. A method is given to calculate the average of a product of components of one unit vector or of two orthogonal unit vectors randomly oriented in an n-dimensional space.

Appendices

A.1. Numerical Evidence in Favor of Conjectures 1.2.1 and 1.2.2

In Chapter 1.2 we referred to the extended numerical experience about random matrices. Here are some details about what we know.

The first large scale use of a computer to study random matrices is due to Porter and Rosenzweig (1960a). In the late fifties they generated and diagonalized a few tens of thousands of real symmetric matrices with random elements. In particular, they studied the following probability densities for the matrix elements H_{ij}, $i \leq j$:

$$(1) \qquad P(H_{ij}) = \begin{cases} \dfrac{1}{2}, & \text{if } |H_{ij}| \leq 1, \\ 0, & \text{otherwise;} \end{cases}$$

$$(2) \qquad P(H_{ij}) = \frac{1}{2}[\delta(H_{ij}+1) + \delta(H_{ij}-1)],$$

$$(3) \qquad P(H_{ij}) = (2\pi)^{-1/2} \exp\left(-H_{ij}^2/2\right).$$

They found that (for $N \times N$ matrices) the eigenvalue density always converged to the "semicircle" law

$$\begin{aligned} \sigma(x) &= (1/\pi)\sqrt{2N - x^2}, \qquad \text{if } \left|x^2\right| \leq 2N, \\ &= 0, \qquad \text{otherwise,} \end{aligned} \tag{A.1.1}$$

and the spacing probability density for consecutive eigenvalues converged always to the same curve which is nearly the "Wigner surmise"

$$p_W(s) = (\pi s/2) \exp\left(-\pi s^2/4\right), \qquad s = S/\langle S \rangle. \tag{A.1.2}$$

The convergence was rapid, it was reached for matrices of order $\approx 20 \times 20$.

The Gaussian probability density, case (3) above, received later an analytical treatment, which covers a large part of this book. This is yet the only case where we know for certain all correlation and cluster functions.

Brody, French, and coworkers in the sixties generated matrices of a different random nature. They considered a finite number of shell model states (in nuclear physics), which were partially filled with nucleons. The two nucleon forces were taken random having some prescribed probability density. The shell model Hamiltonians so generated were thus matrices the elements of which depended on a small number of random variables; most of the matrix elements were *not* statistically independent. Diagonalizing a large number of such matrices, they found that the eigenvalue density is *not* a "semicircle," but a Gaussian centered at the origin. However, the surprising thing is that the spacing probability density for consecutive eigenvalues tends always to the *same* curve which is nearly the "Wigner surmise," Eq. (A.1.2).

Similar computations have since been carried out for quantum chaotic systems, billiards of various shapes and boundary conditions. The eigenvalue density is again something else. However, the spacing probability density for consecutive eigenvalues in the absence of the magnetic field is the same old curve approximated by the "Wigner sumise," Eq. (A.1.2); in the presence of a strong magnetic field this curve is different and corresponds to that derived for the Gaussian unitary ensemble, Chapter 5.

Thus empirical evidence tells us that if the matrices are real symmetric and their $N(N+1)/2$ linearly independent elements depend on independent random variables almost as numerous, then the eigenvalue density tends to the "semicircle" law, Eq. (A.1.1), as the order $N \longrightarrow \infty$. If the matrix elements depend on a smaller number, say about N, of independent random variables, then the eigenvalue density is something else and *not* a "semicircle." But the spacing probability density for consecutive eigenvalues is always the same for large N. Chaotic systems in a magnetic field correspond to matrices which are not real, and have a different spacing probability density, that of the Gaussian unitary ensemble.

A.2. The Probability Density of the Spacings Resulting from a Random Superposition of n Unrelated Sequences of Energy Levels

In Section 1.4 we said that for a simple sequence of levels (i.e., levels having the same spin and parity) the probability density for the nearest neighbor spacings is nicely approximated by the Wigner surmise. For a mixed sequence this probability density results from a random superposition of its constituent simple sequences as follows.

Let ρ_i be the level density in the ith sequence and $p_i(\rho_i S)\,\rho_i dS$, the probability that a spacing in the ith sequence will have a value between S and $S + \mathrm{d}S$. Because $p_i(x)$ is normalized and the level density is the inverse of the mean spacing, we have

$$\int_0^\infty p_i(x)dx = 1, \qquad \int_0^\infty x\, p_i(x)dx = 1. \tag{A.2.1}$$

Let $\Psi_i(x)$ and $E_i(x)$ be defined by

$$\Psi_i(x) = \int_0^x p_i(y)dy = 1 - \int_x^\infty p_i(y)dy = 1 - \int_0^\infty p_i(x+y)dy \tag{A.2.2}$$

and

$$\begin{aligned} E_i(x) &= \int_x^\infty [1 - \Psi_i(y)]\,dy = \int_x^\infty \left[\int_0^\infty p_i(y+z)dz\right] dy \\ &= \int_0^\infty \int_0^\infty p_i\,(x+y+z)\,dydz, \end{aligned} \tag{A.2.3}$$

so that $\Psi_i(\rho_i S)$ is the probability that a spacing in the ith sequence is less than or equal to S and $E_i(\rho_i S)$ is the probability that a given interval of length S will not contain any of the levels belonging to the sequence i.

Consider the system resulting from the superposition of n sequences. The total density is

$$\rho = \sum_i \rho_i. \tag{A.2.4}$$

Let $P(\rho S)\rho\, \mathrm{d}S$ be the probability that a spacing will lie between S and $S + \mathrm{d}S$. Analogously to (A.2.2) and (A.2.3), we introduce the functions $\Psi(x)$ and $E(x)$ by

$$\Psi(x) = \int_0^x P(y)dy = 1 - \int_0^\infty P(x+y)dy \tag{A.2.5}$$

and

$$E(x) = \int_x^\infty [1 - \Psi(y)]\, dy = \int_0^\infty \int_0^\infty P(x+y+z)dy\, dz. \tag{A.2.6}$$

From the observation that $E(\rho S)$ is the probability that a given interval of length S will not contain any of the levels and the randomness of the superposition we have

$$E(\rho S) = \prod_i E_i\left(\rho_i S\right). \tag{A.2.7}$$

Introducing the fractional densities

$$f_i = \frac{\rho_i}{\rho}, \qquad \sum_i f_i = 1 \tag{A.2.8}$$

and the variable $x = \rho S$, we have

$$E(x) = \prod_i E_i\left(f_i x\right). \tag{A.2.9}$$

By differentiating (A.2.9) twice, we obtain

$$P(x) = \frac{d^2 E}{dx^2} = E(x)\left\{ \sum_i f_i^2 \frac{p_i\left(f_i x\right)}{E_i\left(f_i x\right)} + \left[\sum_i f_i \frac{1 - \Psi_i\left(f_i x\right)}{E_i\left(f_i x\right)}\right]^2 - \sum_i \left(f_i \frac{1 - \Psi_i\left(f_i x\right)}{E_i\left(f_i x\right)}\right)^2 \right\}, \tag{A.2.10}$$

which was the purpose of this appendix.

We now consider three special cases.

1. Let the levels in each of the sequences be independent of one another so that $p_i(x) = e^{-x}$. In this case

$$1 - \Psi_i(x) = E_i(x) = e^{-x} \tag{A.2.11}$$

and (A.2.10) yields $P(x) = e^{-x}$, which verifies the obvious fact that the random superposition of sequences of independent random levels produces a sequence of independent random levels.

2. Let all fractional densities be equal to $1/n$ and take the limit as n goes to ∞. Let $x = ny$ so that

$$P(ny) = [E(y)]^n \left\{ \frac{1}{n} \frac{p(y)}{E(y)} + \left(1 - \frac{1}{n}\right) \left[\frac{1 - \Psi(y)}{E(y)}\right]^2 \right\}. \tag{A.2.12}$$

From (A.2.1), (A.2.2), and (A.2.3) we have

$$\Psi(0) = 0, \qquad E(0) = 1, \qquad E'(0) = -1. \tag{A.2.13}$$

Therefore, taking the limit as $n \longrightarrow \infty$, $y \longrightarrow 0$, whereas $ny = x$ is fixed, we see that the terms in the square brackets tend to 1, while from

$$E(y) \approx E(0) + yE'(0) + \cdots = 1 - y + \cdots ;$$

keeping only the first term,

$$[E(y)]^n \approx (1 - y)^n = \left(1 - \frac{x}{n}\right)^n \longrightarrow e^{-x}. \tag{A.2.14}$$

This is a verification of the heuristic reasoning that if the number n of the sequences to be superimposed is large, a level belonging to a sequence will almost certainly be followed by a level of another sequence and these two levels will be independent, whatever $p(x)$ may be.

3. For the "Wigner surmise"

$$p(x) = \frac{\pi}{2} x \exp\left(-\frac{\pi}{4} x^2\right) \tag{A.2.15}$$

we have

$$1 - \Psi(x) = \exp\left(-\frac{\pi}{4} x^2\right), \quad E(x) = 1 - \frac{2}{\sqrt{\pi}} \int_0^{x\sqrt{\pi}/2} e^{-y^2} dy. \tag{A.2.16}$$

Because $p(0) = 0$, we have

$$P(0) = 1 - \sum_{i=1}^{n} f_i^2 \tag{A.2.17}$$

and, in particular, $P(0) \neq 0$.

4. For the Gaussian ensembles the functions $\Psi(x)$ and $E(x)$ are tabulated in Appendix A.15.

A.3. Some Properties of Hermitian, Unitary, Symmetric, or Self-Dual Matrices

For completeness we collect here a few properties of matrices which we need in Chapters 2, 3, and 9. A proof of them can be found in any standard text on matrix theory; for example, Mehta (1989).

A.3.1. A matrix S can be diagonalized by a unitary matrix (i.e., given S, one can find a U with $U^\dagger S U = D$, U unitary, D diagonal), if and only if $SS^\dagger = S^\dagger S$.

Thus Hermitian or unitary matrices can be diagonalized by a unitary matrix. The diagonal elements of D are the eigenvalues of S; they are real for Hermitian S and of the form $\exp(i\theta_j)$, θ_j real, for unitary S.

A.3.2. Any unitary matrix S can be written as $S = \exp(iH)$, where H is Hermitian. Moreover, if S is symmetric, then H is symmetric; if S is self-dual, then H is self-dual.

For if $S = U^\dagger \exp(i\theta) U$, with U unitary and θ diagonal and real, then with $H = U^\dagger \theta U$ one has $S = \exp(iH)$. The symmetry or self-duality of S evidently implies that of H.

A.3.3. Any unitary symmetric matrix S can be written, in many ways, as $S = UU^T$, where U is unitary. Any unitary self-dual matrix S can be written, in many ways, as $S = UU^R$, where U is unitary. The transpose (resp. dual) of U is denoted by U^T (resp. U^R).

For if $S = \exp(iH)$, then write $U = \exp(iH/2)R$, where R is any real orthogonal (resp. quaternion real symplectic) matrix; for example, $R = I$, the unit matrix.

A.3.4. A symmetric Hermitian or a symmetric unitary matrix can be diagonalized by a real orthogonal matrix. A self-dual Hermitian or a self-dual unitary matrix can be diagonalized by a quaternion real symplectic matrix.

For if $A = U^{\dagger}DU = A^T$ with U unitary and D diagonal, then $U = U^*$, and U is also real, i.e., U is real orthogonal. Similarly, if $A = U^{\dagger}DU = A^R$, with U unitary and D diagonal, then U is quaternion real, i.e., U is quaternion real symplectic.

A.3.5. Any antisymmetric Hermitian or antisymmetric unitary matrix S can be reduced to the block diagonal form with a real orthogonal matrix; i.e., given an antisymmetric Hermitian or antisymmetric unitary matrix S, one can find an R with $R^T SR = D$, R real orthogonal, and D block diagonal, $R^T R = 1$, $D_{2j-1,2j} = -D_{2j,2j-1} = \mu_j$, all other elements of D being zero. When S is Hermitian, μ_j are real and when S is unitary, μ_j have the form $\exp(i\theta_j)$.

A.3.6. Any anti-self-dual Hermitian or anti-self-dual unitary matrix can be reduced to the block diagonal form with a quaternion real symplectic matrix.

A.4. Counting the Dimensions of $T_{\beta G}$ and $T'_{\beta G}$ (Chapter 3) and of $T_{\beta C}$ and $T'_{\beta C}$ (Chapter 8)

When we require that two of the eigenvalues be equal, we drop a number of parameters needed to specify a certain two-dimensional subspace, the subspace of these two equal eigenvalues. However, this degenerate eigenvalue is itself one real parameter. Thus, if $f(N, \beta)$ is the number of independent real parameters needed to specify a particular matrix from the ensemble $E_{\beta G}$, the number needed to specify a matrix from the ensemble $E_{\beta G}$ with two equal eigenvalues is

$$f(N, \beta) - f(2, \beta) + 1. \tag{A.4.1}$$

In other words if the number of dimensions of the space $T_{\beta G}$ is $f(N, \beta)$, that of the space $T'_{\beta G}$ is $f(N, \beta) - f(2, \beta) + 1$.

Now to specify a matrix from any of the ensembles $E_{\beta G}$ we need specify only the matrix elements H_{ij} with $i \le j$. The diagonal elements are real and therefore require N real parameters for their specification. The off-diagonal elements H_{ij} with $i < j$ are $N(N-1)/2$ in number and they need $N(N-1)\beta/2$ real parameters. Thus,

$$f(N, \beta) = N + \frac{1}{2}N(N-1)\beta. \tag{A.4.2}$$

By inserting $\beta = 1, 2$, or 4 into (A.4.2) and (A.4.1) we get the dimensions of $T_{\beta G}$ and $T'_{\beta G}$.

To count the dimensions of the spaces $T_{\beta C}$ and $T'_{\beta C}$ we must find the corresponding numbers $f(N, \beta)$, and for this purpose it is sufficient to consider matrices in the neighborhood of unity. Let us then have

$$S = 1 + iA,$$

where A is infinitesimal. Since S is unitary,

$$S^\dagger S \equiv \left(1 - iA^\dagger\right)(1 + iA) = 1$$

or, up to terms linear in A,

$$A = A^\dagger; \tag{A.4.3}$$

A is then Hermitian. If, in addition, S is symmetric (self-dual), then A is symmetric (self-dual). Thus the number of independent real parameters needed to specify a symmetric unitary, self-dual unitary, or unitary matrix S is the same as that needed to specify a symmetric Hermitian, self-dual Hermitian, or Hermitian matrix A, respectively. Thus the dimensions of $T_{\beta C}$ and $T_{\beta G}$ are equal, and hence also those of $T'_{\beta C}$ and $T'_{\beta G}$.

A.5. An Integral over the Unitary Group, Equation (3.5.1) or Equation (14.3.1)

We want to show that

$$\int \exp\left(-\frac{1}{2t}\operatorname{tr}\left(A - UBU^{-1}\right)^2\right) dU$$

$$= c\left(\Delta(\vec{a})\Delta(\vec{b})\right)^{-1} \det\left[\exp\left(-\frac{1}{2t}\left(a_j - b_k\right)^2\right)\right], \tag{A.5.1}$$

where the integral is taken over the group of $n \times n$ unitary matrices U. Here A is an $n \times n$ Hermitian matrix with eigenvalues $a_1, \ldots, a_n$ and $\Delta(\vec{a})$ is the product of their differences, $\Delta(\vec{a}) = \prod_{1 \le j < k \le n} (a_j - a_k)$; similarly the Hermitian matrix B has eigenvalues $b_1, \ldots, b_n$ and $\Delta(\vec{b})$ is the product of their differences. The determinant on the right-hand side is $n \times n$ with $j, k = 1, ..., n$; and the constant c,

$$c = t^{n(n-1)/2} \prod_{j=1}^{n} j!,$$

is neither needed nor evaluated here.

Instead of the above we will derive the equivalent formula

$$\begin{aligned}
&\int \exp\left(-\frac{1}{2t}\operatorname{tr}(A-B)^2\right) f(\vec{b})dB \\
&= \int \exp\left(-\frac{1}{2t}\operatorname{tr}(A-B)^2\right) f(\vec{b})\Delta^2(\vec{b})d\vec{b}dU \\
&= c'n! \int \exp\left(-\frac{1}{2t}\sum_{j=1}^{n}(a_j-b_j)^2\right) f(\vec{b})\frac{\Delta(\vec{b})}{\Delta(\vec{a})}d\vec{b} \qquad \text{(A.5.2)} \\
&= c' \int \left(\Delta(\vec{a})\Delta(\vec{b})\right)^{-1} \det\left[\exp\left(-\frac{1}{2t}(a_j-b_k)^2\right)\right] \\
&\quad \times f(\vec{b})\Delta^2(\vec{b})d\vec{b}, \qquad \text{(A.5.3)}
\end{aligned}$$

valid for an arbitrary symmetric function $f(\vec{b})$ of the eigenvalues of B. Here

$$dB = \prod_j dB_{jj} \prod_{j<k} d\left(\text{Re } B_{jk}\right) d\left(\text{Im } B_{jk}\right), \qquad \text{(A.5.4)}$$

$$d\vec{b} = \prod_{j=1}^{n} db_j, \qquad \text{(A.5.5)}$$

all the integrals are taken from $-\infty$ to ∞, and c' is another constant. The second equality (A.5.3) is easier since under the integral sign one can replace

$$n!\ \Delta(b) \exp\left(-\frac{1}{2t}\sum_j (a_j-b_j)^2\right)$$

by the sum over all permutations P of the indices 1, 2, ..., n, as

$$\Delta(b)\sum_P(-1)^P \exp\left(-\frac{1}{2t}\sum_j (a_j - b_{Pj})^2\right)$$

$$= \Delta(b) \det\left[\exp\left(-\frac{1}{2t}\sum_j (a_j - b_k)^2\right)\right]_{j,k=1,\ldots,n}$$

To see the first equality (A.5.2), let us recall that the partial differential equation (PDE)

$$\frac{\partial F}{\partial t} = \frac{1}{2}\sum_j D_j \frac{\partial^2 F}{\partial x_j^2} \tag{A.5.6}$$

with the initial condition $F(\vec{x};0) = f(\vec{x})$, $\vec{x} \equiv (x_1, ...x_n)$, has the unique solution (*cf.* Morse and Feshbach, 1953)

$$F(\vec{x};t) = \int K(\vec{x},\vec{y};t)f(\vec{y})d\vec{y}, \quad d\vec{y} = dy_1 \cdots dy_n, \tag{A.5.7}$$

where

$$K(\vec{x},\vec{y};t) = \left(\prod_i (2\pi D_i t)^{-1/2}\right) \exp\left(-\sum_j (x_j - y_j)^2 /2D_j t\right). \tag{A.5.8}$$

If we take

$$K(A,B;t) = 2^{-n/2}(\pi t)^{-n^2/2}\exp[-\mathrm{tr}(A-B)^2/2t], \tag{A.5.9}$$

$$\mathrm{tr}(A-B)^2 = \sum_j (A_{jj} - B_{jj})^2 + 2\sum_{j<k}\left((\mathrm{Re}\ A_{jk} - \mathrm{Re}\ B_{jk})^2 + (\mathrm{Im}\ A_{jk} - \mathrm{Im}\ B_{jk})^2\right), \tag{A.5.10}$$

then

$$F(A;t) = \int K(A,B;t) f(B) dB, \tag{A.5.11}$$

satisfies the PDE.

$$\frac{\partial F}{\partial t} = \frac{1}{2} \nabla_A^2 F, \tag{A.5.12}$$

$$\nabla_A^2 = \sum_j \frac{\partial^2}{\partial A_{jj}^2} + \frac{1}{2} \sum_{j<k} \left(\frac{\partial^2}{\partial \left(\text{Re } A_{jk}\right)^2} + \frac{\partial^2}{\partial \left(\text{Im } A_{jk}\right)^2} \right), \tag{A.5.13}$$

with the initial conditon $F(A;0) = f(A)$.

Hermitian matrices A and B can be diagonalized by unitary matrices,

$$A = U_A a U_A^{-1}, \quad B = U_B b U_B^{-1}, \quad a = [a_j \delta_{jk}], \quad b = [b_j \delta_{jk}].$$

Changing variables as in Section 3.3, we get (*cf.* Morse and Feshbach, 1953, Chapter 1 end)

$$\nabla_A^2 = (\Delta(\vec{a}))^{-2} \sum_j \frac{\partial}{\partial a_j} [\Delta(\vec{a})]^2 \frac{\partial}{\partial a_j} + \nabla_{U_A}^2.$$

Now

$$\begin{aligned}
&\frac{1}{\Delta(\vec{a})} \sum_j \frac{\partial^2}{\partial a_j^2} \Delta(\vec{a}) f - (\Delta(\vec{a}))^{-2} \sum_j \frac{\partial}{\partial a_j} \left(\Delta(\vec{a})\right)^2 \frac{\partial f}{\partial a_j} \\
&= \frac{1}{\Delta(\vec{a})} \sum_j \frac{\partial}{\partial a_j} \left(\Delta(\vec{a}) \sum_k{}' \left(a_j - a_k\right)^{-1} \right) f \\
&= \sum_j \left(\left(\sum_k{}' \frac{1}{a_j - a_k} \right)^2 - \sum_k{}' \left(\frac{1}{a_j - a_k} \right)^2 \right) f \\
&= \sum_j \sum_{k \neq p}{}' \left(a_j - a_k\right)^{-1} \left(a_j - a_p\right)^{-1} f,
\end{aligned}$$

where a prime on the summation means that the sum does not contain any singular term. Now the last sum is equal to $P(\vec{a}) f / \Delta(\vec{a})$, where $P(\vec{a})$ is some polynomial in $a_1, ..., a_n$ of degree at least two less than that of $\Delta(\vec{a})$. But $P(\vec{a})/\Delta(\vec{a})$ is symmetric in $a_1, ..., a_n$, so that $P(\vec{a})$ is antisymmetric in $a_1, ..., a_n$ and hence divisible by $\Delta(\vec{a})$. The conclusion is that $P(\vec{a})$ is identically zero. Hence,

$$\nabla_A^2 = (\Delta(\vec{a}))^{-1} \sum_j \frac{\partial^2}{\partial a_j^2} \Delta(\vec{a}) + \nabla_{U_A}^2, \tag{A.5.14}$$

$$F(\vec{a}, U_A; t) = \text{const} \times \int \exp\left(-\frac{1}{2t} \operatorname{tr}\left(a - UBU^{-1}\right)^2\right) \times f\left(\vec{b}, U_B\right) \Delta^2(\vec{b}) d\vec{b} dU_B, \tag{A.5.15}$$

where $U = U_A^{-1} U_B$, and $\Delta^2(\vec{b})$ comes from the Jacobian (*cf.* Section 3.3). If $f(\vec{b}, U_B) = f(\vec{b})$ is independent of U_B, then so is $F(A)$ of U_A, and

$$F(\vec{a}; t) = \text{const} \times \int \exp\left(-\frac{1}{2t} \operatorname{tr}\left(a - UBU^{-1}\right)^2\right) f(\vec{b}) \Delta^2(\vec{b}) d\vec{b} dU, \tag{A.5.16}$$

$$F(\vec{a}; 0) = f(\vec{a}), \tag{A.5.17}$$

satisfies the PDE.

$$\frac{\partial}{\partial t}[\Delta(\vec{a}) F(\vec{a}; t)] = \frac{1}{2} \sum_j \frac{\partial^2}{\partial a_j^2} [\Delta(\vec{a}) F(\vec{a}; t)], \tag{A.5.18}$$

with the initial condition $\Delta(\vec{a}) F(\vec{a}; 0) = \Delta(\vec{a}) f(\vec{a})$. Therefore, $\Delta(\vec{a}) F(\vec{a}; t)$ is given by

$$\Delta(\vec{a}) F(\vec{a}; t) = (2\pi t)^{-n/2} \int \exp\left(-\frac{1}{2t} \sum_{j=1}^n (a_j - b_j)^2\right) \Delta(\vec{b}) f(\vec{b}) d\vec{b}. \tag{A.5.19}$$

Comparing (A.5.16) and (A.5.19) we get

$$\int \exp\left(-\frac{1}{2t} \operatorname{tr}\left(a - UBU^{-1}\right)^2\right) f(\vec{b}) \Delta^2(\vec{b}) d\vec{b} dU = \text{const} \times \int \exp\left(-\frac{1}{2t} \sum_j (a_j - b_j)^2\right) \frac{\Delta(\vec{b})}{\Delta(\vec{a})} f(\vec{b}) d\vec{b} \tag{A.5.20}$$

which is Eq. (A.5.2).

More generally, Eq. (A.5.1) holds for normal matrices A and B. Recall that a matrix M is called normal if $MM^+ = M^+M$; such matrices can be diagonalized by a unitary matrix.

A.6. The Minimum Value of W. Proof of Equation (4.1.6).

To get the minimum value of the potential energy W we present Stieltjes' (1914) ingenious arguments. The existence of a minimum is clear. Let the points $x_1, ..., x_N$ make

$$W = \frac{1}{2}\sum_{1}^{N} x_j^2 - \sum_{1\le i<j\le N} \ln|x_i - x_j| \tag{A.6.1}$$

a minimum; then

$$0 = -\frac{\partial W}{\partial x_j} \equiv -x_j + \sum_{i(\neq j)} \frac{1}{x_j - x_i}. \tag{A.6.2}$$

Consider the polynomial

$$g(x) = (x - x_1)(x - x_2)\cdots(x - x_N), \tag{A.6.3}$$

which has $x_1, x_2, \cdots, x_N$ as its zeros. Logarithmic differentiation gives

$$\sum_{i(\neq j)} \frac{1}{x - x_i} = \frac{g'(x)}{g(x)} - \frac{1}{x - x_j} = \frac{(x - x_j)g'(x) - g(x)}{(x - x_j)g(x)}. \tag{A.6.4}$$

Near $x = x_j$ the left hand side is continuous, while the right-hand side takes the indeterminate form 0/0. Taking the limit $x \to x_j$ one gets from l'Hopital's rule on two differentiations:

$$\begin{aligned}\sum_{i(\neq j)} \frac{1}{x_j - x_i} &= \lim_{x\to x_j} \frac{(x - x_j)g'(x) - g(x)}{(x - x_j)g(x)} \\ &= \lim_{x\to x_j} \frac{(x - x_j)g''(x)}{(x - x_j)g'(x) + g(x)} = \frac{g''(x_j)}{2g'(x_j)}.\end{aligned} \tag{A.6.5}$$

so that (A.6.2) can be written as

$$g''(x_j) - 2x_j g'(x_j) = 0. \tag{A.6.6}$$

This means that the polynomial

$$g''(x) - 2xg'(x)$$

of order N has its zeros at $x_1, ..., x_N$ and therefore it must be proportional to $g(x)$. Comparing the coefficents of x^N, we see that

$$g''(x) - 2xg'(x) + 2Ng(x) = 0. \tag{A.6.7}$$

The polynomial solution of this differential equation is uniquely determined to be the Hermite polynomial of order N:

$$H_N(x) = N! \sum_{m=0}^{[N/2]} \frac{(-1)^m (2x)^{N-2m}}{m!(N-2m)!}. \tag{A.6.8}$$

The discriminant of $H_N(x)$ (Szegö, 1959) is

$$\prod_{1 \le i < j \le N} (x_i - x_j)^2 = 2^{-N(N-1)/2} \prod_{j=1}^{N} j^j, \tag{A.6.9}$$

and from (A.6.8) we get

$$\sum_1^N x_j^2 = N(N-1)/2. \tag{A.6.10}$$

Thus, the minimum value of W is

$$W_0 = \frac{1}{4} N(N-1)(1 + \ln 2) - \frac{1}{2} \sum_1^N j \ln j. \tag{A.6.11}$$

A.7. Relation between R_n, T_n and $E(n; 2\theta)$. Equivalence of Equations (5.1.3) and (5.1.4)

With $R_n(x_1, ..., x_n)$ and $T_n(x_1, ..., x_n)$ defined by Eqs. (5.1.2) and (5.1.3), let us write

$$r_0 = 1, \tag{A.7.1}$$

$$r_n = \int_{-\infty}^{\infty} \cdots \int_{-\infty}^{\infty} R_n(x_1, ..., x_n) \prod_1^n [a(x_i) dx_i], \quad n \geq 1, \tag{A.7.2}$$

$$t_n = \int_{-\infty}^{\infty} \cdots \int_{-\infty}^{\infty} T_n(x_1, ..., x_n) \prod_1^n [a(x_i) dx_i], \quad n \geq 1, \tag{A.7.3}$$

and the generating functions

$$R(z) = \sum_{n=0}^{\infty} \frac{r_n}{n!} z^n \equiv 1 + \sum_{n=1}^{\infty} \frac{r_n}{n!} z^n, \tag{A.7.4}$$

$$T(z) = \sum_{n=1}^{\infty} (-1)^{n-1} \frac{t_n}{n!} z^n, \tag{A.7.5}$$

where the function $a(x)$ is arbitrary. The numbers r_n and t_n may be related by the use of Eq. (5.1.3). In fact,

$$\begin{aligned} t_n &= \sum_G (-1)^{n-m} (m-1)! \, \frac{n!}{G_1! \cdots G_m!} \, \frac{1}{m!} \, r_{G_1} r_{G_2} \cdots r_{G_m} \\ &= \sum_G (-1)^{n-m} \, \frac{n!}{m} \prod_{j=1}^{m} \frac{r_{G_j}}{G_j!}, \end{aligned} \tag{A.7.6}$$

the sum being taken over all partitions G of n:

$$n = G_1 + \cdots + G_m, \quad G_j \geq 1, \quad m \geq 1. \tag{A.7.7}$$

We now relate the functions $T(z)$ and $R(z)$:

$$\begin{aligned} T(z) &= \sum_{n=1}^{\infty} \frac{(-1)^{n-1}}{n!} t_n z^n \\ &= \sum_{n=1}^{\infty} \sum_{G(n)} \frac{(-1)^{m-1}}{m} \prod_{j=1}^{m} \frac{r_{G_j}}{G_j!} z^{G_j}. \end{aligned} \tag{A.7.8}$$

As we are summing finally over all integers n, the restriction

$$\sum_{j=1}^{m} G_j = n \tag{A.7.9}$$

may be removed; we then have

$$\begin{aligned} T(z) &= \sum_{m,G_1,\ldots,G_m=1}^{\infty} \frac{(-1)^{m-1}}{m} \prod_{j=1}^{m} \frac{r_{G_j}}{G_j!} z^{G_j} \\ &= \sum_{m=1}^{\infty} \frac{(-1)^{m-1}}{m} \left(\sum_{\ell=1}^{\infty} \frac{r_\ell}{\ell!} z^\ell \right)^m \\ &= \ln R(z); \end{aligned} \tag{A.7.10}$$

or

$$R(z) = \exp[T(z)]. \tag{A.7.11}$$

Expanding the exponential and equating coefficients of z^n on both sides, we get

$$\begin{aligned} \frac{r_n}{n!} &= \text{coefficient of } z^n \text{ in } \sum_{m=1}^{n} \frac{1}{m!} \left(\sum_{\ell=1}^{n} \frac{(-1)^{\ell-1}}{\ell!} t_\ell z^\ell \right)^m \\ &= \sum_G \frac{1}{m!} \prod_{j=1}^{m} \left(\frac{(-1)^{G_j-1}}{G_j!} t_{G_j} \right); \quad \sum_{j=1}^{m} G_j = n, \end{aligned} \tag{A.7.12}$$

or

$$r_n = \sum_G \frac{(-1)^{n-m}}{m!} \frac{n!}{G_1!G_2!\cdots G_m!} t_{G_1} t_{G_2} \cdots t_{G_m}. \tag{A.7.13}$$

In view of the arbitrariness of $a(x)$, this relation is identical to Eq. (5.1.4).

If we take $a(x)$ as the characteristic function of the interval $(-\theta, \theta)$:

$$a(x) = \begin{cases} 1, & \text{if} \quad |x| \leq \theta, \\ 0, & \text{otherwise}, \end{cases} \tag{A.7.14}$$

then the probability that the interval $(-\theta, \theta)$ contains exactly n levels can be written as

$$E(n; 2\theta) = \frac{N!}{n!(N-n)!} \int_{-\infty}^{\infty} \cdots \int_{-\infty}^{\infty} dx_1 \cdots dx_N \times \prod_{i=1}^{n} a(x_i) \prod_{j=n+1}^{N} [1 - a(x_j)] \times P_N(x_1, \cdots, x_N). \tag{A.7.15}$$

Expanding the product $\prod [1 - a(x_j)]$ and regrouping similar terms we get

$$\begin{aligned} E(n; 2\theta) &= \frac{1}{n!} \sum_{j=n}^{N} \frac{(-1)^{j-n}}{(j-n)!} \int_{-\infty}^{\infty} \cdots \int_{-\infty}^{\infty} R_j(x_1, ..., x_j) \prod_{i=1}^{j} a(x_i) dx_i \\ &= \frac{(-1)^n}{n!} \sum_{j=n}^{N} \frac{(-1)^j}{(j-n)!} r_j \\ &= \frac{(-1)^n}{n!} \left(\frac{d}{dz} \right)^n \sum_{j=0}^{N} \frac{(-1)^j}{j!} r_j z^j \Bigg|_{z=1} . \end{aligned} \tag{A.7.16}$$

In terms of the generating functions this becomes

$$\begin{aligned} E(n; 2\theta) &= \frac{1}{n!} \left(-\frac{d}{dz} \right)^n R(-z)|_{z=1} \\ &= \frac{1}{n!} \left(-\frac{d}{dz} \right)^n \exp[T(-z)]|_{z=1} . \end{aligned} \tag{A.7.17}$$

Substituting the expression for T_n, Eq. (5.2.15),

$$T_{n2}(x_1, ..., x_n) = \sum K_N(x_1, x_2) K_N(x_2, x_3) \cdots K_N(x_n, x_1) \quad \text{(A.7.18)}$$

in (A.7.3), we get

$$\begin{aligned} t_{n2} &= (n-1)! \int_{-\theta}^{\theta} \cdots \int_{-\theta}^{\theta} K_N(x_1, x_2) \\ &\qquad \times K_N(x_2, x_3) \cdots K_N(x_n, x_1) dx_1 \cdots dx_n \\ &= (n-1)! \sum_{j=0}^{N-1} \lambda_j^n. \end{aligned} \quad \text{(A.7.19)}$$

so that

$$T_2(z) = \sum_{j=1}^{\infty} \frac{(-1)^{j-1}}{j} z^j \sum_i \lambda_i^j = \sum_i \ln\,(1 + z\lambda_i). \quad \text{(A.7.20)}$$

Hence from (A.7.17)

$$E(n; 2\theta) = \frac{1}{n!} \left(-\frac{d}{dz} \right)^n \prod_i (1 - z\lambda_i) \bigg|_{z=1}. \quad \text{(A.7.21)}$$

Taking the limit $N \to \infty$, we can replace $K_N(x, y)$ by $Q(\xi, \eta)$ and $E(n; 2\theta)$ by $E_2(n; s)$.

Had we taken the expression of T_n for orthogonal, symplectic or the noninvariant Gaussian ensemble, we would find the corresponding expression for $E(n; s)$. For example, for the noninvariant Gaussian ensemble,

$$T_n(x_1, ..., x_n) = \frac{1}{2} \mathrm{tr} \sum \phi(x_1, x_2) \phi(x_2, x_3) \cdots \phi(x_n, x_1), \quad \text{(A.7.22)}$$

where $\phi(x, y)$ is the 2×2 matrix, Eq. (14.1.14). The integral equation

$$\int_{-\theta}^{\theta} \phi(x, y) F(y) dy = \lambda F(x) \quad \text{(A.7.23)}$$

has N distinct eigenvalues λ_j, each occuring twice, so that t_n is again given by Eq. (A.7.19), and $E(n; 2\theta)$ by Eq. (A.7.21). In the limit $N \to \infty$ one replaces ϕ by σ, Eq. (14.1.34),

$$\sigma(x,y) = \begin{bmatrix} \frac{\sin(x-y)\pi}{(x-y)\pi} & D(x-y) \\ J(x-y) & \frac{\sin(x-y)\pi}{(x-y)\pi} \end{bmatrix}, \tag{A.7.24}$$

$$D(r) = -\frac{1}{\pi}\int_0^\pi t\,\sin(rt)\exp(2\rho^2 t^2)dt, \tag{A.7.25}$$

$$J(r) = -\frac{1}{\pi}\int_\pi^\infty \frac{\sin(rt)}{t}\exp(-2\rho^2 t^2)dt. \tag{A.7.26}$$

Equation (A.7.21) is now

$$E(n;s) = \frac{1}{n!}\left(-\frac{d}{dz}\right)^n \prod_{i=0}^{\infty} (1 - z\lambda_i)\big|_{z=1}, \tag{A.7.27}$$

with λ_i the distinct eigenvalues of

$$\int_{-t}^{t} \sigma(x,y)F(y)dy = \lambda F(x). \tag{A.7.28}$$

A hierarchical relation between R_n and R_{n+1} may be obtained by considering α^2 in Chapter 14 as the "time" variable, constructing a Brownian motion model as in Chapter 8 and solving the partial differential equation so obtained. For more details see French et al. (1988), page 229, Eq. (72).

A.8. Relation between $E(n;s)$, $F(n;s)$, and $p(n;s)$: Equations (5.1.16), (5.1.17), and (5.1.18)

Let $E(0;x)$ be the probability that an interval of length x is empty of eigenvalues. Then $E(0;x+\delta x)$ is the probability that the interval $\delta x + x$ is empty, and $E(0;x) - E(0;x+\delta x)$ is the probability that the interval x is empty and δx is not empty. The probability that δx will contain more than one eigenvalue is of second or higher order in δx. Therefore, taking the limit $\delta x \to 0$ and keeping only the first order terms, we get $-[dE(0;x)/dx]\delta x$ as the probability that x is empty and δx contains one eigenvalue. This is equivalent to $F(0;x) = -dE(0;x)/dx$. By a similar argument we obtain $p(0;x) = -dF(0;x)/dx$:

$$\bullet \overset{\delta x}{-} \bullet \overset{x}{-} \bullet .$$

Let $E(n;x)$ be the probability that the interval x contains exactly n eigenvalues. Increasing x by δx in $E(n;x)$ and subtracting it from $E(n;x)$ we find that either one of the eigenvalues in x moves in δx or a new eigenvalue appears in δx. Therefore,

$$-\frac{dE(n;x)}{dx}\,\delta x = -\text{prob}\begin{pmatrix} x & n-1 \\ \delta x & 1 \end{pmatrix} + \text{prob}\begin{pmatrix} x & n \\ \delta x & 1 \end{pmatrix},$$

where prob $\begin{pmatrix} x & r \\ \delta x & s \end{pmatrix}$ means the probability that x and δx contain r and s eigenvalues respectively. This is equivalent to Eq. (5.1.16b). A similar argument gives Eq. (5.1.16c).

The inverse relations, Eqs. (5.1.17) and (5.1.18), are easy consequences of Eqs. (5.1.16) and can be derived step by step starting at $n = 0$.

The quantity $B_1'(t)$ or $E_1'(s)$ of Section 10.4 is the probability that the interval $s = 2t$ contains either zero or one eigenvalue, i.e., $E_1'(x) = E_1(0;x) + E_1(1;x)$. So

$$\frac{d^2}{dx^2}\left(E_1(0;x) + E_1'(x)\right) = \frac{d^2}{dx^2}[2E_1(0;x) + E_1(1;x)] = p(1;x),$$

i.e., $p(1;x)\delta x$ is the probability that the interval between two eigenvalues containing exactly one eigenvalue inside it has a length btween x and $x + \delta x$. In other words it is the probability density for the next nearest neighbor spacings.

A.9. The Limit of $\sum_0^{N-1} \varphi_j^2(x)$. Equation (5.2.17)

The dominant term in $\sum_0^{N-1} \varphi_j^2(x)$ may be obtained with ease by a physical argument.

The $\varphi_j(x)$ is the normalized oscillator function, so that $\varphi_j^2(x)dx$ gives the probability that an oscillator in the jth state is found in the interval $(x, x+dx)$. Consider N oscillators, one each in the states $0, 1, ..., N-1$, so that when N is large, $\sum_0^{N-1} \varphi_j^2(x)$ is the density of the particles at the point x. The particles are fermions and the temperature is zero, for there is not more than one particle in each state and all states up to a certain energy (Fermi energy) are filled. The Fermi momentum corresponding to this maximum energy can be obtained from the differential equation satisfied by $\varphi_{N-1}(x)$.

$$\hbar^2 \frac{d^2}{dx^2} \varphi_{N-1}(x) + \hbar^2 \left(2N - 1 - x^2\right) \varphi_{N-1}(x) = 0,$$

so that

$$-p_F^2 + \left(2N - 1 - x^2\right) \hbar^2 = 0.$$

Because our oscillators are one-dimensional, their density is given by

$$\sigma(x) \approx \frac{1}{(2\pi\hbar)} \int_{-p_F}^{p_F} dp = \frac{1}{2\pi\hbar} 2p_F, \qquad p \leq p_F.$$

From the last two equations we get

$$\begin{aligned} \sigma(x) \quad &= \frac{1}{\pi} \left(2N - x^2\right)^{1/2}, \qquad && x^2 \leq 2N, \\ &= 0, && x^2 \geq 2N. \end{aligned}$$

This is the dominant term. Terms of the next lower order cannot be obtained from physical arguments alone. To get further information about $\sigma(x)$ we may write from the formula of Christoffel-Darboux (Bateman, 1953)

$$\sum_0^{N-1} \varphi_j^2(x) = N \ \varphi_N^2(x) - [N(N+1)]^{1/2} \ \varphi_{N-1}(x)\varphi_{N+1}(x),$$

and use the known (Erdelyi, 1960) asymptotic behavior of the functions $\varphi_{N-1}(x)$, $\varphi_N(x)$, and $\varphi_{N+1}(x)$ for the various intervals of x.

The level density $\sigma_N(x)$ for the Gaussian orthogonal ensemble contains two extra terms, namely, $\sqrt{N/2}\varphi_{N-1}(x) \int_{-\infty}^{\infty} \varepsilon(x-t)\varphi_N(t)dt$ and

$\varphi_{2m}(x)/\int_{-\infty}^{\infty} \varphi_{2m}(t)dt$. In the limit $N \to \infty$, their contribution is negligible.

A.10. The Limits of $\sum_0^{N-1} \varphi_j(\boldsymbol{x})\varphi_j(\boldsymbol{y})$, etc. Sections 5.2, 6.4, and 6.5

The Christoffel–Darboux formula gives (Bateman, 1953)

$$\sum_0^{N-1} \varphi_j(x)\varphi_j(y) = \left(\frac{1}{2}N\right)^{1/2} \left[\frac{\varphi_N(x)\varphi_{N-1}(y) - \varphi_N(y)\varphi_{N-1}(x)}{x-y}\right]. \quad \text{(A.10.1)}$$

Let $N = 2m$, $2m^{1/2}x = \pi t\xi$, $2m^{1/2}y = \pi t\eta$ and let us take the limit $m \longrightarrow \infty$, $x \longrightarrow 0$, $y \longrightarrow 0$; whereas ξ and η are finite. Using the formula (Bateman, 1953)

$$\lim(-1)^m m^{1/4}\varphi_{2m}(x) = \pi^{-1/2}\cos \pi t\xi,$$

$$\lim(-1)^m m^{1/4}\varphi_{2m+1}(x) = \pi^{-1/2}\sin \pi t\xi,$$

we get

$$\lim \sum_0^{2m-1} \varphi_j(x)\varphi_j(y) = \frac{2m^{1/2}}{\pi}\frac{\sin \pi\xi t \cos \pi\eta t - \cos \pi\xi t \sin \pi\eta t}{\pi\xi t - \pi\eta t} = \frac{2\sqrt{m}}{\pi}\frac{\sin(\xi-\eta)\pi t}{(\xi-\eta)\pi t} \quad \text{(A.10.2)}$$

which is Eq. (5.2.20) or Eq. (5.3.13). To obtain the limit of

$$K_m(x,y) = \sum_0^{m-1} \varphi_{2j}(x)\varphi_{2j}(y)$$

we observe that $K_m(x,y)$ is the even part in x of Eq. (A.10.2).

$$K_m(x,y) = \frac{1}{2}\sum_0^{2m-1} \left[\varphi_j(x)\varphi_j(y) + \varphi_j(-x)\varphi_j(y)\right].$$

Therefore

$$\lim K_m(x,y) = \frac{1}{2}\frac{2\sqrt{m}}{\pi}\left[\frac{\sin(\xi-\eta)\pi}{(\xi-\eta)\pi} + \frac{\sin(\xi+\eta)\pi}{(\xi+\eta)\pi}\right] \tag{A.10.3}$$

$$= \frac{2\sqrt{m}}{\pi}\bar{Q}(\xi,\eta), \quad \text{say.} \tag{A.10.4}$$

Also

$$\lim \sum_0^{m-1} \varphi_{2j}(x)\varphi'_{2j}(y) = \lim \frac{\partial}{\partial y}\sum_0^{m-1} \varphi_{2j}(x)\varphi_{2j}(y)$$

$$= \frac{2\sqrt{m}}{\pi}\frac{\partial}{\partial \eta} \lim K_m(x,y)$$

$$= \left(\frac{2\sqrt{m}}{\pi}\right)^2 \frac{\partial}{\partial \eta}\bar{Q}(\xi,\eta) \tag{A.10.5}$$

and

$$\lim \sum_0^{m-1} \varphi_{2j}(x)\int_0^y \varphi_{2j}(z)dz = \lim \int_0^y K_m(x,z)dz$$

$$= \int_0^\eta \bar{Q}(\xi,\zeta)\mathrm{d}\zeta. \tag{A.10.6}$$

By writing similar equations for

$$\lim \sum_0^{m-1} \varphi'_{2j}(x)\varphi_{2j}(y) \quad \text{and} \quad \lim \sum_0^{m-1} \varphi_{2j}(y) \int_0^x \varphi_{2j}(z)\mathrm{d}z$$

and combining, we obtain Eqs. (6.4.8) and (6.4.9).

A.11. The Fourier Transforms of the Two-Point Cluster Functions

The functions $s(r)$, $(d/dr)[s(r)]$, and $\int_0^r s(z)dz$ are given only for positive values of r by Eqs. (5.2.20), (6.4.8), and (6.4.9); for negative values of r they are defined by the statement that they are even functions of r. Therefore, writing $F[f(x)]$ for the Fourier transform of $f(x)$,

$$\begin{aligned} F[s(r)] &= \int_{-\infty}^{\infty} e^{2\pi ikr} s(r)dr = 2\int_0^{\infty} \cos(2\pi\,|k|\,r)\frac{\sin\pi r}{\pi r}\, dr \\ &= \int_0^{\infty} [\sin{(2\,|k|+1)}\,\pi r - \sin{(2\,|k|-1)}\,\pi r]\, \frac{dr}{\pi r} \\ &= \frac{1}{2}\, [\mathrm{sgn}\,(2\,|k|+1) - \mathrm{sgn}\,(2\,|k|-1)] \\ &= \begin{cases} 1, & \text{if} \quad |k| < 1/2, \\ 1/2, & \text{if} \quad |k| = 1/2, \\ 0, & \text{if} \quad |k| > 1/2; \end{cases} \end{aligned} \tag{A.11.1}$$

$$\begin{aligned} F\left[s^2(r)\right] &\equiv \int_{-\infty}^{\infty} e^{2\pi ikr} s^2(r)dr = 2\int_0^{\infty} \cos(2\pi\,|k|\,r)\frac{\sin^2\pi r}{\pi^2 r^2}\, dr \\ &= \int_0^{\infty} \cos{(2\pi\,|k|\,r)}\, \frac{1-\cos(2\pi r)}{\pi^2 r^2} dr \\ &= \int_0^{\infty} \left(2\pi^2 r^2\right)^{-1} [2\cos{(2\pi\,|k|\,r)} - \cos(2\,|k|+2)\pi r] \\ &\qquad - \cos{(2\,|k|-2)}\,\pi r]\, dr. \end{aligned}$$

Now

$$\int_0^\infty r^{-2}(\cos ar - \cos br)dr = \int_a^b d\lambda \int_0^\infty r^{-1}\sin \lambda r \, dr$$

$$= \frac{\pi}{2}\int_a^b d\lambda \ \mathrm{sgn}\lambda = \frac{\pi}{2}\left(|b| - |a|\right),$$

so that

$$\begin{aligned} F\left[s^2(r)\right] &= \left(2\pi^2\right)^{-1}\frac{\pi}{2}\, 2\pi\left[(|k|+1) + \big|(|k|-1)\big| - 2\ |k|\right] \\ &= \frac{1}{2}\left[(1-|k|) + \big|(1-|k|)\big|\right] \\ &= \begin{cases} 1-|k|, & \text{if } |k|<1, \\ 0, & \text{if } |k|>1. \end{cases} \end{aligned} \tag{A.11.2}$$

By partial integration we have

$$\begin{aligned} F\left[\frac{d}{dr}s(r)\right] &= 2\int_0^\infty \cos\left(2\pi\,|k|\,r\right)\frac{d}{dr}\frac{\sin \pi r}{\pi r}dr \\ &= 2\Big\{-1 + |k|\int_0^\infty r^{-1}[\cos(2\,|k|-1)\pi r \\ &\qquad - \cos(2\,|k|+1)\pi r]dr\Big\} \\ &= 2\left(-1 + |k|\int_{(2|k|-1)\pi}^{(2|k|+1)\pi} d\lambda \int_0^\infty dr\, \sin\,\lambda r\right). \end{aligned}$$

Now

$$\int_0^\infty dr \sin \lambda r = \lim_{\alpha \to 0} \int_0^\infty e^{-\alpha r} \sin \lambda r \, dr$$

$$= \lim_{\alpha \to 0} \frac{\lambda}{\alpha^2 + \lambda^2} = \frac{1}{\lambda},$$

so that

$$F\left(\frac{d}{dr}s(r)\right) = 2\left(-1 + |k| \ln \frac{(2|k|+1)}{(2|k|-1)}\right). \tag{A.11.3}$$

Also by partial integration

$$F\left\{\left[\int_0^r s(z)dz\right]\left[\frac{d}{dr}s(r)\right]\right\} = -F\left[s^2(r)\right] + 4\pi |k|$$

$$\times \int_0^\infty dr \sin(2\pi |k| r) \frac{\sin \pi r}{\pi r}\left[\int_0^r s(z)dz\right]$$

and

$$2\int_0^\infty r^{-1} \sin(2\pi |k| r) \sin \pi r \left(\int_0^r \frac{\sin \pi x}{x} dx\right) dr$$

$$= \int_{(2|k|-1)\pi}^{(2|k|+1)\pi} dz \int_0^\infty dr \sin zr \left(\int_0^r \frac{\sin \pi x}{x} dx\right)$$

$$= \int_{(2|k|-1)\pi}^{(2|k|+1)\pi} dz \int_0^\infty dr \frac{\sin \pi r}{r} \frac{\cos zr}{z}$$

$$= \int_{(2|k|-1)\pi}^{(2|k|+1)\pi} dz \frac{1}{2z} \left[\operatorname{sgn}(\pi + z) + \operatorname{sgn}(\pi - z)\right]$$

$$= \begin{cases} -\dfrac{1}{2} \ln |(2|k|-1)|, & \text{if } |k| < 1, \\ 0, & \text{if } |k| > 1, \end{cases}$$

so that

$$F\left\{\left[\int_0^r s(z)dz\right]\left[\frac{d}{dr}s(r)\right]\right\}$$

$$= \begin{cases} -1+|k|-|k|\ln|(2|k|-1)|, & \text{if} \quad |k|<1, \\ 0, & \text{if} \quad |k|>1. \end{cases} \tag{A.11.4}$$

Combining Eqs. (A.11.1)–(A.11.4), we get all the Fourier transforms quoted in Chapters 5, 6, and 7.

A.12. Some Applications of Gram's Formula

We have used Gram's result repeatedly, e.g., in Sections 5.3, 6.8, 10.2, and 13.2. It may be stated as follows (*cf.* Mehta, 1989).

Let v_i, $i = 1, 2, ..., m$, be m vectors and let $v_{i\nu}$, $\nu = 1, 2, ..., n$, be their components along some basis. Form the scalar products

$$b_{ij} = (v_i, v_j) = \sum_{\nu=1}^{n} v_{i\nu} v_{j\nu}^*, \quad i, j = 1, 2, ..., m;$$

then

$$\det \begin{bmatrix} b_{11} & b_{12} & \dots & b_{1m} \\ b_{21} & b_{22} & \dots & b_{2m} \\ \dots & \dots & \dots & \dots \\ b_{m1} & b_{m2} & \dots & b_{mm} \end{bmatrix}$$

$$= \frac{1}{m!} \sum_{\nu_1,\nu_2,\dots,\nu_m} \left| \det \begin{bmatrix} v_{1\nu_1} & \dots & v_{1\nu_m} \\ \dots & \dots & \dots \\ v_{m\nu_m} & & v_{m\nu_m} \end{bmatrix} \right|^2 \tag{A.12.1}$$

where on the right-hand side we sum over all possible ways of choosing $\nu_1, \nu_2, ..., \nu_m$ among $1, 2, ..., n$. The summation over ν may be finite, denumerable, or continuously infinite, the summation sign being replaced by an integration over a suitable measure.

If we make the correspondence

$$v_{i\nu} \longrightarrow \varphi_{i-1}(x), \quad \sum_{\nu} \longrightarrow \int_{\text{out}} dx = \left(\int_{-\infty}^{-\theta} + \int_{\theta}^{\infty} \right) dx$$

the scalar products become

$$\begin{aligned} b_{ij} &= \sum_{\nu} v_{i\nu} v_{j\nu}^* \\ &\longrightarrow \left(\int_{-\infty}^{-\theta} + \int_{\theta}^{\infty} \right) \varphi_{i-1}(x) \varphi_{j-1}(x) dx \\ &= \delta_{ij} - \int_{-\theta}^{\theta} \varphi_{i-1}(x) \varphi_{j-1}(x) dx = g_{ij} \end{aligned}$$

and Eq. (A.12.1) yields Eq. (5.3.3). If we make the correspondence

$$v_{i\nu} \longrightarrow \varphi_{2i-2}(x), \quad \sum_{\nu} \longrightarrow \int_{\theta}^{\infty} dx,$$

then the scalar products become

$$b_{ij} = \sum_{\nu} v_{i\nu} v_{j\nu}^* \longrightarrow \int_{\theta}^{\infty} \varphi_{2i-2}(x) \varphi_{2j-2}(x) dx,$$

and Eq. (A.12.1) gives Eq. (6.8.2). If we make the correspondence

$$v_{i\nu} \longrightarrow \sqrt{2} \mathrm{e}^{-(1/2)y^2} \frac{y^{2i+1/2}}{(y^2 + \theta^2)^{1/4}}, \quad \sum_{\nu} \longrightarrow \int_0^{\infty} dy,$$

the scalar products become

$$b_{ij} = \sum_{\nu} v_{i\nu} v_{j\nu}^* \longrightarrow \int_0^{\infty} dy \; 2e^{-y^2} \frac{y^{2i+2j+1}}{(y^2 + \theta^2)^{1/2}} = \eta_{i+j}$$

and Eq. (A.12.1) yields Eq. (6.8.9). If we make the correspondence

$$v_{j\nu} \longrightarrow (2\pi)^{-1/2} e^{ip\theta}, \qquad \sum_{\nu} \longrightarrow \int_{\alpha}^{2\pi-\alpha} d\theta,$$

the scalar products become

$$b_{jk} = (2\pi)^{-1} \int_{\alpha}^{2\pi-\alpha} d\theta e^{i(p-q)\theta},$$

and we get Eq. (10.2.3).

A.13. Power Series Expansions of Eigenvalues, of Spheroidal Functions, and of Various Probabilities

The first few terms in the power series expansions of the functions $f_j(x)$, Eq. (5.3.17); λ_j, Eq. (5.3.18); $E_\beta(n;s)$, Eqs. (5.3.27), (5.4.30), (6.5.19), (6.6.13), (6.7.21); and $p_\beta(n;s)$, Eqs. (5.4.31), (6.6.14), (6.6.18), (6.6.19), (6.7.22) are reproduced here.

$$\begin{aligned}
\lambda_0 &= s - \frac{\pi^2}{36}s^3 + \frac{23\pi^4}{32400}s^5 - \frac{79\pi^6}{5715360}s^7 + O(s^9), \\
\lambda_1 &= \frac{\pi^2}{36}s^3 - \frac{\pi^4}{1200}s^5 + \frac{41\pi^6}{2940000}s^7 + O(s^9), \\
\lambda_2 &= \frac{\pi^4}{8100}s^5 - \frac{\pi^6}{2857680}s^7 + O(s^9), \\
\lambda_3 &= \frac{\pi^6}{4410000}s^7 + O(s^9), \\
\lambda_j &= O(s^{2j+1}),
\end{aligned}$$

$$f_0(x) = \left(\frac{1}{2}\right)^{1/2} \left[\left(1 - \frac{\pi^4 s^4}{12960}\right) + \left(-\frac{\pi^2 s^2}{36} + \frac{\pi^4 s^4}{4536}\right) P_2(x) + \frac{\pi^4 s^4}{8400} P_4(x) + O(s^6) \right],$$

$$f_1(x) = \left(\frac{3}{2}\right)^{1/2} \left[\left(1 - \frac{3\pi^4 s^4}{140000}\right) P_1(x) + \left(-\frac{\pi^2 s^2}{100} - \frac{\pi^4 s^4}{45000}\right) P_3(x) + \frac{\pi^4 s^4}{35280} P_5(x) + O(s^6) \right],$$

$$f_2(x) = \left(\frac{5}{2}\right)^{1/2} \left[\left(\frac{\pi^2 s^2}{180} - \frac{\pi^4 s^4}{22680}\right) + \left(1 - \frac{545\pi^4 s^4}{6223392}\right) P_2(x) + \left(-\frac{3\pi^2 s^2}{490} - \frac{23\pi^4 s^4}{384160}\right) P_4(x) + O(s^6) \right]$$

$$f_j(x) = \left(\frac{2j+1}{2}\right)^{1/2} P_j(x) + O(s^2),$$

where $P_j(x)$ are the Legendre polynomials,

$$E_1(0;s) = 1 - s + \frac{\pi^2 s^3}{36} - \frac{\pi^4 s^5}{1200} + \frac{\pi^4 s^6}{8100} + \frac{\pi^6 s^7}{70560} - \frac{\pi^6 s^8}{264600} - \frac{\pi^8 s^9}{6531840} + O(s^{10}),$$

$$E_1(1;s) = s - \frac{\pi^2 s^3}{18} + \frac{\pi^4 s^5}{600} - \frac{\pi^4 s^6}{8100} - \frac{\pi^6 s^7}{35280} + \frac{\pi^6 s^8}{264600} + O(s^9),$$

$$E_1(2;s) = \frac{\pi^2 s^3}{36} - \frac{\pi^4 s^5}{1200} - \frac{\pi^4 s^6}{8100} + O(s^7);$$

$$E_2(0;s) = 1 - s + \frac{\pi^2 s^4}{36} - \frac{\pi^4 s^6}{675} + \frac{\pi^6 s^8}{17640} - \frac{\pi^6 s^9}{291600} - \frac{\pi^8 s^{10}}{637875} + O(s^{11}),$$

$$E_2(1;s) = s - \frac{\pi^2 s^4}{18} + \frac{2\pi^4 s^6}{675} - \frac{\pi^6 s^8}{8820} + \cdots,$$

$$E_2(2;s) = \frac{\pi^2 s^4}{36} - \frac{\pi^4 s^6}{675} + \cdots;$$

$$E_4(0;s) = E_1(0;2s) + \frac{1}{2}E_1(1;2s) = 1 - s + \frac{8\pi^4 s^6}{2025} - \frac{16\pi^6 s^8}{33075} + \cdots,$$

$$E_4(n;s) = E_1(2n;2s) + \frac{1}{2}E_1(2n-1;2s) + \frac{1}{2}E_1(2n+1;2s), \quad n \geq 1;$$

$$p_1(0;s) = \frac{\pi^2 s}{6} - \frac{\pi^4 s^3}{60} + \frac{\pi^4 s^4}{270} + \frac{\pi^6 s^5}{1680} - \frac{\pi^6 s^6}{4725} - \frac{\pi^8 s^7}{90720} + \cdots,$$

$$p_1(1;s) = \frac{\pi^4 s^4}{270} - \frac{\pi^6 s^6}{4725} + \cdots,$$

$$p_1(2;s) = \frac{\pi^8 s^8}{1764000} + \cdots,$$

$$p_2(0;s) = \frac{\pi^2 s^2}{3} - \frac{2\pi^4 s^4}{45} + \frac{\pi^6 s^6}{315} - \frac{\pi^6 s^7}{4050} - \frac{2\pi^8 s^8}{14175} + \cdots,$$

$$p_2(1;s) = \frac{\pi^6 s^7}{4050} + \cdots,$$

$$p_2(2;s) = \frac{\pi^{12} s^{14}}{5358150000} + \cdots;$$

$$p_4(0;s) = \frac{16\pi^4 s^4}{135} - \frac{128\pi^6 s^6}{4725} + \cdots;$$

$$\begin{aligned} B_1(t;y) &= 1 - \frac{\pi^2}{24}\left(t^2 + y^2\right) + \frac{\pi^4}{1920}\left(t^4 + 6t^2y^2 + y^4\right) \\ &\quad + \frac{\pi^4 t^3}{5400}\left(t^2 - 5y^2\right) + \cdots, \\ B_2(t;y) &= 1 - \frac{2}{9}\pi^2 t\left(3y^2 + t^2\right) \\ &\quad + \frac{4}{225}\pi^4 t\left(5y^4 + 10y^2t^2 + t^4\right) + \cdots; \end{aligned}$$

$$\begin{aligned} \mathcal{B}_1(x_1, x_2) &= 1 - \frac{\pi^2}{12}\left(x_1^2 + x_2^2\right) + \frac{\pi^4}{240}\left(x_1^4 + x_2^4\right) \\ &\quad - \frac{\pi^4}{1350}\left(x_1^5 - 5x_1^2x_2^2\left(x_1 + x_2\right) + x_2^5\right) + \cdots, \\ \mathcal{B}_2(x_1, x_2) &= 1 - \frac{\pi^2}{9}\left(x_1^3 + x_2^3\right) + \frac{2\pi^4}{225}\left(x_1^5 + x_2^5\right) + \cdots; \end{aligned}$$

$$\mathcal{P}_1(x_1, x_2) = \frac{\pi^4}{45} x_1 x_2 \left(x_1 + x_2\right) + \cdots.$$

A.14. Numerical Tables of $\lambda_j(s)$, $b_j(s)$, and $E_\beta(n;s)$ for $\beta = 1, 2$, and 4

TABLE A.14.1. EIGENVALUES λ_j OF THE INTEGRAL EQUATION (5.3.14), FOR $0 \le j \le 8$, $0 \le s \le 5$

s	$\lambda_0(s)$	$\lambda_1(s)$	$\lambda_2(s)$	$\lambda_3(s)$	$\lambda_4(s)$	$\lambda_5(s)$	$\lambda_6(s)$	$\lambda_7(s)$	$\lambda_8(s)$
0.127	0.12676	0.00056							
0.255	0.25019	0.00444	0.00001						
0.382	0.36724	0.01463	0.00010						
0.509	0.47534	0.03355	0.00041						
0.637	0.57258	0.06279	0.00124	0.00001					
0.891	0.73072	0.15395	0.00650	0.00010					
1.146	0.84107	0.28243	0.02186	0.00056	0.00001				
1.401	0.91141	0.43134	0.05555	0.00222	0.00004				
1.655	0.95288	0.57972	0.11544	0.00697	0.00020				
1.910	0.97583	0.70996	0.20514	0.01820	0.00071	0.00002			
2.165	0.98793	0.81233	0.32102	0.04102	0.00214	0.00007			
2.419	0.99409	0.88538	0.45226	0.08154	0.56681	0.00022	0.00001		
2.674	0.99715	0.93336	0.58414	0.14510	0.01339	0.00065	0.00002		
2.928	0.99864	0.96278	0.70298	0.23361	0.02866	0.00172	0.00007		
3.183	0.99936	0.97987	0.79992	0.34356	0.05602	0.00418	0.00019	0.00001	
3.438	0.99970	0.98937	0.87226	0.46601	0.10050	0.00932	0.00052	0.00002	
3.692	0.99986	0.99450	0.92218	0.58890	0.16613	0.01950	0.00128	0.00006	
3.947	0.99994	0.99719	0.95442	0.70071	0.25388	0.03781	0.00295	0.00015	0.00001
4.202	0.99997	0.99859	0.97415	0.79357	0.36019	0.06844	0.00639	0.00038	0.00002
4.456	0.99999	0.99929	0.98571	0.86457	0.47705	0.11572	0.01306	0.00091	0.00005
4.711	1.00000	0.99965	0.99226	0.91499	0.59392	0.18282	0.02520	0.00202	0.00012
4.966		0.99983	0.99588	0.94863	0.70068	0.27007	0.04598	0.04286	0.00028
5.093		0.99988	0.99701	0.96055	0.74790	0.32028	0.06078	0.00613	0.00042

TABLE A.14.2. NUMERICAL VALUES OF $b_j(s) \equiv f_{2j}(1) \int_{-1}^{1} f_{2j}(x)dx / \int_{-1}^{1} f_{2j}^2(x)dx$, FOR $0 \leq j \leq 3$, $0 \leq s \leq 5$. THE $f_j(x)$ ARE THE SPHEROIDAL FUNCTIONS, SOLUTIONS OF EQ. (5.3.14)

s	b_0	b_1	b_2	b_3	b_4
0.000	1.00000				
0.127	0.99556	0.00444			
0.255	0.98230	0.01768	0.00002		
0.382	0.96040	0.03949	0.00011		
0.509	0.93021	0.06946	0.00033		
0.637	0.89225	0.10694	0.00081		
0.891	0.79630	0.20060	0.00309	0.00002	
1.146	0.68148	0.31011	0.00835	0.00007	
1.401	0.55958	0.42183	0.01836	0.00023	
1.655	0.44242	0.52179	0.03516	0.00063	0.00001
1.910	0.33878	0.59892	0.06082	0.00147	0.00002
2.165	0.25291	0.64676	0.09721	0.00308	0.00005
2.419	0.18520	0.66324	0.14552	0.00592	0.00011
2.674	0.13369	0.64973	0.20569	0.01064	0.00025
2.928	0.09549	0.61021	0.27574	0.01803	0.00051
3.183	0.06766	0.55075	0.35146	0.02912	0.00099
3.438	0.04765	0.47883	0.42662	0.04505	0.00181
3.692	0.03339	0.40225	0.49401	0.06709	0.00317
3.947	0.02331	0.32783	0.54689	0.09649	0.00532
4.202	0.01622	0.26039	0.58023	0.13423	0.00862
4.456	0.01126	0.20250	0.59141	0.18074	0.01353
4.711	0.00779	0.15483	0.58028	0.23549	0.02065
4.966	0.00539	0.11679	0.54896	0.29657	0.03067
5.093	0.00447	0.10101	0.52694	0.32847	0.03703

TABLE A.14.3. THE PROBABILITIES $E_1(n; s)$ OF HAVING n EIGENVALUES OF A REAL SYMMETRIC MATRIX IN AN INTERVAL s FOR $0 \leq n \leq 7$, $0 \leq s \leq 5$

s	$E_1(0; s)$	$E_1(1; s)$	$E_1(2; s)$	$E_1(3; s)$	$E_1(4; s)$	$E_1(5; s)$	$E_1(6; s)$	$E_1(7; s)$
0.000	1.00000							
0.127	0.87324	0.12620	0.00056					
0.255	0.74980	0.24576	0.00444					
0.382	0.63270	0.35267	0.01460	0.00004				
0.509	0.52445	0.44200	0.03336	0.00019				
0.573	0.47425	0.47893	0.04644	0.00039				
0.637	0.42689	0.51031	0.06209	0.00071				
0.891	0.26753	0.57844	0.14928	0.00473	0.00001			
1.146	0.15546	0.56171	0.26444	0.01823	0.00016			
1.401	0.08367	0.48373	0.38194	0.04971	0.00096			
1.655	0.04167	0.37567	0.47249	0.10612	0.00402	0.00002		
1.910	0.01920	0.26555	0.51451	0.18780	0.01279	0.00014		
2.165	0.00818	0.17178	0.50144	0.28523	0.03269	0.00068		
2.419	0.00322	0.10204	0.44206	0.38032	0.06982	0.00254	0.00002	
2.674	0.00117	0.05577	0.35500	0.45211	0.12814	0.00773	0.00009	
2.928	0.00039	0.02808	0.26096	0.48437	0.20603	0.01978	0.00039	
3.183	0.00012	0.01304	0.17617	0.47130	0.29444	0.04352	0.00140	0.00001
3.438	0.00003	0.00559	0.10946	0.41885	0.37802	0.08372	0.00428	0.00005
3.692	0.00001	0.00221	0.06270	0.34136	0.43954	0.14273	0.01126	0.00020
3.947		0.00081	0.03314	0.25590	0.46572	0.21791	0.02581	0.00071
4.202		0.00027	0.01618	0.17684	0.45179	0.30039	0.05228	0.00223
4.456		0.00008	0.00730	0.11282	0.40273	0.37638	0.09453	0.00607
4.711		0.00002	0.00304	0.06652	0.33077	0.43087	0.15385	0.01458
4.966		0.00001	0.00117	0.03628	0.25083	0.45252	0.22692	0.03117
5.093			0.00071	0.02602	0.21209	0.44948	0.26607	0.04373

TABLE A.14.4. THE PROBABILITIES $E_2(n; s)$ OF HAVING n EIGENVALUES OF A RANDOM HERMITIAN MATRIX IN AN INTERVAL s FOR $0 \leq n \leq 7$, $0 \leq s \leq 5$

s	$E_2(0; s)$	$E_2(1; s)$	$E_2(2; s)$	$E_2(3; s)$	$E_2(4; s)$	$E_2(5; s)$	$E_2(6; s)$	$E_2(7; s)$
0.000	1.00000							
0.127	0.87275	0.12718	0.00007					
0.255	0.74647	0.25242	0.00111					
0.382	0.62344	0.37115	0.00541					
0.509	0.50685	0.47700	0.01614	0.00001				
0.637	0.40008	0.56327	0.03661	0.00004				
0.891	0.22632	0.65684	0.11610	0.00074				
1.146	0.11149	0.63644	0.24674	0.00533				
1.401	0.04747	0.52729	0.40249	0.02270	0.00005			
1.655	0.01739	0.37809	0.53689	0.06717	0.00046			
1.910	0.00547	0.23561	0.60522	0.15101	0.00270			
2.165	0.00147	0.12759	0.58709	0.27268	0.01114	0.00002		
2.419	0.00034	0.05991	0.49506	0.40982	0.03468	0.00019		
2.674	0.00007	0.02433	0.36473	0.52458	0.08518	0.00111		
2.928	0.00001	0.00852	0.23518	0.58041	0.17104	0.00482	0.00001	
3.183		0.00257	0.13262	0.56033	0.28814	0.01627	0.00007	
3.438		0.00066	0.06528	0.47461	0.41494	0.04411	0.00040	
3.692		0.00015	0.02796	0.35367	0.51765	0.09868	0.00188	
3.947		0.00003	0.01040	0.23199	0.56466	0.18595	0.00695	0.00002
4.202			0.00335	0.13381	0.54181	0.29997	0.02093	0.00013
4.456			0.00093	0.06772	0.45897	0.41949	0.05222	0.00066
4.711			0.00022	0.02999	0.34379	0.51330	0.10998	0.00271
4.966			0.00005	0.01159	0.22769	0.55320	0.19836	0.00908
5.093			0.00002	0.00684	0.17687	0.54838	0.25230	0.01549

TABLE A.14.5. PROBABILITIES $E_4(n;s)$ OF HAVING n EIGENVALUES OF A RANDOM SELF-DUAL HERMITIAN MATRIX IN AN INTERVAL s FOR $0 \leq n \leq 7$, $0 \leq s \leq 6$

s	$E_4(0;s)$	$E_4(1;s)$	$E_4(2;s)$	$E_4(3;s)$	$E_4(4;s)$	$E_4(5;s)$	$E_4(6;s)$	$E_4(7;s)$
0.000	1.00000							
0.127	0.87268	0.12732						
0.255	0.74545	0.25445	0.00010					
0.382	0.61903	0.37996	0.00100					
0.509	0.49565	0.49940	0.00495					
0.637	0.37938	0.60461	0.01600					
0.891	0.18829	0.73217	0.07950	0.00003				
1.146	0.07208	0.71061	0.21663	0.00068				
1.401	0.02067	0.56458	0.40825	0.00650				
1.655	0.00438	0.36923	0.59322	0.03317	0.00001			
1.910	0.00068	0.19603	0.69624	0.10687	0.00019			
2.165	0.00008	0.08238	0.67242	0.24321	0.00191			
2.419	0.00001	0.02672	0.53883	0.42302	0.01143			
2.674		0.00656	0.35802	0.59051	0.04488	0.00004		
2.928		0.00120	0.19485	0.67862	0.12487	0.00045		
3.183		0.00016	0.08520	0.64929	0.26207	0.00328		
3.438		0.00002	0.02922	0.51973	0.43513	0.01590	0.00001	
3.692			0.00772	0.34691	0.59068	0.05459	0.00009	
3.947			0.00155	0.19083	0.66723	0.13961	0.00078	
4.202			0.00023	0.08513	0.63202	0.27789	0.00472	
4.456			0.00003	0.03021	0.50335	0.44626	0.02015	0.00001
4.711				0.00832	0.33602	0.59211	0.06339	0.00016
4.966				0.00176	0.18533	0.65888	0.15285	0.00117
5.220				0.00019	0.07777	0.62056	0.29517	0.00630
5.475				0.00004	0.03048	0.48899	0.45617	0.02430
5.730					0.00856	0.32485	0.59459	0.07176
5.984					0.00190	0.18029	0.65110	0.16510
6.239					0.00033	0.08162	0.60330	0.30687
6.366					0.00012	0.05087	0.54700	0.38653

A.15. Numerical Values of $E_\beta(0; s)$, $\Psi_\beta(s)$, and $p_\beta(0; s)$ for $\beta = 1$ and 2 and $s \leq 3.7$

s	$E_1(0;s)$	$E_2(0;s)$	$\Psi_1(s)$	$\Psi_2(s)$	$p_1(0;s)$	$p_2(0;s)$
0.000	1.00000	1.00000				
0.064	0.93641	0.93634	0.00333	0.00028	0.10431	0.01326
0.127	0.87324	0.87275	0.01323	0.00223	0.20620	0.05221
0.191	0.81090	0.80937	0.02948	0.00742	0.30346	0.11438
0.255	0.74980	0.74647	0.05173	0.01721	0.39414	0.19592
0.318	0.69028	0.68436	0.07949	0.03267	0.47657	0.29186
0.382	0.63270	0.62344	0.11220	0.05455	0.54944	0.39649
0.446	0.57734	0.56413	0.14922	0.08321	0.61178	0.50384
0.509	0.52445	0.50685	0.18986	0.11864	0.66295	0.60804
0.573	0.47425	0.45204	0.23339	0.16045	0.70266	0.70370
0.637	0.42689	0.40008	0.27908	0.20796	0.73095	0.78628
0.764	0.34112	0.30596	0.37411	0.31607	0.75477	0.89937
0.891	0.26753	0.22632	0.46963	0.43360	0.73965	0.93351
1.019	0.20589	0.16171	0.56114	0.55059	0.69350	0.89288
1.146	0.15546	0.11149	0.64529	0.65853	0.62542	0.79483
1.273	0.11515	0.07411	0.71986	0.75156	0.54452	0.66302
1.401	0.08367	0.04747	0.78376	0.82693	0.45891	0.52079
1.528	0.05962	0.02929	0.83680	0.88453	0.37514	0.38659
1.655	0.04167	0.01739	0.87956	0.92621	0.29791	0.27194
1.783	0.02856	0.00994	0.91307	0.95482	0.23010	0.18166
1.910	0.01920	0.00547	0.93862	0.97350	0.17304	0.11543
2.037	0.01265	0.00289	0.95759	0.98510	0.12679	0.06986
2.165	0.00818	0.00147	0.97133	0.99197	0.09058	0.04032
2.292	0.00518	0.00072	0.98103	0.99585	0.06314	0.02221
2.419	0.00322	0.00034	0.98772	0.99794	0.04295	0.01168
2.546	0.00196	0.00015	0.99221	0.99902	0.02853	0.00588
2.674	0.00117	0.00007	0.99517	0.99955	0.01851	0.00283
2.801	0.00069	0.00003	0.99706	0.99980	0.01173	0.00130
2.928	0.00039	0.00001	0.99825	0.99992	0.00727	0.00057
3.056	0.00022		0.99898	0.99997	0.00440	0.00024
3.183	0.00012		0.99942	0.99999	0.00261	0.00010
3.310	0.00007		0.99968	1.00000	0.00151	0.00004
3.438	0.00003		0.99982		0.00085	0.00001
3.565	0.00002		0.99991		0.00047	
3.692	0.00001		0.99995		0.00026	

A.16. Proof of Equations (5.5.12)–(5.5.14) and (10.4.20)

Consider the integral equation

$$\lambda g(x) = \int_{-t}^{t} K(x,y)g(y)dy \tag{A.16.1}$$

whose solutions can always be chosen to be either even, $g(-x) = g(x)$, or odd, $g(-x) = -g(x)$. The set of even solutions, $g_{2n}(x)$, labeled by even subscripts, consists of all the solutions of an integral equation obtained from Eq. (A.16.1) when $K(x,y)$ there is replaced by

$$K_+(x,y) = \frac{1}{2}\left\{K(x,y) + K(-x,y)\right\}. \tag{A.16.2}$$

Similarly, the set of odd solutions $g_{2n+1}(x)$ of Eq. (A.16.1), labeled by odd subscripts, consists of all the solutions of an integral equation with the kernel

$$K_-(x,y) = \frac{1}{2}\left\{K(x,y) - K(-x,y)\right\}. \tag{A.16.3}$$

Let us denote the Fredholm determinants of $K(x,y)$, $K_+(x,y)$, and $K_-(x,y)$, by $D(t)$, $D_+(t)$, and $D_-(t)$, respectively. We have

$$\begin{aligned} D(t) &= \det[1-K] = \prod_{n=0}^{\infty}(1-\lambda_n) \\ &= \sum_{n=0}^{\infty}\frac{(-1)^n}{n!}\int\cdots\int dx_1\cdots dx_n \det\left[K\left(x_i,x_j\right)\right]_{i,j=1,\ldots,n}, \end{aligned} \tag{A.16.4}$$

and similar expressions for $D_+(t)$ and $D_-(t)$. In Eq. (A.16.4) and in what follows the integrals will be understood to be taken from $-t$ to t.

We want to show that

$$\frac{d^2D}{dt^2} \equiv \frac{d^2}{dt^2}(D_+D_-) = 4\left(\frac{dD_+}{dt}\right)\left(\frac{dD_-}{dt}\right), \tag{A.16.5}$$

and

$$\frac{D_-}{D_+} = 1 + \sum_{n=0}^{\infty} \frac{\lambda_{2n}}{1-\lambda_{2n}}\, g_{2n}(t) \int g_{2n}(x)dx \tag{A.16.6}$$

where the g_{2n} are normalized even solutions of Eq. (A.16.1):

$$\int g_{2n}^2(x)dx = 1; \tag{A.16.7}$$

provided the kernel $K(x,y)$ is an even function of the difference of its arguments:

$$K(x,y) = K(x-y) = K(y-x). \tag{A.16.8}$$

Proof. (i) Let $Q(x,y)$ be any of the three kernels K, K_+ or K_- and $\Phi(t)$ be the corresponding Fredholm determinant

$$\Phi(t) = \sum_{n=0}^{\infty} \frac{(-1)^n}{n!} \int \cdots \int dx_1 \cdots dx_n \det\left[Q\left(x_i, x_j\right)\right]_{i,j=1,\ldots,n}. \tag{A.16.9}$$

Differentiating the above equation with respect to t we have

$$\begin{aligned}\Phi'(t) &= \sum_{n=1}^{\infty} \frac{(-1)^n}{n!}\, n \int \cdots \int dx_1 \cdots dx_{n-1} \left\{G_{n-1}(t) + G_{n-1}(-t)\right\} \\ &= -2\sum_{n=0}^{\infty} \frac{(-1)^n}{n!} \int \cdots \int dx_1 \cdots dx_n\, G_n(t),\end{aligned} \tag{A.16.10}$$

where

$$G_n(t) = \det \begin{bmatrix} Q(t,t) & Q(t,x_j) \\ Q(x_i,t) & Q(x_i,x_j) \end{bmatrix}_{i,j=1,\ldots,n}, \tag{A.16.11}$$

and in arriving at $G_n(-t) = G_n(t)$ we have used the relations (A.16.8). A comparison of Eq. (A.16.10), (A.16.11) with the expansions of the "resolvent" and the "minors" in the Fredholm theory of integral equations will show that

$$\frac{\Phi'(t)}{\Phi(t)} = -2\left(\frac{Q}{1-Q}\right)(t,t), \tag{A.16.12}$$

with

$$\begin{aligned}
&\left(\frac{Q}{1-Q}\right)(x,y) \\
&\quad = \frac{\text{minor } (x,y) \text{ in } (1-Q)}{\det(1-Q)} \\
&\quad = \sum_{n=0}^{\infty} \int \cdots \int dx_1 \cdots dx_n \; Q(x,x_1)\, Q(x_1,x_2) \cdots Q(x_n,y)\,.
\end{aligned} \tag{A.16.13}$$

Applying this result to the even case we get an expression for D'_+/D_+ as

$$\begin{aligned}
&-2 \sum_{n=0}^{\infty} \int \cdots \int dx_1 \cdots dx_n \; K_+(t,x_1) \cdots K_+(x_n,t) \\
&\quad = -2 \sum_{n=0}^{\infty} \int \cdots \int dx_1 \cdots dx_n \; K(t,x_1) \cdots K(x_{n-1},x_n) K_+(x_n,t).
\end{aligned} \tag{A.16.14}$$

In a similar way

$$\frac{D'_-}{D_-} = -2 \sum_{n=0}^{\infty} \int \cdots \int dx_1 \cdots dx_n \; K(t,x_1) \cdots K(x_{n-1},x_n)\, K_-(x_n,t). \tag{A.16.15}$$

Adding and subtracting the last two equations we get

$$\begin{aligned} a(t) + b(t) &= -2 \sum_{n=0}^{\infty} \int \cdots \int dx_1 \cdots dx_n K(t, x_1) \cdots K(x_n, t) \\ &= -2 \left(\frac{K}{1-K} \right) (t, t), \end{aligned} \tag{A.16.16}$$

$$\begin{aligned} a(t) - b(t) &= -2 \sum_{n=0}^{\infty} \int \cdots \int dx_1 \cdots dx_n K(t, x_1) \cdots K(x_n, -t) \\ &= -2 \left(\frac{K}{1-K} \right) (t, -t), \end{aligned} \tag{A.16.17}$$

where for convenience we have put

$$a(t) = \frac{D'_+}{D_+}, \qquad b(t) = \frac{D'_-}{D_-}. \tag{A.16.18}$$

Differentiating Eq. (A.16.16) once more, we get

$$a' + b' = A_1 + A_2 + B_1 + B_2 \tag{A.16.19}$$

where

$$\begin{aligned} A_1 = & -2 \sum_{n=1}^{\infty} \sum_{j=1}^{n} \int \cdots \int dx_1 \cdots dx_{j-1} dx_{j+1} \cdots dx_n \\ & \times K(t, x_1) \cdots K(x_{j-1}, t) K(t, x_{j+1}) \cdots K(x_n, t), \end{aligned} \tag{A.16.20}$$

$$\begin{aligned} A_2 = & -2 \sum_{n=1}^{\infty} \sum_{j=1}^{n} \int \cdots \int dx_1 \cdots dx_{j-1} dx_{j+1} \cdots dx_n \\ & \times K(t, x_1) \cdots K(x_{j-1}, -t) K(-t, x_{j+1}) \cdots K(x_n, t), \end{aligned} \tag{A.16.21}$$

$$B_1 = -2 \sum_{n=0}^{\infty} \int \cdots \int dx_1 \cdots dx_n \, \frac{\partial K(t, x_1)}{\partial t} K(x_1, x_2) \cdots K(x_n, t), \tag{A.16.22}$$

and

$$B_2 = -2\sum_{n=0}^{\infty}\int\cdots\int dx_1\cdots dx_n\; K(t,x_1)\cdots K\left(x_{n-1},x_n\right)\frac{\partial K(x_n,t)}{\partial t}. \tag{A.16.23}$$

Since $K(x,y)$ depends only on $x-y$, we can in B_1 replace $\partial K(t,x_1)/\partial t$ by $-\partial K(t,x_1)/\partial x_1$ and integrate partially with respect to x_1. Next in the integral

$$\int\cdots\int dx_1\cdots dx_n\; K\left(t,x_1\right)\frac{\partial K(x_1,x_2)}{\partial x_1}K\left(x_2,x_3\right)\cdots K\left(x_n,t\right)$$

we replace $\partial K\left(x_1,x_2\right)/\partial x_1$ by $-\partial K\left(x_1,x_2\right)/\partial x_2$ and integrate partially with respect to x_2, and so on till the partial derivation is pushed to the extreme right. These step-by-step partial integrations give finally

$$B_1 = -A_1 + A_2 - B_2. \tag{A.16.24}$$

Equations (A.16.19), (A.16.20), (A.16.21), (A.16.24), and (A.16.17) give therefore

$$a' + b' = 2A_2 = -(a-b)^2, \tag{A.16.25}$$

which in view of Eq. (A.16.18) is the relation (A.16.5).

(ii) Provided the summation and integration can be interchanged, relation (A.16.6) can be written as

$$\frac{D_-}{D_+} = 1 + \int\left(\frac{K}{1-K}\right)(t,x)dx, \tag{A.16.26}$$

one has only to expand the resolvant in terms of the normalized eigenfunctions

$$\left(\frac{K}{1-K}\right)(t,x) = \sum_{n=0}^{\infty}\frac{\lambda_n}{1-\lambda_n}\,g_n(t)g_n(x) \tag{A.16.27}$$

and note that the odd functions contribute nothing on integration. We will prove relation (A.16.6) in the form (A.16.26).

Let us calculate the logarithmic derivatives with respect to t of the two sides of Eq. (A.16.26).

$$\left(\frac{D_+}{D_-}\right)\frac{d}{dt}\left(\frac{D_-}{D_+}\right) = b - a = 2\left(\frac{K}{1-K}\right)(t,-t) \tag{A.16.28}$$

from Equations (A.16.18) and (A.16.17). Also the derivative of the right-hand side is

$$\begin{aligned}
&\frac{d}{dt}\left\{1+\sum_{n=1}^{\infty}\int\cdots\int dx_1\cdots dx_n K(t,x_1)\cdots K\left(x_{n-1},x_n\right)\right\} \\
&\quad=\sum_{n=1}^{\infty}\sum_{j=1}^{n}\int\cdots\int dx_1\cdots dx_{j-1}dx_{j+1}\cdots dx_n \\
&\qquad\times\{K(t,x_1)\cdots K(x_{j-1},t)K(t,x_{j+1})\cdots K(x_{n-1},x_n) \\
&\qquad+K(t,x_1)\cdots K(x_{j-1},-t)K(-t,x_{j+1})\cdots K(x_{n-1},x_n)\} \\
&\qquad+\sum_{n=1}^{\infty}\int\cdots\int dx_1\cdots dx_n\frac{\partial K(t,x_1)}{\partial t}K\left(x_1,x_2\right)\cdots K\left(x_{n-1},x_n\right).
\end{aligned} \tag{A.16.29}$$

In the last line of the above equation one can again shift the partial derivation to the extreme right by successively replacing $\partial K(x,y)/\partial x$ by $-\partial K(x,y)/\partial y$ and integrating by parts. The expression in the last line of (A.16.29) is therefore

$$\begin{aligned}
&\sum_{n=1}^{\infty}\sum_{j=1}^{n}\int\cdots\int dx_1\cdots dx_{j-1}dx_{j+1}\cdots dx_n \\
&\qquad\times\left[-K\left(t,x_1\right)\cdots K\left(x_{j-1},t\right)K\left(t,x_{j+1}\right)\cdots K\left(x_{n-1},x_n\right)\right. \\
&\qquad\left.+\,K\left(t,x_1\right)\cdots K\left(x_{j-1},-t\right)K\left(-t,x_{j+1}\right)\cdots K\left(x_{n-1},x_n\right)\right].
\end{aligned} \tag{A.16.30}$$

Also

$$\sum_{n=1}^{\infty} \sum_{j=1}^{n} \int \cdots \int dx_1 \cdots dx_{j-1} dx_{j+1} \cdots dx_n$$

$$\times -K(t, x_1) \cdots K(x_{j-1}, -t) K(-t, x_{j+1}) \cdots K(x_{n-1}, x_n)$$

$$= \left(\frac{K}{1-K}\right)(t, -t) \left\{1 + \int dx \left(\frac{K}{1-K}\right)(-t, x)\right\} \tag{A.16.31}$$

and

$$\int dx \left(\frac{K}{1-K}\right)(-t, x) = \int dx \left(\frac{K}{1-K}\right)(t, -x)$$

$$= \int dx \left(\frac{K}{1-K}\right)(t, x). \tag{A.16.32}$$

Putting together Eqs. (A.16.28)–(A.16.32) we see that the logarithmic derivatives of the two sides of (A.16.26) are equal. In addition, Eq. (A.16.26) is obviously valid for $t = 0$. Thus, Eq. (A.16.26) is valid for all t.

Making a change of scale, $\xi = xt$, $\eta = yt$, the integral equation (5.3.14) takes the form (A.16.1) with

$$K(x, y) = \frac{\sin(x-y)\pi}{(x-y)\pi}.$$

(iii) The formal maniplulations encountered in this appendix are valid of course only when the various expansions are uniformly convergent so that term by term differentiation is allowed. This will be so if we take the above kernel K. So that relations (A.16.5) and (A.16.6) are the same as the Eqs. (5.5.14) and (10.4.20).

The relations (A.16.5) and (A.16.6) were circulated as conjectures. This appendix is based on a letter of Gaudin which he wrote in response to those conjectures.

A.17. Correlation Functions the Hard Way

Reading Sections 6.3–6.4, 7.1–7.2, 14.1, or 15.2 it appears miraculous that everything works so well with quaternion matrices and one may wonder how to guess the correct form in cases it works. When the joint probability density function contains the power 2 of $\Delta(x) = \prod_{i<j} |x_i - x_j|$, we have at our disposal Gram's formula and the theory of orthogonal polynomials with its powerful results. The same is not yet the case with quasi-orthogonal polynomials related to the power 1 or 4 of $\Delta(x)$, or to Pfaffians and quaternion matrices. So what one usually does is to compute the two point correlation function in an elementary, though tedious, way keeping as near as possible to the classical theory of orthogonal polynomials, and then try to guess the appropriate quaternion matrix. This was the path followed in all the cases worked out as yet, especially the Gaussian ensembles of Chapter 14. In this appendix we explain this pedestrian way of deriving the correlation functions.

Let $P(x_1, ..., x_N)$ be the joint probability density for the eigenvalues (or levels) $x_1, ..., x_N$. If we can evaluate

$$\left\langle \prod_{i=1}^{N} u(x_i) \right\rangle = \int \cdots \int P(x_1, ..., x_N) \prod_{i=1}^{N} u(x_i) d\mu(x_i) \tag{A.17.1}$$

in a closed form, with $u(x)$ an arbitrary function, then we will obtain all the correlation functions by functional differentiation. But this is difficult. Fortunately, as Dyson pointed out, we do not need to know the integral in any greater detail than its power series expansion when u is near unity. So let us write $u(x) = 1 + a(x)$ and treat $a(x)$ as a small quantity. When $P(x_1, ..., x_N)$ contains $\Delta(x)$ as a factor, one can use the method of integration over alternate variables to modify the unfavorable symmetry before expressing the integral (A.17.1) as a Pfaffian. When $P(x_1, ..., x_N)$ contains the fourth power of $\Delta(x)$, then this fourth power is a confluent alternant with each variable occuring in two columns and the theory of Pfaffians can be used directly. When $P(x_1, ..., x_N)$ contains a product of $\Delta(x)$ and another Pfaffian, a similar procedure works (*cf.* Pandey and Mehta (1983) for details). One can then expand this Pfaffian in a power series in $a(x)$ and compute the first few terms by brute force. This was the way the two-level correlation functions were first calculated in all the cases, before expressing $P(x_1, ..., x_N)$ as the determinant of an $N \times N$ self-dual quaternion matrix satisfying the conditions of Theorem 6.2.1.

A.18. Use of Pfaffians in Some Multiple Integrals. Proof of Equations (6.5.6), (14.3.6)–(14.3.8), and (A.27.6)

Consider an antisymmetric matrix $[a_{ij}]$; $i, j = 1, 2, ..., n$; $a_{ij} = -a_{ji}$. If its order n is odd, it is easy to see that its determinant is zero. On the other hand if n is even, it is known that its determinant is a perfect square (see, e.g., Mehta, 1989); the square root is

$$(\det [a_{ij}])^{1/2} = \frac{1}{(n/2)!} \sum \pm a_{i_1 i_2} a_{i_3 i_4} \cdots a_{i_{n-1} i_n}, \tag{A.18.1}$$

where the summation is extended over all permutations $(i_1, i_2, ..., i_n)$ of $(1, 2, ..., n)$, with the restrictions $i_1 < i_2$, $i_3 < i_4$, ..., $i_{n-1} < i_n$, and the sign is plus or minus depending on whether the permutation

$$\begin{pmatrix} 1 & 2 & \dots & n \\ i_1 & i_2 & \dots & i_n \end{pmatrix}$$

is even or odd. The expression (A.18.1) is known as the "Pfaffian."

Now consider evaluating the integrals (see de Bruijn, 1955)

$$I_1 = \int \cdots \int \det [\, A_i(x_j) \qquad B_i(x_j) \,] \, d\mu(x_1) \cdots d\mu(x_m), \tag{A.18.2}$$

and

$$I_2 = \int \cdots \int \det \begin{bmatrix} \alpha_0 & \beta(x_j) & \gamma(x_j) \\ \alpha_i & A_i(x_j) & B_i(x_j) \end{bmatrix} d\mu(x_1) \cdots d\mu(x_m), \tag{A.18.3}$$

$i = 1, ..., 2m$; $j = 1, ..., m$. The integrands in I_1 and I_2 are determinants of orders $2m \times 2m$ and $(2m+1) \times (2m+1)$, respectively; each variable occurs in two columns and $d\mu(x)$ is a suitable measure. Define

$$\rho_j = \int [\beta(x)B_j(x) - A_j(x)\gamma(x)]d\mu(x), \tag{A.18.4}$$

$$a_{ij} = \int [A_i(x)B_j(x) - A_j(x)B_i(x)]d\mu(x), \tag{A.18.5}$$

so that

$$a_{ij} = -a_{ji}. \tag{A.18.6}$$

Expanding the integrand in Eq. (A.18.2) and integrating independently over all the variables we make the following observations:

1. The integral I_1 is a sum of terms, each being a product of m numbers a_{ij};

2. The indices of the various a_{ij} occuring in any of the above terms are all different. In totality they are all the indices from 1 to $2m$.

3. We may restrict i to be less than j in each of a_{ij} occuring in I_1; for if i is not less than j we may replace a_{ij} by $-a_{ji}$.

4. The coefficient of the term $a_{i_1 i_2} a_{i_3 i_4} \cdots a_{i_{2m-1} i_{2m}}$ in I_1 is $+1$ or -1, depending on whether the permutation

$$\begin{pmatrix} i_1 & i_2 & \ldots & i_{2m} \\ 1 & 2 & \ldots & 2m \end{pmatrix}$$

is even or odd.

From these observations and Eq. (A.18.1) we conclude that

$$I_1 = m! \left(\det \left[a_{ij} \right]_{i,j=1,2,\ldots,2m} \right)^{1/2}. \tag{A.18.7}$$

Similarly, the integral I_2 is

$$I_2 = m! \text{ Pf } M = m! \ (\det M)^{1/2}, \tag{A.18.8}$$

where M is an $(2m+2) \times (2m+2)$ antisymmetric matrix with elements

$$\begin{aligned} M_{1j} &= -M_{j1} = \alpha_{j-2}, & j &= 2, ..., 2m+2, \\ M_{2j} &= -M_{j2} = \rho_{j-2}, & j &= 3, ..., 2m+2, \\ M_{jk} &= a_{j-2,k-2}, & 3 &\leq j,k \leq 2m+2, \\ M_{jj} &= 0, & j &= 1, ..., 2m+2. \end{aligned} \tag{A.18.9}$$

Equations (A.18.7) and (A.18.8) are general results. If $a_{ij} = 0$, for $i+j$ even, then the matrix $[a_{ij}]$ has a checker board structure, every alternate element being zero; and there are simplifications. In this case I_1 is an $m \times m$ determinant and I_2 is a sum of two $(m+1) \times (m+1)$ determinants.

$$I_1 = m! \det \left[a_{2i-1,2j}\right]_{i,j=1,\dots,m}, \tag{A.18.10}$$

and

$$I_2 = m! \det \begin{bmatrix} 0, & \rho_{2j} \\ \alpha_{2i-1}, & a_{2i-1,2j} \end{bmatrix} + m! \det \begin{bmatrix} \alpha_0, & -\alpha_{2j} \\ \rho_{2i-1}, & a_{2i-1,2j} \end{bmatrix}. \tag{A.18.11}$$

The integrand in Eq. (6.5.5) is also a determinant where each variable occurs in two columns. By parity argument

$$a_{2i+1,2j+1} = \int_{\text{out}} [F_{2i}(x)\varphi_{2j}(x) - F_{2j}(x)\varphi_{2i}(x)]dx = 0, \tag{A.18.12}$$

$$a_{2i,2j} = \int_{\text{out}} \left(F_{2i-1}(x)\varphi'_{2j-2}(x) - F_{2j-1}(x)\varphi'_{2i-2}(x)\right) dx = 0, \tag{A.18.13}$$

while

$$\begin{aligned} a_{2i+1,2j+2} &= \int_{\text{out}} \left(F_{2i}(x)\varphi'_{2j}(x) - F_{2j+1}(x)\varphi_{2i}(x) \right) dx \\ &= \int_{y<x,\text{out}} \int \left(\varphi_{2i}(y)\varphi'_{2j}(x) - \varphi'_{2j}(y)\varphi_{2i}(x) \right) dx \\ &= -2 \int_{\text{out}} \varphi_{2i}(x)\varphi_{2j}(x)dx = -2 \left(\delta_{ij} - \int_{-\theta}^{\theta} \varphi_{2i}(x)\varphi_{2j}(x)dx \right) \\ &= -2\, \bar{g}_{ij}. \end{aligned} \tag{A.18.14}$$

From Eqs. (A.18.2), (A.18.10), and (A.18.14) we get Eq. (6.5.6). Similarly, from (A.18.11) we get (6.7.12).

A.19. Determinants of the Forms $[\delta_{ij} - u_i v_j]$ and $[A + u_i - u_j - \text{sgn}(i-j)]$

The determinant in Eq. (6.6.12) is of the form $\det[\delta_{ij} - u_i v_j]$. To evaluate such a determinant it is convenient to border it with an extra row and column. Thus

$$\begin{aligned} &\det \begin{bmatrix} 1-u_1v_1 & -u_1v_2 & \dots & -u_1v_n \\ -u_2v_1 & 1-u_2v_2 & \dots & -u_2v_n \\ \dots & \dots & \dots & \dots \\ -u_nv_1 & -u_nv_2 & \dots & 1-u_nv_n \end{bmatrix} \\ &= \det \begin{bmatrix} 1 & 0 & 0 & \dots & 0 \\ u_1 & 1-u_1v_1 & -u_1v_2 & \dots & -u_1v_n \\ u_2 & -u_2v_1 & 1-u_2v_2 & \dots & -u_1v_n \\ \dots & \dots & \dots & \dots & \dots \\ u_n & -u_nv_1 & -u_nv_2 & \dots & 1-u_nv_n \end{bmatrix} \\ &= \det \begin{bmatrix} 1 & v_1 & v_2 & \dots & v_n \\ u_1 & 1 & 0 & \dots & 0 \\ u_2 & 0 & 1 & \dots & 0 \\ \dots & \dots & \dots & \dots & \dots \\ u_n & 0 & 0 & \dots & 1 \end{bmatrix} \\ &= 1 - \sum_{j=1}^{n} u_j v_j. \end{aligned} \tag{A.19.1}$$

The cofactor b_{ij} of the element (i, j) in $\det[\delta_{ij} - u_i v_j]$ is easy to calculate by the same bordering

$$b_{ij} = \delta_{ij}\left(1 - \sum_{k=1}^{n} u_k v_k\right) + u_j v_i. \tag{A.19.2}$$

To calculate $\det[A + u_i - u_j - \mathrm{sgn}(i-j)]$, $i, j = 1, ..., n$, appearing in Eq. (6.3.35), we border it again with two rows and two columns; thus

$$\begin{aligned}
&\det[A + u_i - u_j - \mathrm{sgn}(i-j)] \\
&= \det\begin{bmatrix} 1 & 0 & 0 \\ A & 1 & -A + u_j \\ -A - u_i & 0 & A + u_i - u_j - \mathrm{sgn}(i-j) \end{bmatrix} \\
&= \det\begin{bmatrix} 1 & 0 & 1 \\ A & 1 & u_j \\ -u_i & 1 & -\mathrm{sgn}(i-j) \end{bmatrix} \\
&= \det\begin{bmatrix} 0 & 1 & 1 \\ -1 & A & u_j \\ -1 & -u_i & -\mathrm{sgn}(i-j) \end{bmatrix} \\
&= \det\begin{bmatrix} 0 & 1 & 1 \\ -1 & 0 & u_j \\ -1 & -u_i & -\mathrm{sgn}(i-j) \end{bmatrix} \\
&\quad + A \det\begin{bmatrix} 0 & 1 \\ -1 & -\mathrm{sgn}(i-j) \end{bmatrix}
\end{aligned}$$

The last two determinants are of anti-symmetric matrices; their orders differ by one. If n is odd, the first determinant is zero and if n is even, the second one is zero. The non-zero determinant is the square of its Pfaffian. Thus,

$$\det\left[A + u_i - u_j - \operatorname{sgn}(i-j)\right] = A, \qquad \text{if } n \text{ is odd,}$$

$$= (1 + u_1 - u_2 + \cdots + u_{n-1} - u_n)^2, \qquad \text{if } n \text{ is even.}$$

A.20. Power Series Expansion of I_m (θ), Equation (6.8.12)

Expanding the integrand in (6.8.10) in powers of θ and integrating, we get

$$\eta_i(\theta) = \xi_{2i} - \frac{1}{2}\theta^2 \xi_{2i-2} + \frac{3}{8}\theta^4 \xi_{2i-4} + \cdots, \tag{A.20.1}$$

where

$$\xi_{2i} = 2\int_0^\infty e^{-y^2} y^{2i} dy = \Gamma\left(i + \frac{1}{2}\right).$$

Taking terms only up to θ^4, we put the expansion (A.20.1) in the determinant (6.8.9). We see that in writing the determinant as a sum of several terms many of them vanish because two rows are proportional. Thus we may write

$$I_m(\theta) = \text{const} \times \left(a_0 - \frac{1}{2}\theta^2 a_1 + \frac{3}{8}\theta^4 (a_2 - a_3) + \cdots\right) \tag{A.20.2}$$

where

$$a_0 = \det\left[\xi_{2i+2j}\right]_{i,j=1,\ldots,m-1},$$

$$a_1 = \det\begin{bmatrix} \xi_2 & \xi_4 & \cdots & \xi_{2m-2} \\ \xi_6 & \xi_8 & \cdots & \xi_{2m+2} \\ \cdots & \cdots & \cdots & \cdots \\ \xi_{2m} & \xi_{2m+2} & \cdots & \xi_{4m-4} \end{bmatrix},$$

$$a_2 = \det\begin{bmatrix} \xi_0 & \xi_2 & \cdots & \xi_{2m-4} \\ \xi_6 & \xi_8 & \cdots & \xi_{2m+2} \\ \cdots & \cdots & \cdots & \cdots \\ \xi_{2m} & \xi_{2m+2} & \cdots & \xi_{4m-4} \end{bmatrix},$$

$$a_3 = \det\begin{bmatrix} \xi_2 & \xi_4 & \cdots & \xi_{2m-2} \\ \xi_4 & \xi_6 & \cdots & \xi_{2m} \\ \xi_8 & \xi_{10} & \cdots & \xi_{2m+4} \\ \cdots & \cdots & \cdots & \cdots \\ \xi_{2m} & \xi_{2m+2} & \cdots & \xi_{4m-4} \end{bmatrix}.$$

The evaluation of determinants whose elements are gamma functions is almost as easy as those whose elements are the successive factorials. Taking out all the common factors, one may reduce these determinants to the triangular form by simple operations. Thus we obtain

$$a_0 = \frac{2}{\sqrt{\pi}(m-1)!} \prod_{j=1}^{2m-1} \Gamma\left(1 + \frac{1}{2}j\right), \quad a_1 = \frac{2}{3}(m-1)a_0,$$

and

$$a_2 = \frac{2}{3}m(m-1)a_0, \quad a_3 = \frac{2}{45}(m-1)(m-2)a_0.$$

Putting these values in Eq. (A.20.2) we get Eq. (6.8.12).

A.21. Proof of the Inequalities (6.8.15)

Let $u_i = \theta^2/y_i^2$ so that $u_i \geq 0$. The second inequality in (6.8.15)

$$\prod_1^n (1 + u_i)^{-1/2} \leq 1$$

is immediate, for each factor in the product lies between 0 and 1.

The first inequality can be proved by induction. Suppose that

$$1 - \frac{1}{2}\sum_{i=1}^{r} u_i \leq \prod_{i=1}^{r} (1+u_i)^{-1/2}$$

is true for $1 \leq r \leq n$ and let us prove then that

$$1 - \frac{1}{2}\sum_{i=1}^{n+1} u_i \leq \prod_{i=1}^{n+1} (1+u_i)^{-1/2}.$$

Let $\sum_1^{n+1} u_i \leq 2$, for otherwise the left-hand side will be negative and therefore smaller than the right-hand side, which is positive. Thus

$$\left(1 - \frac{1}{2}\sum_1^{r} u_i\right)\left(1 - \frac{1}{2}\sum_{r+1}^{n+1} u_i\right)$$

$$\leq \prod_1^{r} (1+u_i)^{-1/2} \prod_{r+1}^{n+1} (1+u_i)^{-1/2}, \quad r \leq n,$$

for both quantities in the product on both sides are positive. Therefore we have

$$\prod_{i=1}^{n+1} (1+u_i)^{-1/2} \geq 1 - \frac{1}{2}\sum_{i=1}^{n+1} u_i + \left(\frac{1}{2}\sum_1^{r} u_i\right)\left(\frac{1}{2}\sum_{r+1}^{n+1} u_i\right)$$

$$\geq 1 - \frac{1}{2}\sum_{i=1}^{n+1} u_i.$$

Also it is easy to verify that

$$1 - \frac{1}{2}u_1 \leq (1+u_1)^{-1/2},$$

and the proof is complete.

A.22. Proof of Equations (9.1.11) and (9.2.11)

Let dM and dM' be connected by a similarity transformation

$$dM' = A\, dM A^{-1}, \tag{A.22.1}$$

where A is nonsingular. We now show that the Jacobian

$$J = \frac{\partial \left(dM'_{ij}\right)}{\partial \left(dM_{ij}\right)} \tag{A.22.2}$$

is unity.

Considering the various matrix elements dM'_{ij} as components of a single vector (and similarly for dM_{ij}), we can write Eq. (A.22.1) as

$$dM'_{ij} = \sum_{k,\ell} A_{ik} A^{-1T}_{j\ell}\, dM_{k\ell}$$

or

$$dM' = \left(A \times A^{-1T}\right) dM, \tag{A.22.3}$$

where the direct or the Kronecker product $(A \times B)$ is defined by the equation

$$(A \times B)_{ij,k\ell} = A_{ik} B_{j\ell}.$$

The Jacobian (A.22.2) is thus seen to be equal to the determinant of $\left(A \times A^{-1T}\right)$.

Now it can be easily verified that if P and Q are matrices of the order $(n \times n)$, whereas R and S are of the order $(m \times m)$,

$$(P \times R) \cdot (Q \times S) = (P \cdot Q) \times (R.S), \tag{A.22.4}$$

where a dot means ordinary matrix multiplication. From Eq. (A.22.4) we obtain

$$(R \times P) = (R \times 1_n) \cdot (1_m \times P),$$

where 1_r is the $(r \times r)$ unit matrix. Taking determinants on both sides of this equation we have

$$\det(R \times P) = (\det\, R)^n (\det\, P)^m. \tag{A.22.5}$$

Thus

$$J = \frac{\partial \left(dM'_{ij}\right)}{\partial \left(dM_{ij}\right)} = \det \left(A \times A^{-1T}\right) = \left[(\det\ A)\left(\det\ A^{-1}\right)\right]^{N} = 1,$$

which establishes the result we wanted.

A.23. Proof of Theorems 10.9.1 and 10.9.2

We start by proving a few lemmas. At first glance they may look unrelated, but their relevance will become clearer as we proceed.

Lemma A.23.1. *Given a sequence of numbers* $\nu_1,\ \nu_2,\ \ldots,\ \nu_N$, *let* $\Delta\left(\nu_a, \nu_b, \nu_c, \ldots\right)$ *be defined by*

$$\Delta\left(\nu_a, \nu_b, \nu_c, \ldots\right) = \left[\nu_a\left(\nu_a + \nu_b\right)\left(\nu_a + \nu_b + \nu_c\right)\cdots\right]^{-1} \tag{A.23.1}$$

We then have the identity

$$\sum_{k=0}^{N}(-1)^k \Delta\left(\nu_k, \nu_{k-1}, \ldots, \nu_1\right)\Delta\left(\nu_{k+1}, \ldots, \nu_N\right) = 0. \tag{A.23.2}$$

It is assumed here and in the following that whenever the corresponding index does not exist the relevant factor is unity. For example, in Eq. (A.23.2) the $k = 0$ term is $\Delta\left(\nu_1, \nu_2, \ldots, \nu_N\right)$ and the $k = N$ term is $(-1)^N \Delta\left(\nu_N, \nu_{N-1}, \ldots, \nu_1\right)$.

Proof. Consider the left-hand side as a function of, say, ν_1. Its possible singularities are at the points $\nu_1 = \left(\nu_2 + \cdots + \nu_\ell\right)$ and at $\nu_1 = 0$. At $\nu_1 = -\left(\nu_2 + \cdots + \nu_\ell\right)$ two terms in the sum become singular,

$k = 0$ and $k = \ell$:

$$\Delta\left(\nu_1, \nu_2, ..., \nu_\ell, ..., \nu_N\right) + (-1)^\ell \Delta\left(\nu_\ell, ..., \nu_1\right) \Delta\left(\nu_{\ell+1}, ..., \nu_N\right)$$

$$= [\nu_1\left(\nu_1 + \nu_2\right) \cdots \left(\nu_1 + \cdots + \nu_{\ell-1}\right)$$

$$\times \left(s_\ell + \nu_{\ell+1}\right) \cdots \left(s_\ell + \nu_{\ell+1} + \cdots + \nu_N\right) s_\ell]^{-1}$$

$$+ (-1)^\ell [\nu_\ell\left(\nu_\ell + \nu_{\ell-1}\right) \cdots \left(\nu_\ell + \cdots + \nu_2\right)$$

$$\times \nu_{\ell+1}\left(\nu_{\ell+1} + \nu_{\ell+2}\right)\left(\nu_{\ell+1} + ... + \nu_N\right) s_\ell]^{-1}, \tag{A.23.3}$$

where $s_\ell = \nu_1 + \nu_2 + \cdots + \nu_\ell$. The coefficients of s_ℓ^{-1} in the two terms are equal and opposite at $s_\ell = 0$. To verify this, replace ν_ℓ in the second term by $-\left(\nu_1 + \cdots + \nu_{\ell-1} - s_\ell\right)$, $\left(\nu_\ell + \nu_{\ell-1}\right)$ by $-\left(\nu_1 + \cdots + \nu_{\ell-2} - s_\ell\right), ...,$ and $\left(\nu_\ell + \nu_{\ell-1} + \cdots + \nu_2\right)$ by $-\left(\nu_1 - s_\ell\right)$, so that at $s_\ell = 0$ factors in the two terms may be compared.

This holds for all ℓ, including $\ell = 1$; that is, the point $\nu_1 = 0$. The original function of ν_1 is therefore nowhere singular and so must be a polynomial in ν_1. Because every term in the sum Eq. (A.23.2) contains at least one linear factor in ν_1 in the denominator, the polynomial mentioned above must be identically zero.

Lemma A.23.2. *Let $\mathcal{A}$ be the antisymmetrization operator, $\mathcal{A} = \sum_P \epsilon_P P$, where P is a permutation of the variables $\nu_1, \nu_2, ..., \nu_N$ and ϵ_p is its parity. We have the identity*

$$(-1)^N \mathcal{A}\, \Delta\left(\nu_1, \nu_2, ..., \nu_N\right) = \left(\nu_1 + \nu_2 + \cdots + \nu_N\right)^{-1} \times \sum_{k=1}^{N} (-1)^k \mathcal{A}\, \Delta\left(\nu_1, ..., \nu_{k-1}, \nu_{k+1}, ..., \nu_N\right). \tag{A.23.4}$$

Proof. The right-hand side is antisymmetric in all of the variables $\nu_1, \nu_2, ..., \nu_N$ and contains $N(N-1)! = N!$ terms. It also contains

$$\Delta(\nu_1, ..., \nu_N) = (\nu_1 + \cdots + \nu_N)^{-1} \Delta(\nu_1, ..., \nu_{N-1})$$

with the coefficient $(-1)^N$. Therefore, the two sides of (A.23.4) are equal.

Lemma A.23.3. *The antisymmetrized* $\Delta(\nu_1, \nu_2, ..., \nu_N)$ *is*

$$\mathcal{A}\,\Delta(\nu_1, \nu_2, ..., \nu_N) = \prod_{k=1}^{N} \nu_k^{-1} \prod_{1 \le j < k \le N} \left[(\nu_k - \nu_j)(\nu_k + \nu_j)^{-1} \right]. \tag{A.23.5}$$

We prove (A.23.5) by induction. Let it then be valid for $N-1$. Using Eq. (A.23.5) on the right-hand side of the identity (A.23.4) and reducing to a common denominator, we note that the degree of this common denominator is $1 + N + N(N-1)/2$, and by counting factors in each term of the sum we find that the degree of the common numerator is $1 + N(N-1)/2$. Because the right-hand side of Eq. (A.23.4) is antisymmetric, it vanishes if $\nu_j = \nu_k$; the numerator therefore must contain a factor $(\nu_j - \nu_k)$. On taking out all such factors that are $N(N-1)/2$ in number, we are left with a linear factor symmetric in all the $\nu_1, \nu_2, ..., \nu_N$. This linear factor can only by $(\nu_1 + \nu_2 + \cdots + \nu_N)$.

A constant multiplier is still undecided. For this we suppose that $\nu_1 \ll \nu_2 \ll \cdots \ll \nu_N$ and compare the dominant term on each side of Eq. (A.23.5). This term is $(\nu_1 \nu_2 \cdots \nu_N)^{-1}$. Finally, (A.23.5) is true for $N = 2$, as can be easily verified.

Lemma A.23.4. *The integral*

$$\int_{\varphi_1}^{\varphi_2} d\theta_N \int_{\varphi_1}^{\theta_N} d\theta_{N-1} \cdots \int_{\varphi_1}^{\theta_2} d\theta_1 \, \exp\left[i(\nu_1\theta_1 + \cdots + \nu_N\theta_N)\right]$$

$$\equiv \int\int d\theta_1 \cdots d\theta_N \, \exp\left[i(\nu_1\theta_1 + \cdots + \nu_N\theta_N)\right] \tag{A.23.6}$$

over the region $\varphi_1 \le \theta_1 \le \theta_2 \le \cdots \le \theta_N \le \varphi_2$ *is equal to*

$$\sum_{k=0}^{N} i^{2k-N} \exp\left[i\left(\nu_1 + \cdots + \nu_k\right)\varphi_1 + i\left(\nu_{k+1} + \cdots + \nu_N\right)\varphi_2\right]$$

$$\times \Delta\left(\nu_k, ..., \nu_1\right) \Delta\left(\nu_{k+1}, ..., \nu_N\right). \qquad \text{(A.23.7)}$$

The proof is by induction. Assume that

$$\int\int d\theta_1 \cdots d\theta_{N-1} \exp\left[i\left(\nu_1\theta_1 + \cdots + \nu_{N-1}\theta_{N-1}\right)\right]$$

over the region $\varphi_1 \le \theta_1 \le \theta_2 \le \cdots \le \theta_N$ is

$$\sum_{k=0}^{N-1} i^{2k-(N-1)} \exp[i\left(\nu_1 + \cdots + \nu_k\right)\varphi_1 + i\left(\nu_{k+1} + \cdots + \nu_{N-1}\right)\theta_N]$$

$$\times \Delta\left(\nu_k, ..., \nu_1\right) \Delta\left(\nu_{k+1}, ..., \nu_{N-1}\right). \qquad \text{(A.23.8)}$$

Therefore, the integral (A.23.6) is

$$\sum_{k=0}^{N-1} i^{2k-N+1} e^{i(\nu_1+\cdots+\nu_k)\varphi_1} \left[e^{i(\nu_{k+1}+\cdots+\nu_N)\varphi_2} - e^{i(\nu_{k+1}+\cdots+\nu_N)\varphi_1}\right]$$

$$\times i^{-1}\left(\nu_{k+1} + \cdots + \nu_N\right)^{-1} \Delta\left(\nu_k, \nu_{k-1}, ..., \nu_1\right) \Delta\left(\nu_{k+1}, ..., \nu_{N-1}\right)$$

$$= \sum_{k=0}^{N-1} i^{2k-N} \exp\left[i\left(\nu_1 + \cdots + \nu_k\right)\varphi_1 + i\left(\nu_{k+1} + \cdots + \nu_N\right)\varphi_2\right]$$

$$\times \Delta\left(\nu_k, \nu_{k-1}, ..., \nu_1\right) \Delta\left(\nu_{k+1}, ..., \nu_N\right)$$

$$- \sum_{k=0}^{N-1} i^{2k-N} e^{i(\nu_1+\ldots+\nu_N)\varphi_1}$$

$$\times \Delta\left(\nu_k, \nu_{k-1}, ..., \nu_1\right) \Delta\left(\nu_{k+1}, ..., \nu_N\right), \qquad \text{(A.23.9)}$$

where we have used the identity

$$(\nu_{k+1} + \cdots + \nu_N)^{-1} \Delta(\nu_{k+1}, ..., \nu_{N-1}) = \Delta(\nu_{k+1}, ..., \nu_N).$$

By Lemma A.23.1, the second sum in Eq. (A.23.9) is the $k = N$ term in Eq. (A.23.7).

Finally, for $N = 1$,

$$\begin{aligned} \int_{\varphi_1}^{\varphi_2} \exp(i\nu_1\theta_1) d\theta_1 &= \frac{1}{i\nu_1} \left(e^{i\nu_1\varphi_2} - e^{i\nu_1\varphi_1}\right) \\ &= i^{-1} e^{i\nu_1\varphi_2} \Delta(\nu_1) + i^{2-1} e^{i\nu_1\varphi_1} \Delta(\nu_1). \end{aligned} \quad \text{(A.23.10)}$$

Thus, we have proved the lemma.

PROOF OF THEOREM 10.9.1

From Lemmas A.23.3 and A.23.4,

$$\begin{aligned} I &= \mathcal{A} \int\int d\theta_1 \cdots d\theta_{2m} \exp[i(\nu_1\theta_1 + \cdots + \nu_{2m}\theta_{2m})] \\ &= \mathcal{A} \sum_{k=0}^{2m} i^{2k-2m} \exp[i(\nu_1 + \cdots + \nu_k) 2\varphi_1 \\ &\qquad + i(\nu_{k+1} + \cdots + \nu_{2m}) 2\varphi_2] \\ &\quad \times \Delta(\nu_k, \nu_{k-1}, ..., \nu_1) \Delta(\nu_{k+1}, ..., \nu_{2m}), \\ &= \sum_{k=0}^{2m} (-1)^{k-m} \sum_{(\alpha)} \exp[i(\nu_{\alpha_1} + \cdots + \nu_{\alpha_k}) 2\varphi_1 \\ &\qquad + i\left(\nu_{\alpha_{k+1}} + \cdots + \nu_{\alpha_{2m}}\right) 2\varphi_2] \\ &\quad \times (-1)^{\Sigma_1^k(\alpha_j - j)} \mathcal{A} \, \Delta\left(\nu_{\alpha_k}, \nu_{\alpha_{k-1}}, ..., \nu_{\alpha_1}\right) \end{aligned}$$

$$\times \mathcal{A}\,\Delta\left(\nu_{\alpha_{k+1}}, \nu_{\alpha_{k+2}}, ..., \nu_{\alpha_{2m}}\right)$$

$$= \sum_{k=0}^{2m} (-1)^{k-m} \sum_{(\alpha)} \exp\left(2i\varphi_1 \sum_{1}^{k} \nu_{\alpha_j} + 2i\varphi_2 \sum_{k+1}^{2m} \nu_{\alpha_j}\right)$$

$$\times (-1)^{\Sigma_1^k (\alpha_j - j)}$$

$$\times (-1)^{k(k-1)/2} \mathcal{A}\,\Delta\left(\nu_{\alpha_1}, \nu_{\alpha_2}, ..., \nu_{\alpha_k}\right)$$

$$\times \mathcal{A}\,\Delta\left(\nu_{\alpha_{k+1}}, \nu_{\alpha_{k+2}}, ..., \nu_{\alpha_{2m}}\right), \tag{A.23.11}$$

where $\Sigma(\alpha)$ means a summation with the conditions $\alpha_1 < \alpha_2 < \cdots < \alpha_k$, $\alpha_{k+1} < \alpha_{k+2} < \cdots < \alpha_{2m}$ over all permutations $\alpha_1, \alpha_2, ..., \alpha_{2m}$ of the indices $1, 2, ..., 2m$.

Let us call $\alpha_1, \alpha_2, ..., \alpha_k$ the "occupied" states and $\alpha_{k+1}, \alpha_{k+2}, ..., \alpha_{2m}$, the "unoccupied" states. We define an "occupation number" ϵ_k, which is $+1$ for the occupied states, $\alpha_1, \alpha_2, ..., \alpha_k$, and zero for the unoccupied states, $\alpha_{k+1}, \alpha_{k+2}, ..., \alpha_{2m}$. With this device we can extend the summation over all indices $1, 2, ..., 2m$. For example,

$$\sum_{j=1}^{k} \alpha_j = \sum_{j=1}^{2m} j\epsilon_j, \qquad \sum_{j=1}^{k} \nu_{\alpha_j} = \sum_{j=1}^{2m} \epsilon_j \nu_j, \qquad \text{etc.} \tag{A.23.12}$$

Using (A.23.5), we obtain from (A.23.11)

$$I = \sum_{\epsilon_1, ..., \epsilon_{2m}} (-1)^{m + \Sigma j\epsilon_j} \exp\left(2i \sum \epsilon_j \nu_j \varphi_1 + 2i \sum \left(1 - \epsilon_j\right) \nu_j \varphi_2\right)$$

$$\times \left(\nu_1 \cdots \nu_{2m}\right)^{-1}$$

$$\times \prod_{1 \le j < k \le 2m} \left[\left(\nu_k - \nu_j\right)\left(\nu_k + \nu_j\right)^{-1}\right]^{(1-\epsilon_k)(1-\epsilon_j) + \epsilon_k \epsilon_j}, \tag{A.23.13}$$

for $\epsilon_k \epsilon_j + \left(1 - \epsilon_k\right)\left(1 - \epsilon_j\right)$ is unity when both states j, k are occupied, or when both are unoccupied, and $\epsilon_k \epsilon_j + \left(1 - \epsilon_k\right)\left(1 - \epsilon_j\right)$ is zero otherwise.

A more symmetrical form is obtained by introducing the "spin" that takes the value +1 for the occupied states and −1 for the unoccupied states

$$s_k = 2\left(\epsilon_k - 1/2\right). \tag{A.23.14}$$

Equation (A.23.13) gives then

$$\begin{aligned} I = &\sum_{s_1,\ldots,s_{2m}} (-1)^{m+\Sigma j(1+s_j)/2} \\ &\times \exp\left(i(\varphi_1+\varphi_2)\sum_j \nu_j + i(\varphi_1-\varphi_2)\sum_j s_j\nu_j \right) \\ &\times \left(\nu_1\cdots\nu_{2m}\right)^{-1} \sum_{1\le j<k\le 2m} \left(\left(\nu_k-\nu_j\right)\left(\nu_k+\nu_j\right)^{-1}\right)^{(1+s_js_k)/2}, \end{aligned} \tag{A.23.15}$$

$$(-1)^{\Sigma j(1+s_j)/2} = \prod_j (-s_j)^j, \tag{A.23.16}$$

and

$$\begin{aligned} &\sum_{1\le j<k\le 2m} \left(\left(\nu_k-\nu_j\right)\left(\nu_k+\nu_j\right)^{-1}\right)^{(1+s_js_k)/2} \\ &\quad = \sum_{1\le j<k\le 2m} \left(\left(\nu_k-\nu_j s_j s_k\right)\left(\nu_k+\nu_j\right)^{-1}\right) \\ &\quad = \prod_{1\le j<k\le 2m} \left[\left(\nu_k+\nu_j\right)^{-1}\left(\nu_k s_k - \nu_j s_j\right)\right] \prod_k s_k^{k-1}. \end{aligned} \tag{A.23.17}$$

Substituting (A.23.16) and (A.23.17) into (A.23.15), we get

$$\begin{aligned} I = &(-1)^m \exp\left[i\left(\varphi_1+\varphi_2\right)\sum_j \nu_j \right] \\ &\times (-1)^{m(2m+1)} \left(\nu_1\cdots\nu_{2m}\right)^{-1} \prod_{1\le j<k\le 2m} \left(\nu_k+\nu_j\right)^{-1} \end{aligned}$$

$$\times \sum_{s_1,...,s_{2m}} \prod_{1\le j<k\le 2m} (\nu_k s_k - \nu_j s_j)$$

$$\times \exp\left[i(\varphi_1 - \varphi_2)\sum_j s_j\nu_j \right] \prod_j s_j$$

$$= \exp\left[i(\varphi_1 + \varphi_2)\sum_j \nu_j \right] (\nu_1 ... \nu_{2m})^{-1} \prod_{1\le j<k\le 2m} (\nu_k + \nu_j)^{-1}$$

$$\times \sum_{s_1,...,s_{2m}} \exp\left[i(\varphi_1 - \varphi_2)\sum_j s_j\nu_j \right] \det\left[s_j^k \nu_j^{k-1} \right]_{j,k=1,...,2m}. \tag{A.23.18}$$

The summation over s_j in Eq. (A.23.18) is over all possible choices $s_j = \pm 1$. By absorbing the exponential factor $\exp[i(\varphi_1 - \varphi_2)\, s_j\nu_j]$ in the jth row of the determinant in Eq. (A.23.18), we see that each row depends only on a single s_j. The summation over s_j can therefore be carried out in each row, independently of the others.

$$\sum_{s_j=\pm 1} \exp\left[i(\varphi_1 - \varphi_2)\, s_j\nu_j \right] s_j^{2k} \nu_j^{2k-1} = 2\nu_j^{2k-1} \cos \nu_j(\varphi_2 - \varphi_1), \tag{A.23.19}$$

$$\sum_{s_j=\pm 1} \exp\left[i(\varphi_1 - \varphi_2)\, s_j\nu_j \right] s_j^{2k+1} \nu_j^{2k} = -2i\nu_j^{2k} \sin \nu_j(\varphi_2 - \varphi_1), \tag{A.23.20}$$

Theorem 10.9.1 is now evident.

PROOF OF THEOREM 10.9.2

In Theorem 10.9.1 take the limits $-\nu_{2j-1} = \nu_{2j} = j - 1/2$. Because so many factors vanish in the numerator and the denominator, it is best to add the $(2j)$th row to the $(2j-1)$th, divide the sum by $\nu_{2j-1} + \nu_{2j}$, and then take the limits $\nu_{2j-1} + \nu_{2j} \longrightarrow 0$. Finally put $\nu_{2j} = j - 1/2$. This procedure gives

$$\lim \prod_{j=1}^{m} (\nu_{2j-1} + \nu_{2j})^{-1} \det \left[\nu_j^{2k-2} \sin \nu_j x \quad \nu_j^{2k-1} \cos \nu_j x \right] = D^T(x), \tag{A.23.21}$$

where $D^T(x)$ is the transpose of $D(x)$, Eq. (10.9.7). Also

$$\nu_1 \cdots \nu_{2m} = (-1)^m \sum_{j=1}^{m} (j - 1/2)^2 = (-1)^m \left[\frac{(2m)!}{2^{2m} m!} \right]^2, \tag{A.23.22}$$

$$\nu_1 + \cdots + \nu_{2m} = 0, \tag{A.23.23}$$

and a straightforward, though lengthy, calculation gives

$$\begin{aligned}
&\prod_{1 \le j < k \le 2m} (\nu_j + \nu_k) \prod_{j=1}^{m} (\nu_{2j-1} + \nu_{2j})^{-1} \\
&= \prod_{1 \le j < k \le m} (\nu_{2j-1} + \nu_{2k-1}) \\
&\quad \times (\nu_{2j-1} + \nu_{2k}) (\nu_{2j} + \nu_{2k-1}) (\nu_{2j} + \nu_{2k}) \\
&= \prod_{1 \le j < k \le m} (\nu_{2j} + \nu_{2k})^2 (-\nu_{2j} + \nu_{2k})^2 \\
&= \prod_{1 \le j < k \le m} (j + k - 1)^2 (k - j)^2 \\
&= \prod_{j=1}^{m-1} \left(\frac{(m + j - 1)!}{(2j - 1)!} \right)^2 (j!)^2 = \left[\prod_{j=1}^{m-1} (2j)! \right]^2 .
\end{aligned} \tag{A.23.24}$$

Substituting from Eqs. (A.23.21), (A.23.22), (A.23.23), and (A.23.24) into (10.9.5)–(10.9.6), we obtain Eqs. (10.9.5), (10.9.6) and (10.9.7).

A.24. Proof of the Inequality (11.1.5)

Consider all $N(N-1)/2$ chords that join the N points $A_1, A_2, ..., A_N$ which lie on the unit circle. For definiteness let the angle variables of these points be in increasing order. Moreover, in the subsequent argument, whenever the index of any point exceeds N we subtract a multiple of N so that it is one of the numbers $1, 2, ..., N$. We want to maximize the product of the lengths of all the chords.

Let us divide the set of chords into a number of classes. In the first class we put the chords A_1A_2, A_2A_3, ..., $A_{N-1}A_N$, A_NA_1. In the second one we put A_1A_3, A_3A_5, A_5A_7, ... and so on until we get back to A_1. If A_2 is left out, we construct a separate class of A_2A_4, A_4A_6, ... until we get back to A_2. In the next class we put the chords A_1A_4, A_4A_7, A_7A_{10}, ... until A_1 is repeated. If A_2 is left out, we construct a separate class A_2A_5, A_5A_8, ... Similarly for A_3. And so on, until all the chords are exhausted. We maximize the product of the lengths of the chords belonging to a particular class. It is conceivable that the maximization conditions are different for different classes. If this occurs, we will really be in trouble.

Notice that if the points P_1, P_2 are fixed, but the point P_3 varies over the upper arc, the product of the chords P_1P_3, P_2P_3 is maximum when the two chords are equal. (See Figure A.24.1.) From this it follows that the product of the chords belonging to any one class is maximum when the points A_1, A_2, ..., A_N lie at the vertices of a regular polygon. This condition is the same for any of the classes and the above mentioned trouble does not arise.

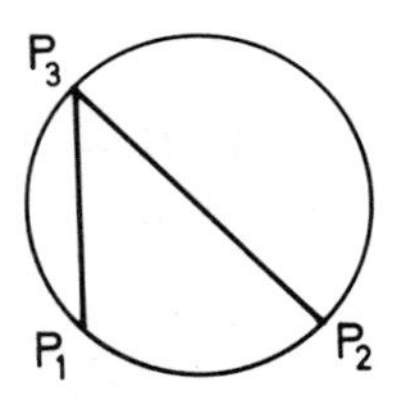

FIG. A.24.1. When P_1 and P_2 are fixed and P_3 varies over the upper arc of the circle, then the maximum of the product of P_1P_3 and P_2P_3 is attained when $P_1P_3 = P_2P_3$.

A.25. Good's Proof of Equation (11.1.11)

For non-negative integers $a_1, a_2, ..., a_n$ let us write

$$F(\vec{x}; \vec{a}) = \prod_{i \neq j=1}^{n} \left(1 - \frac{x_j}{x_i}\right)^{a_j},$$

and

$$M(\vec{a}) = \frac{(a_1 + \cdots + a_n)!}{a_1! \cdots a_n!}.$$

When $F(\vec{x}; \vec{a})$ is expanded in positive and negative powers of $x_1, x_2, ..., x_n$, let the constant term be $G(\vec{a})$. We will show that $G(\vec{a}) = M(\vec{a})$.

Consider the polynomial

$$P(x) = \sum_{j=1}^{n} \prod_{i=1}^{n}{}' \frac{(x - x_i)}{(x_j - x_i)},$$

where the prime on the product means that the term $i = j$ is not taken. This polynomial, of degree $n - 1$, takes the value 1 at $x = x_1$, $x = x_2$,..., $x = x_n$; therefore it is identically equal to 1. Put $x = 0$ to get

$$\sum_{j=1}^{n} \prod_{i=1}^{n}{}' \left(1 - \frac{x_j}{x_i}\right)^{-1} = 1.$$

Multiplying $F(\vec{x}; \vec{a})$ by this function, we see that if $a_j > 0$, for every $j = 1, 2, ..., n$, then

$$F(\vec{x}; \vec{a}) = \sum_{j=1}^{n} F(\vec{x}; a_1, ..., a_{j-1}, a_j - 1, a_{j+1}, ..., a_n),$$

so that

$$G(\vec{a}) = \sum_{j=1}^{n} G(a_1, ..., a_{j-1}, a_j - 1, a_{j+1}, ..., a_n). \qquad \text{(A.25.1)}$$

If some $a_j = 0$, then x_j occurs only in negative powers in $F(\vec{x}; \vec{a})$, and $G(\vec{a})$ is then equal to the constant term in

$$F(x_1, ..., x_{j-1}, x_{j+1}, ...x_n; a_1, ..., a_{j-1}, a_{j+1}, ..., a_n);$$

that is

$$G(\vec{a}) = G(a_1, ..., a_{j-1}, a_{j+1}, ..., a_n), \quad \text{if} \quad a_j = 0. \tag{A.25.2}$$

Also obviously

$$G\left(\vec{0}\right) = 1. \tag{A.25.3}$$

The recursive equations (A.25.1)–(A.25.3) determine uniquely $G(\vec{a})$. But $M(\vec{a})$ satisfies these three equations. Therefore, $G(\vec{a}) = M(\vec{a})$.

A.26. Some Recurrence Relations and Integrals Used in Chapter 14

In this appendix we collect some recurrence relations and identities often used in Chapter 14 and in later appendices.

A.26.1.

The harmonic oscillator functions

$$\begin{aligned}\varphi_j(x) &= \left(2^j j!\sqrt{\pi}\right)^{-1/2} \exp(x^2/2) \left(-\frac{d}{dx}\right)^j \exp(-x^2) \\ &= \left(2^j j!\sqrt{\pi}\right)^{-1/2} \exp(-x^2/2) H_j(x),\end{aligned}$$

with $H_j(x)$, the Hermite polynomials, obey the orthonormality

$$\int_{-\infty}^{\infty} \varphi_i(x)\varphi_j(x)dx = \delta_{ij} \tag{A.26.1}$$

and the recurrence relations

$$\sqrt{2}\varphi_j'(x) = \sqrt{j}\varphi_{j-1}(x) - \sqrt{j+1}\varphi_{j+1}(x), \tag{A.26.2}$$

$$\sqrt{2}x\varphi_j(x) = \sqrt{j}\varphi_{j-1}(x) + \sqrt{j+1}\varphi_{j+1}(x). \tag{A.26.3}$$

These equalities follow from those for Hermite polynomials (Bateman 1953; Szegö 1939),

$$H_j'(x) = 2j\ H_{j-1}(x),$$

$$2xH_j(x) = H_{j+1}(x) + 2j\ H_{j-1}(x).$$

A.26.2.

The functions $\psi_j(x)$ and $A_j(x)$ defined by Eqs. (14.1.27) and (14.1.28) satisfy the equations

$$\sqrt{2}\psi_j(x) = \left(\frac{1+\alpha^2}{1-\alpha^2}\right)^j \left[\sqrt{j}(1-\alpha^2)\varphi_{j-1}(x) - \sqrt{j+1}(1+\alpha^2)\varphi_{j+1}(x)\right] \quad \text{(A.26.4)}$$

$$\sqrt{2}\varphi_j(x) = \left(\frac{1+\alpha^2}{1-\alpha^2}\right)^j \left[\sqrt{j}(1-\alpha^2)A_{j-1}(x) - \sqrt{j+1}(1+\alpha^2)A_{j+1}(x)\right] \quad \text{(A.26.5)}$$

These relations can be easily derived from Eqs. (A.26.2) and (A.26.3).

A.26.3.

The orthonormality relations involving $\psi_j(x)$, $A_j(x)$, and $\varphi_j(x)$ are

$$\int_{-\infty}^{\infty} \varphi_i(x)\varphi_j(x)dx = \delta_{ij}\ , \quad \text{(A.26.1)}$$

$$\int_{-\infty}^{\infty} \psi_i(x)A_j(x)dx = -\delta_{ij}\ , \quad \text{(A.26.6)}$$

$$\int_{-\infty}^{\infty} \psi_i(x)\varphi_j(x)dx = \int_{-\infty}^{\infty} A_i(x)\varphi_j(x)dx = 0\ , \quad i+j \quad \text{even.} \quad \text{(A.26.7)}$$

The second is obtained by partial integration and the third by a parity argument. Note that $\varphi_j(x)$ has the parity of j, while $\psi_j(x)$ and $A_j(x)$ have parities opposite to that of j.

A.26.4.

The recurrence relations

$$\psi_{2j}(x)A_{2j}(y) = -\varphi_{2j+1}(x)\varphi_{2j+1}(y) + (1-\alpha^2) \times \left[\left(\frac{1+\alpha^2}{1-\alpha^2}\right)^{2j} \sqrt{j}\varphi_{2j-1}(x)A_{2j}(y) - \left(\frac{1+\alpha^2}{1-\alpha^2}\right)^{2j+2} \sqrt{j+1}\varphi_{2j+1}(x)A_{2j+2}(y)\right], \tag{A.26.8}$$

$$\psi_{2j+1}(x)A_{2j+1}(y) = -\varphi_{2j}(x)\varphi_{2j}(y) + (1-\alpha^2) \times \left[\left(\frac{1+\alpha^2}{1-\alpha^2}\right)^{2j} \sqrt{j}\varphi_{2j}(x)A_{2j-1}(y) - \left(\frac{1+\alpha^2}{1-\alpha^2}\right)^{2j+2} \sqrt{j+1}\varphi_{2j+2}(x)A_{2j+1}(y)\right], \tag{A.26.9}$$

$$\varphi_{2j}(x)A_{2j}(y) + A_{2j+1}(x)\varphi_{2j+1}(y) = (1-\alpha^2)\left[\left(\frac{1+\alpha^2}{1-\alpha^2}\right)^{2j} \sqrt{j}A_{2j-1}(x)A_{2j}(y) - \left(\frac{1+\alpha^2}{1-\alpha^2}\right)^{2j+2} \sqrt{j+1}A_{2j+1}(x)A_{2j+2}(y)\right], \tag{A.26.10}$$

$$\psi_{2j}(x)\varphi_{2j}(y) + \varphi_{2j+1}(x)\psi_{2j+1}(y)$$

$$= (1-\alpha^2)\Bigg[\left(\frac{1+\alpha^2}{1-\alpha^2}\right)^{2j}\sqrt{j}\varphi_{2j-1}(x)\varphi_{2j}(y) - \left(\frac{1+\alpha^2}{1-\alpha^2}\right)^{2j+2}\sqrt{j+1}\varphi_{2j+1}(x)\varphi_{2j+2}(y)\Bigg] \tag{A.26.11}$$

can be derived from Eqs. (A.26.4) and (A.26.5). These relations are useful to verify the equivalence of expressions (14.1.16) and (14.1.17), of (14.1.18) and (14.1.19) and of (14.1.21) and (14.1.22).

A.26.5.

We need the integral

$$\int_{-\infty}^{\infty}\varphi_{2j}(y)\exp\left(-\frac{1}{2}\alpha^2y^2\right)dy = \left[2\pi/\left(1+\alpha^2\right)\right]^{1/2}\frac{(2j)!}{j!}\left[2^{2j}(2j)!\sqrt{\pi}\right]^{-1/2}\left(\frac{1-\alpha^2}{1+\alpha^2}\right)^{j} \tag{A.26.12}$$

To prove this equation we use the generating function

$$\exp\left(-y^2+2yz-z^2\right) = \sum_{j=0}^{\infty}\frac{z^j}{j!}\left(-\frac{d}{dy}\right)^j\exp(-y^2) = \sum_{j=0}^{\infty}\frac{z^j}{j!}\exp\left(-\frac{1}{2}y^2\right)\left(2^j j!\sqrt{\pi}\right)^{1/2}\varphi_j(y) \tag{A.26.13}$$

to write

$$\sum_{j=0}^{\infty} \frac{z^j}{j!} \left(2^j j! \sqrt{\pi}\right)^{1/2} \int_{-\infty}^{\infty} \varphi_j(y) \exp\left(-\frac{1}{2}\alpha^2 y^2\right) dy$$
$$= \int_{-\infty}^{\infty} \exp\left[-\frac{1}{2}\left(1+\alpha^2\right) y^2 + 2yz - z^2\right] dy$$
$$= \exp\left[z^2\left(\frac{1-\alpha^2}{1+\alpha^2}\right)\right] \times \int_{-\infty}^{\infty} \exp\left[-\frac{1+\alpha^2}{2}\left(y - \frac{2}{1+\alpha^2} z\right)^2\right] dy$$
$$= \left(\frac{2\pi}{1+\alpha^2}\right)^{1/2} \sum_{j=0}^{\infty} \frac{z^{2j}}{j!} \left(\frac{1-\alpha^2}{1+\alpha^2}\right)^j .$$

Equating coefficients of z^{2j} on both sides we get Eq. (A.26.12).

A.26.6.

We need the convolution integrals

$$\int_{-\infty}^{\infty} g(x,y)\psi_j(y)dy = \varphi_j(x), \tag{A.26.14}$$

$$\int_{-\infty}^{\infty} g(x,y)\varphi_j(y)dy = A_j(x), \tag{A.26.15}$$

with $g(x,y)$ given by Eq. (14.1.23). By partial integration (A.26.14) is equivalent to

$$\int_{-\infty}^{\infty} \exp\left[-\frac{1}{2}\alpha^2\left(x^2+y^2\right) - \left(\frac{1-\alpha^4}{4\alpha^2}\right)(x-y)^2\right] \varphi_j(y)dy$$
$$= \frac{2\alpha\sqrt{\pi}}{1+\alpha^2}\left(\frac{1-\alpha^2}{1+\alpha^2}\right)^j \varphi_j(x). \tag{A.26.16}$$

To prove this we expand both sides of the identity

$$\int_{-\infty}^{\infty} \exp\left[2yz - z^2 - \frac{1}{2}\left(1+\alpha^2\right)y^2 - \left(\frac{1-\alpha^4}{4\alpha^2}\right)(x-y)^2\right] dy$$

$$= \frac{2\alpha\sqrt{\pi}}{1+\alpha^2} \exp\Bigg[-\left(\frac{1-\alpha^2}{1+\alpha^2}\right)^2 z^2 + 2\left(\frac{1-\alpha^2}{1+\alpha^2}\right) zx$$

$$-\frac{1}{2}(1-\alpha^2)x^2\Bigg] \qquad \text{(A.26.17)}$$

in powers of z and use the generating function (A.26.13).

Equation (A.26.15) can be rewritten as

$$\left(\frac{1+\alpha^2}{1-\alpha^2}\right)^{j+1/2} \int_{-\infty}^{\infty} f(x-y)\exp\left(-\frac{1}{2}\alpha^2y^2\right)\varphi_j(y)dy$$

$$= \int_{-\infty}^{\infty} \exp\left(\frac{1}{2}\alpha^2y^2\right)\varphi_j(y)\operatorname{sgn}(x-y)dy.$$

This last equation is true, since the derivatives of both sides are equal, Eq. (A.26.16), and the equality holds at one point $x = \infty$. In fact at $x = \infty$, $f(x-y) = \operatorname{sgn}(x-y) = 1$, and we have to make sure that

$$\left(\frac{1+\alpha^2}{1-\alpha^2}\right)^{j+1/2} \int_{-\infty}^{\infty} \exp\left(-\frac{1}{2}\alpha^2y^2\right)\varphi_j(y)dy$$

$$= \int_{-\infty}^{\infty} \exp\left(\frac{1}{2}\alpha^2y^2\right)\varphi_j(y)dy. \qquad \text{(A.26.18)}$$

For j odd, both sides are zero and the equality is evident. For j even, we use Eq. (A.26.12) and the one obtained by changing the sign of α^2 in it. Taking ratios we get Eq. (A.26.18).

A.26.7. RECURRENCE AND ORTHONORMALITY RELATIONS BETWEEN $\mathbf{\Psi_j(x)}$, $\mathbf{A}_j(x)$, AND $\varphi_\mathbf{j}(\mathbf{x})$.

The function $\mathbf{\Psi}_j(x)$ and $\mathbf{A}_j(x)$, Eqs. (14.2.20) and (14.2.21), are obtained from $\psi_j(x)$ and $A_j(x)$ Eqs. (14.1.27) and (14.1.28), by changing the sign of α^2. So from (A.26.4)–(A.26.7) we get

$$\sqrt{2}\mathbf{\Psi}_j(x) = \left(\frac{1-\alpha^2}{1+\alpha^2}\right)^j \times \left[\sqrt{j}(1+\alpha^2)\varphi_{j-1}(x) - \sqrt{j+1}(1-\alpha^2)\varphi_{j+1}(x),\right], \tag{A.26.19}$$

$$\sqrt{2}\varphi_j(x) = \left(\frac{1-\alpha^2}{1+\alpha^2}\right)^j \times \left[\sqrt{j}(1+\alpha^2)\mathbf{A}_{j-1}(x) - \sqrt{j+1}(1-\alpha^2)\mathbf{A}_{j+1}(x),\right], \tag{A.26.20}$$

$$\int_{-\infty}^{\infty} \mathbf{\Psi}_i(x)\mathbf{A}_j(x)dx = -\delta_{ij}, \tag{A.26.21}$$

$$\int_{-\infty}^{\infty} \mathbf{\Psi}_i(x)\varphi_j(x)dx = \int_{-\infty}^{\infty} \mathbf{A}_i(x)\varphi_j(x)dx = 0, \quad i+j \quad \text{even.} \tag{A.26.22}$$

A.26.8.

We also need the convolution integrals

$$\int_{-\infty}^{\infty} \mathbf{g}(x,y)\varphi_j(y)dy = \mathbf{\Psi}_j(x), \tag{A.26.23}$$

$$\int_{-\infty}^{\infty} \mathbf{g}(x,y)\mathbf{A}_j(y)dy = \varphi_j(x). \tag{A.26.24}$$

Equation (A.26.23) can be established from the definitions of $\mathbf{g}(x,y)$, $\mathbf{\Psi}_j(x)$, Eqs. (14.2.18), (14.2.20), and using equations (A.26.16), (A.26.2) and (A.26.3). For Eq. (A.26.24) we have from the definitions of $\mathbf{g}(x,y)$ and $\mathbf{A}_j(x)$, Eqs. (14.2.18), (14.2.21)

$$\begin{aligned}\int_{-\infty}^{\infty} \mathbf{g}(x,y)\mathbf{A}_j(y)dy = &-(1+\alpha^2)(\frac{1-\alpha^4)}{4\alpha^3}\sqrt{\pi}\left(\frac{1+\alpha^2}{1-\alpha^2}\right)^j \\ &\times \int\int_{-\infty}^{\infty} dydz \times \varepsilon(y-z)\varphi_j(z) \\ &\times \exp\left[-\frac{1}{2}\alpha^2(x^2+z^2)\right](x-y) \\ &\times \exp\left(-\frac{1-\alpha^4}{4\alpha^2}(x-y)^2\right),\end{aligned}$$

with $2\varepsilon(x) = \text{sgn}(x)$, Eq. (14.1.29). Changing the variable y to $t = x - y$, one can integrate over t,

$$\begin{aligned}&\frac{1-\alpha^4}{-2\alpha^2}\int_{-\infty}^{\infty} \varepsilon(x-z-t)t \exp\left(-\frac{1-\alpha^4}{4\alpha^2}t^2\right) dt \\ &= \exp\left(-\frac{1-\alpha^4}{4\alpha^2}(x-z)^2\right).\end{aligned}$$

Now using Eq. (A.26.16) one gets Eq. (A.26.24).

A.27. Normalization Integral, Equation (14.1.11)

The constant in Eq. (14.1.11) being fixed by the condition

$$\int_{-\infty}^{\infty} \cdots \int_{-\infty}^{\infty} p(x_1, ..., x_N)dx_1 \cdots dx_N = 1 \tag{A.27.1}$$

with $p(x_1, ..., x_N)$ given by Eq. (14.1.7), we shall evaluate the integral

$$C_N^{-1} = \int_{-\infty}^{\infty} \cdots \int_{-\infty}^{\infty} dx_1 \cdots dx_N \exp\left(-\frac{1+\alpha^2}{2}\sum x_i^2\right) \Delta(x)\text{Pf}\,[F_{ij}], \tag{A.27.2}$$

with

$$\Delta(x) = \prod_{i<j} (x_i - x_j), \tag{14.1.8}$$

and F_{ij} defined by Eq. (14.1.9) and (14.1.9′).

The Pfaffian and $\Delta(x)$ are alternating functions of the variables $x_1, ..., x_N$. The $\Delta(x)$ can be written as a determinant

$$\exp\left(-\sum x_i^2/2\right) \Delta(x) = \prod_0^{N-1} \left(2^{-j} j! \sqrt{\pi}\right)^{1/2} \det\left[\varphi_{i-1}(x_j)\right]_{i,j=1,\ldots,N}, \tag{A.27.3}$$

where $\varphi_i(x)$ is the normalised "oscillator function" given by Eq. (14.1.26). The number of terms in the Pfaffian is $(2m)!/(2^m m!)$ with $N = 2m$ or $N = 2m - 1$.

It is convenient to discuss separately the cases N even and N odd. Let us first take the case $N = 2m$ even. From the symmetry and what we said above, one has

$$\begin{aligned} C_N^{-1} = &\prod_0^{N-1} \left(2^{-j} j! \sqrt{\pi}\right)^{1/2} \int_{-\infty}^{\infty} \cdots \int_{-\infty}^{\infty} dx_1 \cdots dx_N \\ &\times \exp\left(-\frac{1}{2}\alpha^2 \sum x_i^2\right) \\ &\times \operatorname{Pf}\left[f\left(x_i - x_j\right)\right] \det\left[\varphi_{i-1}(x_j)\right] \end{aligned} \tag{A.27.4}$$

$$\begin{aligned} = &\prod_0^{N-1} \left(2^{-j} j! \sqrt{\pi}\right)^{1/2} \frac{(2m)!}{2^m m!} \int_{-\infty}^{\infty} \cdots \int_{-\infty}^{\infty} dx_1 \cdots dx_N \\ &\times \exp\left(-\frac{1}{2}\alpha^2 \sum x_i^2\right) \\ &\times \prod_{j=1}^{m} f\left(x_{2j} - x_{2j-1}\right) \det\left[\varphi_{i-1}(x_j)\right]. \end{aligned} \tag{A.27.5}$$

For the last integral, we have from the theory of Pfaffians (Mehta, 1989 or Appendix A.18)

$$C_N^{-1} \prod_0^{N-1} \left(2^{-j} j! \sqrt{\pi}\right)^{-1/2} \left(\frac{(2m)!}{2^m m!}\right)^{-1} = m! 2^m \mathrm{Pf}\left[a_{ij}\right]_{i,j=0,1,\ldots,2m-1} \tag{A.27.6}$$

where for $i, j = 0, 1, \ldots, 2m-1$,

$$\begin{aligned} a_{ij} &= \frac{1}{2} \int_{-\infty}^{\infty} \int_{-\infty}^{\infty} \exp\left[-\frac{1}{2}\alpha^2\left(x^2+y^2\right)\right] \\ &\qquad \times f(y-x)\left[\varphi_i(x)\varphi_j(y) - \varphi_j(x)\varphi_i(y)\right] dx dy \\ &= \int_{-\infty}^{\infty} \int_{-\infty}^{\infty} \exp\left[-\frac{1}{2}\alpha^2\left(x^2+y^2\right)\right] \\ &\qquad \times f(y-x)\varphi_i(x)\varphi_j(y) dx dy. \end{aligned} \tag{A.27.7}$$

From parity one sees that $a_{ij} = 0$ if $i+j$ is even. Thus, the Pfaffian reduces to a determinant

$$\mathrm{Pf}\left[a_{ij}\right]_{i,j=0,1,\ldots,2m-1} = \det\left[a_{2i,2j+1}\right]_{i,j=0,1,\ldots,m-1}. \tag{A.27.8}$$

The determinant is not changed if we add to any row (column) a constant multiple of another row (column). To choose convenient multiples, we observe that using the identity (A.26.4)

$$\begin{aligned} &\sqrt{2}\frac{d}{dy}\left[\varphi_j(y)\exp\left(-\frac{1}{2}\alpha^2 y^2\right)\right] \\ &\quad = \exp\left(-\frac{1}{2}\alpha^2 y^2\right) \\ &\qquad \times \left[\sqrt{j}(1-\alpha^2)\varphi_{j-1}(y) - \sqrt{j+1}(1+\alpha^2)\varphi_{j+1}(y)\right], \end{aligned} \tag{A.27.9}$$

an integration by parts in Eq. (A.27.7) gives

$$\frac{1}{\sqrt{2}}\left[\sqrt{j}(1-\alpha^2)a_{i,j-1}-\sqrt{j+1}(1+\alpha^2)a_{i,j+1}\right]$$

$$= -\frac{2}{\sqrt{\pi}}\left(\frac{1-\alpha^4}{4\alpha^2}\right)^{1/2}$$

$$\times \int_{-\infty}^{\infty}\int_{-\infty}^{\infty} \exp\bigg(-\frac{1}{2}\alpha^2\left(x^2+y^2\right) - \frac{1-\alpha^4}{4\alpha^2}(x-y)^2\bigg)\varphi_i(x)\varphi_j(y)dxdy$$

$$= -2\left[(1-\alpha^2)/(1+\alpha^2)\right]^{j+1/2}\delta_{ij}. \tag{A.27.10}$$

In the last step we made use of Eqs. (A.26.16) and (A.26.1). Thus

$$a_{2i,2j+1} - \left(\frac{2j}{2j+1}\right)^{1/2}\left(\frac{1-\alpha^2}{1+\alpha^2}\right)a_{2i,2j-1}$$

$$= \frac{2}{1+\alpha^2}\left(\frac{2}{2j+1}\right)^{1/2}\left(\frac{1-\alpha^2}{1+\alpha^2}\right)^{2j+1/2}\delta_{ij}$$

$$= b_{ij}, \quad \text{say}, \tag{A.27.11}$$

and

$$\det\left[a_{2i,2j+1}\right] = \det\left[b_{ij}\right]$$

$$= \prod_{j=0}^{m-1}\left[\frac{2}{1+\alpha^2}\left(\frac{2}{2j+1}\right)^{1/2}\left(\frac{1-\alpha^2}{1+\alpha^2}\right)^{2j+1/2}\right]. \tag{A.27.12}$$

From Eqs. (A.27.6), (A.27.8), and (A.27.12) we finally have for $N = 2m$,

$$C_N^{-1} = \prod_0^{N-1} \left(2^{-j} j! \sqrt{\pi}\right)^{1/2} N! \prod_0^{m-1} \left[\frac{2}{1+\alpha^2} \left(\frac{2}{2j+1}\right)^{1/2} \times \left(\frac{1-\alpha^2}{1+\alpha^2}\right)^{2j+1/2} \right] \tag{A.27.13}$$

which is Eq. (14.1.11), since

$$\begin{aligned} \prod_0^{N-1} \left(\sqrt{\pi} j!\right)^{1/2} &= \prod_0^{m-1} (\pi (2j)! (2j+1)!)^{1/2} \\ &= \prod_0^{m-1} \left[(2j+1)^{1/2} \pi^{1/2} \Gamma(2j+1) \right] \\ &= \prod_0^{m-1} \left[(2j+1)^{1/2} 2^{2j} \Gamma(j+\tfrac{1}{2}) \Gamma(j+1) \right] \\ &= 2^{m(m-1)} \prod_0^{m-1} (2j+1)^{1/2} \prod_1^{2m} \left(\frac{2}{j} \Gamma(1+\tfrac{1}{2}j) \right). \end{aligned} \tag{A.27.14}$$

When $N = 2m - 1$ is odd, we have similarly

$$\begin{aligned} C_N^{-1} & \prod_0^{N-1} \left(2^{-j} j! \sqrt{\pi}\right)^{-1/2} \\ &= \frac{(2m)!}{2^m m!} \int_{-\infty}^{\infty} \cdots \int_{-\infty}^{\infty} dx_1 \cdots dx_N \exp\left(-\frac{1}{2}\alpha^2 \sum x_i^2\right) \\ &\quad \times \prod_{j=1}^{m-1} f\left(x_{2j} - x_{2j-1}\right) \det\left[\varphi_{i-1}(x_j)\right] \\ &= \frac{(2m)!}{2^m m!} 2^{m-1} (m-1)! \mathrm{Pf}\left[a_{ij}\right]_{i,j=0,1,\ldots,2m-1}, \end{aligned} \tag{A.27.15}$$

where now for $i, j = 0, 1, ..., 2m-2$, a_{ij} is given by Eq. (A.27.7), and

$$a_{i,2m-1} = -a_{2m-1,i} = \int_{-\infty}^{\infty} \exp\left(-\frac{1}{2}\alpha^2 x^2\right) \varphi_i(x) dx,$$

$$i = 0, 1, ..., 2m-2, \quad \text{(A.27.16)}$$

$$a_{2m-1,2m-1} = 0. \quad \text{(A.27.17)}$$

Again we have $a_{ij} = 0$, if $i + j$ is even, so that

$$\text{Pf}\,[a_{ij}]_{i,j=0,1,...,2m-1} = \det\,[a_{2i,2j+1}]_{i,j=0,1,...,m-1}\,. \quad \text{(A.27.18)}$$

We can again, without changing the value of the determinant, replace $a_{2i,2j+1}$ by b_{ij}, where for $j = 0, 1, ..., m-2$ and $i = 0, 1, ..., m-1$ the b_{ij} is given by Eq. (A.27.11), while for $i = 0, 1, ..., m-1$

$$\begin{aligned} b_{i,m-1} &= \int_{-\infty}^{\infty} \exp\left(-\frac{1}{2}\alpha^2 x^2\right) \varphi_{2i}(x) dx \\ &= \left(\frac{2\pi}{1+\alpha^2}\right)^{1/2} [2^{2i}(2i)!\sqrt{\pi}]^{-1/2} \frac{(2i)!}{i!} \left(\frac{1-\alpha^2}{1+\alpha^2}\right)^i \quad \text{(A.27.19)} \end{aligned}$$

[Eq. (A.26.12)].

Collecting the results for $N = 2m-1$, one has

$$\begin{aligned} C_N^{-1} = & \prod_0^{N-1} \left(2^{-j} j! \sqrt{\pi}\right)^{1/2} N! \\ & \times \prod_0^{m-2} \left[\frac{2}{1+\alpha^2} \left(\frac{2}{2j+1}\right)^{1/2} \left(\frac{1-\alpha^2}{1+\alpha^2}\right)^{2j+1/2}\right] \\ & \times \left(\frac{2\pi}{1+\alpha^2}\right)^{1/2} [2^{2m-2}(2m-2)!\sqrt{\pi}]^{-1/2} \\ & \times \frac{(2m-2)!}{(m-1)!} \left(\frac{1-\alpha^2}{1+\alpha^2}\right)^{m-1}, \end{aligned} \quad \text{(A.27.20)}$$

which on similar manipulations can be seen to be Eq. (14.1.11) with $N = 2m - 1$.

A.28. Another Normalization Integral, Equation (14.2.9)

The constant in Eq. (14.2.9) being fixed by the normalization condition

$$\int_{-\infty}^{\infty} \cdots \int_{-\infty}^{\infty} \mathbf{p}(x_1, ..., x_{2N})\, dx_1 \cdots dx_{2N} = 1,$$

with $\mathbf{p}(x_1, ..., x_{2N})$ given by Eq. (14.2.7), we need the integral

$$\mathbf{C}_N^{-1} = \int_{-\infty}^{\infty} \cdots \int_{-\infty}^{\infty} dx_1 \cdots dx_{2N} \exp\left(-\frac{1+\alpha^2}{2} \sum_1^{2N} x_i^2\right) \times \Delta(x)\ \mathrm{Pf}\left[\mathbf{F}(x_i - x_j)\right],$$

$$\mathbf{F}(x) = x \exp\left[-\left(1-\alpha^4\right) x^2/4\alpha^2\right].$$

We follow the reasoning of Appendix A.27 for the case of an even number of variables and write

$$\mathbf{C}_N^{-1} \prod_0^{2N-1} \left(2^{-j} j! \sqrt{\pi}\right)^{-1/2} = (2N)!\ \mathrm{Pf}\left[a_{ij}\right]_{i,j=0,1,\ldots,2N-1} = (2N)!\ \det\left[a_{2i,2j+1}\right]_{i,j=0,1,N-1}$$

where

$$a_{ij} = \int \int_{-\infty}^{\infty} \mathbf{F}(y-x) \exp\left[-\frac{1}{2}\alpha^2\left(x^2+y^2\right)\right] \varphi_i(x)\varphi_j(y) dx\, dy$$
$$= \int \int_{-\infty}^{\infty} (y-x)\varphi_i(x)\varphi_j(y) \times \exp\left(-\frac{1}{2}\alpha^2\left(x^2+y^2\right) - \frac{1-\alpha^4}{4\alpha^2}(x-y)^2\right) dx\, dy.$$

Now from Eqs. (A.26.3), (A.26.16), and (A.26.1) one gets after a little algebra

$$a_{2i,2j+1} = \frac{2\alpha^3\sqrt{2\pi}}{\left(1+\alpha^2\right)^2} \left(\frac{1-\alpha^2}{1+\alpha^2}\right)^{2i} \left(\sqrt{2i+1}\ \delta_{ij} + \sqrt{2i}\ \delta_{i,j+1}\right).$$

Thus

$$\mathbf{C}_N^{-1} = \prod_0^{2N-1} \left(2^{-j} j! \sqrt{\pi}\right)^{1/2} (2N)! \\ \times \prod_0^{N-1} \left[\frac{2\alpha^3 \sqrt{2\pi}}{(1+\alpha^2)^2} \left(\frac{1-\alpha^2}{1+\alpha^2} \right)^{2j} \sqrt{2j+1} \right]$$

which is the same as Eq. (14.2.9).

A.29. Joint Probability Density as a Determinant of a Self-Dual Quaternion Matrix. Section 14.4, Equations (14.4.2) and (14.4.5)

Equation (14.4.2) will be verified separately for the cases N even and N odd. The manipulations are similar to those of Chapter 6.

When N is even, we observe that the $2N \times 2N$ matrix

$$G = \begin{bmatrix} S_N(x_i, x_j) & D_N(x_i, x_j) \\ I_N(x_i, x_j) & S_N(x_j, x_i) \end{bmatrix}_{i,j=1,\ldots,N,} \tag{A.29.1}$$

can be written as a product of two rectangular matrices of orders $2N \times N$ and $N \times 2N$, respectively:

$$G = G_1 G_2,$$

$$G_1 = \begin{bmatrix} \varphi_{2k}(x_i) & -\psi_{2k}(x_i) \\ -A_{2k}(x_i) & \varphi_{2k}(x_i) \end{bmatrix}_{\substack{i=1,2,\ldots,N, \\ k=0,1,\ldots,N/2-1}},$$

$$G_2 = \begin{bmatrix} \varphi_{2k}(x_j) & \psi_{2k}(x_j) \\ A_{2k}(x_j) & \varphi_{2k}(x_j) \end{bmatrix}_{\substack{k=0,1,\ldots,N/2-1 \\ j=1,2,\ldots,N}}. \tag{A.29.2}$$

The rank of G_1 or of G_2 is at most N; it is in fact N, since the $\varphi_{2k}(x)$ is an orthonormal sequence. The rank of G, the product of G_1 and G_2, is therefore at most N (see, e.g., Mehta 1989); in fact we know that it is N, since its first N rows are linearly independent. Therefore, the last N rows of G are linear combinations of its first N rows and *vice versa.* The ordinary determinant of the $2N \times 2N$ matrix

$$[\phi(x_i, x_j)] = \begin{bmatrix} S_N(x_i, x_j) & D_N(x_i, x_j) \\ J_N(x_i, x_j) & S_N(x_j, x_i) \end{bmatrix}_{i,j=1,\dots,N}$$

is therefore not changed if we subtract from its last N rows the corresponding rows of G :

$$\begin{aligned} \det[\phi(x_i, x_j)] &= \det \begin{bmatrix} S_N(x_i, x_j) & D_N(x_i, x_j) \\ g(x_i, x_j) & 0 \end{bmatrix} \\ &= (-1)^N \det[g(x_i, x_j)] \det[D_N(x_i, x_j)]. \end{aligned} \quad \text{(A.29.3)}$$

But

$$\begin{aligned} \det[g(x_i, x_j)] = 2^{-N} &\left(\frac{1+\alpha^2}{1-\alpha^2} \right)^{N/2} \\ &\times \exp\left(-\alpha^2 \sum x_i^2 \right) \det[f(x_i - x_j)] \end{aligned} \quad \text{(A.29.4)}$$

and

$$\begin{aligned} \det[D_N(x_i, x_j)] &= \det[\varphi_{2k}(x_i) \quad \psi_{2k}(x_i)] \begin{bmatrix} \psi_{2k}(x_j) \\ -\varphi_{2k}(x_j) \end{bmatrix} \\ &= (\det[\varphi_{2k}(x_i) \quad \psi_{2k}(x_i)])^2 . \end{aligned} \quad \text{(A.29.5)}$$

Now ψ_{2k} being a linear combination of φ_{2k-1} and φ_{2k+1}, Eq. (A.26.4), we can replace the columns $\psi_0, \psi_2, \dots, \psi_{2N-2}$ successively by $\varphi_1, \varphi_3, \dots, \varphi_{2N-1}$

in the determinant. The result is

$$\begin{aligned}\det\left[D_N(x_i,x_j)\right] &\propto \{\det\left[\varphi_{j-1}(x_i)\right]\}^2 \\ &\propto \exp\left(-\sum x_i^2\right)\left\{\det\left[x_i^{j-1}\right]\right\}^2 \\ &\propto \exp\left(-\sum x_i^2\right)\left[\Delta(x)\right]^2 . \end{aligned} \tag{A.29.6}$$

From Eqs. (A.29.3), (A.29.4), and (A.29.6) we get

$$\det\left[\phi(x_i,x_j)\right] \propto \exp\left(-(1+\alpha^2)\sum_1^N x_i^2\right)\det\left[f(x_i-x_j)\right]\left[\Delta(x)\right]^2 ,$$

which in view of Eq. (14.1.7) is

$$\det\left[\phi(x_i,x_j)\right] \propto \left[p(x_1,\ldots,x_N)\right]^2 . \tag{A.29.7}$$

(Note that here the matrix is written in its partitioned form, its elements being ordinary numbers, not quaternions, and the determinant is the usual one.)

When $N=2m+1$ odd, the method used above fails at one point: the $N\times N$ matrix

$$D_N(x_i,x_j) = \left[\varphi_{2k+1}(x_i) \quad -\psi_{2k+1}(x_i)\right]\begin{bmatrix}\psi_{2k+1}(x_j)\\ \varphi_{2k+1}(x_j)\end{bmatrix} \tag{A.29.8}$$

of rank $N-1$ has a zero determinant, while its multiplying factor becomes infinite due to the extra terms ξ_N and μ_N.

Instead of G consider the $2N\times 2N$ matrix

$$G_\delta = \begin{bmatrix}\varphi_{2k+1}(x_i) & -\psi_{2k+1}(x_i) & \delta\varphi_{2m}(x_i)\\ -A_{2k+1} & \varphi_{2k+1}(x_i) & (c\delta)^{-1}\exp\left(-\alpha^2x_i^2/2\right)\end{bmatrix}$$

$$\times\begin{bmatrix}\varphi_{2k+1}(x_j) & \psi_{2k+1}(x_j)\\ A_{2k+1}(x_j) & \varphi_{2k+1}(x_j)\\ (c\delta)^{-1}\exp\left(-\alpha^2x_j^2/2\right) & \delta\varphi_{2m}(x_j)\end{bmatrix} \tag{A.29.9}$$

where δ is arbitrary and

$$c = \int_{-\infty}^{\infty} \varphi_{2m}(t)\exp\left(-\alpha^2 t^2/2\right) dt. \tag{A.29.10}$$

Thus

$$\mathrm{G}_\delta = \begin{bmatrix} S_N(x_i, x_j) + \xi_N(x_i, x_j) & D_N(x_i, x_j) + \delta^2 \varphi_{2m}(x_i)\varphi_{2m}(x_j) \\ I_N(x_i, x_j) + (c\delta)^{-2}\exp\left[-\alpha^2\left(x_i^2 + x_j^2\right)/2\right] & S_N(x_j, x_i) + \xi_N(x_j, x_i) \end{bmatrix} \tag{A.29.11}$$

with ξ_N, S_N, I_N and D_N given by Eqs. (14.1.15), (14.1.17), (14.1.22), and (14.1.19). The rank of G_δ is N. The determinant of the $2N \times 2N$ matrix

$$[\phi_\delta(x_i,\, x_j)] = \begin{bmatrix} S_N(x_i, x_j) + \xi_N(x_i, x_j) & D_N(x_i, x_j) + \delta^2 \varphi_{2m}(x_i)\varphi_{2m}(x_j) \\ J_N(x_i, x_j) & S_N(x_j, x_i) + \xi_N(x_j, x_i) \end{bmatrix}, \tag{A.29.12}$$

is not changed if we subtract from its last N rows the corresponding rows of G_δ; the resulting determinant factorizes,

$$\begin{aligned} \det\left[\phi_\delta(x_i, x_j)\right] = {} & \det\left[D_N(x_i, x_j) + \delta^2 \varphi_{2m}(x_i)\varphi_{2m}(x_j)\right] \\ & \times (-1)^N \det[g(x_i, x_j) \\ & - (c\delta)^{-2}\exp\left\{-\alpha^2\left(x_i^2 + x_j^2\right)/2\right\} \\ & + \mu(x_i, x_j) - \mu(x_j, x_i)]. \end{aligned} \tag{A.29.13}$$

The first factor is

$$\begin{aligned} \delta^2\{\det[&\varphi_1(x_i), \psi_1(x_i); \varphi_3(x_i), \psi_3(x_i); \ldots; \\ &\varphi_{2m-1}(x_i), \psi_{2m-1}(x_i); \varphi_{2m}(x_i)]\}^2. \end{aligned}$$

We can replace the last but one column by a linear combination of the last two columns. Choosing this combination properly, Eq. (A.26.4), ψ_{2m-1} can be replaced by φ_{2m-2}. Then the column ψ_{2m-3} can be replaced by the column φ_{2m-4}, and so on. Thus, the column ψ_{2k+1} is replaced by φ_{2k} for $k = 0, 1, ..., m-1$, the determinant being multiplied by a constant. Thus the first factor in Eq. (A.29.13) is proportional to

$$\delta^2 \left\{ \det \left[\varphi_{j-1} \left(x_i \right) \right]_{i,j=1,2,\ldots,2m+1} \right\}^2 \propto \delta^2 \exp \left(-\sum x_i^2 \right) \left[\Delta(x) \right]^2 . \tag{A.29.14}$$

The second factor in Eq. (A.29.13) is proportional to

$$\exp \left(-\alpha^2 \sum x_i^2 \right) \det \left[f \left(x_i - x_j \right) - k\delta^{-2} + k' \left(h \left(x_i \right) - h \left(x_j \right) \right) \right] \tag{A.29.15}$$

where

$$h(x) = \int_{-\infty}^{\infty} \exp \left(\alpha^2 t^2 / 2 \right) \varepsilon(x-t) \varphi_{2m}(t) dt, \tag{A.29.16}$$

$2\varepsilon(x) = \text{sgn } x$, Eq. (14.1.29), and k and k' are certain constants. But

$$\det \left[f(x_i - x_j) - k\delta^{-2} + k'(h(x_i) - h(x_j)) \right]$$
$$= \det \begin{bmatrix} f \left(x_i - x_j \right) - k\delta^{-2} + k' \left(h \left(x_i \right) - h \left(x_j \right) \right) & 0 & k\delta^{-2} - k' h \left(x_i \right) \\ k\delta^{-2} + k' h \left(x_j \right) & 1 & -k\delta^{-2} \\ 0 & 0 & 1 \end{bmatrix} \tag{A.29.17}$$

$$= \det \begin{bmatrix} f \left(x_i - x_j \right) & -k' h \left(x_i \right) & 1 \\ k' h \left(x_j \right) & -k\delta^{-2} & 1 \\ -1 & -1 & 0 \end{bmatrix}$$

$$= -k\delta^{-2} \det \begin{bmatrix} f \left(x_i - x_j \right) & 1 \\ -1 & 0 \end{bmatrix} + \det \begin{bmatrix} f \left(x_i - x_j \right) & -k' h \left(x_i \right) & 1 \\ k' h \left(x_j \right) & 0 & 1 \\ -1 & -1 & 0 \end{bmatrix} \tag{A.29.18}$$

The second determinant in Eq. (A.29.18) above, being that of an anti-symmetric matrix of odd order $N+2$, is zero.

Thus the second factor of Eq. (A.29.13) is proportional to

$$\delta^{-2}\exp\left(-\alpha^2\sum x_i^2\right)\det\begin{bmatrix} f\left(x_i-x_j\right) & 1 \\ -1 & 0 \end{bmatrix}. \tag{A.29.19}$$

From Eqs. (A.29.13), (A.29.14), and (A.29.19) on taking the limit $\delta \longrightarrow 0$, we get for N odd,

$$\begin{aligned} \det\left[\phi\left(x_i,x_j\right)\right] &\propto \left[\exp\left(-\left(1+\alpha^2\right)\sum x_i^2/2\right)\Delta(x)\right]^2 \det\left[F_{ij}\right] \\ &\propto \left[p\left(x_1,...,x_N\right)\right]^2 \end{aligned} \tag{A.29.20}$$

in view of Eq. (14.1.7). The value Eq. (A.29.10) of the constant c is not required for this appendix; it will be needed in Appendix A.30.

Thus whether N is even or odd, the determinant of the $2N\times 2N$ matrix $\left[\phi\left(x_i,x_j\right)\right]$ is proportional to $\left\{p\left(x_1,...,x_N\right)\right\}^2$, Eq. (A.29.7) or (A.29.20). Therefore, from Eq. (6.2.6), considering $\phi(x_i,x_j)$ as a quaternion element of an $N\times N$ self-dual quaternion matrix

$$\det\left[\phi\left(x_i,x_j\right)\right] \propto p\left(x_1,...,x_N\right). \tag{A.29.21}$$

The constant of proportionality is fixed by the normalization, Theorem 6.2.1 applied N times and the fact that

$$\int_{-\infty}^{\infty}\phi(x,x)dx = \int_{-\infty}^{\infty}\left[S_N(x,x)+\xi_N(x,x)\right]dx = N. \tag{A.29.22}$$

Arguments similar to the even N case given above show that the ordinary $2N\times 2N$ determinant

$$\det\left[\mathbf{\Phi}\left(x_i,x_j\right)\right] \propto \left[\mathbf{p}\left(x_1,...,x_{2N}\right)\right]^2, \tag{A.29.23}$$

which is Eq. (14.4.5) except for the constant. This constant is again determined by normalization, Theorem 6.2.1, and the fact that

$$\int_{-\infty}^{\infty}\mathbf{\Phi}(x,x)dx = \int_{-\infty}^{\infty}\mathbf{S}_N(x,x)dx = 2N. \tag{A.29.24}$$

A.30. Verification of Equation (14.4.3)

We will verify the equation

$$\int_{-\infty}^{\infty} \phi(x,y)\phi(y,z)dy = \phi(x,z) + \tau\phi(x,z) - \phi(x,z)\tau, \qquad (14.4.3)$$

where ϕ is defined by Eq. (14.1.14) and

$$\tau = \frac{1}{2}\begin{bmatrix} 1 & 0 \\ 0 & -1 \end{bmatrix}. \qquad (A.30.1)$$

We will also verify the same equation when ϕ is replaced by $\mathbf{\Phi}$, Eq. (14.2.10), and τ by $-\tau$.

Writing for brevity $F^{\dagger}(x,y) = F(y,x)$ and

$$F * G \equiv \int_{-\infty}^{\infty} F(x,y)G(y,z)dy, \qquad (A.30.2)$$

Eq. (14.4.3) above is

$$\begin{aligned} \phi * \phi &\equiv \begin{bmatrix} S_N + \xi_N & D_N \\ J_N & S_N^{\dagger} + \xi_N^{\dagger} \end{bmatrix} * \begin{bmatrix} S_N + \xi_N & D_N \\ J_N & S_N^{\dagger} + \xi_N^{\dagger} \end{bmatrix} \\ &= \begin{bmatrix} S_N + \xi_N & 2D_N \\ 0 & S_N^{\dagger} + \xi_N^{\dagger} \end{bmatrix}, \end{aligned} \qquad (A.30.3)$$

where S_N, ξ_N, D_N, and J_N are defined by (14.1.17), (14.1.15), (14.1.19), and (14.1.20). This amounts to verifying that

$$\begin{aligned} (S_N + \xi_N) * (S_N + \xi_N) &= S_N + \xi_N, \\ D_N * J_N = 0 &= J_N * (S_N + \xi_N), \\ (S_N + \xi_N) * D_N &= D_N, \end{aligned} \qquad (A.30.4)$$

and, as a consequence,

$$\begin{aligned} (S_N + \xi_N)^{\dagger} * (S_N + \xi_N)^{\dagger} &= (S_N + \xi_N)^{\dagger}, \\ J_N * D_N = 0 &= (S_N + \xi_N)^{\dagger} * J_N, \\ D_N * (S_N + \xi_N)^{\dagger} &= D_N. \end{aligned} \qquad (A.30.5)$$

For the verification of Eqs. (A.30.4) we will repeatedly use results of orthonormality and convolution integrals of Appendix A.26, Sections A.26.3 and A.26.6. Thus from the expressions (14.1.17), (14.1.19), and (14.1.22) and the equations of Appendix A.26.3, we have

$$S_N * S_N = S_N, \qquad S_N * D_N = D_N,$$

$$D_N * I_N = S_N, \qquad I_N * S_N = I_N, \tag{A.30.6}$$

while from Appendix A.26.6, we have

$$g * S_N = \; - I_N, \qquad D_N * g = -S_N. \tag{A.30.7}$$

For N odd we also need

$$\xi_N * \xi_N = \xi_N, \qquad \mu_N^\dagger * \xi_N = \mu_N^\dagger, \tag{A.30.8}$$

$$g * \xi_N = \mu_N^\dagger, \tag{A.30.9}$$

and

$$S_N * \xi_N, \quad \xi_N * S_N, \quad \xi_N * D_N,$$

$$\mu_N * \xi_N \quad \left(\mu_N - \mu_N^\dagger\right) * S_N, \quad D_N * \left(\mu_N - \mu_N^\dagger\right). \tag{A.30.10}$$

Those in Eq. (A.30.8) are easily verified, and Eq. (A.30.9) is a consequence of Eq. (A.26.15). For Eqs. (A.30.10) we need to know

$$\int_{-\infty}^{\infty} \varphi_{2j+1}(y)\varphi_{2m}(y)dy, \; \int_{-\infty}^{\infty} \psi_{2j+1}(y)A_{2m}(y)dy,$$

$$\int_{-\infty}^{\infty} \varphi_{2j+1}(y)\exp\left(-\alpha^2 y^2/2\right) dy, \; \int_{-\infty}^{\infty} A_{2m}(y)\varphi_{2m}(y)dy, \tag{A.30.11}$$

and

$$\int_{-\infty}^{\infty} A_{2j+1}(y)\varphi_{2m}(y)dy, \int_{-\infty}^{\infty} \varphi_{2j+1}(y)A_{2m}(y)dy, \quad (A.30.12)$$

$$\int_{-\infty}^{\infty} \exp\left(-\alpha^2 y^2/2\right) \psi_{2j+1}(y)dy. \quad (A.30.13)$$

Integrals (A.30.11) are zero by parity. For Eq. (A.30.12) we note that A_{2j+1} is a linear combination of φ_{2j} and A_{2j-1}, Eq. (A.26.5), and hence of $\varphi_{2j}, \varphi_{2j-2}, ..., \varphi_2, \varphi_0$. Similarly, φ_{2j+1} is a linear combination of $\psi_{2j}, \psi_{2j-2}, ..., \psi_2, \psi_0$ by Eq. (A.26.4). Thus, A_{2j+1} is orthogonal to φ_{2m} and φ_{2j+1} is orthogonal to A_{2m}, for $j < m$ (see Appendix A.26.3). The integrand in Eq. (A.30.13) is a perfect derivative, Eq. (14.1.27), of a quantity vanishing at both ends.

Therefore, all the integrals in Eqs. (A.30.11)–(A.30.13) are zero and so are those in Eq. (A.30.10).

Thus, we have verified Eq. (14.4.3) in all the cases for $\phi(x, y)$ given by Eq. (14.1.14).

The verification of Eq. (14.4.3) with ϕ replaced by Φ, Eq. (14.2.10), and τ replaced by $-\tau$ is similar. From the expressions (14.2.11), (14.2.13), (14.2.16), and (14.2.18) together with Eqs. (A.26.1), (A.26.21)–(A.26.24) we have

$$\begin{aligned} &\mathbf{S}_N * \mathbf{S}_N = \mathbf{S}_N, \quad \mathbf{S}_N * \mathbf{D}_N = \mathbf{D}_N, \quad \mathbf{I}_N * \mathbf{S}_N = \mathbf{I}_N, \\ &\mathbf{S}_N * \mathbf{g} = -\mathbf{D}_N, \quad \mathbf{D}_N * \mathbf{I}_N = \mathbf{S}_N, \quad \mathbf{g} * \mathbf{I}_N = \mathbf{S}_N, \end{aligned}$$

or

$$\Phi * \Phi \equiv \begin{bmatrix} \mathbf{S}_N & \mathbf{K}_N \\ \mathbf{I}_N & \mathbf{S}_N^\dagger \end{bmatrix} * \begin{bmatrix} \mathbf{S}_N & \mathbf{K}_N \\ \mathbf{I}_N & \mathbf{S}_N^\dagger \end{bmatrix} = \begin{bmatrix} \mathbf{S}_N & 0 \\ 2\mathbf{I}_N & \mathbf{S}_N^\dagger \end{bmatrix} = \Phi + \Phi\tau - \tau\Phi,$$

with τ given by Eq. (A.30.1).

A.31. The Limits of $J_N(x, y)$ and $D_N(x, y)$ as $N \to \infty$. Asymptotic Forms of $J(r; \rho)$ and $D(r; \rho)$. Sections 14.1 and 14.2

The limits of $I_N(x, y)$ and $D_N(x, y)$ for $N \to \infty$ can be obtained by taking the Fourier transform of Eq. (A.30.7). Writing $F(f(x))$ for the

Fourier transform of $f(x)$, as in Appendix A.11, we have (Bateman, 1954, Vol. 1), for $b > 0$,

$$\begin{aligned} F(\mathrm{erf}(bx)) &= 2i \int_0^\infty \sin(2\pi kx)\mathrm{erf}(bx)dx \\ &= \frac{i}{\pi k} \exp(-\pi^2 k^2/b^2), \end{aligned} \tag{A.31.1}$$

and

$$\begin{aligned} F\left(\frac{\sin\,(bx)}{x}\right) &= 2\int_0^\infty \cos(2\pi kx)\,\frac{\sin(bx)}{x}\,dx \\ &= \begin{cases} \pi, & \text{if} \quad 2\pi\,|k| < b, \\ \pi/2, & \text{if} \quad 2\pi\,|k| = b, \\ 0, & \text{if} \quad 2\pi\,|k| > b. \end{cases} \end{aligned} \tag{A.31.2}$$

which along with Eq. (A.30.7) give for the Fourier transform of I_N

$$F(I_N) \approx \frac{i}{2\pi k} \exp(-8\pi^2\rho^2k^2) \begin{cases} 1, & \text{if} \quad |k| < 1/2, \\ 1/2, & \text{if} \quad |k| = 1/2, \\ 0, & \text{if} \quad |k| > 1/2. \end{cases} \tag{A.31.3}$$

An inverse Fourier transform then gives

$$\begin{aligned} I_N(x,y) &\approx -\frac{1}{\pi}\int_0^r dt \int_0^\pi dk\, \cos(kt)\, \exp(-2\rho^2k^2) \\ &= -\frac{1}{\pi}\int_0^\pi dk\, \frac{\sin(kr)}{k}\, \exp(-2\rho^2k^2), \end{aligned} \tag{A.31.4}$$

so that

$$J_N(x,y) \approx J(r;\rho) = -\frac{1}{\pi}\int_0^r dt \int_\pi^\infty dk\, \cos(kt)\, \exp(-2\rho^2k^2). \tag{A.31.5}$$

Similarly from the second equality in Eq. (A.30.7) we have

$$D_N(x,y) \approx [R_1(x)]^2 D(r;\rho), \tag{A.31.6}$$

$$\begin{aligned} D(r;\rho) &= \frac{d}{dr}\left(\frac{1}{\pi}\int_0^{\pi} dk\, \cos(kr)\, \exp(2\rho^2k^2)\right) \\ &= -\frac{1}{\pi}\int_0^{\pi} dk\; k\, \sin(kr)\, \exp(2\rho^2k^2). \end{aligned} \tag{A.31.7}$$

Integration gives

$$\begin{aligned} &\int_\pi^\infty \frac{\sin(kr)}{k}\exp(-2\rho^2k^2)dk \\ &= \exp(-r^2/8\rho^2)\text{Im}\int_\pi^\infty \exp\left(-2\rho^2\left(k - ir/(4\rho^2)\right)^2\right) dk/k \\ &= \exp[-r^2/(8\rho^2)]\text{Im}\Bigg[\exp\left(-2\rho^2\left(k-\frac{ir}{4\rho^2}\right)^2\right) \\ &\qquad \left(-\frac{1}{4\rho^2k[k - ir/(4\rho^2)]} + \cdots\right)\Bigg]_\pi^\infty \\ &= \exp\left(-2\rho^2\pi^2\right)\,\text{Im}\; e^{i\pi r}\left(\frac{-1}{\pi\,(4\pi\rho^2 - ir)} + \cdots\right) \\ &= \frac{1}{\pi}\exp\left(-2\rho^2\pi^2\right)\left(\frac{r\,\cos(\pi r) + 4\pi\rho^2\sin(\pi r)}{16\pi^2\rho^4 + r^2} + \cdots\right). \end{aligned} \tag{A.31.8}$$

Similarly

$$\begin{aligned} &\int_0^\pi k\, \sin(kr)\exp(2\rho^2k^2)dk \\ &\quad = \pi\, \exp(2\pi^2\rho^2)\left(\frac{r\,\cos(\pi r) - 4\pi\rho^2\sin(\pi r)}{16\pi^2\rho^4 + r^2} + \cdots\right). \end{aligned} \tag{A.31.9}$$

A.32. Evaluation of the Integral (15.1.9) for Complex Matrices

In the exposition given here and in Appendices A.34 and A.37 we follow the method of Ginibre (1965). Appendix A.35 gives an alternative simpler proof due to Dyson. We start with the following proposition.

Any complex nonsingular $N \times N$ matrix X can be expressed in one and only one way as

$$X = UYV, \tag{A.32.1}$$

where U is a unitary matrix, Y is a triangular matrix with all diagonal elements equal to unity, $y_{ij} = 0,\ i > j;\ y_{ii} = 1$, and V is a diagonal matrix with real positive diagonal elements.

Proof. Given $X = [x_{ij}]$, we solve the homogeneous linear equations in u_{rj}:

$$\begin{aligned} \sum_{j=1}^{N} u_{rj} x_{ji} &= 0, \qquad i < r, \\ \sum_{j=1}^{N} u_{rj} u_{ij}^{*} &= 0, \qquad i > r, \end{aligned} \tag{A.32.2}$$

successively for $r = N, N-1, ..., 1$. Because the number of unknowns is always one greater than the number of equations, the u_{rj} for a fixed r are not all zero. We may normalize them to satisfy

$$\sum_{j=1}^{N} u_{rj} u_{rj}^{*} = 1, \tag{A.32.3}$$

without disturbing the equalities (A.32.2). Thus, we have found a unitary matrix $U_1 = [u_{rj}]$ such that $Y_1 = U_1 X$ is triangular $(Y_1)_{ij} = 0, i > j$. Because X is nonsingular, all diagonal elements of Y_1 are different from zero. Writing the diagonal elements of Y_1 in the polar form, $(Y_1)_{jj} = v_j \exp(i\theta_j)$, we construct two diagonal matrices, one unitary, U_2, with diagonal elements $(U_2)_{jj} = \exp(i\theta_j)$, and the other positive definite, V, with diagonal elements $V_{jj} = v_j$. Putting $Y = U_2^\dagger Y_1 V^{-1}$ and $U = U_1^\dagger U_2$, we see that $X = UYV$, where U, Y, and V have the properties required in the proposition. Next let $X = UYV = U'Y'V'$; then $U'^\dagger U = Y'V'V^{-1}Y^{-1}$ is unitary (left-hand side) as well as triangular with real positive diagonal elements (right-hand side). Thus $U'^\dagger U$ and $Y'V'V^{-1}Y^{-1}$ are unit matrices. A comparison of the diagonal elements on the two sides of $YV = Y'V'$ now gives $V = V'$. The decomposition $X = UYV$ is therefore unique.

Using the fact that $S = XEX^{-1}$, $U^\dagger U = 1$, and $EV = VE$, we can write

$$\operatorname{tr}\left(S^\dagger S\right) = \operatorname{tr}\left[E^\dagger X^\dagger X E \left(X^\dagger X\right)^{-1}\right] = \operatorname{tr}\left[E^\dagger Y^\dagger Y E \left(Y^\dagger Y\right)^{-1}\right], \tag{A.32.4}$$

$$dA = X^{-1}dX = V^{-1}Y^{-1}\left(U^{-1}dU\right)YV + V^{-1}Y^{-1}dYV + V^{-1}dV. \tag{A.32.5}$$

The volume element $\prod_{i\neq j} dA_{ij}^{(0)} dA_{ij}^{(1)}$ needed in Eq. (15.1.9) is the quotient of the volume element $\prod_{ij} dA_{ij}^{(0)} dA_{ij}^{(1)}$ by that of the set of all complex diagonal matrices. We put aside the quantities that do not depend on the eigenvalues, for they give only multiplicative constants. All of these constants can be adjusted in the final normalization. From Eq. (A.32.5) and the structure of Y and V we see that

$$\prod_{i,j} dA_{ij}^{(0)} dA_{ij}^{(1)} = \prod_{i<j} \left(Y^{-1}dY\right)_{ij}^{(0)} \left(Y^{-1}dY\right)_{ij}^{(1)} a, \tag{A.32.6}$$

where a depends only on U and V. We replace $\prod_{i\neq j} dA_{ij}^{(0)} dA_{ij}^{(1)}$ in Eq. (15.1.9) with $\prod_{i<j} \left(Y^{-1}dY\right)_{ij}^{(0)} \left(Y^{-1}dY\right)_{ij}^{(1)}$ and calculate

$$\int \exp\left[-\operatorname{tr}\left(E^\dagger H E H^{-1}\right)\right] \prod_{i<j} \left(Y^{-1}dY\right)_{ij}^{(0)} \left(Y^{-1}dY\right)_{ij}^{(1)}, \tag{A.32.7}$$

where

$$H = Y^\dagger Y. \tag{A.32.8}$$

The matrix H is Hermitian. Any of its upper left diagonal block of size n is obtained from the upper left diagonal block of Y of the same size: $H_n = Y_n^\dagger Y_n$. Therefore, for every n, det $H_n = 1$, and the diagonal elements H_{nn} are successively and uniquely determined once the off-diagonal elements are given. Thus we need $N(N-1)$ real parameters to specifiy H, the same number needed to specify Y. We can further convince ourselves that the correspondence of Y and H is one to one. However, we do not need this last result, for we are omitting the constants anyway. Now we make a change of variables. First, because det $Y = 1$,

$$\prod_{i<j} \left(Y^{-1}dY\right)_{ij}^{(0)} \left(Y^{-1}dY\right)_{ij}^{(1)} = \prod_{i<j} dY_{ij}^{(0)} dY_{ij}^{(1)}. \tag{A.32.9}$$

Next, we take $H_{ij}^{(0)}, H_{ij}^{(1)}$ for $i < j$ as independent variables. The superscripts (0) and (1) denote, as always, the real and the imaginary parts. From

$$H_{ij} = Y_{ij} + \sum_{k<i} Y_{ki}^* Y_{kj}, \qquad i < j,$$

one can easily calculate the Jacobian of the transformation from Y to H, it being unity. The integral (15.1.9) is

$$J = C \int \exp\left[-\text{tr}\left(E^\dagger H E H^{-1}\right)\right] \prod_{i<j} dH_{ij}^{(0)} dH_{ij}^{(1)}, \tag{A.32.10}$$

where C is a constant.

The integration over H is done in N steps. At every step we integrate over the variables of the last column and thus decrease by one the size of the matrix, whose structure remains the same. For this we need the recursion relation Eq. (A.32.17) derived below.

Let $H' = Y_n^\dagger Y_n$, $E' = [z_i \delta_{ij}]_{i,j=1,2,\ldots,n}$, be the relevant matrices of order n and H, E be those obtained from H', E' by removing the last row and last column. Let the greek indices run from 1 to n, and the latin indices from 1 to $n-1$. Let $\Delta'_{\alpha\beta}$ be the cofactor of $H'_{\alpha\beta}$ in H' and Δ_{ij}, the cofactor of H_{ij} in H. Let $g_i = H'_{in}$. Because det H' = det $H = 1$, we have

$$\Delta'_{\alpha\beta} = \left(H'^{-1}\right)_{\beta\alpha}, \quad \Delta_{ij} = \left(H^{-1}\right)_{ji}. \tag{A.32.11}$$

Expanding det H', Δ'_{in}, Δ'_{ij} by the last row and last column, we have

$$1 = H'_{nn} - \sum_{i,j} g_i^* g_j \Delta_{ji}, \tag{A.32.12}$$

$$\Delta'_{in} = -\sum_{\ell} \Delta_{i\ell} g_\ell^*, \tag{A.32.13}$$

$$\Delta'_{ij} = H'_{nn}\Delta_{ij} - \sum_{\ell,k} g_k^* g_\ell \Delta_{ij}^{\ell k}, \tag{A.32.14}$$

where $\Delta_{ij}^{\ell k}$ is the cofactor obtained from H by removing the ith and ℓth rows and the jth and kth columns. Sylvester's theorem (*cf.* Mehta, 1989) expresses $\Delta_{ij}^{\ell k}$ in terms of Δ_{rs}

$$\Delta_{ij}^{\ell k} = \Delta_{ij}\Delta_{\ell k} - \Delta_{ik}\Delta_{\ell j}. \tag{A.32.15}$$

In writing Eq. (A.32.15), we have replaced det H by unity on the left-hand side. Let

$$\phi_n = \text{tr}\left(E'^{\dagger} H' E' H'^{-1}\right) = \sum_{\alpha,\beta} z_\alpha^* z_\beta H'_{\alpha\beta} \Delta'_{\alpha\beta}. \tag{A.32.16}$$

Separating the last row and last column and making use of Eqs. (A.32.11) to (A.32.15), we get, after some simplification,

$$\phi_n = |z_n|^2 + \phi_{n-1} + \left\langle g^* \left| H^{-1}\left(E^\dagger - z_n^*\right) H \left(E - z_n\right) H^{-1} \right| g \right\rangle, \tag{A.32.17}$$

where

$$\langle g^* |B| g\rangle = \sum_{i,j} g_i^* B_{ij} g_j. \tag{A.32.18}$$

Substituting (A.32.17) for $n = N$ in Eq. (A.32.10), we get

$$J = C\, e^{-|z_N|^2} \int e^{-\phi_{N-1}} \prod_{l \le i < j \le N-1} dH_{ij}^{(0)} dH_{ij}^{(0)}$$

$$\times \int \exp\left[-\left\langle g^* \left| H^{-1}\left(E^\dagger - z_N^*\right) H \left(E - z_N\right) H^{-1} \right| g \right\rangle\right]$$

$$\times \prod_{l \le i \le N-1} dg_i^{(0)} dg_i^{(1)}, \tag{A.32.19}$$

The last integral is immediate and gives

$$\pi^{N-1} \left\{\det\left[H^{-1}\left(E^\dagger - z_N^*\right) H \left(E - z_N\right) H^{-1}\right]\right\}^{-1}$$

$$= \pi^{N-1} \prod_{i=1}^{N-1} \left|z_i - z_N\right|^{-2}. \tag{A.32.20}$$

The process can be repeated N times and we finally get

$$J = C \exp\left(-\sum_1^N |z_i|^2\right) \prod_{1 \le i < j \le N} \left|z_i - z_j\right|^{-2} \tag{A.32.21}$$

where C is a new constant.

A.33. A Few Remarks About the Eigenvalues of a Quaternion Real Matrix and Its Diagonalization

A quaternion-real matrix S is one whose elements are real quaternions (*cf.* Chapter 2). If we replace the elements of an $N \times N$ quaternion matrix S by their 2×2 matrix representation (2.4.3), we get a $2N \times 2N$ matrix $C(S)$ with complex elements. A real quaternion is represented by a 2×2 matrix of the form

$$\begin{bmatrix} a & -b^* \\ b & a^* \end{bmatrix},$$

so that the matrix $C(S)$ has the form

$$C(S) = \begin{bmatrix} a_{ij} & -b^*_{ij} \\ b_{ij} & a^*_{ij} \end{bmatrix} . \tag{A.33.1}$$

The matrix $C(S)$ has $2N$ (complex or c-) eigenvalues and at least one (complex or c-) eigenvector belonging to each distinct eigenvalue. If $[x_i \ , \ y_i]^T$ is a c-eigenvector of $C(S)$ belonging to the c-eigenvalue α,

$$\sum_j \begin{bmatrix} a_{ij} & -b^*_{ij} \\ b_{ij} & a^*_{ij} \end{bmatrix} \begin{bmatrix} x_j \\ y_j \end{bmatrix} = \alpha \begin{bmatrix} x_i \\ y_i \end{bmatrix} \tag{A.33.2}$$

$$\alpha = \alpha_0 + i\alpha_1; \qquad \alpha_0, \quad \alpha_1 \text{ real} , \tag{A.33.3}$$

that is,

$$\begin{aligned} \sum_j \left(a_{ij}x_j - b^*_{ij}y_j\right) &= \alpha x_i, \\ \sum_j \left(b_{ij}x_j + a^*_{ij}y_j\right) &= \alpha y_i . \end{aligned} \tag{A.33.2$'$}$$

Then, taking the complex conjugate of these equations and changing the order in which they are written, we see that

$$\sum_j \begin{bmatrix} a_{ij} & -b^*_{ij} \\ b_{ij} & a^*_{ij} \end{bmatrix} \begin{bmatrix} -y^*_j \\ x^*_j \end{bmatrix} = \alpha^* \begin{bmatrix} -y^*_i \\ x^*_i \end{bmatrix} , \tag{A.33.4}$$

that is $[-y^*_i \ , \ x^*_i]^T$ is another c-eigenvector belonging to the c-eigenvalue α^*. We now write Eqs. (A.33.2) and (A.33.4) together:

$$\sum_j \begin{bmatrix} a_{ij} & -b^*_{ij} \\ b_{ij} & a^*_{ij} \end{bmatrix} \begin{bmatrix} x_j & -y^*_j \\ y_j & x^*_j \end{bmatrix} = \begin{bmatrix} x_i & -y^*_i \\ y_i & x^*_i \end{bmatrix} \begin{bmatrix} \alpha & 0 \\ 0 & \alpha^* \end{bmatrix} \tag{A.33.5}$$

or, in the quaternion notation,

$$Sx = x\alpha , \quad \alpha = \alpha_0 + \alpha_1 e_1 . \tag{A.33.6}$$

We see that the eigenvectors and eigenvalues of S are real quaternions. Moreover, the (quaternion or q-) eigenvalue does not contain the e_2 and e_3 parts that give rise to off-diagonal terms in its 2×2 matrix representation. Thus, the quaternion α in Eq. (A.33.6) may be identified with the complex number α in Eq. (A.33.3).

We say that two quaternions λ_1 and λ_2 are essentially distinct if the equation $\lambda_1 \mu = \mu \lambda_2$ implies $\mu = 0$. If $x_1, x_2, ..., x_r$ are q-eigenvectors belonging to the essentially distinct q-eigenvalues $\lambda_1, \lambda_2, ..., \lambda_r$, then $x_1, x_2, ..., x_r$ are right linearly independent; that is the right linear (vector) equation

$$x_1 c_1 + x_2 c_2 + \cdots + x_r c_r = 0 \tag{A.33.7}$$

implies

$$c_1 = c_2 = \cdots = c_r = 0 \ .$$

A proof can be supplied by induction.

However, the right linear independence of a set of q-vectors does not necessarily lead to their left linear independence, as may be seen from the following example. The vectors

$$x_1 = \begin{bmatrix} 1 \\ \\ e_1 \end{bmatrix} , \quad x_2 = \begin{bmatrix} e_2 \\ \\ -e_3 \end{bmatrix} \tag{A.33.8}$$

are right linearly independent, but they are left linearly dependent. Thus, we are still far from the diagonalization of S by purely quaternion means.

However, we may again use the intermediatory of the matrix $C(S)$. If all the c-eigenvalues z_1, z_1^*, ..., z_N, z_N^* of $C(S)$ are distinct, none of them being real, the complex matrix X whose columns are the eigenvectors of $C(S)$ belonging to these c-eigenvalues, is nonsingular and

$$X^{-1} C(S) X = E \ , \tag{A.33.9}$$

where E is diagonal with diagonal elements $z_1, z_1^*, ..., z_N, z_N^*$. It is easy to be convinced that when X^{-1} is re-expressed as an $N \times N$ quaternion matrix all its elements will be real quaternions. Therefore, if all the N (quaternion) eigenvalues of S are essentially distinct, a quaternion real matrix X exists such that

$$S = XEX^{-1} \ , \tag{A.33.10}$$

where E is diagonal and quaternion real.

A.34. Evaluation of the Integral (15.2.9)

As in Appendix A.32, we decompose the $2N \times 2N$ matrix X into the unique product $X = UYV$, where U is unitary, Y is triangular with unit diagonal elements $Y_{ij} = 0$, $i > j$; $Y_{ii} = 1$, and V is diagonal with real positive elements. Moreover, because X has the form

$$\begin{bmatrix} a_{ij} & -b_{ij}^* \\ \\ b_{ij} & a_{ij}^* \end{bmatrix}, \tag{A.34.1}$$

the U, Y, and V all have the same form. In particular, $Y_{2i-1,2i} = -Y_{2i,2i-1}^* = 0$, and $V_{2i-1,2i-1} = V_{2i,2i}$.

In fact any matrix A having the form (A.34.1) satisfies the relation $ZA = A^*Z$, where Z is given by Eq. (2.4.1). And conversely, if $ZA = A^*Z$, then A has the form (A.34.1). From $ZX = X^*Z$ one sees that

$$U^T ZU = Y^* VZV^{-1}Y^{-1} \tag{A.34.2}$$

is unitary and antisymmetric (the left-hand side) and has nonzero elements only in the 2×2 blocks along the principal diagonal (comparison of elements on the two sides). One also has V_{ii} real and positive. Thus, $U^T ZU = Z$. Substituting this in Eq. (A.34.2), we get $Y^{*-1}ZY = VZV^{-1} = Z$. Thus U, Y and V all have the form of (A.34.1).

If we let $H = Y^\dagger Y$, then, because $H^\dagger = H$ and Y has the form (A.34.1), $H_{2i-1,2i-1} = H_{2i,2i} = h_i$ and $H_{2i-1,2i} = H_{2i,2i-1} = 0$. Moreover, h_i is completely determined by the condition $\det H_{2i-1} = \det Y_{2i-1}^\dagger Y_{2i-1} = 1$. Thus we may consider H_{ij}, $i < j$ and $(i,j) \neq (2k-1, 2k)$ as independent complex variables. As in Appendix A.32, we change from the volume element

$$\prod_{i<j} \prod_{\lambda=0}^{1} dA_{2i-1,2j}^{(\lambda)} dA_{2i,2j}^{(\lambda)} \quad \text{to} \quad \prod dY_{ij}^{(0)} dY_{ij}^{(1)}$$

and finally to $\prod dH_{ij}^{(0)} dH_{ij}^{(1)}$, where the product $\prod$ over the elements of dY or dH are taken over all $i < j$ except the pairs $(i,j) = (2k-1, 2k)$.

To calculate the integral

$$\int \exp\left[-(1/2)\operatorname{tr}\left(E^\dagger HEH^{-1}\right)\right] \prod dH_{ij}^{(0)} dH_{ij}^{(1)}, \tag{A.34.3}$$

a recurrence relation similar to Eq. (A.32.17) is needed. Let H', E' denote matrices of order $2n$; H'', E'', their upper left diagonal blocks of order $(2n-1)$, and H, E, their upper left diagonal blocks of order $2n-2$. The cofactors of H', H'', and H are denoted, respectively, by Δ', Δ'', and Δ. Let

$$g = (g_i)\,, \qquad g' = (g'_i)\,, \tag{A.34.4}$$

where

$$g_i = H'_{i,2n-1}, \qquad g'_i = H'_{i,2n}, \quad i = 1, 2, ..., 2n-2,$$

$$g_{2n-1} = g_{2n} = g'_{2n-1} = g'_{2n} = 0. \tag{A.34.5}$$

Then

$$g'_{2i} = g^*_{2i-1}, \qquad g'_{2i-1} = -g^*_{2i},$$

or

$$g' = -Zg^*, \tag{A.34.6}$$

where Z is given by Eq. (2.4.1). Using the facts

$$H'_{2n-1,2n-1} = H'_{2n,2n} = h_n, \qquad H'_{2n-1,2n} = H'_{2n,2n-1} = 0, \tag{A.34.7}$$

and

$$\det H' = \det H'' = \det H = 1, \tag{A.34.8}$$

we get by expanding according to the last row and last column

$$1 = h_n - \sum_{i,j} g^*_i g_j \Delta_{ji}, \tag{A.34.9}$$

$$\Delta''_{2n-1,\ell} = \sum_k g_k \Delta_{k\ell}, \tag{A.34.10}$$

$$\begin{aligned} \Delta''_{ij} &= h_n \Delta_{ij} - \sum_{k,\ell} g^*_k g_\ell \Delta^{\ell k}_{ij} \\ &= \Delta_{ij} + \sum_{k,\ell} g^*_k g_\ell \Delta_{ik} \Delta_{\ell j}, \end{aligned} \tag{A.34.11}$$

where in the last step of Eq. (A.34.11) we have used Eqs. (A.34.9) and (A.32.15). The matrices H', H'', and H are positive definite and so are their inverses. In particular,

$$\sum_{i,j} f_i^* \Delta_{ij} f_j > 0 \tag{A.34.12}$$

if not all f_i are zero. By equating to zero the expansion of a determinant whose first $2n - 1$ rows are identical with those of H' and whose last row is identical with the last but one of H', we get

$$\begin{aligned} 0 &= \sum_i g_i^* \Delta'_{2n,i} + h_n \Delta'_{2n,2n-1} \\ &= \sum_{i,j} g_i^* g_j' \Delta''_{ji} + h_n \sum_j g_j' \Delta''_{j,2n-1}, \end{aligned} \tag{A.34.13}$$

which on making use of Eqs. (A.34.9), (A.34.10), and (A.34.11) gives

$$0 = \sum_{j,p} g_j' g_p^* \Delta_{jp} \left(1 + \sum_{i,q} g_i^* \Delta_{qi} g_q \right). \tag{A.34.14}$$

In view of Eq. (A.34.12), this is equivalent to

$$0 = \sum_{j,p} g_j' \Delta_{jp} g_p^* = \left(\sum_{j,p} g_j'^* \Delta_{pj} g_p \right)^* \tag{A.34.15}$$

or

$$\sum_j \Delta''_{j,2n-1} g_j' = \left(H''^{-1} g' \right)_{2n-1} = 0. \tag{A.34.15$'$}$$

Next we put

$$\phi_n = \frac{1}{2} \mathrm{tr} \left(E'^\dagger H' E' H'^{-1} \right) \tag{A.34.16}$$

and apply Eq. (A.32.17) twice to get

$$\phi_n = |z_n|^2 + \phi_{n-1} + \frac{1}{2} \langle g^* | U | g \rangle + \frac{1}{2} \langle g'^* | V | g' \rangle . \tag{A.34.17}$$

The notation is that of Eq. (A.32.18)

$$\langle f^* |B| f\rangle = \sum_{i,j} f_i^* B_{ij} f_i, \tag{A.34.18}$$

where

$$U = H^{-1}\left(E^\dagger - z_n^*\right) H \left(E - z_n\right) H^{-1}, \tag{A.34.19}$$

$$V = H''^{-1}\left(E''^\dagger - z_n\right) H'' \left(E'' - z_n^*\right) H''^{-1}. \tag{A.34.20}$$

From Eq. (A.34.15) we see that V is essentially equal to U.

$$\langle g'^* |V| g'\rangle = \langle g'^* |U| g'\rangle . \tag{A.34.21}$$

Last, from Eq. (A.34.6), $U^\dagger = U$, and $UZ = ZU^*$, Z given by (2.4.1), we have

$$\langle g'^* |U| g'\rangle = \langle g^* |U| g\rangle . \tag{A.34.22}$$

Collecting Eqs. (A.34.17), (A.34.21) and (A.34.22), the recurrence relation becomes

$$\phi_n = |z_n|^2 + \phi_{n-1} + \langle g^* |U| g\rangle \tag{A.34.23}$$

where U is given by Eq. (A.34.19).

The rest of the integration is identical to that in Appendix A.32.

A.35. Another Proof of Equations (15.1.10) and (15.2.10)

A shorter proof of Eqs. (15.1.10) and (15.2.10) due to Dyson is as follows.

Consider first the case when S is a matrix with complex elements. Given S, one can always find a unitary matrix U such that $U^\dagger SU = T$ is triangular, i.e., $T_{jk} = 0$ if $j > k$ (*cf.* Mehta, 1989). The diagonal elements $T_{jj} = z_j$ of T are the eigenvalues of S. The matrix S has N^2 complex elements, so that the number of real parameters entering S is $2N^2$. The number of real parameters entering U is N^2 and that entering T is $N(N+1)$. So the number of real parameters in U and T seem to exceed those in S by N. Actually, S does not determine U and T uniquely. If we replace U by UV where V is any unitary diagonal matrix, then $(UV)^\dagger S(UV) = V^\dagger TV$ remains triangular. We will use this freedom below to impose N conditions on the variations of U.

Differentiating $S = UTU^\dagger$, we get

$$\begin{aligned} dS &= U(dT + U^\dagger dU\; T - T\; U^\dagger dU)U^\dagger \\ &= U\; dA\; U^\dagger, \end{aligned} \tag{A.35.1}$$

$$\begin{aligned} dA &= dT + U^\dagger dU\; T - T\; U^\dagger dU \\ &= dT + i(dH\; T - T\; dH), \end{aligned} \tag{A.35.2}$$

where dT is triangular and $dH = -iU^\dagger dU$ is Hermitian. Let us restrict the variations in U so that

$$dH_{jj} = -i(U^\dagger dU)_{jj} = 0, \qquad j = 1,\ 2,\ ...,\ N. \tag{A.35.3}$$

In terms of its components Eq. (A.35.2) reads

$$\begin{aligned} dA_{jk} &= dT_{jk} + i(T_{kk} - T_{jj})dH_{jk} \\ &\quad + i\sum_{\ell<k} dH_{j\ell}T_{\ell k} - i\sum_{j<\ell} T_{j\ell}dH_{\ell k}. \end{aligned} \tag{A.35.4}$$

Let us order the indices (j,k), $1 \le j,\ k \le N$, so that (j_1,k_1) preceeds (j_2,k_2) if either $j_1 > j_2$ or $j_1 = j_2$ and $k_1 < k_2$. If $j \le k$ we take $T_{j,k}$ and if $j > k$ we take $H_{j,k}$ as the variable. In other words the variables $A_{j,k}$, $(T_{j,k},\ H_{j,k})$ are arranged as

$$A_{N,1},\ A_{N,2},\ ...,\ A_{N,N},\ A_{N-1,1},\ ...,\ A_{N-1,N},\ ...,\ A_{1,1},\ ...,\ A_{1,N};$$

$$H_{N,1},\ H_{N,2},\ ...,\ H_{N,N-1},\ T_{N,N},\ H_{N-1,1},\ ...,\ H_{N-1,N-2};$$

$$T_{N-1,N-1},\ T_{N-1,N},\ H_{N-2,1},\ ...,\ T_{2,N},\ T_{1,1},\ T_{1,2},\ ...,\ T_{1,N}.$$

With this ordering, the elements above the main diagonal in the Jacobian matrix $\partial(A_{jk})/\partial(H_{jk},T_{jk})$ are all zero, and the diagonal elements are either $i(T_{kk} - T_{jj})$ or 1. Therefore,

$$\begin{aligned}\mu(dS) &= \prod_{j,k} dS_{jk} dS^*_{jk} = \prod_{j,k} dA_{jk} dA^*_{jk} \\ &= \prod_{j<k} |T_{jj} - T_{kk}|^2 \, dH_{jk} dH^*_{jk} \prod_{j\le k} dT_{jk} dT^*_{jk} \\ &= \prod_{j<k} |z_j - z_k|^2 \prod_j dz_j dz^*_j \prod_{j<k} dT_{jk} dT^*_{jk} dH_{jk} dH^*_{jk}, \quad \text{(A.35.5)}\end{aligned}$$

where $z_j = T_{jj}$ are the eigenvalues of S. Also

$$\operatorname{tr}(S^\dagger S) = \operatorname{tr}(T^\dagger T) = \sum_j |z_j|^2 + \sum_{j<k} |T_{jk}|^2, \quad \text{(A.35.6)}$$

$$\int \exp\left(-\sum_{j<k} |T_{jk}|^2\right) \prod_{j<k} dT_{jk} dT^*_{jk} = \pi^{N(N+1)/2}, \quad \text{(A.35.7)}$$

and

$$\int \prod_{j<k} dH_{jk} dH^*_{jk} = \Omega_U \Big/ \int \prod_j dH_{jj} = \Omega_U/(2\pi)^N, \quad \text{(A.35.8)}$$

where Ω_U is the volume of the unitary group. Collecting the results one gets

$$\int \exp[-\operatorname{tr}(S^\dagger S)] \mu(dS) = \int P_c(z_1, ..., z_N) \prod_j dz_j dz^*_j, \quad \text{(A.35.9)}$$

with

$$P_c(z_1, ..., z_N) = \text{const} \times \prod_{j<k} |z_j - z_k|^2 \exp\left(-\sum_j |z_j|^2\right). \quad \text{(A.35.10)}$$

When S has quaternion real elements, then U is symplectic, T is triangular quaternion real, diagonal elements of T do not contain the e_2 or e_3 quaternion units, the eigenvalues of S are $x_j \pm iy_j$ if $T_{jj} = x_j + e_1 y_j$, idH is anti-self-dual and the proof given above works with minor modifications.

When S is real and has only real eigenvalues, all distinct, then U is real orthogonal, T is real triangular, dH is real symmetric and the proof given above is again valid. If S is real and has some complex eigenvalues, then $S = UTU^\dagger$ with U real orthogonal is not possible and the proof given here fails.

A.36. Proof of Equation (15.2.38)

Let us put

$$\psi(x, y) = (2\pi)^{-1/2} \sum_{k=0}^{\infty} I_{k+1/2}(x) \left[y^{k+1/2} - y^{-(k+1/2)} \right]. \tag{A.36.1}$$

Differentiating Eq. (A.36.1) with respect to x and using the relation

$$I'_\nu(x) = \frac{1}{2} \left[I_{\nu+1}(x) + I_{\nu-1}(x) \right], \tag{A.36.2}$$

we have

$$\begin{aligned} \frac{\partial \psi}{\partial x} &= (2\pi)^{-1/2} \frac{1}{2} \sum_{k=0}^{\infty} \left[I_{k+3/2}(x) + I_{k-1/2}(x) \right] \\ &\quad \times \left[y^{k+1/2} - y^{-(k+1/2)} \right] \\ &= (2\pi)^{-1/2} \frac{1}{2} \sum_{k=0}^{\infty} I_{k+1/2}(x) \left[y^{k+1/2} - y^{-(k+1/2)} \right] \left(y + y^{-1} \right) \\ &\quad + (2\pi)^{-1/2} \frac{1}{2} \left(y^{1/2} - y^{-1/2} \right) \left[I_{1/2}(x) + I_{-1/2}(x) \right] \\ &= \frac{1}{2} \left(y + y^{-1} \right) \psi(x, y) + (2\pi)^{-1/2} \\ &\quad \times \frac{1}{2} \left(y^{1/2} - y^{-1/2} \right) \left(\frac{2}{\pi x} \right)^{1/2} e^x, \end{aligned} \tag{A.36.3}$$

where we have used the fact that

$$I_{1/2}(x) + I_{-1/2}(x) = \left(\frac{2}{\pi x} \right)^{1/2} (\sinh x + \cosh x). \tag{A.36.4}$$

The differential equation (A.36.3) can be immediately solved to give us

$$\psi(x,y) = \frac{1}{2\pi}\exp\left[\frac{1}{2}\left(y+y^{-1}\right)x\right]\left(y^{1/2}-y^{-1/2}\right)$$
$$\times\int_0^x \exp\left[x' - \frac{1}{2}\left(y+y^{-1}\right)x'\right]\frac{dx'}{\sqrt{x'}}. \qquad \text{(A.36.5)}$$

Changing the integration variable from x' to $t = 1 - x'/x$, we get

$$\psi(x,y) = \frac{1}{2\pi}e^{x}\left(y^{1/2}-y^{-1/2}\right)\sqrt{x}$$
$$\times\int_0^1 \exp\left[\frac{1}{2}\left(y^{1/2}-y^{-1/2}\right)^2 xt\right]\frac{dt}{\sqrt{1-t}} \qquad \text{(A.36.6)}$$

Finally putting $x = zz^*$ and $y = z^*/z$ we have

$$\psi(x,y) = \phi(z,z^*). \qquad \text{(A.36.7)}$$

and Eq. (A.36.6) gives Eq. (15.2.38).

A.37. The Case of Random Real Matrices

For $N \times N$ real matrices the linear measure is

$$\mu(dS) = \prod_{i,j=1}^{N} dS_{ij}.$$

The procedure leading to Eq. (15.1.4) is the same as in the complex case, and instead of Eq. (15.1.8) we have

$$\mu(dS) = \prod_{i\neq j}|z_i - z_j|\,dA_{ij}, \qquad dA = X^{-1}dX,$$

where X diagonalizes the matrix S and $z_1, z_2, \ldots, z_N$ are the distinct eigenvalues of S. We then have to evaluate

$$\int e^{-\operatorname{tr}(S^\dagger S)}\prod_{i\neq j}dA_{ij}.$$

In case all the eigenvalues of a real matrix S are real, the corresponding eigenvectors can all be taken to be real and the real X that satisfies $S = XEX^{-1}$ can be written uniquely in the form $X = UYV$ (as in Appendix A.32), where U, Y, and V are real, U is unitary, hence orthogonal, Y is triangular with unit diagonal elements, and V is diagonal positive. As in Appendix A.32, we can derive the recurrence relation

$$\phi_n = z_n^2 + \phi_{n-1} + \langle g \left| H^{-1} \left(E - z_n \right) H \left(E - z_n \right) H^{-1} \right| \rangle ,$$

where $\phi_n = \operatorname{tr}\left(E' H' E' H'^{-1} \right)$, E' is an $n \times n$ real diagonal matrix with diagonal elements $z_1, z_2, ..., z_n$, $H' = Y_n^{\dagger} Y_n$, and Y_n is an $n \times n$ real triangular matrix with unit diagonal elements. The integral

$$\int \exp\left[-\operatorname{tr}\left(EHEH^{-1} \right) \right] \prod_{i<j} dH_{ij}$$

is immediate and comes out to be proportional to

$$\prod_{i<j} \left[\left(z_i - z_j \right)^2 \right]^{-1/2} .$$

The joint probability density of the eigenvalues is therefore

$$P\left(z_1, ..., z_N \right) = C \exp\left(-\sum_1^N z_i^2 \right) \prod_{i<j} \left| z_i - z_j \right| \tag{A.37.1}$$

in the case when all the eigenvalues are real. Equation (A.37.1) has the same form as Eq. (3.1.18).

In case some of the eigenvalues are complex, this procedure can still be carried out for the real eigenvalues. One is then left with the evaluation of an integral

$$\int e^{-\operatorname{tr}\left(S^{\dagger} S \right)} d\mu\left(H \right)$$

where S is a real matrix, none of whose eigenvalues is real. And this integral is not easy. For details the reader may refer to the original paper of Ginibre (1965).

A.38. Variance of the Number Statistic. Section 16.1

The variance of the number statistic is most easily calculated via the two level correlation or cluster function. However, we will compute it also via the n-level spacing functions; it will serve as an illustration and as a check of the consistency.

If $R_1(x)$ and $R_2(x, y)$ are the one and two level correlation functions, then

$$\langle n^2 \rangle = \int_{-L}^{L} \int_{-L}^{L} [R_1(x)\delta(x-y) + R_2(x,y)] dx dy. \tag{A.38.1}$$

Measuring the distances in terms of the local mean spacing $D = R_1^{-1}(x)$, one has

$$\begin{aligned} (\delta n)^2 \equiv \langle n^2 \rangle - \langle n \rangle^2 &= \int_{-s/2}^{s/2} \int_{-s/2}^{s/2} [\delta(x-y) - Y(x-y)] dx dy. \\ &= s - 2 \int_0^s (s-\xi) Y(\xi) d\xi, \end{aligned} \tag{A.38.2}$$

where $2L = sD$, and $Y(x-y)$ is the two level cluster function. Substituting the expressions of Y from Chapters 5, 6 and 7, we get the variance of n for the various ensembles.

Unitary: $\beta = 2$,

$$\begin{aligned} Y(\xi) &= \left(\frac{\sin(\pi\xi)}{\pi\xi} \right)^2, \\ (\delta n)_2^2 &= s - \frac{2}{\pi^2} \int_0^{\pi s} (\pi s - \xi) \left(\frac{\sin \xi}{\xi} \right)^2 d\xi \\ &= \frac{2s}{\pi} \int_{\pi s}^{\infty} \left(\frac{\sin \xi}{\xi} \right)^2 d\xi + \frac{1}{\pi^2} \int_0^{2\pi s} \frac{1 - \cos \xi}{\xi} d\xi \\ &= \frac{1}{\pi^2} [\ln(2\pi s) + \gamma + 1] + O(s^{-1}), \end{aligned} \tag{A.38.3}$$

where γ is Euler's constant, $\gamma = 0.5772...$.

Orthogonal: $\beta = 1$,

$$Y(\xi) = \left(\frac{\sin(\pi\xi)}{\pi\xi}\right)^2 + \frac{d}{d\xi}\left(\frac{\sin(\pi\xi)}{\pi\xi}\right)\int_\xi^\infty \frac{\sin(\pi t)}{\pi t}dt,$$

$$(\delta n)_1^2 = s - \frac{2}{\pi^2}\int_0^{\pi s}(\pi s - \xi)$$

$$\times \left(\left(\frac{\sin\xi}{\xi}\right)^2 + \frac{d}{d\xi}\left(\frac{\sin\xi}{\xi}\right)\int_\xi^\infty \frac{\sin t}{t}dt\right)d\xi$$

$$= \frac{4s}{\pi}\int_{\pi s}^\infty \left(\frac{\sin\xi}{\xi}\right)^2 d\xi + \frac{2}{\pi^2}\int_0^{2\pi s}\frac{1-\cos\xi}{\xi}d\xi$$

$$-\frac{1}{4} + \frac{1}{\pi^2}\left(\int_{\pi s}^\infty \frac{\sin t}{t}dt\right)^2.$$

$$= \frac{2}{\pi^2}\left(\ln(2\pi s) + \gamma + 1 - \frac{\pi^2}{8}\right) + O(s^{-1}). \tag{A.38.4}$$

Symplectic: $\beta = 4$,

$$Y(\xi) = \left(\frac{\sin(2\pi\xi)}{2\pi\xi}\right)^2 - \int_0^\xi \frac{\sin(2\pi t)}{2\pi t}dt\frac{d}{d\xi}\frac{\sin(2\pi\xi)}{2\pi\xi},$$

$$(\delta n)_4^2 = s - \frac{2}{(2\pi)^2}\int_0^{2\pi s}(2\pi s - \xi)$$

$$\times \left(\left(\frac{\sin\xi}{\xi}\right)^2 - \int_0^\xi \frac{\sin t}{t}dt\frac{d}{d\xi}\frac{\sin\xi}{\xi}\right)$$

$$= s - \frac{1}{\pi^2}\int_0^{2\pi s}(2\pi s - \xi)\left(\frac{\sin\xi}{\xi}\right)^2 d\xi$$

$$+ \frac{1}{4\pi^2}\left(\int_0^{2\pi s}\frac{\sin t}{t}dt\right)^2$$

$$= \frac{1}{2\pi^2}\left(\ln(4\pi s) + \gamma + 1 + \frac{\pi^2}{8}\right) + O(s^{-1}). \tag{A.38.5}$$

A.38.1. AVERAGES OF THE POWERS OF n VIA n-LEVEL SPACINGS

Consider a distribution of points or levels on a straight line. Let these levels be the eigenvalues of a random matrix taken from one of the three ensembles, unitary, orthogonal or symplectic. The probability $E(n,s)$ that a randomly chosen interval of length s contains exactly n levels obviously must satisfy

$$\sum_{n=0}^{\infty} E(n,s) = 1. \tag{A.38.6}$$

It is convenient to choose the length scale so that the average spacing is unity. The mean value of n is then given by

$$\sum_{n=0}^{\infty} nE(n,s) = s. \tag{A.38.7}$$

And we want to compute the mean values of powers of n. For the three ensembles studied in Chapters 5, 6, and 7, we have the following expressions of $E(n,s)$.

Unitary ensemble, $\beta = 2$,

$$E_2(0,s) = \prod_j (1-\lambda_j) \equiv \prod_j (1+x_j)^{-1}, \tag{A.38.8}$$

$$E_2(n,s) = E_2(0,s) \sum_{(i)} x_{i_1} \cdots x_{i_n}. \tag{A.38.9}$$

Orthogonal ensemble, $\beta = 1$,

$$E_1(0,s) = \prod_j (1-\lambda_{2j}) \equiv \prod_j (1+y_j)^{-1}, \tag{A.38.10}$$

$$E_1(2n,s) = E_1(0,s). \sum_{(i)} y_{i_1} ... y_{i_n} \left(1 - (b_{i_1} + ... + b_{i_n})\right), \tag{A.38.11}$$

$$E_1(2n-1,s) = E_1(0,s). \sum_{(i)} y_{i_1} ... y_{i_n} \left(b_{i_1} + ... + b_{i_n}\right). \tag{A.38.12}$$

Symplectic ensemble, $\beta = 4$,

$$E_4(0, s/2) = E_1(0, s)\left(1 + \frac{1}{2}\sum_i b_i y_i\right), \tag{A.38.13}$$

$$E_4(n, s/2) = E_1(0, s)\left(\sum_{(i)} y_{i_1} \cdots y_{i_n}\left(1 - \frac{1}{2}\left(b_{i_1} + \cdots + b_{i_n}\right)\right)\right.$$
$$\left. + \frac{1}{2}\sum_{(i)} y_{i_1} \cdots y_{i_{n+1}}\left(b_{i_1} + \cdots + b_{i_{n+1}}\right)\right), \tag{A.38.14}$$

where

$$x_i = \lambda_i\left(1 - \lambda_i\right)^{-1}, \quad y_i = x_{2i} = \lambda_{2i}\left(1 - \lambda_{2i}\right)^{-1}, \tag{A.38.15}$$

$$b_i = f_{2i}(1)\int_{-1}^{1} f_{2i}(x)dx \Big/ \int_{-1}^{1} f_{2i}^2(x)dx, \tag{A.38.16}$$

λ_i and $f_i(x)$ are, respectively, the eigenvalues and eigenfunctions of the integral equation

$$\lambda_i f_i(x) = \int_{-1}^{1} K(x - y) f_i(y) dy, \tag{A.38.17}$$

with the kernel

$$K(x) = \frac{\sin(\pi s x/2)}{\pi x}. \tag{A.38.18}$$

The sums or products in Eqs. (A.38.8)–(A.38.14) are taken over all integers with the restrictions $0 \le j < \infty$, $0 \le i_1 < i_2 < i_3 < \cdots$.

The $f_i(x)$ are known as spheroidal functions. They are even or odd, respectively, when the index is even or odd,

$$f_i(-x) = (-1)^i f_i(x). \tag{A.38.19}$$

They are orthogonal and may be normalized,

$$\int_{-1}^{1} f_i(x) f_j(x) dx = \delta_{ij}. \tag{A.38.20}$$

One may then write

$$K(x-y) = \sum_i \lambda_i f_i(x) f_i(y), \tag{A.38.21}$$

so that

$$\sum_i \lambda_i = \int_{-1}^{1} K(0)dx = s. \tag{A.38.22}$$

We will need the following identities:

$$\begin{aligned} 1 + \sum_n \sum_{(i)} \xi_{i_1} \cdots \xi_{i_n} &\equiv 1 + \sum_i \xi_i + \sum_{i<j} \xi_i \xi_j + \sum_{i<j<k} \xi_i \xi_j \xi_k + \cdots \\ &= \prod_i (1 + \xi_i), \end{aligned} \tag{A.38.23}$$

$$\begin{aligned} \sum_n \sum_{(i)} (b_{i_1} + \cdots + b_{i_n}) \xi_{i_1} \cdots \xi_{i_n} &\equiv \sum_i b_i \xi_i + \sum_{i<j} (b_i + b_j) \xi_i \xi_j + \cdots \\ &= \prod_i (1 + \xi_i) \sum_j \frac{b_j \xi_j}{1 + \xi_j}. \end{aligned} \tag{A.38.24}$$

One can prove these identities by recurrence. One can deduce others by replacing ξ_i by $z\xi_i$, differentiating several times with respect to z and finally setting $z = 1$. For example, from Eq. (A.38.24) we deduce

$$\begin{aligned} &\sum_n n \sum_{(i)} (b_{i_1} + \cdots + b_{i_n}) \xi_{i_1} \cdots \xi_{i_n} \\ &\equiv \sum_i b_i \xi_i + 2 \sum_{i<j} (b_i + b_j) \xi_i \xi_j + \cdots \\ &= \prod_i (1 + \xi_i) \left(\sum_j \frac{b_j \xi_j}{(1 + \xi_j)^2} + \sum_j \frac{b_j \xi_j}{1 + \xi_j} \sum_k \frac{\xi_k}{1 + \xi_k} \right). \end{aligned} \tag{A.38.25}$$

Setting all b_i equal to 1, we get from Eqs. (A.38.24) and (A.38.25) other identities. For later convenience we write one of them,

$$\sum_n n^2 \sum_{(i)} \xi_{i_1} \cdots \xi_{i_n}$$

$$\equiv \sum_i \xi_i + 2^2 \sum_{i<j} \xi_i \xi_j + 3^2 \sum_{i<j<k} \xi_i \xi_j \xi_k + \cdots$$

$$= \prod_i (1+\xi_i) \left(\sum_j \frac{\xi_j}{(1+\xi_j)^2} + \left(\sum_j \frac{\xi_j}{1+\xi_j} \right)^2 \right). \quad \text{(A.38.26)}$$

A.38.2. UNITARY ENSEMBLE

This case is the simplest. Take $\xi_i = x_i$ in Eqs. (A.38.23), (A.38.24), and (A.38.26) with all $b_i = 1$. This gives, with Eq. (A.38.22), Eqs. (A.38.6) and (A.38.7) and

$$\sum_n n^2 E_2(n,s) = \sum_i \lambda_i (1-\lambda_i) + \left(\sum_i \lambda_i \right)^2, \quad \text{(A.38.27)}$$

so that the variance of n is

$$(\delta n)_2^2 = \langle n^2 \rangle_2 - \langle n \rangle^2 = \sum_i \lambda_i (1-\lambda_i). \quad \text{(A.38.28)}$$

A.38.3. ORTHOGONAL ENSEMBLE

Take $\xi_i = y_i$ and b_i defined by Eqs. (A.38.15) and (A.38.16). Now Eqs. (A.38.11), (A.38.12), (A.38.23), (A.38.25), and (A.38.26) give in addition to Eq. (A.38.6)

$$\sum_n n E_1(n,s) = 2 \sum_i \lambda_{2i} - \sum_i b_i \lambda_{2i}, \quad \text{(A.38.29)}$$

and

$$\sum_n n^2 E_1(n,s) = \left(\sum_i (2\lambda_{2i} - b_i)\right)^2 + 4\sum_i \lambda_{2i}(1-\lambda_{2i})(1-b_i) + \left(\sum_i b_i\lambda_{2i}\right)\left(1-\sum_i b_i\lambda_{2i}\right). \quad (A.38.30)$$

From Eqs. (A.38.19) and (A.38.21) one obtains

$$\sum_i b_i\lambda_{2i} = \sum_i \lambda_i f_i(1)\int_{-1}^{1} f_i(x)dx = \int_{-1}^{1} K(1-x)dx, \quad (A.38.31)$$

and

$$\sum_i (\lambda_{2i} - \lambda_{2i+1}) = \sum_i \lambda_i \int_{-1}^{1} f_i(x)f_i(-x)dx = \int_{-1}^{1} K(2x)dx. \quad (A.38.32)$$

After some simplifications, one has therefore

$$\sum_i b_i\lambda_{2i} = \sum_i (\lambda_{2i} - \lambda_{2i+1}) = \frac{1}{\pi}\int_0^{\pi s} \frac{\sin x}{x} dx. \quad (A.38.33)$$

From Eqs. (A.38.29), (A.38.33), and (A.38.22) we get (A.38.7). Equations (A.38.29) and (A.38.30) give the variance of n,

$$(\delta n)_1^2 = \langle n^2\rangle_1 - \langle n\rangle^2 = 4\sum_i \lambda_{2i}(1-\lambda_{2i})(1-b_i) + \left(\sum_i b_i\lambda_{2i}\right)\left(1-\sum_i b_i\lambda_{2i}\right). \quad (A.38.34)$$

A.38.4. SYMPLECTIC ENSEMBLE

Take $\xi_i = y_i$ and b_i, Eqs. (A.38.15) and (A.38.16). Equations (A.38.13), (A.38.14), (A.38.22)–(A.38.24), (A.38.26), and (A.38.33) give in addition to Eqs. (A.38.6) and (A.38.7),

$$\sum_n n^2 E_4(n, s/2)$$

$$= E_1(0,s)\left(\sum_n n^2 \sum_{(i)} y_{i_1} \cdots y_{i_n} - \frac{1}{2}\sum_n (2n-1)\sum_{(i)} (b_{i_1} + \cdots + b_{i_n})\, y_{i_1} \cdots y_{i_n}\right)$$

$$= \sum_i \lambda_{2i}(1-\lambda_{2i}) + \left(\sum_i \lambda_{2i}\right)^2 - \sum_i b_i \lambda_{2i}(1-\lambda_{2i}) - \sum_i b_i \lambda_{2i} \sum_j \lambda_{2j} + \frac{1}{2}\sum_i b_i \lambda_{2i}. \tag{A.38.35}$$

The variance of n is therefore

$$(\delta n)_4^2 = \sum_n n^2 E_4(n, s/2) - \left(\sum_i \lambda_{2i} - \frac{1}{2}\sum_i b_i \lambda_{2i}\right)^2$$

$$= \sum_i \lambda_{2i}(1-\lambda_{2i})(1-b_i) + \frac{1}{2}\sum_i b_i \lambda_{2i}\left(1 - \frac{1}{2}\sum_j b_j \lambda_{2j}\right). \tag{A.38.36}$$

To see that Eqs. (A.38.28), (A.38.34), and (A.38.36) are identical to Eqs. (A.38.3)–(A.38.5), one has to evaluate the various sums.

Equations (A.38.20) and (A.38.21) give

$$\sum_i \lambda_i^2 = \int_{-1}^{1}\int_{-1}^{1} K^2(x-y)\,dx\,dy = 2\int_0^1 \left(\frac{\sin(\pi s \xi)}{\pi \xi}\right)^2 (1-\xi)\,d\xi. \tag{A.38.37}$$

So that with Eq. (A.38.22),

$$\sum_i \lambda_i(1-\lambda_i) = \frac{2s}{\pi}\int_{\pi s}^{\infty}\left(\frac{\sin x}{x}\right)^2 dx + \frac{1}{\pi^2}\int_0^{2\pi s}\frac{1-\cos x}{x}dx$$
$$= \frac{1}{\pi^2}\left(\ln(2\pi s)+\gamma+1+\frac{\sin(2\pi s)}{2\pi s}+O(s^{-2})\right). \tag{A.38.38}$$

From Eqs. (A.38.22) and (A.38.33), one has

$$\sum_i \lambda_{2i} = \frac{1}{2}\sum_i \lambda_i + \frac{1}{2}\sum_i(\lambda_{2i}-\lambda_{2i+1}) = \frac{s}{2}+\frac{1}{2\pi}\int_0^{\pi s}\frac{\sin x}{x}dx$$
$$= \frac{s}{2}+\frac{1}{4}-\frac{1}{2\pi^2 s}\cos(\pi s)+O(s^{-2}). \tag{A.38.39}$$

From

$$K(x-y)+K(x+y) = 2\sum_i \lambda_{2i} f_{2i}(x) f_{2i}(y) \tag{A.38.40}$$

one gets

$$4\sum_i \lambda_{2i}^2 = \int_{-1}^{1}\int_{-1}^{1}\left(\frac{\sin((x-y)\pi s/2)}{(x-y)\pi}+\frac{\sin((x+y)\pi s/2)}{(x+y)\pi}\right)^2 dxdy$$
$$= 2\sum_i \lambda_i^2 + \frac{4}{\pi^2}\int_0^{\pi s} d\xi \int_0^{\pi s-\xi} d\eta\, \frac{\sin \xi}{\xi}\,\frac{\sin \eta}{\eta}$$
$$= 2\sum_i \lambda_i^2 + 2\left(\frac{1}{\pi}\int_0^{\pi s}\frac{\sin \xi}{\xi}\,d\xi\right)^2, \tag{A.38.41}$$

and

$$
\begin{aligned}
\sum_i b_i \lambda_{2i}^2 &= \int_{-1}^{1}\int_{-1}^{1} \frac{\sin[(1-y)\pi s/2]}{(1-y)\pi}\, \frac{\sin[(x-y)\pi s/2]}{(x-y)\pi} dx dy \\
&= \frac{1}{\pi^2}\int_0^{\pi s} d\xi \int_{\xi-\pi s}^{\xi} d\eta\, \frac{\sin\,\xi}{\xi}\, \frac{\sin\,\eta}{\eta} \\
&= \frac{1}{2}\left(\frac{1}{\pi}\int_0^{\pi s} \frac{\sin\,\xi}{\xi}\, d\xi\right)^2 + \frac{1}{2}\left(\frac{1}{\pi}\int_0^{\pi s} \frac{\sin\,\xi}{\xi}\, d\xi\right)^2 \\
&= \frac{1}{2}\left(\frac{1}{2} - \frac{\cos(\pi s)}{\pi^2 s}\right)^2 \\
&\quad + \frac{1}{\pi^2}\int_0^{\pi s} d\xi \int_0^{\pi s-\xi} d\eta\, \frac{\sin\,\xi}{\xi}\, \frac{\sin\,\eta}{\eta} + O(s^{-2}).
\end{aligned} \tag{A.38.42}
$$

From Eqs. (A.38.39), (A.38.41), (A.38.33), and (A.38.42) one gets, therefore,

$$
\begin{aligned}
&4\sum_i \lambda_{2i}(1-\lambda_{2i})(1-b_i) \\
&\quad = 4\sum_i \left(\lambda_{2i} - \lambda_{2i}^2 - b_i\lambda_{2i} + b_i\lambda_{2i}^2\right) \\
&\quad = \frac{2}{\pi^2}\left(\ln(2\pi s) + \gamma + 1 - \frac{\pi^2}{4} + O(s^{-2})\right).
\end{aligned} \tag{A.38.43}
$$

Substituting these asymptotic expressions in Eqs. (A.38.28), (A.38.34), and (A.38.36), we get finally

$$
(\delta n)_2^2 = \frac{1}{\pi^2}\left(\ln(2\pi s) + \gamma + 1\right) + O(s^{-1}), \tag{A.38.44}
$$

$$
(\delta n)_1^2 = \frac{2}{\pi^2}\left(\ln(2\pi s) + \gamma + 1 - \frac{\pi^2}{8}\right) + O(s^{-1}), \tag{A.38.45}
$$

and

$$(\delta n)_4^2 = \frac{1}{2\pi^2}\left(\ln(4\pi s) + \gamma + 1 + \frac{\pi^2}{8}\right) + O(s^{-1}), \tag{A.38.46}$$

taking care of the fact that for the symplectic ensemble E_4 was given at $s/2$ and not at s.

For comparison with the Poisson process when levels have no correlations, we note that

$$Y(\xi) = 0, \quad E_0(n,s) = \frac{s^n}{n!}\, e^{-s}, \quad \text{and} \quad (\delta n)_0^2 = s. \tag{A.38.47}$$

A.39. Optimum Linear Statistic. Section 16.1

The general linear statistic

$$W = \sum_i f(E_i), \tag{A.39.1}$$

has the mean value

$$\langle W \rangle = D^{-1} \int_{-L}^{L} f(x)dx \tag{A.39.2}$$

and the variance

$$\begin{aligned} V_W &= \langle W \rangle^2 - \langle W \rangle^2 \\ &= \int_{-L}^{L} \int_{-L}^{L} [R_2(x,y) + R_1(x)\delta(x-y) - R_1(x)R_1(y)] f(x) f(y) dx dy \\ &= \int_{-s/2}^{s/2} \int_{-s/2}^{s/2} f(xD) f(yD) [\delta(x-y) - Y(x-y)] dx dy \end{aligned} \tag{A.39.3}$$

where $R_1(x) = D^{-1}$ is the level density, $2L = sD$, and $Y(x)$ is the two level cluster function. Going to the Fourier transforms,

$$f(x) = \int_{-\infty}^{\infty} e^{2\pi ikx}\varphi(k)dk, \tag{A.39.4}$$

$$\varphi(k) = \int_{-\infty}^{\infty} e^{-2\pi ikx}f(x)dx, \tag{A.39.5}$$

$$\delta(x) = \int_{-\infty}^{\infty} e^{-2\pi ikx}dk, \quad Y(x) = \int_{-\infty}^{\infty} e^{-2\pi ikx}b(k)dk, \tag{A.39.6}$$

we have

$$V_W = D^{-1}\int_{-\infty}^{\infty} \varphi(k)\varphi(-k)[1 - b(Dk)]dk, \tag{A.39.7}$$

where $b(k)$, the Fourier transform of $Y(x)$, is given by Eqs. (5.2.22), (6.4.17), and (7.2.9) for the three ensembles, respectively.

If $f(x)$ is a smooth function, $\varphi(k)$ is large only for values of k of the order of L^{-1}, and the whole of the integral (A.39.7) comes from values of k in this neighborhood. Therefore, the approximation

$$b(k) = 1 - \frac{2}{\beta}|k| \tag{A.39.8}$$

may be used with an error of the order of $D/2L = s^{-1}$,

$$\begin{aligned} V_W &\cong \frac{2}{\beta}\int_{-\infty}^{\infty} \varphi(k)\varphi(-k)\,|k|\,dk \\ &= -\frac{2}{\beta\pi^2}\int_{-L}^{L}\int_{-L}^{L} f'(x)f'(y)\,\ln|x-y|\,dxdy. \end{aligned} \tag{A.39.9}$$

Setting $f(x) = g(x/L)$, the figure of merit $\Phi_W = V_W/\langle W\rangle^2$ is

$$\Phi_W = \frac{4}{\pi^2 s^2\beta}\frac{\int_{-1}^{1}\int_{-1}^{1} g'(x)g'(y)\ln|x-y|\,dxdy}{\left(\int_{-1}^{1} g'(x)dx\right)^2}. \tag{A.39.10}$$

This is minimum if $g^{'}(x)$ satisfies the integral equation

$$\int_{-1}^{1} g^{'}(y)\ln|x-y|\,dy = x, \tag{A.39.11}$$

i.e., when $g(x) = (1-x^2)^{1/2}/\pi$ (*cf.* Section 4.2). The optimum statistic is therefore

$$W = \sum_{i=1}^{n} \left(1-(E_i/L)^2\right)^{1/2}, \tag{A.39.12}$$

with $\langle W\rangle = \pi L/(2D)$ and $V_W = 1/(2\beta)$.

Note that Eqs. (A.39.3) and (A.39.7) are exact, while the approximations (A.39.8), (A.39.9) and the following are valid only for smooth functions.

A.40. Mean Value of Δ. Section 16.2

The best straight line fitting an observed staircase graph is chosen by least square deviation

$$\begin{aligned} \Delta &= \min_{A,B} \frac{1}{2L}\int_{-L}^{L}\left(N(E)-\frac{A}{D}E-B\right)^2 dE \\ &= \min_{A,B} \frac{1}{s}\int_{-s/2}^{s/2}[n(x)-Ax-B]^2 dx \end{aligned} \tag{A.40.1}$$

where D is the mean spacing, $2L = sD$, $E = xD$ and $N(E) = n(x)$. Equating to zero the partial derivatives with respect to A and B fixes the values of A, B giving this minimum

$$A = \int xn(x)dx\Big/\int x^2 dx \equiv \frac{12}{s}n_1, \tag{A.40.2}$$

$$B = \int n(x)dx\Big/\int dx = n_0, \tag{A.40.3}$$

with

$$n_0 = \frac{1}{s}\int n(x)dx, \quad n_1 = \frac{1}{s^2}\int xn(x)dx. \tag{A.40.4}$$

Here and in what follows, all integrals are taken from $-s/2$ to $s/2$, unless explicitly indicated otherwise. Substituting the values of A, B in (A.40.1) we get

$$\Delta = \frac{1}{s}\int n^2(x)dx - 12n_1^2 - n_0^2. \tag{A.40.5}$$

Taking averages

$$\langle\Delta\rangle = \frac{1}{s}\int \left\langle n^2(x)\right\rangle dx - 12\left\langle n_1^2\right\rangle - \left\langle n_0^2\right\rangle \tag{A.40.6}$$

Now

$$\left\langle n_1^2\right\rangle = \frac{1}{s^4}\int\int xy\left\langle n(x)n(y)\right\rangle dxdy, \tag{A.40.7}$$

$$\left\langle n_0^2\right\rangle = \frac{1}{s^2}\int\int \left\langle n(x)n(y)\right\rangle dxdy, \tag{A.40.8}$$

and

$$\langle n(x)n(y)\rangle = \int_{-s/2}^{x} d\xi \int_{-s/2}^{y} d\eta \left\{R_1(\xi)\delta(\xi-\eta) + R_2(\xi,\eta)\right\} \tag{A.40.9}$$

where $R_1(\xi)$ is the level density and $R_2(\xi,\eta)$ is the two-level correlation function. As we are measuring distances in terms of the mean spacing D,

$$R_1(\xi) = 1\ , \qquad R_2(\xi,\eta) = 1 - Y(\xi-\eta), \tag{A.40.10}$$

where $Y(r)$ is the two-level cluster function given by Eq. (5.2.20), (6.4.14), or (7.2.6);

$$Y(r) = \left(\frac{\sin(\pi r)}{\pi r}\right)^2, \qquad \beta = 2, \tag{A.40.11}$$

$$Y(r) = \left(\frac{\sin(\pi r)}{\pi r}\right)^2 + \frac{d}{dr}\left(\frac{\sin(\pi r)}{\pi r}\right)\int_r^\infty \frac{\sin(\pi t)}{\pi t}dt, \quad \beta = 1, \tag{A.40.12}$$

and

$$Y(r) = \left(\frac{\sin(2\pi r)}{2\pi r}\right)^2 - \frac{d}{dr}\left(\frac{\sin(2\pi r)}{2\pi r}\right)\int_0^r \frac{\sin(2\pi t)}{2\pi t}dt, \quad \beta = 4. \tag{A.40.13}$$

If one imposes $\xi \geq \eta$, then because of the symmetry one should multiply by 2, except for the delta function. Introducing the variables $u = \xi - \eta$, $v = \xi$, integrations over all variables except u are elementary. Thus, for any function $f(x)$

$$\begin{aligned}
&\int\int dx dy \int_{-s/2}^{x} d\xi \int_{-s/2}^{y} d\eta f(\xi - \eta) \\
&= \int\int d\xi d\eta \left(\frac{s}{2} - \xi\right)\left(\frac{s}{2} - \eta\right) f(\xi - \eta) \\
&= 2\int_0^s du \int_{u-s/2}^{s/2} dv \left(\frac{s}{2} - v\right)\left(\frac{s}{2} - v + u\right) f(u) \\
&= 2\int_0^s du \int_0^{s-u} dv v(v+u) f(u) \\
&= \frac{1}{3}\int_0^s du\,(s-u)^2\,(2s+u) f(u),
\end{aligned} \tag{A.40.14}$$

$$\begin{aligned}
&\int dx \int_{-s/2}^{x} d\xi \int_{-s/2}^{x} d\eta f(\xi - \eta) \\
&= 2\int d\xi \int_{-s/2}^{\xi} d\eta \left(\frac{s}{2} - \xi\right) f(\xi - \eta) \\
&= 2\int_0^s du \int_0^{s-u} dv . v f(u) \\
&= \int_0^s du (s-u)^2 f(u)
\end{aligned} \tag{A.40.15}$$

and

$$\int\int dxdy \int_{-s/2}^{x} d\xi \int_{-s/2}^{y} d\eta \; xyf(\xi-\eta)$$
$$= \frac{1}{4}\int\int d\xi d\eta \left(\frac{s^2}{4}-\xi^2\right)\left(\frac{s^2}{4}-\eta^2\right) f(\xi-\eta)$$
$$= \frac{1}{2}\int_0^s du \int_0^{s-u} dvv\,(s-v)\,(v+u)\,(s-v-u)f(u)$$
$$= \frac{1}{60}\int_0^s du\,(s-u)^3\left(s^2+3su+u^2\right) f(u). \tag{A.40.16}$$

Collecting the results we get after some simplification

$$\langle\Delta\rangle = \frac{1}{15}s^{-4}\int_0^s du(s-u)^3\left(2s^2-9su-3u^2\right)$$
$$\times\left(\frac{1}{2}\delta(u)+1-Y(u)\right)$$
$$= \frac{1}{15}s^{-4}\int_0^s du(s-u)^3\left(2s^2-9su-3u^2\right)\left(\frac{1}{2}\delta(u)-Y(u)\right). \tag{A.40.17}$$

For a random level sequence without correlations (Poisson process), $Y(u)=0$, and

$$\langle\Delta\rangle = \frac{1}{15}s. \tag{A.40.18}$$

For the cases when $Y(u)$ is given by Eqs. (A.40.11)–(A.40.13), we use the asymptotic estimations,

$$\begin{aligned}\int_0^s \left(\frac{\sin(\pi u)}{\pi u}\right)^2 du &= \frac{1}{2} - \int_{\pi s}^{\infty} \left(\frac{\sin x}{x}\right)^2 dx \\ &= \frac{1}{2} - \frac{1}{2\pi s} + O(s^{-2}), \qquad \text{(A.40.19)}\\ \int_0^s u \left(\frac{\sin(\pi u)}{\pi u}\right)^2 du &= \frac{1}{2\pi^2} \int_0^{2\pi s} \frac{1-\cos x}{x} dx \\ &= \frac{1}{2\pi^2}[\ln(2\pi s)+\gamma] + O(s^{-1}), \qquad \text{(A.40.20)}\end{aligned}$$

and

$$\int_0^s u^j \left(\frac{\sin(\pi u)}{\pi u}\right)^2 du = \frac{1}{2\pi^2}\frac{s^{j-1}}{j-1} + O(s^{j-2}), \quad j \geq 2. \qquad \text{(A.40.21)}$$

Changing variables from πu, πt to u, t and integrating by parts, one has

$$\begin{aligned}&\int_0^s \frac{d}{du}\left(\frac{\sin(\pi u)}{\pi u}\right)\int_u^{\infty} \frac{\sin(\pi t)}{\pi t}\, dt\, du \\ &\quad = \frac{1}{\pi}\int_0^{\pi s}\left(\frac{\sin u}{u}\right)\int_u^{\infty}\frac{\sin t}{t}\, dt\, du \\ &\quad = -\frac{1}{2\pi^2 s} + O(s^{-2}), \qquad \text{(A.40.22)}\\ &\int_0^s u\frac{d}{du}\left(\frac{\sin(\pi u)}{\pi u}\right)\int_u^{\infty}\frac{\sin(\pi t)}{\pi t}\, dt\, du \\ &\quad = \frac{1}{\pi^2}\int_0^{\pi s} u\left(\frac{\sin u}{u}\right)\int_u^{\infty}\frac{\sin t}{t}\, dt\, du \\ &\quad = \frac{1}{\pi^2}\int_0^{\pi s} u\left(\frac{\sin u}{u}\right)^2 du - \frac{1}{2\pi^2}\left(\frac{\pi}{2}\right)^2 \\ &\qquad + \int_{\pi s}^{\infty} du \int_u^{\infty} dt \left(\frac{\sin u}{u}\right)\left(\frac{\sin t}{t}\right) \\ &\quad = \frac{1}{2\pi^2}[\ln(2\pi s)+\gamma] - \frac{1}{8} + O(s^{-1}), \qquad \text{(A.40.23)}\\ &\int_0^s u^j \frac{d}{du}\left(\frac{\sin(\pi u)}{\pi u}\right)\int_u^{\infty}\frac{\sin(\pi t)}{\pi t}\, dt\, du \\ &\quad = \frac{1}{2\pi^2}\frac{s^{j-1}}{j-1} + O(s^{j-2}), \quad j \geq 2, \qquad \text{(A.40.24)}\end{aligned}$$

and similarly for integrals containing

$$\frac{d}{du}\left(\frac{\sin(2\pi u)}{2\pi u}\right)\int_0^u \frac{\sin(2\pi t)}{2\pi t}\,dt\,.$$

Substituting these asymptotic estimations in Eqs. (A.40.17) we get finally

$$\langle\Delta\rangle = \frac{1}{\pi^2}\left(\ln(2\pi s)+\gamma-\frac{5}{4}-\frac{\pi^2}{8}\right)+O(s^{-1}),\quad \beta=1, \tag{A.40.25}$$

$$\langle\Delta\rangle = \frac{1}{2\pi^2}\left(\ln(2\pi s)+\gamma-\frac{5}{4}\right)+O(s^{-1}),\quad \beta=2, \tag{A.40.26}$$

$$\langle\Delta\rangle = \frac{1}{4\pi^2}\left(\ln(4\pi s)+\gamma-\frac{5}{4}+\frac{\pi^2}{8}\right)+O(s^{-1}),\quad \beta=4. \tag{A.40.27}$$

The evaluation of $\langle\Delta^2\rangle$ involves 3- and 4-level correlations as well. They are tedious but present no difficulty in principle, specially for the unitary ensemble. The final result is that the variance of Δ, i.e., $\langle\Delta^2\rangle - \langle\Delta\rangle^2$, is a small constant independent of s in the three cases.

A.41. Tables of Functions $\mathcal{B}_\beta(x_1, x_2)$ and $\mathcal{P}_\beta(x_1, x_2)$ for β=1 and 2

TABLE A.41.1. $\mathcal{B}_1(x_1, x_2)$. It is the probability that a distance x_1 on one side and a distance x_2 on the other side of a randomly chosen eigenvalue do not contain any other eigenvalue when the matrix is chosen from the Gaussian orthogonal ensemble. $\mathcal{B}_1(x_1, x_2)$ being symmetric in x_1 and x_2, its values are given only for $x_1 \leq x_2$.

	$\pi x_1/2$	0.	0.1	0.2	0.3	0.4	0.5	0.6	0.7	0.8
$\pi x_2/2$	$x_2 \backslash x_1$	0.	0.064	0.127	0.191	0.255	0.318	0.382	0.446	0.509
0.	0.	1.								
0.1	0.064	0.996673	0.993347							
0.2	0.127	0.986770	0.983448	0.973565						
0.3	0.191	0.970517	0.967204	0.957354	0.941211					
0.4	0.255	0.948273	0.944975	0.935182	0.919149	0.897263				
0.5	0.318	0.920509	0.917236	0.907528	0.891656	0.870019	0.843124			
0.6	0.382	0.887796	0.884559	0.874967	0.859312	0.838005	0.811560	0.780578		
0.7	0.446	0.850776	0.847584	0.838145	0.822766	0.801871	0.775987	0.745717	0.711721	
0.8	0.509	0.810140	0.807006	0.797756	0.782713	0.762317	0.737102	0.707674	0.674690	0.638836
0.9	0.573	0.766610	0.763547	0.754522	0.739878	0.720065	0.695624	0.667162	0.635330	0.600805
1.0	0.637	0.720917	0.717936	0.709174	0.694987	0.675839	0.652271	0.624889	0.594338	0.561279
1.2	0.764	0.625886	0.623106	0.614971	0.601863	0.584259	0.562699	0.537776	0.510106	0.480318
1.4	0.891	0.530374	0.527836	0.520442	0.508593	0.492763	0.473479	0.451307	0.426825	0.400610
1.6	1.019	0.438855	0.436588	0.430019	0.419547	0.405634	0.388783	0.369515	0.348361	0.325840
1.8	1.146	0.354707	0.352729	0.347024	0.337982	0.326038	0.311654	0.295305	0.277461	0.258575

Table A.41.1 $\mathcal{B}_1(x_1,\ x_2)$ (CONTINUED)

	$\pi x_1/2$	0.	0.1	0.2	0.3	0.4	0.5	0.6	0.7	0.8
$\pi x_2/2$	$x_2 \backslash x_1$	0.	0.064	0.127	0.191	0.255	0.318	0.382	0.446	0.509
2.0	1.273	0.280137	0.278449	0.273610	0.265984	0.255970	0.243983	0.230439	0.215745	0.200289
2.2	1.401	0.216245	0.214840	0.210830	0.204550	0.196351	0.186597	0.175643	0.163833	0.151486
2.4	1.528	0.163196	0.162053	0.158809	0.153758	0.147204	0.139454	0.130805	0.121539	0.111912
2.6	1.655	0.120437	0.119529	0.116966	0.112999	0.107883	0.101871	0.095204	0.088105	0.080778
2.8	1.783	0.086932	0.086228	0.084251	0.081208	0.077309	0.072755	0.067736	0.062426	0.056981
3.0	1.910	0.061384	0.060851	0.059361	0.057082	0.054179	0.050809	0.047120	0.043242	0.039291
3.2	2.037	0.042408	0.042014	0.040917	0.039250	0.037139	0.034704	0.032055	0.029288	0.026488
3.4	2.165	0.028670	0.028385	0.027596	0.026405	0.024905	0.023186	0.021328	0.019400	0.017462
3.6	2.292	0.018969	0.018768	0.018214	0.017381	0.016341	0.015155	0.013882	0.012570	0.011259
3.8	2.419	0.012285	0.012145	0.011765	0.011197	0.010491	0.009693	0.008840	0.007967	0.007101
4.0	2.546	0.007787	0.007694	0.007439	0.007060	0.006592	0.006066	0.005508	0.004940	0.004381
4.2	2.674	0.004833	0.004771	0.004604	0.004357	0.004054	0.003715	0.003358	0.002998	0.002645
4.4	2.801	0.002937	0.002897	0.002789	0.002632	0.002440	0.002227	0.002004		
4.6	2.928	0.001747	0.001722	0.001655	0.001557	0.001438				
4.8	3.056	0.0010177	0.0010023	0.0009611						
5.0	3.183	0.0005805								

Table A.41.1. $\mathcal{B}_1(X_1, X_2)$ (CONTINUED)

	$\pi x_1/2$	0.9	1.0	1.2	1.4	1.6	1.8	2.0	2.2	2.4
$\pi x_2/2$	$x_2\backslash x_1$	0.573	0.637	0.764	0.891	1.019	1.146	1.273	1.401	1.528
0.9	0.573	0.564265								
1.0	0.637	0.526374	0.490265							
1.2	0.764	0.449025	0.416818	0.351814						
1.4	0.891	0.373222	0.345188	0.189074	0.235515					
1.6	1.019	0.302445	0.278636	0.231395	0.186831	0.146802				
1.8	1.146	0.239074	0.219347	0.180558	0.144411	0.112341	0.085076			
2.0	1.273	0.184426	0.168479	0.137413	0.108825	0.083782	0.062763	0.045785		
2.2	1.401	0.138895	0.126316	0.102044	0.079993	0.060927	0.045133	0.032545	0.022860	
2.4	1.528	0.102157	0.092475	0.073971	0.057380	0.043223	0.031652	0.022555	0.015650	0.010581
2.6	1.655	0.073401	0.066127	0.052361	0.040180	0.029926	0.021658	0.015247	0.010448	0.006975
2.8	1.783	0.051535	0.046200	0.036203	0.027475	0.020228	0.014465	0.010057	0.006805	
3.0	1.910	0.035365	0.031544	0.024456	0.018352	0.013353	0.009432	0.006476		
3.2	2.037	0.023724	0.021052	0.016145	0.011977	0.008610	0.006007			
3.4	2.165	0.015561	0.013736	0.010419	0.007639	0.005425				
3.6	2.292	0.009982	0.008764	0.006573	0.004762					
3.8	2.419	0.006263	0.005469	0.004055						
4.0	2.546	0.003843	0.003338							

TABLE A.41.2. $\mathcal{P}_1(\mathbf{x_1}, \mathbf{x_2})$. It is the probability density of two consecutive spacings between the eigenvalues of a random matrix taken from the Gaussian orthogonal ensemble; $\mathcal{P}_1(x_1, x_2) = \partial^2 \mathcal{B}_1(x_1, x_2)/\partial x_1 \partial x_2$. $\mathcal{P}_1(x_1, x_2)$ being symmetric in x_1 and x_2, its values are given only for $x_1 \le x_2$.

	$\pi x_1/2$	0.1	0.2	0.3	0.4	0.5	0.6	0.7	0.8	0.9
$\pi x_2/2$	$x_2 \backslash x_1$	0.064	0.127	0.191	0.255	0.318	0.382	0.446	0.509	0.573
0.1	0.064	0.00								
0.2	0.127	0.004	0.00849							
0.3	0.191	0.006	0.01595	0.02793						
0.4	0.255	0.010	0.02474	0.04207	0.06249					
0.5	0.318	0.016	0.03484	0.05842	0.08446	0.11249				
0.6	0.382	0.020	0.04640	0.07560	0.10768	0.14205	0.17725			
0.7	0.446	0.027	0.05811	0.09341	0.13184	0.17185	0.21195	0.25135		
0.8	0.509	0.033	0.07016	0.11175	0.15582	0.20074	0.24554	0.28898	0.32978	
0.9	0.573	0.039	0.08241	0.12966	0.17867	0.22829	0.27713	0.32374	0.36680	0.40534
1.0	0.637	0.044	0.09421	0.14642	0.20014	0.25380	0.30580	0.35468	0.39926	0.43856
1.2	0.764	0.055	0.11484	0.17578	0.23672	0.29583	0.35163	0.40277	0.44805	0.48647
1.4	0.891	0.064	0.13065	0.19708	0.26182	0.32317	0.37963	0.42983	0.47296	0.50824
1.6	1.019	0.070	0.14003	0.20867	0.27400	0.33434	0.38850	0.43541	0.47426	0.50457
1.8	1.146	0.072	0.14273	0.21029	0.27317	0.32998	0.37961	0.42124	0.45444	0.47899
2.0	1.273	0.071	0.13906	0.20287	0.26094	0.31213	0.35569	0.39106	0.41805	0.43672

Table A.41.2. $\mathcal{P}_1(X_1, X_2)$ (CONTINUED)

	$\pi x_1/2$	0.1	0.2	0.3	0.4	0.5	0.6	0.7	0.8	0.9
$\pi x_2/2$	$x_2 \backslash x_1$	0.064	0.127	0.191	0.255	0.318	0.382	0.446	0.509	0.573
2.2	1.401	0.067	0.13015	0.18804	0.23962	0.28404	0.32077	0.34957	0.37045	0.38362
2.4	1.528	0.061	0.11730	0.16800	0.21219	0.24936	0.27919	0.30164	0.31693	0.32547
2.6	1.655	0.054	0.10210	0.14500	0.18161	0.21162	0.23495	0.25177	0.26237	0.26721
2.8	1.783	0.046	0.08600	0.12113	0.15046	0.17392	0.19156	0.20360	0.21045	0.21261
3.0	1.910	0.038	0.07017	0.09808	0.12089	0.13863	0.15146	0.15972	0.16379	0.16417
3.2	2.037	0.030	0.05557	0.07708	0.09426	0.10727	0.11629	0.12169	0.12382	0.12314
3.4	2.165	0.023	0.04274	0.05884	0.07142	0.08067	0.08680	0.09013	0.09100	0.08979
3.6	2.292	0.018	0.03195	0.04369	0.05263	0.05901	0.06301	0.06493	0.06505	0.06371
3.8	2.419	0.013	0.02325	0.03156	0.03775	0.04200	0.04451	0.04554	0.04532	0.04392
4.0	2.546	0.009	0.01647	0.02219	0.02635	0.02911	0.03068	0.03098	0.03054	0.03051
4.2	2.674	0.006	0.01135	0.01520	0.01799	0.01951	0.02032	0.02182	0.01811	
4.4	2.801	0.004	0.00771	0.00998	0.01160	0.01406	0.01099			
4.6	2.928	0.002	0.00499	0.00738	0.00622					
4.8	3.056	0.00	0.00271							
5.0	3.183									

Table A.41.2. $\mathcal{P}_1(X_1, X_2)$ (CONTINUED)

	$\pi x_1/2$	1.0	1.2	1.4	1.6	1.8	2.0	2.2	2.4
$\pi x_2/2$	$x_2 \backslash x_1$	0.637	0.764	0.891	1.019	1.146	1.273	1.401	1.528
1.0	0.637	0.47173							
1.2	0.764	0.51757	0.55629						
1.4	0.891	0.53524	0.56405	0.56133					
1.6	1.019	0.52626	0.54434	0.53206	0.49548				
1.8	1.146	0.49501	0.50292	0.48297	0.44217	0.38796			
2.0	1.273	0.44735	0.44659	0.42166	0.37955	0.32751	0.27196		
2.2	1.401	0.38956	0.38240	0.35501	0.31429	0.26678	0.21795	0.17182	
2.4	1.528	0.32777	0.31640	0.28893	0.25164	0.21015	0.16889	0.13098	0.09848
2.6	1.655	0.26689	0.25345	0.22771	0.19514	0.16032	0.12676	0.09698	0.07311
2.8	1.783	0.21065	0.19685	0.17404	0.14674	0.11861	0.09252	0.07086	
3.0	1.910	0.16137	0.14842	0.12912	0.10713	0.08542	0.06647		
3.2	2.037	0.12009	0.10871	0.09309	0.07615	0.06023			
3.4	2.165	0.08688	0.07745	0.06534	0.05258				
3.6	2.292	0.06120	0.05366	0.04401					
3.8	2.419	0.04184	0.03506						
4.0	2.546	0.02627							

TABLE A.41.3. $\mathcal{B}_2(x_1, x_2)$. Same as in Table A.41.1, but when the matrix is chosen from the Gaussian unitary ensemble.

	$\pi x_1/4$	0.	0.1	0.2	0.3	0.4	0.5	0.6
$\pi x_2/4$	$x_2\backslash x_1$	0.	0.127	0.255	0.382	0.509	0.637	0.764
0.	0.	1.000000						
0.1	0.127	0.997766	0.995532					
0.2	0.255	0.982791	0.980559	0.965609				
0.3	0.382	0.945446	0.943223	0.928374	0.891497			
0.4	0.509	0.881362	0.879168	0.864573	0.828545	0.767460		
0.5	0.637	0.792044	0.789907	0.775812	0.741338	0.683475	0.604792	
0.6	0.764	0.683930	0.681896	0.668613	0.636520	0.583368	0.512099	0.429375
0.7	0.891	0.566398	0.564515	0.552374	0.523477	0.476374	0.414260	0.343398
0.8	1.019	0.449406	0.447720	0.437009	0.411952	0.371839	0.319924	0.261827
0.9	1.146	0.341475	0.340020	0.330935	0.310078	0.277340	0.235820	0.190312
1.0	1.273	0.248436	0.247232	0.239841	0.223213	0.197652	0.165922	0.131900
1.1	1.401	0.173072	0.172116	0.166361	0.153681	0.134609	0.111458	0.087198
1.2	1.528	0.115466	0.114741	0.110456	0.101215	0.087625	0.071506	0.055007
1.3	1.655	0.073788	0.073262	0.070212	0.063780	0.054536	0.043826	0.033125
1.4	1.783	0.045176	0.044812	0.042738	0.038463	0.032460	0.025670	0.019051
1.5	1.910	0.026505	0.026263	0.024917	0.022203	0.018482	0.014374	0.010467
1.6	2.037	0.014904	0.014752	0.013917	0.012272	0.010069	0.007697	0.005497
1.7	2.165	0.008035	0.007942	0.007448	0.006496	0.005250	0.003942	0.002759
1.8	2.292	0.004153	0.004100	0.003820	0.003293	0.002621	0.001932	0.001324
1.9	2.419	0.002059	0.002029	0.001878	0.001599	0.001253	0.000906	0.000608
2.0	2.546	0.000979	0.000963	0.000885	0.000744	0.000573	0.000407	
2.1	2.674	0.000446	0.000439	0.000400	0.000332	0.000251		
2.2	2.801	0.000195	0.000192	0.000173	0.000142			
2.3	2.928	0.000082	0.000080	0.000072				
2.4	3.056	0.000033	0.000032					
2.5	3.183	0.000013						

Table A.41.3. $\mathcal{B}_2(X_1, X_2)$ (CONTINUED)

	$\pi x_1/4$	0.7	0.8	0.9	1.0	1.1	1.2
$\pi x_2/4$	$x_2\backslash x_1$	0.891	1.019	1.146	1.273	1.401	1.528
0.7	0.891	0.271058					
0.8	1.019	0.203681	0.150614				
0.9	1.146	0.145725	0.105911	0.073107			
1.0	1.273	0.099308	0.070862	0.047965	0.030828		
1.1	1.401	0.064492	0.045139	0.029935	0.018830	0.011248	
1.2	1.528	0.039932	0.027394	0.017784	0.010940	0.006386	0.003540
1.3	1.655	0.023586	0.015847	0.010064	0.006051	0.003449	0.001866
1.4	1.783	0.013295	0.008743	0.005429	0.003188	0.001773	
1.5	1.910	0.007155	0.004603	0.002793	0.001601		
1.6	2.037	0.003678	0.002313	0.001371			
1.7	2.165	0.001807	0.001110				
1.8	2.292	0.00084					
1.9	2.419						

TABLE A.41.4. $\mathcal{P}_2(x_1, x_2)$. Same as in Table A.41.2, but when the matrix is chosen from the Gaussian unitary ensemble.

	πx_1	0.1	0.2	0.3	0.4	0.5	0.6	0.7	0.8
πx_2	$x_2 \backslash x_1$	0.032	0.064	0.095	0.127	0.159	0.191	0.223	0.255
0.1	0.032	0.00							
0.2	0.064	0.001	0.0041						
0.3	0.095	0.003	0.0126	0.0354					
0.4	0.127	0.007	0.0275	0.0712	0.13458				
0.5	0.159	0.012	0.0476	0.1161	0.20865	0.30913			
0.6	0.191	0.018	0.0703	0.1629	0.27969	0.39789	0.49321		
0.7	0.223	0.024	0.0915	0.2026	0.33410	0.45773	0.54773	0.58807	
0.8	0.255	0.028	0.1072	0.2280	0.36212	0.47897	0.55414	0.57610	0.54681
0.9	0.286	0.030	0.1148	0.2351	0.36053	0.46110	0.51663	0.52042	0.47881
1.0	0.318	0.030	0.1134	0.2242	0.33246	0.41188	0.44721	0.43670	0.38987
1.1	0.350	0.028	0.1040	0.1989	0.28587	0.34328	0.36137	0.34247	0.29664
1.2	0.382	0.024	0.0891	0.1652	0.23018	0.28806	0.27392	0.25191	0.2117
1.3	0.414	0.020	0.0716	0.1288	0.17412	0.19691	0.19532	0.17436	0.1424
1.4	0.446	0.015	0.0542	0.0946	0.12420	0.13638	0.13136	0.11387	0.0900
1.5	0.477	0.011	0.0386	0.0656	0.08367	0.08926	0.08343	0.07046	
1.6	0.509	0.008	0.0261	0.0430	0.05335	0.05519	0.0506		
1.7	0.541	0.005	0.0166	0.0267	0.03213	0.0329			
1.8	0.573	0.003	0.0101	0.0159	0.0189				
1.9	0.605	0.001	0.0058	0.008					
2.0	0.637	0.00	0.003						

TABLE A.41.4. $\mathcal{P}_2(x_1, x_2)$ (CONTINUED)

	πx_1	0.9	1.0	1.1
πx_2	$x_2 \backslash x_1$	0.286	0.318	0.350
0.9	0.286	0.40681		
1.0	0.318	0.32136	0.24628	
1.1	0.350	0.2372	0.1770	0.1213
1.2	0.382	0.1647	0.1176	
1.3	0.414	0.1066		

Notes

Our knowledge of random matrices is most extensive for the following ensembles.

Gaussian unitary ensemble or GUE. This is the ensemble of $N \times N$ Hermitian matrices with the joint probability density proportional to $\exp\left(-\text{tr } H^2\right)$; i.e., apart from the Hermitian character, the real and imaginary parts of every matrix element is a Gaussian random variable with a common variance. This ensemble is invariant under unitary transformations, and is appropriate to describe systems without time reversal symmetry. The joint probability density of the eigenvalues is proportional to

$$|\Delta(x)|^2 \exp\left(-\sum_{i=1}^{N} x_i^2\right), \tag{N.1}$$

where

$$\Delta(x) = \prod_{1 \le i < j \le N} (x_i - x_j). \tag{N.2}$$

The n-point correlation and cluster functions for any finite n (and finite or infinite N) are known, as is the probability $E_2(n; s)$ of having exactly n eigenvalues in a randomly chosen interval of given length s. (See Chapter 5.)

Gaussian orthogonal ensemble or GOE. This is the ensemble of $N \times N$ real symmetric matrices with the joint probability density proportional to $\exp\left(-\text{tr } H^2/2\right)$; i.e., apart from the symmetry, each matrix element is a Gaussian random variable with the same variance. This ensemble is invariant under orthogonal transformations, and is appropriate to describe systems with time reversal and rotational symmetry, i.e., most of the physical systems found in nature. The joint probability density of the eigenvalues is proportional to

$$|\Delta(x)| \exp\left(\sum_{i=1}^{N} x_i^2/2\right), \tag{N.3}$$

with $\Delta(x)$ given by Eq. (N.2). The n-point correlation and cluster functions as well as the probability $E_1(n;s)$ of having exactly n eigenvalues in a randomly chosen interval of given length s are again explicitely known. (See Chapter 6.)

Gaussian symplectic ensemble or GSE. This is the ensemble of $N \times N$ quaternion self-dual matrices with the joint probability density proportional to $\exp(-2 \text{ tr } H^2)$. This ensemble is invariant under symplectic transformations, and is appropriate to describe systems with time reversal symmetry, half odd integer spin and no rotational symmetry; such systems are rare in nature. The joint probability density of the eigenvalues is proportional to

$$|\Delta(x)|^4 \exp\left(-2\sum_{i=1}^{N} x_i^2\right), \tag{N.4}$$

with $\Delta(x)$ given by Eq. (N.2). All correlation and cluster functions and the probability $E_4(n;s)$ of having exactly n eigenvalues in a randomly chosen interval of given length s are again known. (See Chapters 7 and 10.)

The probability $E_\beta(0;s)$, $\beta = 1$, 2 or 4, of having no eigenvalue in a randomly chosen interval of length s is expressed in terms of two Fredholm determinants or infinite products $D_+(s)$ and $D_-(s)$. Their power series expansions for small s and asymptotic expansions for large s are known. (See Chapters 5, 6, 7, 10 and 12.) Extensive numerical tables of $E_\beta(n;s)$ and their few derivatives, useful for applications have been computed. (See Appendices A.14, A.15, and A.41.)

Circular orthogonal, circular symplectic, and circular unitary ensembles. These are the ensembles of random unitary matrices, which are in addition symmetric, (for orthogoanl), or self-dual (for symplectic), or having no other restricition (for unitary). All local statistical properties, i.e., those extending to a finite number of eigenvalues, are in the limit of infinite matrices identical to those of the corresponding Gaussian ensembles. (See Chapter 10.)

Ensemble of Hermitian matrices, where apart from the Hermitian character, the real parts of the matrix elements are Gaussian random variables with a common variance, so also the imaginary parts, but the common variance of the real parts is not equal to the common variance of the imaginary parts. Here again the n-point correlation and cluster functions for any finite n are known. (See Chapter 14.)

We also know quite a lot about the ensemble of antisymmetric Hermitian matrices (or the ensemble of any complex matrices), whose matrix elements (or their real and imaginary parts) are Gaussian random variables; but these ensembles are marginal for physical applications. (See Chapters 13 and 15.)

In the last few decades, a huge amount of experimental as well as numerical data have been collected and analysed for their statistical properties concerning various systems such as nuclear excitation energies, atomic energies, possible energies of a particle free to move on billiard tables of odd shapes (classically chaotic systems), characteristic ultrasonic frequencies of structural materials like the aluminium beams, imaginary parts of the zeros of the Riemann zeta function on the line $\mathrm{Re}\, z = 1/2$, and so on. Their agreement with the theoretical predictions of various random matrix models, especially the GOE and GUE, is quite convincing. (See Chapters 1 and 16.) But we do not yet quite understand why this should be so. Why the zeros of the Riemann zeta on the line $\mathrm{Re}\, z = 1/2$ should behave locally as the eigenvalues of a matrix from the GUE, or why the possible energies of a classically chaotic system should behave as the eigenvalues of a matrix from the GOE, is not clear.

With the availability of large computers people have generated matrices with random elements. These matrices were either real symmetric or Hermitian. The probability density of the matrix elements, or of their real and imaginary parts, was not necessarily Gaussian. The statistical properties of a few eigenvalues of such matrices were found to be quite universal. Apparently they do not depend on the probability density of the individual matrix elements. And we do not quite understand why this should be so.

Classically, chaotic systems fall into various catagories according to their chaoticity; but we cannot yet make this finer distinction by looking at their quantum mechanical energy spectrum alone. There have been attempts to see whether the eigenvectors or the wave functions of such systems carry any information about their chaotic nature; but these attempts are not very successful yet.

A generalization of the Euler beta integral due to Selberg might be ignored at first sight. But it has deep consequences and unsuspected relations with other branches of mathematics, such as the theory of random matrices and finite groups generated by reflections. (See Chapter 17.) Similarly, an integral over the unitary group, seemingly innocent, is quite unobvious and deep. (See Appendix A.5.) A generalization of this later integral for other Lie algebras is known from Harish Chandra (1957). It reads as follows.

Let G be a compact simple Lie group, L its Lie algebra of order N and rank n, W the Weyl (or Coxeter) group of L, R_+ the set of its positive roots, and $m_i = d_i - 1$ its Coxeter indices. Also for X and Y elements of L let (X, Y) be a bilinear form invariant under G, i.e. $(X_1 + X_2, Y) = (X_1, Y) + (X_2, Y)$, $(X, Y_1 + Y_2) = (X, Y_1) + (X, Y_2)$, and $(gX, gY) = (X, Y)$ for $g \in G$. Then

$$\int_{g \in G} \exp\left(c\left(X, gYg^{-1}\right)\right) dg$$
$$= \text{const} \sum_{w \in W} \varepsilon_w \exp\left(c\left(X, wY\right)\right) \Big/ \prod_{\alpha \in R_+} (\alpha, X)(\alpha, Y) \quad \text{(N.5)}$$

where ε_w is the parity of w and the

$$\text{const} = c^{-(N-n)/2} \prod_{\alpha \in R_+} \frac{|\alpha|^2}{2} \prod_{i=1}^{n} m_i! \quad \text{(N.6)}$$

But this generalization is not good enough. For example, one knows the integral

$$\int \exp\left(c\, \text{tr}(A - QBQ^{-1})^2\right) dQ \quad \text{(N.7)}$$

over the group of all real orthogonal matrices Q, when A and B are real antisymmetric matrices, but *not* when A and B are real symmetric matrices.

Chapter 1

The experimental information about slow neutron resonances in various nuclei was collected in the 1960s and 1970s mainly by a few groups of workers, the most extensive being that of the Columbia University group of Camarda et al. A good recent analysis of all this data was done by French, et al. (1985) and O. Bohigas, et al. (1982, 1985). For nuclear level densities see Bethe (1937), Lang and Lecouteur (1954) and Cameron (1956). For the Hardy–Ramanujan formula, see Andrews (1976). The level spacing law (1.5.1) was proposed by E. P. Wigner and appeared in print in Canadian Mathematical Congress Proceedings, University of Toronto Press, Toronto (1957). This and other important papers on the subject before 1965, together with a detailed introductory article by Porter himself can be found in Porter (1965). The possibility of choosing arbitrary global properties with disregard to the local ones was suggested by Balian (1968). For some interesting material about quantum chaos, see the papers in *Quantum chaos and statistical nuclear physics*, Proceedings of the 2nd International Conference on Quantum Chaos and 4th International Colloquium on Statistical Nuclear Physics, Cuernavaca, Mexico, 1986, edited by Seligman and Nishioka, Springer Verlag (1986). A good review article about quantum chaos is Eckhardt (1988). About small metallic particles see the review article by R. Kubo in *Polarization, matière et rayonnement*, Presses Universitaires de France, Paris (1970). For a critical discussion of various reaction width distributions, strength functions, nuclear reaction theories, experimental situation about small metallic particles and many other interesting topics see the thorough and exhaustive review article by Brody et al. (1981). About the zeros of the zeta functions see Montgomery (1973, 1975), Titchmarsh (1951), Davenport and Heilbronn (1936), Potter and Titchmarsh (1935), Cassels (1961), and Odlyzko (1987, 1988).

Chapter 2

Section 2.2 is based on Wigner (1959), 2.3 to 2.5 are largely based on Dyson (1962-I), and Section 2.6 on Porter and Rosenzweig (1960a). All these papers are reproduced in Porter (1965).

Chapter 3

Sections 3.1 to 3.3 are largely based on a paper of Wigner (1962). Section 3.6 is based on Balian (1968).

Chapter 4

Section 4.2 is based on Wigner (1957).

Chapter 5

Section 5.2 is largely based on Dyson (1970), Section 5.3 on Kahn (1963), and Section 5.4 on Mehta and des Cloizeaux (1972). For ergodic properties, see Pandey (1979).

Chapter 6

The use of quaternion matrices for correlation and cluster functions described in Sections 6.2–6.4 was discovered by Dyson (1970) for the circular ensembles and were adapted to the Gaussian case by Mehta (1971). The method of integration on alternate variables is essentally due to a remark of Gaudin (personal communication). Sections 6.5 and 6.6 are taken from Mehta and des Cloizeaux (1972), while Section 6.8 is from Mehta (1960). Effetof (1982) used anticommuting or super-variables to rederive the two-level correlation function for the three Gaussian ensembles, orthogonal, unitary, and symplectic. A similar, perhaps more readable, method using integrations over commuting and anti-commuting variables to rederive the k-point correlation function for the Gaussian unitary ensemble can be found in T. Guhr's recent preprint "Dyson's correlation functions and graded symmetry," Max Planck Inst. für Kernphysik, Heidelberg.

Chapter 8

The papers of Uhlenbeck, Ornstein, and Wang and other papers on the mathematics of Brownian motion can be found in the collection of Nelson Wax (1954). This chapter owes much to Dyson (1962).

Chapter 9

This chapter is largely based on Dyson (1962a-I).

Chapter 10

Sections 10.1, 10.3, and 10.5 are based on Dyson (1970), Sections 10.2 and 10.4 on Dyson (1962a-III), Section 10.6 on Mehta and Dyson (1963-V), Section 10.8 on Dyson (1962b), and Section 10.9 on unpublished notes of Wigner (about 1960).

Chapter 11

Sections 11.1 to 11.4 are based on Dyson (1962a-I and -II).

Chapter 12

Section 12.1 is based on des Cloizeaux and Mehta (1973), Section 12.2 on Widom (1971) and as noted in the text Sections 12.3 to Sections 12.6 are copied from Dyson (1976).

Chapter 13

This chapter is based on Mehta and Rosenzweig (1968).

Chapter 14

This chapter resumes the three papers of Pandey and Mehta (1983). Pandey and Shukla recently studied an ensemble of unitary matrices with unequal mean absolute square values of its symmetric and anti-symmetric parts. This ensemble is not invariant under unitary transformations, it is intermediate between the orthogoanl and unitary circular ensembles studied in Chapters 9, 10, and 11. The results are identical with those for the Hermitian matrices studied in Section 14.1. Similarly, they studied an ensemble of unitary matrices intermediate between symplectic and unitary circular ensembles.

Chapter 15

Section 15.1 is based on Ginibre (1965). Jancovici (1981) considered the case when the power in Eq. (15.1.10) is $2+\varepsilon$ instead of 2, and computed the first term in the expansion of the two-point function $R_2(z_1, z_2)$ in powers of ε. Section 15.2 is based on Ginibre (1965) and Mehta and Srivastava (1966).

Chapter 16

The number variance, μ_2, gained much popularity when Odlyzko (1987) found that for the Riemann zeta zeros it does not rise as required by the theory of random matrices (i.e., GUE), and when Berry's suggestion (1988) of an ad hoc formula reproduced quite well its numerical variations. For the least square statistic, three ways of choosing the straight line originally were considered; two of them, called Δ_1 and Δ_2, had one free parameter each, while the remaining one, Δ_3, depended on two parameters. Experience showed Δ_3 to be better than either Δ_1 or Δ_2. So we suggest dropping the subscript 3.

Chapter 17

As indicated in the text, Sections 17.2 and 17.3 follow Selberg (1944) and Aomoto (1987), respectively. Not knowing Selberg's work, equality (17.6.7) has harassed many people for years; even today no argument is known which proves only that without passing through Eq. (17.1.3) or Eq. (17.6.5). Section 17.9 is based on Macdonald (1982). Section 17.10 follows Askey and Richards (1989).

It is instructive to recall the computer proof of Eq. (17.9.1) due to Garvan (1989). Denoting the two sides of Eq. (17.9.1) by $f(\gamma)$ and $F(\gamma)$, respectively, it will be sufficient to show that

$$\frac{f(\gamma+1)}{f(\gamma)} = \frac{F(\gamma+1)}{F(\gamma)}. \tag{N.8}$$

Since starting with the evident equality $f(0) = F(0)$, the above equation will imply $f(\gamma) = F(\gamma)$ for any positive integer γ and then Carlson's theorem will extend its validity for complex γ.

Now $F(\gamma+1)/F(\gamma)$ is a polynomial in γ of degree $\sum_{i=1}^{n} d_i = N + n$. From Aomoto's argument, Section 17.8, one can show that $f(\gamma+1)/f(\gamma)$ is also a polynomial in γ of degree $N + n$. If two polynomials each of degree $N + n$ are equal at $N + n + 1$ distinct values of γ, then they are identically equal. Garvan could verify the equality of the polynomials with the help of a computer for small values of N and n, i.e., for the groups H_3 and F_4.

All these integrals and constant term identities have their several so called q-extensions, either proved or conjectured. See Macdonald (1982) or Morris (1982) and references given there.

Chapter 18

This chapter resumes the thesis of Bronk (1962); see also his two aricles in *J. Math. Phys.* **5** (1964) and **6** (1965).

Appendices

A.1. Almost all of the numerical computations corroborating the Conjectures 1.2.1 and 1.2.2 were carried out in the early days of the random matrix hypothesis when very little was known analytically. Such numerical studies are lacking for the case when the matrix element probability densities do not have all their moments finite. For example, when

$$P(H) = \prod_{i \le j} \frac{a}{\pi} \left(H_{ij}^2 + a^2\right)^{-1} \tag{N.9}$$

even the second moment of H_{ij} is infinite, and we do not know whether the level density is the "semi circle" or whether the spacing probability density resembles the "Wigner surmise."

A.5. The integral in Eq. (14.3.1) or (A.5.1) is over the group of unitary matrices U. It is equivalent to Eq. (N.5) for the compact Lie group A_n. The same integral over other compact Lie groups is of great interest. D. Altschular and C. Itzykson have recently rederived a formule due to Harish Chandra. But, as we said earlier, this is not good enough for an analytic treatment of, say, a fixed diagonal real matrix perturbed by a random real symmetric matrix.

A.14–A.15. The tables of $\lambda_j(s)$ and $E_\beta(n;s)$ for $\beta = 1$ and 2 first appeared in Mehta and des Cloizeaux (1972). The values of $p_\beta(0;s)$ given here are more precise than those appearing in the first edition of this book, since formulas (5.4.32) and (6.6.18) were used thus avoiding two numerical differentiations.

References

Abramowitz, M., and Stegun, I. A. *Handbook of Mathematical Functions,* Dover, New York (1965).

Al'tshuler, B. L., and Shklovskii, B. I. Repulsion of energy levels and conductivity of small metal samples, *Sov. Phys. JETP* **64** (1986) 127–135.

Andrews, G. E. The theory of partitions, *Encyclopedia of Mathematics and its Applications,* Vol. 2, Addison-Wesley, Reading (1976).

Aomoto, K. Jacobi polynomials associated with Selberg integrals, *SIAM J. Math. Anal.* **18** (1987) 545–549.

Arnold, V. I., and Avez, A. *Problèmes Ergodiques de la Mécanique Classique,* Gauthiers Villars, Paris (1967); English translation, *Ergodic problems in classical mechanics,* Benjamin, New York (1968).

Askey, R. A. Some basic hypergeometric extensions of integrals of Selberg and Andrews, *SIAM J. Math. Anal.* **11** (1980) 938–951.

Askey, R. A., and Richards, D. Selberg's second beta integral and an integral of Mehta, in *Probability, Statistics and Mathematics: Papers in Honor of S. Karlin,* edited by T. W. Anderson, K. B. Athreya, and D. L. Inglehart, pp. 27–39, Academic Press, New York (1989).

Auluck, F. C., and Kothari, D. S. Statistical mechanics and the partitions of numbers, *Proc. Camb. Philos. Soc.* **42** (1946) 272–277.

Balazs, N. L., and Voros, A. Chaos on the pseudosphere, *Phys. Rep.* **143** (1986) 109–240.

Balazs, N. L., and Voros, A. The quantized baker's transformation, *Ann. Phys.* **190** (1989) 1–31.

Balian, R. Random matrices and information theory, *Nuovo Cimento* B 57 (1968) 183–193.

Bateman, H. *Higher Transcendental Functions,* edited by A. Erdelyi, Vol. 2, Chapters 10.10 and 10.13, McGraw Hill, New York (1953).

Bateman, H. Tables of Integral Transforms, edited by A. Erdelyi, Vol. 1, Sections 1.6(1), 2.11(4), McGraw Hill, New York (1954).

Bateman, H. Tables of Integral Transforms, edited by A. Erdelyi, Vol. 2, Section 16.5, McGraw Hill, New York, (1954).

Berry, M. V. Semi-classical Theory of Spectral Rigidity, Proc. Royal Soc. London, **A400** (1985) 229–251.

Berry, M. V. Semi-classical formula for the number variance of the Riemann zeros, *Nonlinearity* **1** (1988) 399–407.

Bethe, H. A. Nuclear physics: nuclear dynamics, theoretical, *Rev. Mod. Phys.* **9** (1937) 69–244.

Bohigas O., and Giannoni, M. J. Chaotic motion and random matrix theories, in *Mathematical and Computational Methods in Nuclear Physics,* edited by J. S. Dehesa et al. (Proceedings, Granada, Spain, 1983), Lecture Notes in Physics Vol. 209, pp. 1–99, Springer Verlag, Berlin (1984a).

Bohigas, O., Giannoni, M. J., and Schmit, C. Characterization of chaotic quantum spectra and universality of level fluctuation laws, *Phys. Rev. Lett.* **52** (198ba) 1–4.

Bohigas, O., Giannoni, M. J., and Schmit, C. Spectral properties of the Laplacian and random matrix theories, *J. Physique Lett.* **45** (1984c) *L1015–L1022.*

Bohigas, O., Giannoni, M. J., and Schmit, C. Spectral fluctuations and chaotic motion, in *Heavy Ion Collisions*, edited by P. Boudre *et al.*, pp. 145–163, Plenum Press (1984d).

Bohigas, O., Haq, R. U., and Pandey, A. Fluctuation properties of nuclear energy levels and widths: comparison of theory with experiment, in *Nuclear Data for Science and Technology,* edited by K.H. Böckhoff, pp. 809–814, Brusseles (1983).

Bohigas, O., Haq, R. U., and Pandey, A. Higher order correlations in spectra of complex systems, *Phys. Rev. Lett.* **54** (1985) 1645–1648 and references given there.

Bohigas, O., Tomsovic, S., and Ullmo, D., Classical transport effects of chaotic levels (to appear).

Bohr, A. On the theory of nuclear fission, in *Proceedings of the First International Conference on Peaceful Uses of Atomic Energy, Geneva, 1955,* Vol. 2, pp. 151–154, Columbia University Press (1956).

Bohr, A., and Mottelson, B. R. *Nuclear Structure,* 2 vols., Benjamin, New York, (1975).

Bouwcamp, C. J. On spheroidal functions of order zero, *J. Math. Phys.* **26** (1947) 79–92.

Brody, T. A. A statistical measure for the repulsion of energy levels, *Nuovo Cimento Lett.* **7** (1973) 482–489.

Brody, T. A., Flores, J., French, J. B., Mello, P. A., Pandey, A., and Wong, S. S. M. Random matrix Physics: spectrum and strength fluctuations, *Rev. Mod. Phys.* **53** (1981) 385–479.

Bronk, B. V. Topics in the theory of random matrices, thesis, Princeton University, 1964a.

Bronk, B. V. Accuracy of the semicircle approximation for the density of eigenvalues of random matrices, *J. Math. Phys.* **5** (1964b) 215–220.

Bronk, B. V. Exponential ensemble for random matrices, *J. Math. Phys.* **6** (1965) 228–237.

Caillol, J.M. Exact results for a two-dimensional one component plasma on a sphere, *J. Phys. Lettres* **42** (1981) L245–L247.

Camarda, H., Desjardins, J. S., Garg, J. B., Haken, G., Havens, W. W. Jr., Liou, H. I., Peterson, J. W., Rahn, F., Rainwater, J., Rosen, J. L., Singh, U. N., Slagowitz, M., Weinchank, S. See the series of papers Neutron resonance spectroscopy I–XIV in Phys. Rev. of the 1960s and 1970s. To find references, see for example, *Phys. Rev. C* **8** (1973) 1833–1836, or *Phys. Rev. C* **11** (1975) 1117–1121, or Neutron resonance spectroscopy, *Phys. Rev. C* **11** (1975) 1117–1121.

Camarda, H. S., and Georgopulos, P. D., Statistical behavior of atomic energy levels: agreement with random matrix theory, *Phys. Rev. Letters* **50** (1983) 492–495.

Cameron, A. G. W. Nuclear level spacings, Can. *J. Phys.* **36** (1956) 1040–1057.

Cassels, J. W. S. Footnote to a Note of Davenport and Heilbronn, J. London Math. Soc. **36** (1961) 177–184.

Chadan, K., and Sabatier, P. C. *Inverse Scattering Problems in Quantum Scattering Theory,* Eqs. (III.5.7) and (IV.1.10), Springer Verlag (1977).

Chandrasekhar, S. Stochastic problems in physics and astronomy, *Rev. Mod. Phys.* **15** (1943) 1–89.

Chevalley, C. *Theory of Lie Groups,* pp. 16–24, Princeton University Press, Princeton, N.J. (1946).

Cipra, B. A. Zeroing in on the zeta function, *Science,* 11 March 1988, pp. 1241–1242; Zeta zero update, *Science,* 3 March 1989, p. 1143.

Cristofori, F., Sona, P. G., and Tonolini, F. The statistics of the eigenvalues of random matrices, *Nucl. Phys.* **78** (1966) 553–556.

Davenport, H., and Heilbronn, H. On the zeros of certain Dirichlet series I, II, *J. London Math. Soc.* **11** (1936) 181–185 and 307–312.

de Bruijn, N. G., On some multiple integrals involving determinants, *J. Indian Math. Soc.*, **19** (1955) 133–151.

Delande, D., and Gay, J. C. Quantum chaos and statistical properties of energy levels: numerical study of the hydrogen atom in a magnetic field, *Phys. Rev. Lett.* **57** (1986) 2006–2009; J. Phys. B **17** (1984) L335ff; **19** (1986) L173ff.

Derrida, B., and Vannimenus, J. A transfer matrix approach to random resitor networks, *J. Phys. A* **15** (1982) L557–L564.

des Cloizeaux, J., and Mehta, M. L. Some asymptotic expressions for prolate spheroidal functions and for the eigenvalues of differential and integral equations of which they are solutions, *J. Math. Phys.* **13** (1972) 1745–1754.

des Cloizeaux, J., and Mehta, M. L. Asymptotic behaviour of spacing distributions for the eigenvalues of random matrices, *J. Math. Phys.* **14** (1973) 1648–1650.

Dieudonné, J. *La Géometrie des Groupes Classiques,* Ergeb. Math., Vol. 5, Springer, Berlin (1955).

Dyson, F. J. The dynamics of a disordered linear chain, *Phys. Rev.* **92** (1953) 1331–1338.

Dyson, F. J. Statistical theory of energy levels of complex systems, I, II, and III, *J. Math. Phys.* **3** (1962a) 140–156, 157–165, 166–175; cited as I, II, and III.

Dyson, F. J. A Brownian motion model for the eigenvalues of a random matrix, *J. Math. Phys.* **3** (1962b) 1191–1198.

Dyson, F. J. The three fold way. Algebraic structure of symmetry groups and ensembles in quantum mechanics, *J. Math. Phys.* **3** (1962c) 1199–1215;

Dyson, F. J. Correlations between the eigenvalues of a random matrix, *Commun. Math. Phys.* **19** (1970) 235–250.

Dyson, F. J. A class of matrix ensembles, *J. Math. Phys.* **13** (1972) 90–97.

Dyson, F. J. Fredholm determinants and inverse scattering problems, *Commun. Math. Phys.* **47** (1976) 171–183.

Dyson, F. J., and Mehta, M. L., Statistical theory of energy levels of complex systems, IV *J. Math. Phys.* **4** (1963) 701–712 cited as IV.

Ebeling, K.J. Statistical properties of random wave fields, in *Physical Acoustics,* edited by W.P. Masson, pp. 233–310, Academic Press, New York (1984).

Eckhardt, B. Quantum mechanics of classically non-integrable systems, *Phys. Rep.* **163** (1988) 205–297.

Effetof, K. B. Statistics of the levels in small metallic particles, *Zh. Eksp. Teor. Fiz.* **83** (1982) 833–847 [English translation, *Sov. Phys. JETP* **56** (1982) 467–475.

Engleman, R. The eigenvalues of a randomly distributed matrix, *Nuovo Cimento* **10** (1958) 615–621.

Erdelyi, A. Asymptotic forms for Laguerre polynomials, *J. Indian Math. Soc., Golden Jubilee Commemoration,* **1907-1908,** (1960) 235–250.

Faddeev, L. D. The inverse problem in quantum theory of scattering, *Usp. Mat. Nauk* **14** (4) (1959) 57–119 [English translation in *J. Math. Phys.* **4** (1963) 72–104].

Fox, D., and Kahn, P. B. Identity of the n-th order spacing distributions for a class of Hamiltonian unitary ensembles, *Phys. Rev.* **134** (1964) B1151–B1192.

French, J. B., and Kota, V. K. B. Nuclear level densities and partition functions with interactions, *Phys. Rev. Letters*, **51** (1983) 2183–2186.

French, J. B., Kota, V. K. B., Pandey, A., and Tomosovic, S. Bounds on time reversal non-invariance in the nuclear Hamiltonian, *Phys. Rev. Lett.* **54** (1985) 2313–2316

French, J. B., Kota, V. K. B., Pandey, A., and Tomosovic, S. Statistical properties of many particle spectra, V and VI, *Ann. Phys.* **181** (1988) 198–234 and 235–260. For earlier papers in this series, see references there.

Fröhlich, H. Die spezifische Wärme der Electronen Kleiner Metallteilchen bei tiefen Temperturen, *Physica* **4** (1937) 406–412.

Gallagar, P. X., and Muller, J. H. Primes and zeros in short intervals, *J. Reine Angewandte Math. (Crelle)* **303-304** (1978) 205–220.

Garg, J. B., Rainwater, J., Peterson, J. S., and Havens, W. W., Jr. Neutron resonance spectroscopy III, Th^{232} and U^{238}, *Phys. Rev.* **134** (1964) B985–B1009.

Garvan, F. G. Some Macdonald-Mehta integrals by brute force, in q-*Series and Partitions*, edited by D. Stanton, IMA volumes in mathematics and its applications, Vol. 18, Springer-Verlag, Berlin, pp. 77–98, (1989).

Gaudin, M. Sur la loi limite de l'éspacement des valeurs propres d'une matrice aléatoire, *Nucl. Phys.* **25** (1961) 447–458.

Gel'fand, I. M., and Levitan, B. M. On the determination of a differential equation by its spectral function, *Izv. Akad. Nauk SSSR, Ser. Mat.* **15** (1951) 309–360 [English translation in *Amer. Math. Soc. Translations* **1** (2) (1955) 253–304.

Ginibre, J. Statistical ensembles of complex, quaternion and real matrices, *J. Math. Phys.* **6** (1965) 440–449.

Good, I. J. Short proof of a conjecture by Dyson, *J. Math. Phys.* **11** (1970) 1884.

Gorkov, L. P., and Eliashberg, G. M. Minute metallic particles in an electro-magnetic field, *Zh. Eksp. Teor. Fiz.* **48** (1965) 1407–1418 [*Sov. Phys. JETP* **21** (1965) 940–947].

Goursat, E. *Cours d'Analyse Mathématique,* Vol. 3, pp. 389, 454, Gauthier Villars, Paris (1956).

Gradshteyn, I. S., and Rhizhik, I. M. *Tables of Integrals, Series and Products,* Academic Press, New York (1965).

Grobe, R., and Haake, F. Quantum distinction of regular and chaotic dissipative motions, *Phys. Rev. Lett.* **61** (1988) 1899–1902.

Grobe, R., and Haake, F. Universality of cubic level repulsion for dissipative quantum chaos, *Phys. Rev. Lett.,* **62** (1989) 2893–2896.

Habsieger, L. Conjecture de Macdonald et q-integral de Selberg-Askey, Thèse, Université Louis Pasteur, Strassbourg, 1987.

Hannay, J. H., and Ozeiro de Almeida, A. M. Periodic orbits and a correlation function for the semi-classical density of states, *J. Phys. A, Maths. and General,* **17** (1984) 3429–3440.

Haq, R. U., Pandey, A., and Bohigas, O. Fluctuation properties of nuclear energy levels: do theory and experiment agree? *Phys. Rev. Lett.* **48** (1982) 1086–1089.

Hardy, G. H., and Ramanujan, S. Asymptotic formulae in combinatory analysis, *Proc. London Math. Soc.* **17** (1918) 75–115.

Harish Chandra, Differential operators on a semi-simple lie algebra, *Amer. J. Math.* **79** (1957) 87–120.

Harer, J., and Zagier, D. The Euler characteristic of the moduli space of curves, *Invent. Math.* **85** (1986) 457–485.

Harvey, J. A., and Hughes, D. J. Spacings of nuclear energy levels, *Phys. Rev.* **109** (1958) 471–479.

Hejhal, D. A., The Selberg trace formula and the Riemann zeta function, *Duke Math. J.* **43** (1976) 441–482.

Hermann, H., Derrida, B., and Vannimenus, J. Superconductivity exponents in two- and three-dimensional percolation, *Phys. Rev. B* **30** (1984) 4080–4082.

Hsu, P. L., *Ann. Eugenics* **9** (1939) 250ff.

Itzykson, C., and Zuber, J. B. Matrix integration and Combinatorics of Modular Groups, preprint SPhT/90-004, Saclay, 91191 Gif-sur-Yvette Cedex, France (1990).

Jancovici, B. Exact results for the two-dimensional one-component plasma, *Phys. Rev. Lett.* **46** (1981) 386–388.

Jimbo, M., Miwa, T., Mori, Y., and Sato, M., Density matrix of impenetrable Bose gas and the fifth Painlevé transcendent, *Jpn. Acad. Ser. A, Math. Sci.* **55** (1979) 317–322; *Physica D* **1** (1980) 80–158.

Jost, R., Über die falschen Nullstellen der Eigenwerte der S-Matrix, *Helv. Phys. Acta* **20** (1947) 256–266.

Kahn, P. B. Energy level spacing distributions, *Nucl. Phys.* **41** (1963) 159–166.

Khinchin, A. I. *Mathematical Foundations of Information Theory,* Dover, New York (1957).

Kisslinger, L. S., and Sorenson, R. A. Pairing plus long range force for single closed shell nuclei, *Kgl. Danske Videnskab. Selskab. Mat.-Fys. Medd.* **32** (9) (1960).

Kramer, H. A., *Proc. Acad. Sci. Amsterdam* **33** (1930) 959ff.

Kubo, R. Electrons in small metallic particles, in *Polarization, Matière et Rayonnement,* volume jubiliaire en l'honneur d'Alfred Kastler, Société Française de Physique, pp. 325–339, Presses Univ. de France (1969).

Landau, L., and Smorodinski, Ya. Lektsii po teori atomnogo yadra, *Gorsar. Izd. Tex.-teoreticheskoi Lit., Moscow* (1955) 92–93.

Lang, J. M. B., and Lecouteur, K. J. Statistics of nuclear levels, *Proc. Phys. Soc. London,* **A67** (1954) 586–600.

Leff, H. S. Statistical theory of energy level spacing distributions for complex spectra, Thesis, State University of Iowa, SUI-63-23, 1963.

Leff, H. S. Systematic characterization of m-th order energy level spacing distributions, *J. Math. Phys.* **5** (1964a) 756–762.

Leff, H. S. Class of ensembles in the statistical theory of energy level spectra, *J. Math. Phys.* **5** (1964b) 763–768.

Levinson, N. On the uniqueness of the potential in a Schrödinger equation for a given asymptotic phase, *Kgl. Danske Vidensk. Selsk. Mat-Fys. Medd.* **25**(9) (1949) 1–29.

Liou, H. I., Camarda, H. S., and Rahn, F. Applications of statistical tests for single level populations to neutron resonance spectroscopy data, *Phys. Rev. C* **5** (1972b) 1002–1015.

Liou, H. I., Camarda, H. S., Wynchank, S., Slagowitz, M., Hacken, G., Rahn, F., and Rainwater, J. Neutron resonance spectroscopy. VIII. The separated isotopes of Erbium: Evidence for Dyson's theory concerning level spacings, *Phys. Rev. C* **5** (1972a) 974–1001.

Macdonald, I. G. *Symmetric Functions and Hall Polynomials,* Clarendon Press, Oxford (1979).

Macdonald, I. G. Some conjectures for root systems, *SIAM J. Math. Anal.* **13** (1982) 988–1004.

Mailly, D., Sanquer, M., Pichard, J. L., and Pari, P. Reduction of quantum noise in a Ga-Al-As/Ga-As heterojunction by a magnetic field: an orthogonal to unitary Wigner statistics transition, *Europhys. Lett.* **8** (1989) 471–476.

Maradudin, A. A., Mazur, P., Montroll, E. W., and Weiss, G. H. Remarks on the vibrations of diatomic lattices, *Rev. Mod. Phys.* **30** (1958) 175–196, Section IV, 186–195.

Marchenko, V. A. Concerning the theory of a differential operator of the second order, *Dokl. Akad. Nauk SSSR* **72** (1950) 457–460; The construction of the potential energy from the phases of the scattered waves, *ibid.* **104** (1955) 695–698.

Mayer, M. G., and Jensen, J. H. D. *Elementary Theory of Nuclear Shell Structure,* Wiley, New York (1955).

McCoy, B., and Wu, T. T. *The Two Dimensional Ising Model,* Harvard University Press, Cambridge, Mass. (1973).

Mehta, M. L. On the statistical properties of the level spacings in nuclear spectra, *Nucl. Phys.* **18** (1960) 395–419.

Mehta, M. L. A note on correlations between eigenvalues of a random matrix, *Commun. Math. Phys.* **20** (1971) 245–250.

Mehta, M. L. Determinants of quaternion matrices, *J. Math. Phys. Sci.* **8** (1974) 559–570.

Mehta, M. L. A note on certain multiple integrals, *J. Math. Phys.* **17** (1976) 2198–2202.

Mehta, M. L. A method of integration over matrix variables, *Commun. Math. Phys.* **79** (1981) 327–340.

Mehta, M. L. Random matrices in nuclear physics and number theory, *Contemp. Math.* **50** (1986) 295–309.

Mehta, M. L. *Matrix Theory,* Editions de Physique, Orsay, France (1989).

Mehta, M. L., and des Cloizeaux, J. The probabilities for several consecutive eigenvalues of a random matrix, *Indian J. Pure Appl. Math.* **3** (1972) 329–351.

Mehta, M. L., and Gaudin, M. On the density of eigenvalues of a random matrix, *Nucl. Phys.* **18** (1960) 420–427.

Mehta, M. L., and Dyson, F. J., Statistical theory of energy levels of complex systems, V, *J. Math. Phys.* **4** (1963) 713–719.

Mehta, M. L., and Mehta, G. C. Discrete Coulomb gas in one dimension: correlation functions, *J. Math. Phys.* **16** (1975) 1256–1258.

Mehta, M. L., and Pandey, A. Spacing distribution for some Gaussian ensembles of Hermitian matrices, *J. Phys. A* **16** (1983a) L601–L606.
Mehta, M. L., and Pandey, A. On some Gaussian ensembles of Hermitian matrices, *J. Phys. A* **16** (1983b) 2655–2684.
Mehta, M. L., and Rosenzweig, N. Distribution laws for the roots of a random antisymmetric Hermitian matrix, *Nucl. Phys.* **A109** (1968) 449–456.
Mehta, M. L., and Srivastava, P. K. Correlation functions for eigenvalues of real quaternion matrices, *J. Math. Phys.* **7** (1966) 341–344.
Mello, P. A. Macroscopic approach to universal conductance fluctuations in disordered metals, *Phys. Rev. Lett.* **11** (1988) 1089–1092.
Mello, P. A., and Pichard, J. L. On maximum entropy approaches to quantum electronic transport, *Phys. Rev. Lett.* (to appear).
Mello, P. A., Akkermans, E., and Shapiro, B. Macroscopic approach to correlations in the electronic transmission and reflection from disordered conductors, *Phys. Rev. Lett.* **61** (1988) 459–462.
Mello, P. A., Pereyra, P., and Kumar, N. Macroscopic approach to multichannel disordered conductors, *Ann. Phys.* **181** (1988) 290–317.
Mon, K. K., and French, J. B. Statistical properties of many particle spectra, *Ann. Phys.* **95** (1975) 90–111.
Monahan, J. E., and Rosenzweig, N. Analysis of the distributions of the spacings between nuclear energy levels, II, *Phys. Rev. C* **5** (1972) 1078–1083.
Montgomery, H. L. The pair correlation of zeros of the zeta function, in *Proceedings of the Symposium on Pure Mathematics 24*, American Mathematical Society, Providence, R.I. (1973) pp. 181–193; Distribution of the zeros of the Riemann zeta function, in *Proceedings of the International Congress of Mathematicians,* Vancouver, B.C., 1974, Vol. 1, pp. 379–381, Canadian Mathematical Congress, Montreal, Quebec (1975).
Moore, C. E. *Atomic Energy Levels,* NBS Circular 467, Washington DC, I (1949), II (1952), III (1958).
Morris, W.G. Constant term identities for finite and affine root systems: Conjectures and theorems, Thesis, University of Wisconsin, Madison, 1982.
Morse, P. M., and Feshbach, H. *Methods of Mathematical Physics,* Chapter 2.4, McGraw-Hill, New York (1953).
Mushkelishvili, N. I., *Singular Integral Equations,* Groningen, Netherlands (1953).
Muttalib, K. A., Pichard, J. L., and Stone, A. D. Random matrix theory and universal statistics for disordered quantum conductors, *Phys. Rev. Lett.* **59** (1987) 2475–2478.
Odlyzko, A. M. On the distribution of spacings between zeros of the zeta function, *Math. Comput.* **48** (1987) 273–308.
Odlyzko, A. M. The 10^{20}-th zero of the Riemann zeta function and 70 million of its neighbours, AT&T Bell Lab preprint (1989).
Odlyzko, A. M., and Schönhage, A. Fast algorithms for multiple evaluations of the Riemann zeta function, *Trans. Amer. Math. Soc.* **309** (1988) 797–809.
Olson, W. H., and Uppulury, V. R. R. in *Probability theory,* Vol. 3 of Proceedings of the Sixth Berkeley Symposium on Mathematical Statistics and Probability, edited by L. M. Le Cam, J. Neyman, and E. L. Scott, p. 615, University of California Press, Berkeley (1972).

Opdam, E. M. Some applications of hypergeometric shift operators, *Invent. Math.* **98** (1989) 1–18, Section 6.5.

Pandey, A. Statistical properties of many particle spectra III: Ergodic behaviour in random matrix ensembles, *Ann. Phys.* **119** (1979) 170–191.

Pandey, A., and Mehta, M. L. Gaussian ensembles of random Hermitian matrices intermediate between orthogonal and unitary ones, *Commun. Math. Phys.* **87** (1983) 449–468.

Pastur, L. A., On the spectrum of random matrices, *Theoretical and Math. Phys.* **10** (1972) 67–74; (Russian original, *Teor. i Matem. Fizika* **10** (1972) 102–112).

Pechukas, P. Distribution of energy eigenvalues in the irregular spectrum, *Phys. Rev. Lett.* **51** (1983) 943–946.

Penner, R. C. The moduli space of a punctured surface and perturbative series, *Bull. Amer. Math. Soc.* **15** (1986) 73–77.

Penner, R. C. Perturbative series and the moduli space of Riemann surfaces, *J. Diff. Geom.* **27** (1988) 35–53.

Porter, C. E. (editor). Statistical theories of spectra: fluctuations, Academic Press, New York (1965).

Porter, C. E., and Rosenzweig, N. Statistical properties of atomic and nuclear spectra, *Ann. Acad. Sci. Fennicae, Serie A, VI Physica* **44** (1960a) 1–66.

Porter, C. E., and Rosenzweig, N. "Repulson of energy levels" in complex atomic spectra, *Phys. Rev.* **120** (1960b) 1698–1714.

Porter, C. E., and Thomas, R. G. Fluctuations of nuclear reaction widths, *Phys. Rev.* **104** (1956) 483–491.

Potter, H. S. A., and Titchmarsh, E. C. The zeros of Epstein zeta functions, *Proc. London Math. Soc.* **39** (2) (1935) 372–384.

Rahn, F., Camarda, H. S., Haken, G., Havens, W. W., Jr., Liou, H. I., Rainwater, J., Slagowitz, M., and Wynchank, S. Neutron resonance spectroscopy, X, *Phys. Rev. C* **6** (1972) 1854–1869.

Riemann, B. *Gesamelte Werke,* Teubner, Leipzig (1876) (reprinted by Dover, New York, 1973).

Robin, L. *Fonctions Sphériques de Legendre et Fonctions Sphéroidales,* Vol. 3, p. 250 formula 255, Gauthier-Villars, Paris (1959).

Rosen, J. L. Neutron resonances in U^{238} Thesis, Columbia University, New York, 1959.

Rosen, J. L., Desjardins, J. S., Rainwater, J., and Havens, W. W., Jr., Slow neutron resonance spectroscopy, *Phys. Rev.* **118** (1960) 687–697.

Rosenzweig, N. in *Statistical Physics,* Brandeis Summer Institute, Vol. 3, Benjamin, New York (1963).

Rosenzweig, N., Monahan, J. E., and Mehta, M. L. Perturbation of statistical properties of nuclear states and transitions by interactions that are odd under time reversal, *Nucl. Phys.* **A109** (1968) 437–448.

Scott, J. M. C. Neutron widths and the density of the nuclear levels, *Philos. Mag.* **45** (1954) 1322–1331.

Selberg, A. Bemerkninger om et multiplet integral, *Norsk Matematisk Tidsskrift,* **26** (1944) 71–78.

Seligman, T. H., and Nishioka, H. (editors). *Quantum Chaos and Statistical Nuclear Physics,* Proceedings of the Second International Conference on Quantum Chaos and the Fourth International Colloquium on Statistical Nuclear Physics,

Cuernavaca, Mexico, 1986, Lecture Notes in Physics, Vol. 263, Springer Verlag (1986).

Seligman, T. H., and Verbaarschot, J. J. M. Quantum spectra of classically chaotic systems without time reversal invariance, *Phys. Lett.* **A108** (1985) 183–187.

Seligman, T. H., and Verbaarschot, J. J. M. Fluctuations of quantum spectra and their semiclassical limit in the transition between order and chaos, *J. Phys. A Math. Gen.* **18** (1985) 2227–2234.

Seligman, T. H., Verbaarschot, J. J. M., and Zirnbauer, M. R. Spectral fluctuation properties of Hamiltonian systems: the transition region between order and chaos, *J. Phys. A Math. Gen.* **18** (1985) 2751–2770.

Shannon, C. E. The mathematical theory of communication, *Bell Syst. Technol. J.* **27** (1948) 379–423, 623–656.

Shannon, C. E., and Weaver, W. The mathematical theory of communication, University of Illinois Press (1962). (This is a reprint of the *Bell Syst. Tech. J.* paper of Shannon cited above.)

Shohat, J. A., and Tamarkin, J. D. *The Problem of Moments,* p. 8, American Mathematical Society Providence, Rhode Island (1943).

Sieber, M., and Steiner, F. Classical and quantum mechanics of a strongly chaotic billiard system, preprint, Hamburg, DESY 89-093.

Slepian, D. Some asymptotic expansions for prolate spheroidal functions, *J. Math. Phys.* **54** (1965) 99–140.

Smythe, W. R. *Static and Dynamic Electricity,* p. 104, problems 29 and 30, McGraw Hill, New York (1950).

Stieltjes, T. J. *Sur Quelques Théorèmes d'Algèbre, Oeuvres Complètes,* vol. 1, p. 440, Noordhoff, Groningen, The Netherlands (1914).

Stratton, J. A., Morse, P. M., Chu, L. J., Little, J. D. C., and Corbato, F. J. *Spheroidal Wave Functions,* MIT Press, Cambridge, Mass (1956).

Szegö, G. *Orthogonal Polynomials,* Sections 6.7 and 6.7.1, pp. 139 and 142, American Mathematical Society, New York (1959).

Titchmarsh, E.C., *Theory of Functions,* p. 186, Oxford University Press, London and New York (1939).

Titchmarsh, E.C. *The Theory of the Riemann Zeta Function,* Chapter 10, Clarendon Press, Oxford (1951).

Tricomi, F. G. Sul comportamento asintotico dei polinomi de Laguerre, *Ann. Mat. Pura Appl.* **28** (1949) 263–289.

Uhlenbeck, G. E., and Ornstein, L. S. On the theory of the Brownian motion, *Phys. Rev.* **36** (1930) 823–841.

Ullah, N. Invariance hypothesis and higher order correlations of Hamiltonian matrix elements, *Nucl. Phys.* **58** (1964) 65–71.

Ullah, N. Asymptotic solution for the Brownian motion of the eigenvalues of a random matrix, *Nucl. Phys.* **78** (1966) 557–560.

Ullah, N. Ensemble average of an arbitrary number of pairs of different eigenvalues using Grassmann integration, *Commun. Math. Phys.* **104** (1986) 693–695.

Ullah, N., and Porter, C. E. Invariance hypothesis and Hamiltonian matrix elements correlations, *Phys. Lett.* **6** (1963) 301–302.

Van Buren, A. L. A fortran computer program for calculating the linear prolate functions, Report 7994, Naval Research Lab., Washington DC, May, 1976.

Vo-Dai, T., and Derome, J. R. Correlations between eigenvalues of random matrices, *Nuovo Cimento* B **30** (1975) 239–253.

Wang, M. C., and Uhlenbeck, G. E. On the theory of the Brownian motion II, *Rev. Mod. Phys.* **17** (1945) 323–342.

Wax, Nelson (editor). Selected topics in noise and stochastic processes, Dover, New York (1954).

Weaver, R. L. Spectral statistics in elastodynamics, *J. Acoust. Soc. Amer.* **85** (1989) 1005–1013.

Weyl, H. Classical groups, Princeton University Press, Princeton, N.J. (1946).

Widom, B. Strong szegö limit theorem on circular arcs, *Indiana Univ. Math. J.* **21** (1971) 277–283.

Wigner, E. P. On the statistical distribution of the widths and spacings of nuclear resonance levels, *Proc. Cambridge Philos. Soc.* **47** (1951) 790–798.

Wigner, E. P. Characteristic vectors of bordered matrices with infinite dimensions, I and II, *Ann. Math.* **62** (1955) 548–564; **65** (1957a) 203–207.

Wigner, E. P. Results and theory of resonance absorption, in Gatlinberg Conference on neutron physics by time of flight, 1956, Oak Ridge Natl. Lab. Report ORNL-2309 (1957b) 59–70.

Wigner, E. P. Statistical properties of real symmetric matrices with many dimensions, in *Canadian Mathematical Congress Proceedings*, pp. 174–184, University of Toronto Press, Toronto, Canada, (1957c). Reproduced in *Statistical theories of spectra: fluctuations*, edited by C.E. Porter, Academic, New York, 1965.

Wigner, E. P. *Group Theory,* Chapter 26, Academic, New York (1959).

Wigner, E. P. Distribution laws for the roots of a random Hermitian matrix (1962), in *Statistical Theories of Spectra: Fluctuations,* edited by C.E. Porter, Academic, New York (1965).

Wilson, K. G. Proof of conjecture by Dyson, *J. Math. Phys.* **3** (1962) 1040–1043.

Zano, N., and Pichard, J. L. Random matrix theory and universal statistics for disordered quantum conductors with spin dependent hopping, *J. Phys.* **49** (1988) 907–920.

Author Index

H

I

J

K

L

M

N

O

P

R

S

T

U

V

W

Z

Subject Index

E

F

G

H

I

J

L